Electrogenerated Chemiluminescence Microscopy and Imaging

Francesco Paolucci · Andrea Fiorani ·
Giovanni Valenti · Massimo Marcaccio
Editors

Electrogenerated Chemiluminescence Microscopy and Imaging

Theory, Instruments, and Applications

Springer

Editors
Francesco Paolucci
Department of Chemistry "Giacomo Ciamician"
University of Bologna
Bologna, Italy

Giovanni Valenti
Department of Chemistry "Giacomo Ciamician"
University of Bologna
Bologna, Italy

Andrea Fiorani
Department of Chemistry
Keio University
Yokohama, Japan

Massimo Marcaccio
Department of Chemistry "Giacomo Ciamician"
University of Bologna
Bologna, Italy

ISBN 978-3-032-14601-4 ISBN 978-3-032-14602-1 (eBook)
https://doi.org/10.1007/978-3-032-14602-1

This Springer imprint is published by the registered company Springer Nature Switzerland AG
The registered company address is: Gewerbestrasse 11, 6330 Cham, Switzerland

Contents

Introduction

Andrea Fiorani, Massimo Marcaccio, Giovanni Valenti, and Francesco Paolucci

Abstract Electrochemiluminescence (ECL) is the light emission from a luminophore excited by homogeneous electron transfer between electrochemically generated radical species. This chapter describes the theoretical fundamentals of ECL, luminophores, mechanisms of ECL generation, and the steps occurred for its development into a powerful electroanalytical technique, owing the orthogonal nature of electrochemical stimulus and light response, which give ECL outstanding sensitivity and resolution. Furthermore, ECL intrinsically surface-confined generation prompted its evolution to an imaging technique, which is the focus of this book.

Keywords Electrogenerated chemiluminescence · Imaging · Microscopy · Electron transfer · Marcus theory · Immunoassay

1 Background

The emission of light from a molecular species (luminophore) which has been excited through a homogeneous electron transfer between two radical precursors in solution produces chemiluminescence (CL) [1–4]. Furthermore, if the radical precursors

A. Fiorani
Department of Chemistry, Keio University, Yokohama, Japan
e-mail: andrea.fiorani@keio.jp

M. Marcaccio · G. Valenti · F. Paolucci (✉)
Department of Chemistry "Giacomo Ciamician", Alma Mater Studiorum—University of Bologna, Bologna, Italy
e-mail: francesco.paolucci@unibo.it

M. Marcaccio
e-mail: massimo.marcaccio@unibo.it

G. Valenti
e-mail: g.valenti@unibo.it

F. Paolucci et al. (eds.), *Electrogenerated Chemiluminescence Microscopy and Imaging*,
https://doi.org/10.1007/978-3-032-14602-1_1

are electrochemically generated through a heterogeneous electron transfer at electrode, this is called electrogenerated chemiluminescence, or electrochemiluminescence (ECL) [5–8]. From this definition, it is possible to identify the three main stages involved in ECL, namely (i) electrochemical, (ii) chemical, and (iii) photophysical reactions.

These three steps introduce a remarkable difference compared to CL. First, ECL is a primarily surface-confined process, limited within the diffusion layer of the reactants electrogenerated near the electrode surface. Second, the emission is solely induced by the electrochemical reaction controlled by the potential program, it means ECL can be finely controlled in time.

These two distinct traits give ECL a high degree of spatiotemporal resolution which was crucial for the development of ECL as imaging method [9–12].

ECL represents an ideal marriage between electrochemical and spectroscopic methods, which attributes its success to the orthogonal stimulus and response, i.e., the decoupling of electrochemical stimulus from the light (analytical) signal that enable very low detection limits in analytical applications, and spatiotemporal resolution in imaging.

2 History of ECL

The first example of light emission induced by electrochemical reactions can be traced back to the 1920s, by electrolysis of Grignard compounds in ether solution [13], or luminol in alkaline solution [14] where the term galvano-luminescence was suggested. However, the investigation of ECL as it is known nowadays began in the mid-1960s with the pioneering works of Hercules [15], Chandross [16], and Bard [17] with polycyclic aromatic hydrocarbons as luminophores in organic solvents. It took about 6 years to report the first example of ECL from an inorganic complex, i.e., tris(2,2′-bipyridine)ruthenium(II), ($Ru(bpy)_3^{2+}$), in acetonitrile by Tokel and Bard in 1972 [18]. Since then, ECL developed from fundamental research to a powerful electroanalytical technique [6, 7]. The main leap forward was the shift from organic to aqueous electrolytes in 1981 by Bard group [19]. However, the breakthrough arrived in 1990 [20] by Leland and Powell of IGEN Incorporated whom proposed the most effective ECL system to date: $Ru(bpy)_3^{2+}$/tri-*n*-propylamine (TPrA) (details in paragraph 7). Although this date corresponds to scientific publications, industrial research was already active from mid-1980s [21, 22].

The ECL clinical diagnostic method was introduced in 1991 [23], leading to IGEN's development of the first ECL clinical analyzer, commercialized in 1994 [24]. Notably, it took 30 years from initial fundamental studies to achieve commercially viable applications (Table 1). ECL research is primarily devoted to analytical applications, particularly for biological targets in clinical diagnostics [6, 7]. However, mechanistic aspects remain crucial for enhancing ECL sensitivity [25, 26].

The history of ECL cannot be told separately from the scientific advancements that Professor Allen J. Bard put forward through the years, and his input in the

Table 1 Timeline of major events in the field of ECL

Luminescence from electrolysis of Grignard compound **1927** [13] and luminol **1929** [14]	ECL of aromatic hydrocarbons **1964** [15–17]
Theory of chemiluminescent electron transfer reactions **1965** [1]	Theory of ECL transients **1966** [28]
Magnetic field effects **1969** [29]	$Ru(bpy)_3{}^{2+}$ ECL **1972** [18]
Oxalate as coreactant **1977** [30]	Aqueous ECL system **1981** [19]
ECL of $Ru(bpy)_3{}^{2+}$ in polymer **1981** [31]	Persulfate as coreactant **1982** [32]
$Ru(bpy)_3{}^{2+}$ as ECL label **1984** [33]	TPrA as coreactant **1990** [20]
ECL for clinical diagnostic **1991** [23]	Commercial immunoassay ECL by IGEN **1994** [24]
Single molecule ECL **1995** [34]	ECL imaging at microelectrode **1998** [35]
Laser action driven by ECL **1998** [36]	Bipolar electrode ECL **2001** [37]
Near-field scanning ECL microscopy **2001** [38]	Si nanocrystals solution ECL **2002** [39]
Mechanism of ECL-beads immunoassay **2002** [40]	ECL of $Ru(bpy)_3{}^{2+}$-doped silica nanoparticles **2006** [41]
ECL imaging of polymer nanoparticles **2008** [42]	ECL imaging of 2002 mechanism **2014** [43]
Mixed-metal complexes annihilation ECL **2015** [44]	In-situ coreactant ECL at boron-doped diamond **2016** [45]
ECL imaging of cells **2017** [46]	Aggregation induced ECL **2017** [47]
2002 mechanism re-evaluation by ECL imaging **2020** [26]	Single biomolecule imaging by ECL **2021** [48]
Single molecule ECL imaging **2021** [49]	Bead-based ECL catalytic enhancement by iridium(III) complex **2024** [50]

development of ECL clinical analysis application. Therefore, as for this definition, “those who, during the preceding year, shall have conferred the greatest benefit on mankind” [27], Professor Bard had deserved the Nobel Prize in Chemistry.

3 Fundamentals of Electrochemistry and Photophysics for ECL

The generation of radical species by electrochemical reactions is the first step to induce ECL, generally by electrochemical methods, such as cyclic voltammetry or chronoamperometry. It is also necessary a certain degree of radical stability, therefore they can diffuse and react inside the diffusion layer, where electrochemical reversible electron transfer is highly desirable, and advantageous to reach higher ECL emission. For this reason, characterization of the electrochemical reactivity is highly necessary. Taking as example the $Ru(bpy)_3{}^{2+}$ (Fig. 1), the most used inorganic complex in ECL, it shows four reversible peaks (1 oxidation, and 3 reductions), and an irreversible

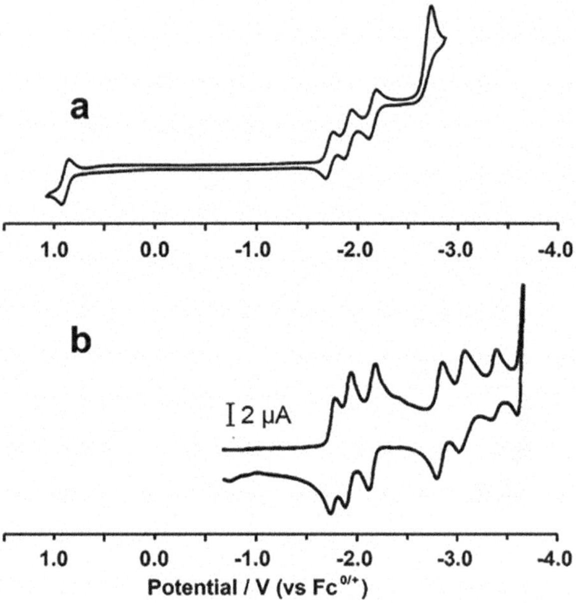

Fig. 1 Cyclic voltammograms of $Ru(bpy)_3{}^{2+}$: **a** in acetonitrile at room temperature, and **b** in dimethylformamide at −54 °C. Adapted with permission from [51], copyright (2022) Elsevier

reduction peak at more extreme negative potentials, leading to complex breakup. This last irreversible peak is actually reversible, and $Ru(bpy)_3{}^{2+}$ can allocate three more electrons, providing the cyclic voltammetry is conducted at low temperature. In ECL, the first oxidation and reduction are generally enough, although generating higher oxidized or reduced species is not precluded, and their use in ECL is limited only by radical stability, or solvent potential window.

More generally, we can express the electrochemical step as follow: an ECL active luminophore (L) is oxidized ($L^{\cdot+}$) and reduced ($L^{\cdot-}$), respectively, at the electrode surface (Fig. 2).

In the diffusion layer, the homogeneous electron transfer between them will afford the excited state of one of them (L*), leaving the other in the fundamental state. We should remark that L is generally the same species, but different molecules can be used, as well.

The emission that is obtained from the luminophore after the excitation by the homogeneous electron transfer can be directly inferred from the photophysics of the molecule which implies the need of a careful characterization of the luminophore photophysical processes before ECL experiments.

The sequence of electrochemical and chemical reactions to reach the excited state described here are referred as annihilation ECL (see paragraph 7).

The chemical reactions leading to the excited state that occur within the diffusion layer between $L^{\cdot+}$ and $L^{\cdot-}$ are represented in Fig. 3.

The *E-route* describes a radiationless decay, ECL is not observed because the energy is not sufficient to reach the excited state. When the energy is sufficient the *T-route* and *S-route* occur with ECL from triplet or singlet excited stated, respectively.

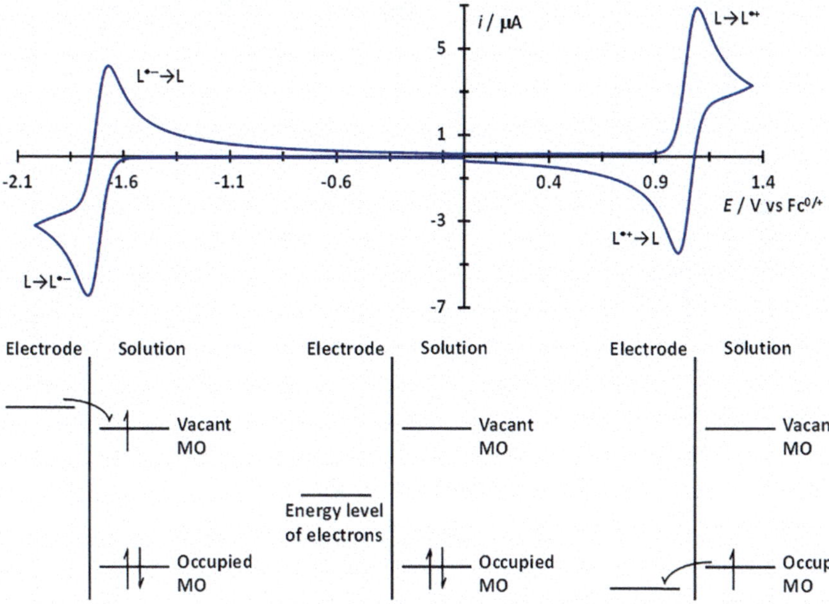

Fig. 2 Oxidation and reduction for a single electron process of a generic luminophore (L): cyclic voltammetry and schematic molecular orbital (MO)

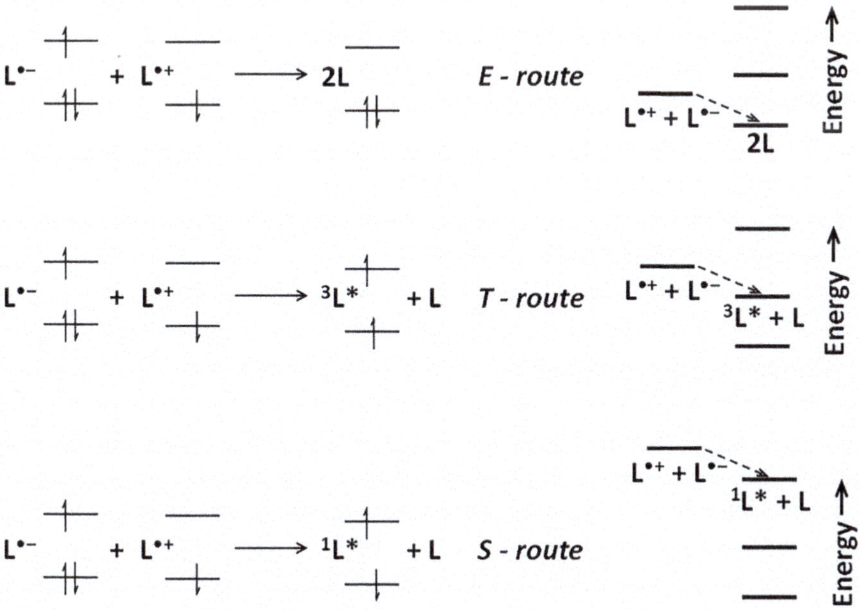

Fig. 3 Schematic representation of ECL pathways to excited state

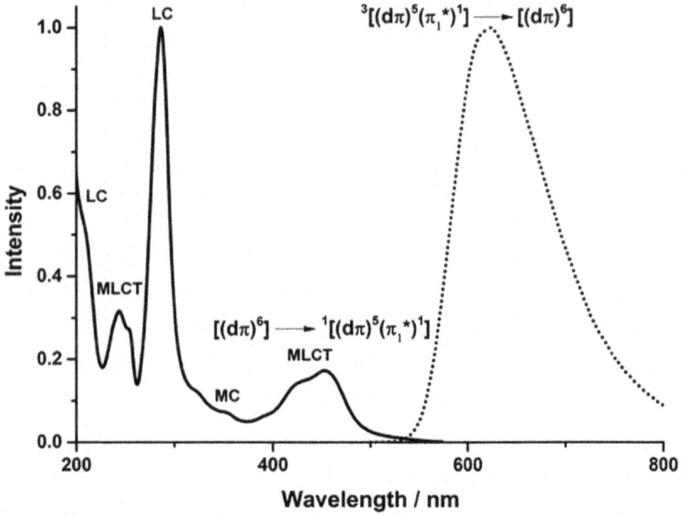

Fig. 4 Spectra of $[Ru(bpy)_3]Cl_2$ in double distilled water: absorption (straight line) and emission (dotted line). Adapted with permission from [51], copyright (2022) Elsevier

In addition to the *S-route*, excitation of the singlet state can be obtained by the process of triplet–triplet annihilation (TTA). Even if the energy available from the electron transfer between the two radicals is not sufficient, the ECL can be emitted from an excited state with higher energy, and the whole process is referred as "energy-deficient" ECL. Spectroscopic analysis of the ECL emission and considerations of the energy requirements to reach the excited state permits to differentiate between the *S-route* and the TTA mechanism.

Considering the $Ru(bpy)_3^{2+}$, the typical ECL orange emission at 610 nm is observed (Fig. 4).

The combination of electrochemical and photophysical process can be conveniently represented as in Fig. 5.

4 Energy Requirements

Taking for granted that the luminophore has an emissive excited state, the energy available in the electron transfer has to be greater than the energy of the excited state to produce ECL. In this case the process is referred as "energy sufficient" ECL. The energy available from the electron transfer reaction can be computed by a combination of formal potentials of the reacting species (Eq. 1–2).

$$\Delta G^0 = -eE^0_{\text{redox}} \tag{1}$$

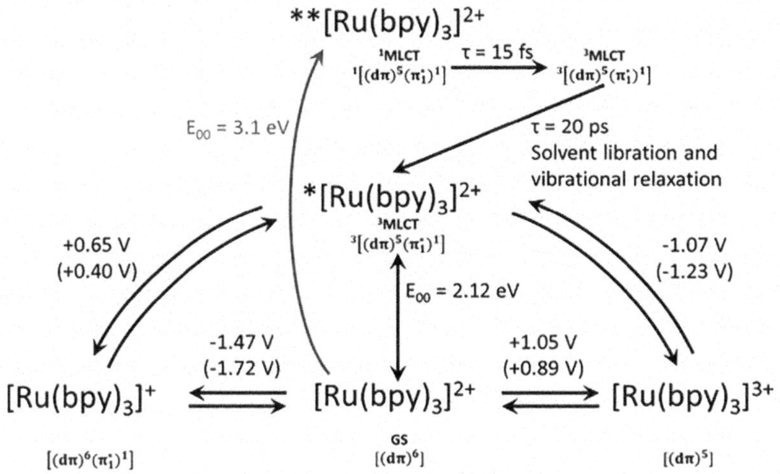

Fig. 5 Ground and excited state redox potentials of $[Ru(bpy)_3]^{2+}$, in water vs Ag/AgCl (3 M NaCl), and in acetonitrile vs Ferrocene (data in parenthesis). Adapted with permission from [51], copyright (2022) Elsevier

Thus

$$-\Delta H^0 = e\left[E^0\left(L^{\cdot+}, L\right) - E^0\left(L, L^{\cdot-}\right)\right] - T\Delta S^0 > E_S = \frac{1239.8}{\lambda_{es}} \quad (2)$$

where e is the elementary charge, $E^0(L^{\cdot+}/L)$ and $E^0(L/L^{\cdot-})$ are the standard potentials (V) of the reacting specie L for oxidation and reduction, respectively, E_S (eV) is the energy emitted from the excited state of the luminophore at the emission wavelength λ_{es} (nm). The term 1239.8 (eV nm) is the constant hc. The term $T\Delta S^0$ is small ($\leq$0.1 eV) at 298 K for typical ECL reaction involving a ΔG^0 of 2–3 eV [52]. If the peak potentials from cyclic voltammetry are used instead of the standard potential an additional term of 0.06 V has to be subtracted. This is the difference between the peak potentials and the standard potentials for a fast, chemically reversible, one-electron electrode process involving oxidized and reduced forms having equal diffusion coefficients. The final correction term is thus 0.16 eV [52].

Neglecting the correction term, because smaller than the usual energy involved, the "energy sufficient" condition can be expressed as (3)

$$\Delta G_{es} = -e\left[E^0\left(L^{\cdot+}, L\right) - E^0\left(L, L^{\cdot-}\right)\right] + E_S \quad (3)$$

where ΔG_{es} is the Gibbs free energy of excited state formation. The formation of the excited stated is thus $\Delta G_{es} < 0$ when the energy involved in the electron transfer is larger than E_s. Equation (3) is particularly useful for the depiction of the ECL "wall of energy sufficiency" (Fig. 6) which combines potentials and emission wavelengths.

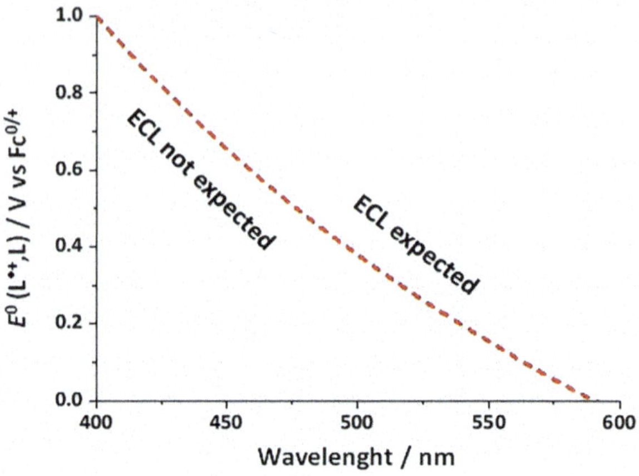

Fig. 6 ECL wall of energy sufficiency for E^0 (L, L$^{\cdot-}$) = − 2.1 V versus $Fc^{0/+}$ computed by Eq. (3). Further details in paragraph 7.1. The red dashed line is $\Delta G_{es} = 0$

5 Theoretical Description of Electrogenerated Chemiluminescence

The energetic requirement only does not explain the formation of the excited state in full. In fact, even if the energy is sufficient to populate the excited state, the formation of the ground state is a competing reaction that can decrease or compromise the excited state formation efficiency because, in principle, both reactions proceed at the same time.

To help rationalize this behavior, the formation of the excited state versus the ground state has to be treated on kinetic grounds, that is, the formation of the excited state is faster than the formation of the ground state, even if the relative driving force is smaller.

The existence of a redox process that is kinetically faster, which is in competition with a higher exergonic redox reaction, is the mechanistic basis explaining why the formation of the excited state is favored instead of the more stable ground state.

This is rationalized in the Marcus theory of electron transfer with the prediction of the "inverted region," where the kinetic constant of electron transfer (k_{ET}) decreases with the driving force. ECL was among the first experimental evidence of inverted region as acknowledged by Marcus in his Nobel lecture [53].

Taking as example the $Ru(bpy)_3^{2+}$, the following treatment outlines the formation of the excited state.

The energy involved in the formation of excited state and ground state (Scheme 1) are computed by Eq. 3.

The potential energy curves for these two states are qualitatively depicted in Fig. 7.

Here, parameter λ is introduced, known as the *reorganization energy*, the energy required to restructure the precursor complex. It represents the free energy change required to distort the configuration of the reactants, from its energy minimum at q_R to the coordinate q_P, which corresponds to the energy minimum of the configuration of the products, without electron transfer.

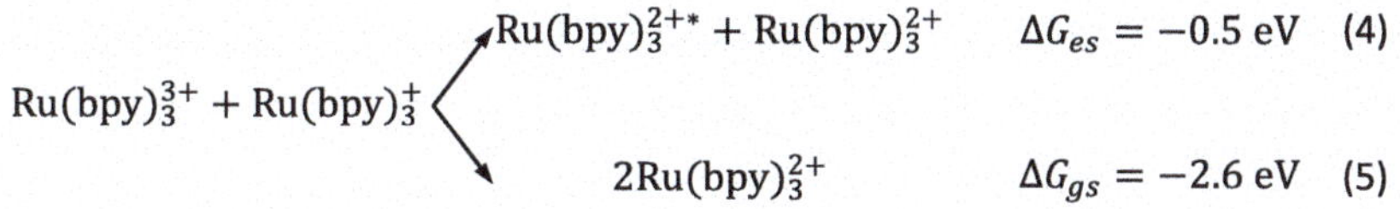

Scheme 1 Redox reaction of $Ru(bpy)_3^{3+}$ with $Ru(bpy)_3^{+}$, and corresponding energy involved

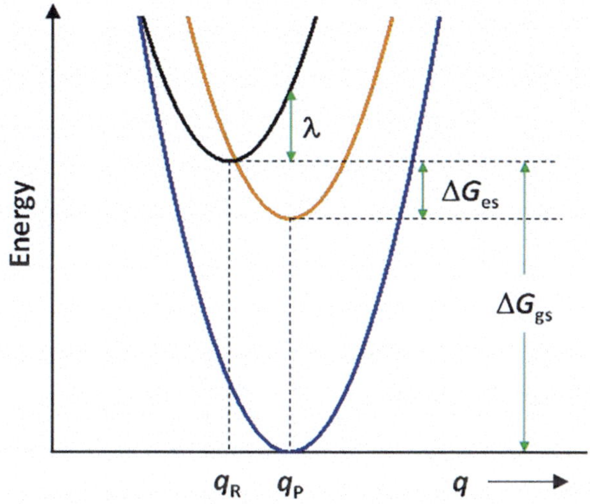

Fig. 7 Qualitative reaction coordinate diagrams for ECL systems containing $Ru(bpy)_3^{2+}$. $2Ru(bpy)_3^{2+}$ (blue), $Ru(bpy)_3^{2+}$ + $Ru(bpy)_3^{2+*}$ (orange), $Ru(bpy)_3^{+}$ + $Ru(bpy)_3^{3+}$ (black). Corresponding λ is estimated to be 0.67 eV in acetonitrile

This energy is the sum of the precursor state of reactants, the inner λ (λ_i), and peripheral elements, mainly the solvent, the outer λ (λ_o).

The constant of electron transfer in the Marcus theory assumes the following form (Eq. 6).

$$k_{ET} = A_{\exp}\left[\frac{-(\Delta G^0 + \lambda)^2}{4\lambda k_B T}\right] \tag{6}$$

The pre-exponential factor A is a combination of the *nuclear frequency factor* (ν_n/s^{-1}), the frequency attempts on the energy barrier, which is associated with changes in bond lengths, rotation about bonds, and solvent motion, and the *electronic transmission coefficient* (κ_{el}), the probability of electron tunneling.

This translates in a k_{ET} larger for the formation of the excited state than for the ground state (Fig. 8).

On the left side of the maximum, there is a normal region, where the Gibbs free energy of activation ($\Delta G^{\ddagger}$) decreases, hence the rate of electron transfer increases, as ΔG^0 becomes more negative. When $\Delta G^0 = -\lambda$, $\Delta G^{\ddagger}$ is zero, and the rate of electron transfer is predicted to be at a maximum. At more negative ΔG^0 values, i.e., for very exergonic reactions, the activation energy becomes larger, and the rate of electron transfer smaller. This is the inverted region, and there are two physical reasons

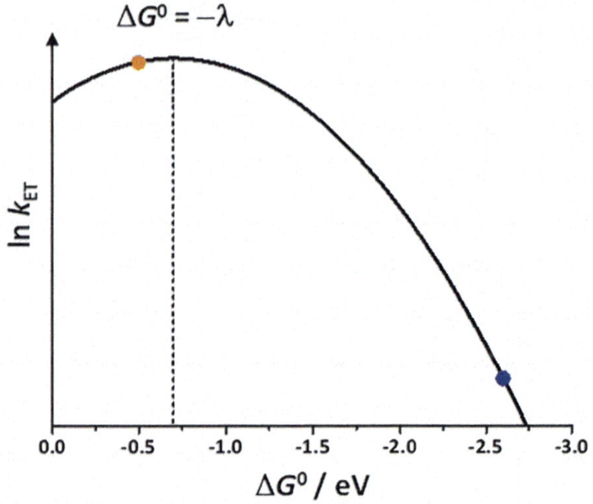

Fig. 8 The logarithm of the rate constants k_{ET} for reactions in Scheme 1 as function of the free energy of reaction ΔG: Eq. 4 (orange dot) and Eq. 5 (blue dot). The maximum of this curve lies at of $\Delta G^0 = -\lambda$

responsible for this effect. First, the products are required to allocate the output energy responding very quickly, however, as the free energy of reaction becomes larger, the probability of such process decreases as ΔG^0 exceeds $-\lambda$. Second, the free energy curve of reactants intersects that of products in their ground states at higher energy than the products electronically excited, so that the energy surfaces no longer allow for adiabatic electron transfer in the inverted region [1, 52, 54].

Because $k_{ET,es} \ggg k_{ET,gs}$, the efficiency of $Ru(bpy)_3{}^{2+}$ excited state formation approaches unity as the temperature decrease down to – 30 °C (Fig. 9).

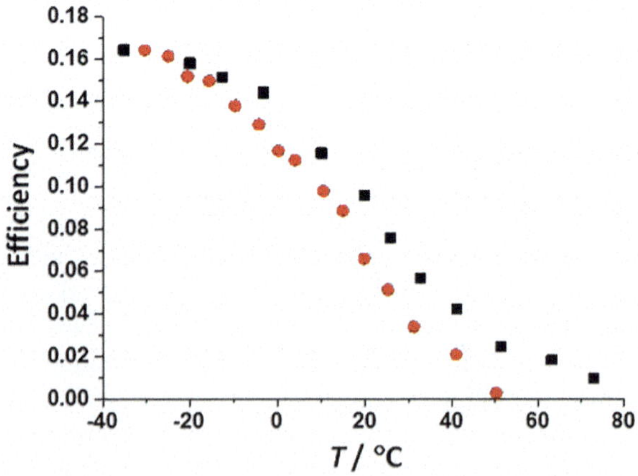

Fig. 9 Variation of the ECL efficiency (red dots) and the luminescence quantum yield (black squares) of $Ru(bpy)_3{}^{2+}$ in acetonitrile as function of the temperature. Adapted with permission from [55], copyright (1979) American Chemical Society

6 ECL Luminophores

Beside electrochemical and photophysical requirements, an ECL emitter should possess efficient ECL also in physiological conditions, binding sites for bioconjugation that shall not affect the photophysical/electrochemical properties, combined with a good solubility in aqueous buffers to permit labelling reaction with biological elements. All these features are extremely important in view of biological assays based on ECL, including ECL imaging.

In this context, ruthenium and iridium complexes play the main role as ECL emitters, displaying rich and stable redox processes, although not always accessible in aqueous electrolyte in case of iridium complexes. Requirements imposed on the ECL emitters restrict the available candidates, which makes ruthenium polypyridine complexes the luminophores of choice in commercial applications of ECL, and generally for bioanalysis.

6.1 Inorganic Complexes

In this category are included coordination complexes based on transition metals, such as ruthenium, iridium, osmium, and platinum. Among all the investigated metals, ruthenium is the forefather and the most widely used, while iridium is gaining more attention for its promising higher efficiency (Fig. 10) [51, 56].

The studies on inorganic complexes as ECL luminophores date back to the 1970s, when Tokel and Bard first reported the ECL from $Ru(bpy)_3^{2+}$ annihilation in acetonitrile [18]. Since then, a large number of different ruthenium-based complexes were investigated, searching for higher ECL efficiency. The $Ru(bpy)_3^{2+}$ is the benchmark within coordination complexes for ECL as reflected by its extensive use in academic research [57–59] and immunoassay technological applications [6–8]. The general

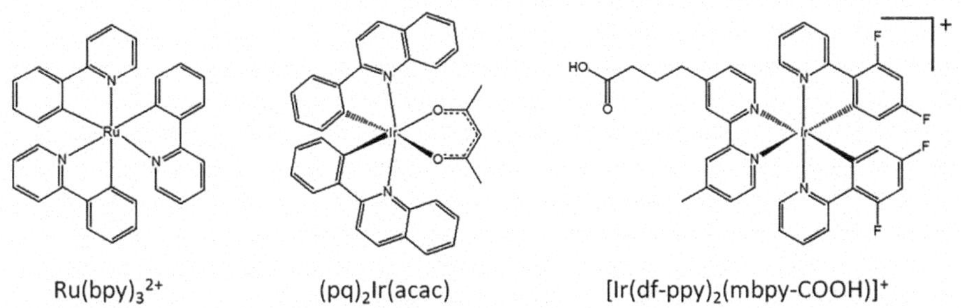

Fig. 10 Examples of metal complexes and organometallic ECL luminophores. 2-phenylquinoline (pq), acetylacetonate (acac), 3,5-difluoro-2-(2-pyridinyl)phenyl (df-ppy), 4-methyl-4′-carboxypropyl-2,2′-bipyridyl (mbpy-COOH)

strategy to tune the ECL properties is ligand modification, or substitution to prepare heteroleptic complexes.

On the other hand, iridium complexes revealed a promising ECL efficiency that prompted intensive research to develop a new class of high emitting iridium-based luminophores for ECL.

However, iridium complexes show generally poor solubility in aqueous solutions, and are prone to oxygen quenching. Therefore, they require careful design to be made soluble while retaining efficient ECL emission. Interestingly, water soluble Ir complexes are increasingly being proposed showing higher ECL efficiency than $Ru(bpy)_3{}^{2+}$.

The nature of the excited state in reaction (4), whether formed by the reduced or oxidized species ($Ru(bpy)_3{}^{+}$ or $Ru(bpy)_3{}^{3+}$), can reveal key mechanistic details.

Electron transfer is enhanced by extended molecular orbital overlap, making the formation of the $Ru(bpy)_3{}^{2+}$ excited state from the oxidized precursor, via ligand-to-ligand electron transfer, kinetically preferred compared to the reduced precursor, which involves metal-to-metal electron transfer.

Studies of mixed inorganic ECL systems with iridium complexes, where the HOMO electron density is delocalized across both ligands and the metal center, provided further mechanistic insights [44]. DFT calculations suggest that the HOMO of $Ir(ppy)_3$ is approximately 50% metal-based, making HOMO/HOMO electron transfer less disfavored when an iridium complex acts as the oxidant compared to ruthenium complexes, which typically have a purely metal-based HOMO.

6.2 Organic Molecules

Polycyclic aromatic hydrocarbons (PAHs), such as 9,10-diphenylanthracene (DPA), rubrene, pyrene, and anthracene, were the initial compounds investigated in ECL studies [60]. High substitution levels facilitate both oxidation and reduction, yielding less reactive radicals, while substantial steric hindrance improves stability by protecting reactive sites, resulting in highly consistent ECL intensity, which is favored by bulky and rigid organic molecules. Fluorene, spirobifluorene, helicenes, and twisted or substituted PAHs are the primary molecules studied in this context (Fig. 11). Excimer formation is commonly observed during annihilation ECL because the charge transfer process requires close proximity of oppositely charged radicals. For example, steric hindrance of dendrons surrounding an emissive pyrene core of sulfurated dendrimers affects the efficiency of excimer formation [61].

Among organic molecules, BODIPY (4,4-difluoro-4-bora-3a,4a-diaza-s-indacene) dye and its derivatives emerged as highly efficient ECL luminophores [62]. Their electrochemistry and ECL are strictly dependent on molecular substitutions which affects the electron density distribution in the BODIPY core.

Completely substituted BODIPY dye shows electrochemically reversible one-electron oxidation and reduction with formation of stable radical ions. On the other

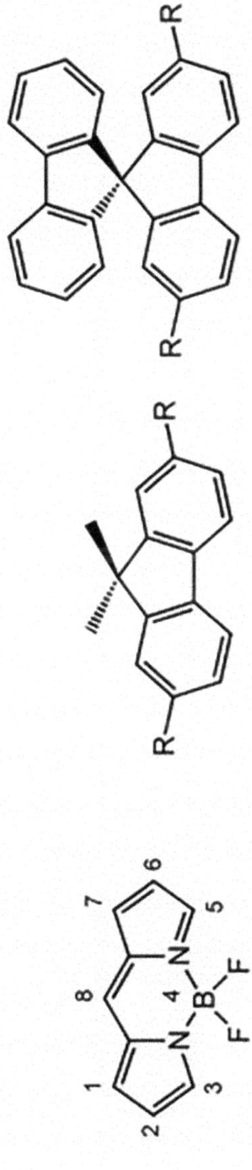

Fig. 11 Examples of organic ECL luminophores. From left to right: 4,4-difluoro-4-bora-3a,4a-diaza-*s*-indacene (BODIPY), fluorene, and spirobifluorene

hand, without functionalization in positions 2, 3, 5, or 6, the radical cation is instable, while without functionalization in position 8, the radical anion is more reactive.

Although organic luminophores typically shows high ECL efficiency, they are necessarily investigated in organic solvents.

Several approaches have been explored to achieve ECL in water, including incorporating organic compounds into core–shell silica-PEG nanoparticles [63], utilizing an emulsion system [64], synthesizing organic nanoparticles [65], or chemically functionalizing DPA with ionic groups [66]. However, these strategies cannot alleviate the solubility issue, in particular if compared with readily available inorganic complexes in view of analytical application. Luminol, which is an organic molecule, but is water soluble, can be functionalized for labeling of biological structures, and it offers more possibility of applications. However, luminol is not catalytically regenerated after the chemical reactions and photon emission making it an intrinsically low emissive molecule [67] (details in paragraph 7).

7 ECL Systems

ECL generation follows two main routes, namely annihilation and coreactant. Annihilation involves the reaction between two radicals, from the same species or different, in a homogeneous electron transfer in solution. Generally, the two radicals are formed by using alternating potential technique, either cyclic voltammetry or chronoamperometry, at the same electrode, or two separated but near electrodes, to generate the reduced and oxidized ion radical species.

The coreactant is sacrificial molecule added in excess concentration compared to the luminophore. Following its oxidation or reduction, the coreactant undergoes a chemical reactions generating a reducing or oxidizing intermediate. The electron transfer between coreactant radical intermediate and the luminophore, previously oxidized or reduced, generates the luminophore in its excited state.

A noteworthy difference is that annihilation ECL requires both oxidation and reduction reactions, while coreactant ECL either oxidation or reduction reaction.

Energy involved in the electron transfer to generate the excited state is generally in the range of 2–3 eV for ECL, which is difficult to reach in water (i.e., generation of ion radicals within water electrolysis potential window). Therefore, annihilation ECL is conducted in organic solvents, or water-containing mixed solvents, while coreactant ECL can proceed easily in aqueous electrolytes, which promoted ECL application in bioanalysis.

During the generation of ECL it is possible to measure the ECL efficiency (Φ^0_{ECL}) as the number of photons emitted for each charge transfer events (Eq. 7)

$$\text{ECL quantum efficiency (\%)} = \frac{\text{total photons}}{\text{total electrons}} \times 100 \quad (7)$$

Because the absolute photon number is not easily accessible, relative ECL efficiency of the luminophore (x) against a standard (st) has been introduced to bypass this technical issue (Eq. 8).

$$\Phi_{ECL} = \frac{\left(\frac{\int \mathrm{ECL}\,\mathrm{d}t}{\int i\,\mathrm{d}t}\right)_x}{\left(\frac{\int \mathrm{ECL}\,\mathrm{d}t}{\int i\,\mathrm{d}t}\right)_{st}} \tag{8}$$

where the integral of ECL is the total luminescence measured by the PMT and the integral of the current is the charge injected during the electrochemical step generating the ECL signal.

The relative ECL intensity can be multiplied by the absolute Φ^0_{ECL}, as measured by Bard and coworkers to be 5% for 1 mM $Ru(bpy)_3{}^{2+}$ in acetonitrile with tetra-n-butylammonium fluoborate at rotating ring-disk electrode [68] or tetra-n-butylammonium perchlorate by double potential step [55], to obtain the absolute Φ^0_{ECL} of the luminophore under investigation. Theoretically, this strategy can be used to compute the efficiency, but while Eq. 8 could be acceptable for the relative assessment of ECL luminophores, it has several critical disadvantages to produce a reliable Φ^0_{ECL}. The ECL is affected by many parameters (mechanism, solvent, degassing, electrode material, electrochemical method, optical setup, and PMT wavelength sensitivity) that is physically impossible to replicate the conditions used in the original measurements of $Ru(bpy)_3{}^{2+}$ Φ^0_{ECL}. This has hampered the comparison of Φ^0_{ECL} from different laboratories, contrary to conventional luminescence systems where the quantum yield is defined [69]. To overcome these issues, Ding and coworkers reported a simple procedure to measure the absolute Φ^0_{ECL} using a photon-counting photomultiplier device (see Chap. 4).

7.1 Coreactant ECL

Because coreactant ECL holds a pivotal role in applications, mainly analytical, and in ECL imaging, it follows a detailed explanation of the different coreactants and their relative mechanisms [70].

Coreactants are molecules which, after oxidation or reduction, undergo a homogeneous follow-up reaction (e.g., deprotonation or splitting) yielding a strongly reducing (or oxidizing) radical intermediates able to provide a high energetic electron transfer reaction with the luminophore to reach the excited state. In this case, ECL needs only a single potential step, both for the coreactant and luminophore. It is possible to distinguish two main coreactant types, if the single potential step involves oxidation or reduction, namely "oxidative-reduction" or "reductive-oxidation" ECL mechanisms, respectively.

ECL with coreactants offers distinct advantages, making it more common than the annihilation method in ECL applications. Firstly, it eliminates the need for electrochemically reversible redox reactions of the luminophore for both oxidation and reduction, significantly expanding the range of viable ECL luminophores. Secondly, it allows the use of solvents with a narrow potential window, such as water, where solvent discharge (oxidation or reduction) might otherwise hinder the formation of one of the annihilation partners. For example, in aqueous buffer solutions, hydrogen evolution can compete with the generation of the reduced luminophore, disrupting the annihilation process. Additionally, using a single oxidation step removes the need to eliminate oxygen to prevent its reduction.

Common coreactants for the "oxidative-reduction" ECL mechanism include amines, with tri-n-propylamine (TPrA) being the most widely used due to its ability to produce high-intensity ECL when paired with $Ru(bpy)_3^{2+}$, alongside oxalate. For the "reductive-oxidation" ECL mechanism, persulfate, hydrogen peroxide, and benzoyl peroxide are employed. Other coreactants, such as amine-related compounds (e.g., amino acids, peptides, nucleic acids, NADH, alkaloids, pharmaceuticals, pesticides, hydrazine), organic acids, and alcohols, can also be used. While these may yield lower ECL signals compared to TPrA, they can be also analyte for direct detection applications. Typically, the luminophore reacts with the coreactant and is regenerated after light emission, except for luminol, which transforms into an unreactive compound, 3-aminophthalate dianion. Thus, ECL systems can be effectively classified based on the unique reaction mechanism of each coreactant.

It will be given particular attention to amines, specifically TPrA, which evidences special chemical reactions able to generate ECL without the direct involvement of the luminophore in redox reactions at the electrode, i.e., $Ru(bpy)_3^{2+}$ is physically immobilized far from the electrode surface, beyond the tunneling distance for electron transfer, < 3 nm [71]. For this special feature, TPrA is the only coreactant that has found applications in commercial clinical analyzers, in conjunction with its widespread use in academic research.

7.1.1 ECL Mechanism by "Oxidative-Reduction"

This group includes coreactants that, upon oxidation, generate highly energetic radicals that reduce the luminophore, such as oxalate and amines.

The first demonstration of coreactant was developed by using oxalate in aqueous solution, reacting with $Ru(bpy)_3^{2+}$ [19], and used for oxalate detection in synthetic urine [72], as shown in Scheme 2.

The mechanism involving TPrA is more complex, but more versatile, making it the only coreactant suitable for developing commercial instruments for clinical analysis [73]. In addition to the first described "oxidative-reduction" mechanism (Scheme 3) reported in 1990 [20], a mechanism involving exclusively TPrA oxidation and a catalytic mechanism also occur. Furthermore, annihilation between $Ru(bpy)_3^{3+}$ and $Ru(bpy)_3^{+}$ is theoretically feasible, as both species can be generated in the presence of TPrA [40].

Scheme 2 ECL mechanism of oxalate and $Ru(bpy)_3{}^{2+}$

$$Ru(bpy)_3^{2+} \longrightarrow Ru(bpy)_3^{3+} + e^-$$
$$Ru(bpy)_3^{3+} + C_2O_4^{2-} \longrightarrow Ru(bpy)_3^{2+} + CO_2 + CO_2^{\bullet -}$$
$$Ru(bpy)_3^{3+} + CO_2^{\bullet -} \longrightarrow Ru(bpy)_3^{2+*} + CO_2$$
$$Ru(bpy)_3^{2+*} \longrightarrow Ru(bpy)_3^{2+} + h\nu$$

Scheme 3 ECL "oxidative-reduction" mechanism of TPrA and $Ru(bpy)_3{}^{2+}$

$$TPrAH^+ \rightleftarrows TPrA + H^+$$
$$TPrA \rightleftarrows TPrA^{\bullet +} + e^-$$
$$TPrA^{\bullet +} \rightleftarrows TPrA^{\bullet} + H^+$$
$$Ru(bpy)_3^{2+} \longrightarrow Ru(bpy)_3^{3+} + e^-$$
$$Ru(bpy)_3^{3+} + TPrA^{\bullet} \longrightarrow Ru(bpy)_3^{2+*} + Pr_2N^+CHCH_2CH_3$$
$$Ru(bpy)_3^{2+*} \longrightarrow Ru(bpy)_3^{2+} + h\nu$$

Scheme 3 outlines the first mechanism studied with TPrA in aqueous solution (Fig. 12a) [20], which paved the way for commercial development of clinical analysis based on ECL [24]. A few years later, a seminal paper from the Bard group revealed the true mechanism driving ECL emission in clinical analyzers [40] demonstrating that it involves only TPrA oxidation (Fig. 12b), without requiring direct oxidation of $Ru(bpy)_3{}^{2+}$ at the electrode (Scheme 4).

The term "heterogeneous" derives from its use in ECL analyzers, where $Ru(bpy)_3{}^{2+}$ is immobilized on microbeads or directly on the electrode [6, 7, 26], preventing its diffusion because it belongs on a different phase (Scheme 4, Fig. 12). In contrast, in the "homogeneous" case, both $Ru(bpy)_3{}^{2+}$ and TPrA are free to diffuse in the solution (Scheme 3, Fig. 12).

The distinction between ECL mechanisms is highlighted in the variation of optimal pH value to reach the highest ECL signal (Fig. 13).

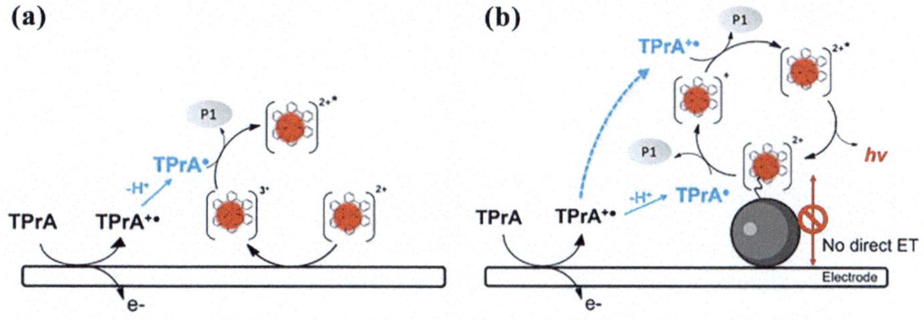

Fig. 12 Schematic representation of electrogenerated chemiluminescence mechanism involving TPrA and $Ru(bpy)_3{}^{2+}$: **a** homogeneous, where both coreactant and luminophore can be oxidized at the electrode surface; **b** heterogeneous where only the coreactant is oxidized. Adapted with permission from [11], copyright (2022) American Chemical Society

Scheme 4 Heterogeneous ECL mechanism of TPrA and $Ru(bpy)_3{}^{2+}$

$$TPrAH^+ \rightleftarrows TPrA + H^+$$
$$TPrA \rightleftarrows TPrA^{\bullet +} + e^-$$
$$TPrA^{\bullet +} \rightleftarrows TPrA^{\bullet} + H^+$$
$$Ru(bpy)_3^{2+} + TPrA^{\bullet} \longrightarrow Ru(bpy)_3^{+} + Pr_2N^+CHCH_2CH_3$$
$$Ru(bpy)_3^{+} + TPrA^{\bullet +} \longrightarrow Ru(bpy)_3^{2+*} + TPrA$$
$$Ru(bpy)_3^{2+*} \longrightarrow Ru(bpy)_3^{2+} + h\nu$$

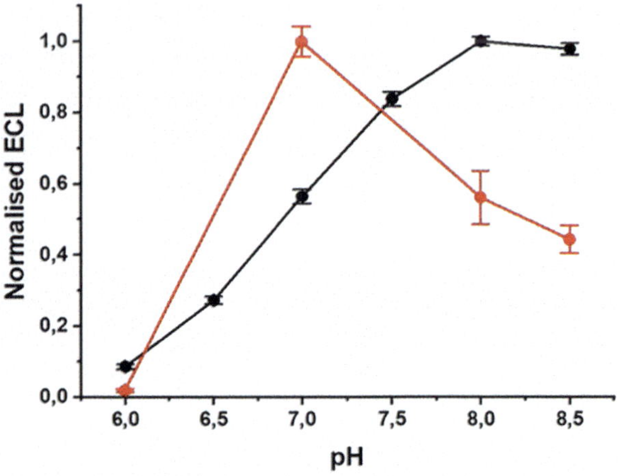

Fig. 13 Normalized ECL as a function of pH: $Ru(bpy)_3{}^{2+}$ free diffusing in solution (black), and $Ru(bpy)_3{}^{2+}$ labeled on 2.8-μm beads deposited on the electrode (red). Adapted with permission from [74], copyright (2023) Royal Society of Chemistry

The overall ECL mechanism's reaction rate is influenced by the deprotonation of the TPrA radical cation to form the TPrA radical, which depends on pH. At high pH, faster deprotonation accelerates the mechanism in Scheme 3 (Fig. 12a), whereas at low pH, slower deprotonation makes the heterogeneous ECL mechanism in Scheme 4 (Fig. 12b) significant for signal emission [75].

Beyond pH effects, the impact of amine radical cation deprotonation was evident when comparing 2-(dibutylamino)ethanol (DBAE) with TPrA. Faster deprotonation with DBAE enhances the ECL signal for the mechanism in Scheme 3 [76], but the emission is nearly suppressed for the heterogeneous ECL mechanism in Scheme 4 when used with microbeads [43].

Finally, a catalytic mechanism is possible in the $Ru(bpy)_3{}^{2+}$/TPrA system, where freely diffusing $Ru(bpy)_3{}^{3+}$ oxidizes TPrA in solution, as shown in Scheme 5.

In 1987, similar chemical reactions were used for detecting aliphatic amines via chemiluminescence, where $Ru(bpy)_3{}^{3+}$ was electrochemically generated and then introduced into the amine sample solution in front of a photomultiplier [77]. This work likely played a crucial role in the development of ECL, as the seminal paper by Leland and Powell was published 3 years later, with IGEN Inc. involved in both studies [20].

Scheme 5 Catalytic ECL mechanism of TPrA and $Ru(bpy)_3{}^{2+}$

$$TPrAH^+ \rightleftarrows TPrA + H^+$$
$$Ru(bpy)_3^{2+} \longrightarrow Ru(bpy)_3^{3+} + e^-$$
$$TPrA + Ru(bpy)_3^{3+} \longrightarrow Ru(bpy)_3^{2+} + TPrA^{\bullet +}$$
$$TPrA^{\bullet +} \rightleftarrows TPrA^{\bullet} + H^+$$
$$Ru(bpy)_3^{3+} + TPrA^{\bullet} \longrightarrow Ru(bpy)_3^{2+*} + Pr_2N^+CHCH_2CH_3$$
$$Ru(bpy)_3^{2+*} \longrightarrow Ru(bpy)_3^{2+} + h\nu$$

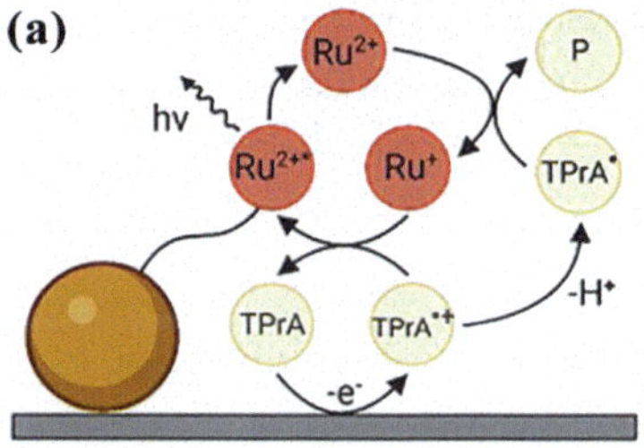

Fig. 14 Schematics of heterogeneous bead-based immunoassay: conventional ECL pathway (**a**) and the enhanced "redox mediated" pathway (**b**). The magnetic microbead is represented by the orange sphere while the $[Ru(bpy)_3]^{2+}$ luminophore and the $[Ir(sppy)_3]^{3-}$ complex are labeled as Ru^{2+} and Ir^{3-}, respectively. Adapted with permission from [50], copyright (2024) Royal Society of Chemistry

This mechanism is typically observed at high $Ru(bpy)_3{}^{2+}$ concentrations (hundreds of μM to mM), [78], offering little practical value for analytical applications.

However, recent studies on the homogeneous oxidation of TPrA by a freely diffusing Ir complex showed that this approach can enhance the ECL signal in a microbead-based immunoassay, a heterogeneous system, through a "redox-mediated" pathway (Fig. 14) [50, 79]. The $[Ir(0)]^{3-}$ complex, oxidized at the electrode to form $[Ir(I)]^{2-}$, can oxidize TPrA, producing $TPrA^{\bullet}$ further from the electrode surface, ultimately boosting the ECL signal by up to 107% [50].

7.1.2 ECL Mechanism by "Reductive-Oxidation"

This group includes coreactants that, upon reduction, generate highly energetic radicals that oxidize the luminophore, primarily peroxides like persulfate [32, 80, 81], hydrogen peroxide [82], and benzoyl peroxide [83]. These coreactants share a common reaction pathway, as depicted for peroxydisulfate in Scheme 6.

The key distinction between $S_2O_8{}^{2-}$ and TPrA lies in the absence of a heterogeneous mechanism for $S_2O_8{}^{2-}$, where the reduction of $Ru(bpy)_3{}^{2+}$ to $Ru(bpy)_3{}^+$ is critical. Experimental evidence showed no ECL emission when $Ru(bpy)_3{}^{2+}$ was

Scheme 6 ECL "reductive-oxidation" mechanism of $S_2O_8^{2-}$ and $Ru(bpy)_3^{2+}$

$$Ru(bpy)_3^{2+} + e^- \longrightarrow Ru(bpy)_3^{+}$$
$$S_2O_8^{2-} + e^- \longrightarrow SO_4^{\bullet -} + SO_4^{2-}$$
$$Ru(bpy)_3^{+} + S_2O_8^{2-} \longrightarrow Ru(bpy)_3^{2+} + SO_4^{\bullet -} + SO_4^{2-}$$
$$Ru(bpy)_3^{+} + SO_4^{\bullet -} \longrightarrow Ru(bpy)_3^{2+*} + SO_4^{2-}$$
$$Ru(bpy)_3^{2+*} \longrightarrow Ru(bpy)_3^{2+} + h\nu$$

labeled on microbeads and $S_2O_8^{2-}$ was used as the coreactant, as confirmed by microscope imaging analysis [84]. This underscores the significant advantage of using TPrA for ECL imaging of cells or large biological entities [9, 26].

7.1.3 Luminol

Luminol, an organic molecule, is utilized in ECL by directly taking advantage of its chemiluminescence (CL) properties [85, 86]. It exhibits a characteristic CL emission centered at 425 nm from the excited 3-aminophthalate dianion (3AP), though the emission can range from approximately 424 to 510 nm depending on the solvent, with ECL typically centered at 440 nm [87, 88]. Luminol reacts with H_2O_2 to produce CL in the presence of a suitable catalyst, such as peroxidase or metal ions, making it compatible with oxidase enzymes that generate H_2O_2 for ECL analytical applications [88]. The ECL mechanism involves luminol oxidation at the electrode, followed by a reaction with H_2O_2, forming an adduct that releases an N_2 molecule to yield the excited 3AP, which emits light (Scheme 7). The ECL emission from luminol, although affected by electrode material, solvent, and pH, is generally observed in the potential range of 0.3–0.5 V (vs. Ag/AgCl) [89–92].

Luminol can be labeled on proteins and microbeads for ECL investigations. However, the irreversible chemical reactions that generate the light make luminol intrinsically low efficient, with a single photon emission each molecule [67], in contrast to inorganic complexes which reversible reactions can afford multiple photon emission from a single luminophore, resembling a catalytic cycle [26].

Scheme 7 Simplified reaction mechanism for ECL of luminol (L) with H_2O_2

$$L \longrightarrow L^{\bullet +} + e^-$$
$$L^{\bullet +} \longrightarrow +L^{\bullet} + H^+$$
$$L^{\bullet} + H_2O_2 \longrightarrow 3AP^* + N_2$$
$$3AP^* \longrightarrow 3AP + h\nu$$

8 ECL Analytical Applications and ECL Microscopy and Imaging

ECL possesses the advantages of electrochemical methods, such as fast measurements in low volume of sample. In particular, it is suitable for the development of point-of-care testing. Thanks to the combination between electrochemical and spectroscopic methods, ECL offers a superior temporal and spatial control on light emission. Because the light emission is triggered by an electrochemical reaction, it brings a decoupling of the stimulus from the analytical signal which greatly suppresses the background. In addition, the signal is intrinsically amplified because the luminophore can afford a sequence of excitation and emission steps. These two last conditions allow to reach very easily low detection limits (pM-fM) and a broad dynamic range [93, 94].

The analytical applications of ECL began as natural consequence of the introduction of coreactant ECL, either for quantification of luminophore or coreactant. Although without immediate sight of application, it resulted in the development of ECL immunoassay when Allen J. Bard and George M. Whitesides deposited the idea of luminophore-labeled proteins in 1985 [21]. Since then, the ECL for clinical diagnostic was proposed 1991 [23], and first clinical analyzer entered the market in 1994 [24]. Today, two analytical platforms based on ECL immunoassays (ECLIA) are available (Fig. 15).

From the previous examples of coreactant mechanism, only TPrA in combination with $Ru(bpy)_3^{2+}$, or related inorganic complex of Ru of Ir, are found in the clinical analysis market because the favorable combination of energetic and mechanistic reactions involved in ECL immunoassay between $Ru(bpy)_3^{2+}$ and TPrA [9, 11, 12, 26].

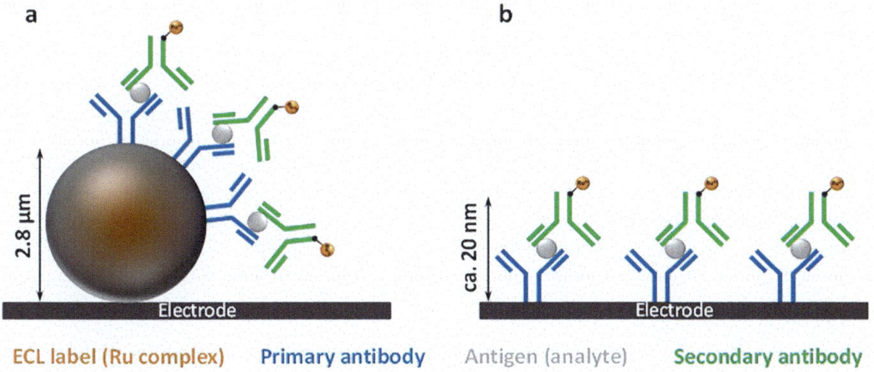

Fig. 15 Schematic representation of ECL sandwich immunoassay: **a** bound on magnetic microbead by Roche Diagnostic, and **b** directly bound onto the electrode by Meso Scale Diagnostics. Primary antibody (blue), antigen (gray), secondary antibody (green), and ruthenium ECL label (orange)

In the wide variety of ECL analytical application, it is appropriate to distinguish between the academic research from clinical analysis. In the former classification any analyte or strategy can be pursued, only limited by human ingenuity. In the latter, the research is directed to immunochemistry for life science, drug discovery and development, and clinical trials or diagnostic. Major analytes are amino acids, drugs, toxins, proteins and immunoglobulins, hormones and DNA. These important biological targets can be in serum, plasma, urine, saliva, or directly on cells.

Another important difference is the wide range of ECL sensing strategies being investigated for analytical applications in academic research. Developing sensors to be used in real setting is still challenging, and the research should prioritize simple and effective systems which are robust and reliable, rather than eccentric or fashionable.

In fact, although ECL sensing is not limited to these two platforms (Fig. 15), when put to test, other strategies are far away from any hypothetical development into commercial analyzers.

ECL microscopy (ECLM) and imaging are the complementary application of ECL as transduction strategy. Because ECL is limited in space and time, imaging the ECL emission add further information on the system under investigation. The spatiotemporal resolution permits to develop multiplexing analysis platforms, locate micro-objects and imaging cells [46, 95], organelles [96], single biomolecule [48] and single entities [97], or visualizing chemical reaction, especially useful in mechanistic investigations [26, 43]. This book is devoted to this research field, and each chapter is meant to provide a comprehensive view of ECLM and imaging, encompassing from fundamental research to foreseeable applications.

References

1. Marcus, R.A.: On the theory of chemiluminescent electron-transfer reactions. J. Chem. Phys. **43**, 2654–2657 (1965)
2. Vasil'ev, R.F.: Chemiluminescence excitation mechanisms. Russ. Chem. Rev. **39**, 529–544 (1970)
3. McCapra, F.: Chemical generation of excited states: the basis of chemiluminescence and bioluminescence. Methods Enzymol. **305**, 3–47 (2000)
4. Zomer, G.: The nature of chemiluminescent reactions in chemiluminescence and bioluminescence. In: Roda, A. (ed.) Chemiluminescence and Bioluminescence: Past, Present and Future, pp. 51–90. RSC Publishing (2010)
5. Bard, A.J. (ed.): Electrogenerated Chemiluminescence. CRC Press, Boca Raton (2004)
6. Richter, M.M.: Electrochemiluminescence (ECL). Chem. Rev. **104**, 3003–3036 (2004)
7. Miao, W.: Electrogenerated chemiluminescence and its biorelated applications. Chem. Rev. **108**, 2506–2553 (2008)
8. Sojic, N. (ed.): Analytical Electrogenerated Chemiluminescence: From Fundamentals to Bioassays. The Royal Society of Chemistry (2020)
9. Knežević, S., Han, D., Liu, B., Jiang, D., Sojic, N.: Electrochemiluminescence microscopy. Angew. Chem. Int. Ed. **63**, e202407588 (2024)
10. Villani, E.: Imaging wireless electrode reactions by bipolar electrochemiluminescence. Electrochemistry **92**, 101007 (2024)

11. Rebeccani, S., Zanut, A., Santo, C.I., Valenti, G., Paolucci, F.: A guide inside electrochemiluminescent microscopy mechanisms for analytical performance improvement. Anal. Chem. **94**, 336–348 (2022)
12. Zanut, A., Fiorani, A., Rebeccani, S., Kesarkar, S., Valenti, G.: Electrochemiluminescence as emerging microscopy techniques. Anal. Bioanal. Chem. **411**, 4375–4382 (2019)
13. Dufford, R.T., Nightingale, D., Gaddum, L.W.: Luminescence of Grignard compounds in electric and magnetic fields, and related electrical phenomena. J. Am. Chem. Soc. **49**, 1858–1864 (1927)
14. Harvey, N.: Luminescence during electrolysis. J. Phys. Chem. **33**, 1456–1459 (1929)
15. Hercules, D.M.: Chemiluminescence resulting from electrochemically generated species. Science **145**, 808–809 (1964)
16. Visco, R.E., Chandross, E.A.: Electroluminescence in solutions of aromatic hydrocarbons. J. Am. Chem. Soc. **86**, 53505351 (1964)
17. Santhanam, K.S.V., Bard, A.J.: Chemiluminescence of electrogenerated 9,10-diphenylanthracene anion radical. J. Am. Chem. Soc. **87**, 139–140 (1965)
18. Tokel, N.E., Bard, A.J.: Electrogenerated chemiluminescence. IX. Electrochemistry and emission from systems containing tris(2,2′-bipyridine)ruthenium(II) dichloride. J. Am. Chem. Soc. **94**, 2862–2863 (1972)
19. Rubinstein, I., Bard, A.J.: Electrogenerated chemiluminescence. 37. Aqueous ecl systems based on tris(2,2′-bipyridine)ruthenium(2^+) and oxalate or organic acids. J. Am. Chem. Soc. **103**, 512–516 (1981)
20. Leland, J.K., Powell, M.J.: Electrogenerated chemiluminescence: an oxidative-reduction type ECL reaction sequence using tripropyl amine. J. Electrochem. Soc. **137**, 3127–3131 (1990)
21. Bard, A.J., Whitesides, G.M.: Luminescent metal chelate labels and means for detection. WO1986002734 (1986)
22. Leland, J.K., Powell, M.J.: Electrochemiluminescent reaction utilizing amine-derived reductant. WO9005296 (1990)
23. Blackburn, G.F., Shah, H.P., Kenten, J.H., Leland, J., Kamin, R.A., Link, J., Peterman, J., Powell, M.J., Shah, A., Talley, D.B., Tyagi, S.K., Wilkins, E., Wu, T.-G., Massey, R.J.: Electrochemiluminescence detection for development of immunoassays and DNA probe assays for clinical diagnostic. Clin. Chem. **37**, 1534–1539 (1991)
24. Yang, H., Leland, J.K., Yost, D., Massey, R.J.: Electrochemiluminescence: a new diagnostic and research tool. Nat. Biotechnol. **12**, 193–194 (1994)
25. Fiorani, A., Valenti, G., Iurlo, M., Marcaccio, M., Paolucci, F.: Electrogenerated chemiluminescence: a molecular electrochemistry point of view. Curr. Opin. Electrochem. **8**, 31–38 (2018)
26. Zanut, A., Fiorani, A., Canola, S., Saito, T., Ziebart, N., Rapino, S., Rebeccani, S., Barbon, A., Irie, T., Josel, H.-P., Negri, F., Marcaccio, M., Windfuhr, M., Imai, K., Valenti, G., Paolucci, F.: Insights into the mechanism of coreactant electrochemiluminescence facilitating enhanced bioanalytical performance. Nat. Commun. **11**, 2668 (2020)
27. https://www.nobelprize.org/about-the-nobel-prize/. Accessed 29 Nov 2024
28. Feldberg, S.W.: Theory of controlled potential electrogeneration of chemiluminescence. J. Am. Chem. Soc. **88**, 390–393 (1966)
29. Faulkner, L.R., Bard, A.J.: Electrogenerated chemiluminescence. IV. Magnetic field effects on the electrogenerated chemiluminescence of some anthracenes. J. Am. Chem. Soc. **91**, 209–210 (1969)
30. Chang, M.-M., Saji, T., Bard, A.J.: Electrogenerated chemiluminescence. 30. Electrochemical oxidation of oxalate ion in the presence of luminescers in acetonitrile solutions. J. Am. Chem. Soc. **99**, 5399–5403 (1977)
31. Rubinstein, I., Bard, A.J.: Polymer films on electrodes. 5. Electrochemistry and chemiluminescence at Nafion-coated electrodes. J. Am. Chem. Soc. **103**, 5007–5013 (1981)
32. White, H.S., Bard, A.J.: Electrogenerated chemiluminescence. 41. Electrogenerated chemiluminescence and chemiluminescence of the $Ru(2,2'bpy)_3{}^{2+}$–$S_2O_8{}^{2-}$ system in acetonitrile-water solutions. J. Am. Chem. Soc. **104**, 6891–6895 (1982)

33. Ege, D., Becker, W.G., Bard, A.J.: Electrogenerated chemiluminescent determination of $Ru(bpy)_3^{2+}$ at low levels. Anal. Chem. **56**, 2413–2417 (1984)
34. Collinson, M.M., Wightman, R.M.: Observation of individual chemical reactions in solution. Science **268**, 1883–1885 (1995)
35. Wightman, R.M., Curtis, C.L., Flowers, P.A., Maus, R.G., McDonald, E.M.: Imaging microelectrodes with high-frequency electrogenerated chemiluminescence. J. Phys. Chem. B **102**, 9991–9996 (1998)
36. Horiuchi, T., Niwa, O., Hatakenaka, N.: Evidence for laser action driven by electrochemiluminescence. Nature **394**, 659–661 (1998)
37. Arora, A., Eijkel, J.C.T., Morf, W.E., Manz, A.: A wireless electrochemiluminescence detector applied to direct and indirect detection for electrophoresis on a microfabricated glass device. Anal. Chem. **73**, 3282–3288 (2001)
38. Zu, Y., Ding, Z., Zhou, J., Lee, Y., Bard, A.J.: Scanning optical microscopy with an electrogenerated chemiluminescent light source at a nanometer tip. Anal. Chem. **73**, 2153–2156 (2001)
39. Ding, Z., Quinn, B.M., Haram, S.K., Pell, L.E., Korgel, B.A., Bard, A.J.: Electrochemistry and electrogenerated chemiluminescence from silicon nanocrystal quantum dots. Science **296**, 1293–1297 (2002)
40. Miao, W., Choi, J.-P., Bard, A.J.: Electrogenerated chemiluminescence 69: the tris(2,2′-bipyridine)ruthenium(II), ($Ru(bpy)_3^{2+}$)/tri-*n*-propylamine (TPrA) system revisited—a new route involving $TPrA^{\cdot+}$ cation radicals. J. Am. Chem. Soc. **124**, 14478–14485 (2002)
41. Zhang, L., Dong, S.: Electrogenerated chemiluminescence sensors using $Ru(bpy)_3^{2+}$ doped in silica nanoparticles. Anal. Chem. **78**, 5119–5123 (2006)
42. Chang, Y.-L., Palacios, R.E., Fan, F.-R.F., Bard, A.J., Barbara, P.F.: Electrogenerated chemiluminescence of single conjugated polymer nanoparticles. J. Am. Chem. Soc. **130**, 8906–8907 (2008)
43. Sentic, M., Milutinovic, M., Kanoufi, F., Manojlovic, D., Arbault, S., Sojic, N.: Mapping electrogenerated chemiluminescence reactivity in space: mechanistic insight into model systems used in immunoassays. Chem. Sci. **5**, 2568–2572 (2014)
44. Kerr, E., Doeven, E.H., Barbante, G.J., Hogan, C.F., Bower, D.J., Donnelly, P.S., Connell, T.U., Francis, P.S.: Annihilation electrogenerated chemiluminescence of mixed metal chelates in solution: modulating emission colour by manipulating the energetics. Chem. Sci. **6**, 472–479 (2015)
45. Irkham, W.T., Fiorani, A., Valenti, G., Paolucci, F., Einaga, Y.: Co-reactant-on-Demand ECL: electrogenerated chemiluminescence by the in situ production of $S_2O_8^{2-}$ at boron-doped diamond electrodes. J. Am. Chem. Soc. **138**, 15636–15641 (2016)
46. Valenti, G., Scarabino, S., Goudeau, B., Lesch, A., Jović, M., Villani, E., Sentic, M., Rapino, S., Arbault, S., Paolucci, F., Sojic, N.: Single cell electrochemiluminescence imaging: from the proof-of-concept to disposable device-based analysis. J. Am. Chem. Soc. **139**, 16830–16837 (2017)
47. Carrara, S., Aliprandi, A., Hogan, C.F., De Cola, L.: Aggregation-induced electrochemiluminescence of platinum(II) complexes. J. Am. Chem. Soc. **139**, 14605–14610 (2017)
48. Liu, Y., Zhang, H., Li, B., Liu, J., Jiang, D., Liu, B., Sojic, N.: Single biomolecule imaging by electrochemiluminescence. J. Am. Chem. Soc. **143**, 17910–17914 (2021)
49. Dong, J., Lu, Y., Xu, Y., Chen, F., Yang, J., Chen, Y., Feng, J.: Direct imaging of single-molecule electrochemical reactions in solution. Nature **596**, 244–249 (2021)
50. Fracassa, A., Santo, C.I., Kerr, E., Knežević, S., Hayne, D.J., Francis, P.S., Kanoufi, F., Sojic, N., Paolucci, F., Valenti, G.: Redox-mediated electrochemiluminescence enhancement for bead-based immunoassay. Chem. Sci. **15**, 1150–1158 (2024)
51. Villani, E., Sakanoue, K., Einaga, Y., Inagi, S., Fiorani, A.: Photophysics and electrochemistry of ruthenium complexes for electrogenerated chemiluminescence. J. Electroanal. Chem. **921**, 116677 (2022)
52. Bard, A.J., Faulkner, L.R., White, H.S.: Electrochemical Methods: Fundamentals and Applications, 3rd edn. Wiley (2022)

53. Marcus, R.A.: Electron transfer reactions in chemistry: theory and experiment (nobel lecture). Angew. Chem. Int. Ed. **32**, 1111–1121 (1993)
54. Marcus, R.A.: Erratum: on the theory of chemiluminescent electron transfer reactions. J. Chem. Phys. **52**, 2803–2804 (1970)
55. Wallace, W.L., Bard, A.J.: Electrogenerated chemiluminescence. 35. Temperature dependence of the ECL efficiency of tris(2,2′-bipyridine)rubidium(2^+) in acetonitrile and evidence for very high excited state yields from electron transfer reactions. J. Phys. Chem. **83**, 1350–1357 (1979)
56. Haghighatbin, M.A., Laird, S.E., Hogan, C.F.: Electrochemiluminescence of cyclometalated iridium (III) complexes. Curr. Opin. Electrochem. **7**, 216–223 (2018)
57. Fiorani, A., Valenti, G., Irkham, P.F., Einaga, Y.: Quantification of electrogenerated chemiluminescence from tris(bipyridine)ruthenium(II) and hydroxyl ions. Phys. Chem. Chem. Phys. **22**, 15413–15417 (2020)
58. Fiorani, A., Eßmann, V., Santos, C.S., Schuhmann, W.: Enhancing electrogenerated chemiluminescence on platinum electrodes through surface modification. ChemElectroChem **7**, 1256–1260 (2020)
59. Kesarkar, S., Valente, S., Zanut, A., Palomba, F., Fiorani, A., Marcaccio, M., Rampazzo, E., Valenti, G., Paolucci, F., Prodi, L.: Neutral dye doped silica nanoparticles for electrogenerated chemiluminescence signal amplification. J. Phys. Chem. C **123**, 5686–5691 (2019)
60. Fiorani, A., Difonzo, M., Rizzo, F., Valenti, G.: Versatile electrochemiluminescent organic emitters. Curr. Opin. Electrochem. **34**, 100998 (2022)
61. Valenti, G., Fiorani, A., Di Motta, S., Bergamini, G., Gingras, M., Ceroni, P., Negri, F., Paolucci, F., Marcaccio, M.: Molecular size and electronic structure combined effects on the electrogenerated chemiluminescence of sulfurated pyrene-cored dendrimers. Chem. Eur. J. **21**, 2936–2947 (2015)
62. Alexander, B., Nepomnyashchii, A.B., Bard, A.J.: Electrochemistry and electrogenerated chemiluminescence of BODIPY dyes. Acc. Chem. Res. **45**, 1844–1853 (2012)
63. Valenti, G., Rampazzo, E., Bonacchi, S., Khajvand, T., Juris, R., Montalti, M., Marcaccio, M., Paolucci, F., Prodi, L.: A versatile strategy for tuning the color of electrochemiluminescence using silica nanoparticles. Chem. Commun. **48**, 4187–4189 (2012)
64. Dick, J.E., Renault, C., Kim, B.K., Bard, A.J.: Electrogenerated chemiluminescence of common organic luminophores in water using an emulsion system. J. Am. Chem. Soc. **136**, 13546–13549 (2014)
65. Omer, K.M., Bard, A.J.: Electrogenerated chemiluminescence of aromatic hydrocarbon nanoparticles in an aqueous solution. J. Phys. Chem. C **113**, 11575–11578 (2009)
66. Natarajan, P., Schmittel, M.: 9,10-diarylanthracenes as stable electrochemiluminescent emitters in water. J. Org. Chem. **77**, 8669–8677 (2012)
67. Fiorani, A., Santo, C.I., Sakanoue, K., Calabria, D., Mirasoli, M., Paolucci, F., Valenti, G., Einaga, Y.: Electrogenerated chemiluminescence from luminol-labelled microbeads triggered by in situ generation of hydrogen peroxide. Anal. Bioanal. Chem. **416**, 7277–7283 (2024)
68. Tokel-Takvoryan, N.E., Hemingway, R.E., Bard, A.J.: Electrogenerated chemiluminescence. XIII. Electrochemical and electrogenerated chemiluminescence studies of ruthenium chelates. J. Am. Chem. Soc. **95**, 6582–6589 (1973)
69. Suzuki, K., Kobayashi, A., Kaneko, S., Takehira, K., Yoshihara, T., Ishida, H., Shiina, Y., Oishic, S., Tobita, S.: Reevaluation of absolute luminescence quantum yields of standard solutions using a spectrometer with an integrating sphere and a back-thinned CCD detector. Phys. Chem. Chem. Phys. **11**, 9850–9860 (2009)
70. Sakanoue, K., Fiorani, A., Irkham, E.Y.: The effect of boron-doping level and surface termination in diamond on electrogenerated chemiluminescence. ACS Appl. Electron. Mater. **3**, 4180–4188 (2021)
71. Imai, K., Valenti, G., Villani, E., Rapino, S., Rampazzo, E., Marcaccio, M., Prodi, L., Paolucci, F.: Numerical simulation of doped silica nanoparticle electrochemiluminescence. J. Phys. Chem. C **119**, 26111–26118 (2015)
72. Rubinstein, I., Martin, C.R., Bard, A.J.: Electrogenerated chemiluminescent determination of oxalate. Anal. Chem. **55**, 1580–1582 (1983)

73. Sakanoue, K., Fiorani, A., Santo, C.I., Irkham, V.G., Paolucci, F., Einaga, Y.: Boron-doped diamond electrode outperforms the state-of-the-art electrochemiluminescence from microbeads immunoassay. ACS Sens. **7**, 1145–1155 (2022)
74. Fiorani, A., Valenti, G., Paolucci, F., Einaga, Y.: Electrogenerated chemiluminescence at boron-doped diamond electrodes. Chem. Commun. **59**, 7900–7910 (2023)
75. Fiorani, A., Han, D., Jiang, D., Fang, D., Paolucci, F., Sojic, N., Valenti, G.: Spatially resolved electrochemiluminescence through a chemical lens. Chem. Sci. **11**, 10496–10500 (2020)
76. Liu, X., Shi, L., Niu, W., Li, H., Xu, G.: Environmentally friendly and highly sensitive Ruthenium(II) tris(2,2′-bipyridyl) electrochemiluminescent system using 2-(Dibutylamino)ethanol as co-reactant. Angew. Chem. Int. Ed. **46**, 421–424 (2007)
77. Noffsinger, J.B., Danielson, N.D.: Generation of chemiluminescence upon reaction of aliphatic amines with tris(2,2′-bipyridine)ruthenium(III). Anal. Chem. **59**, 865–868 (1987)
78. Kanoufi, F., Zu, Y., Bard, A.J.: Homogeneous oxidation of trialkylamines by metal complexes and its impact on electrogenerated chemiluminescence in the trialkylamine/$Ru(bpy)_3{}^{2+}$ system. J. Phys. Chem. B **105**, 210–216 (2001)
79. Fracassa, A., Mariani, C., Fiorani, A., Einaga, Y., Hogan, C.F., Paolucci, F., Sojic, N., Francis, P.S., Valenti, G.: Overcoming kinetic barriers of remote electrochemiluminescence on boron-doped diamond via catalytic coreactant oxidation. Chem. Commun. **61**, 3900–3903 (2025)
80. Villani, E., Valenti, G., Marcaccio, M., Mattarozzi, L., Barison, S., Garoli, D., Cattarin, S., Paolucci, F.: Coreactant electrochemiluminescence at nanoporous gold electrodes. Electrochim. Acta **277**, 168–175 (2018)
81. Fiorani, A., Irkham, V.G., Paolucci, F., Einaga, Y.: Electrogenerated chemiluminescence with peroxydisulfate as a coreactant using boron doped diamond electrodes. Anal. Chem. **90**, 12959–12963 (2018)
82. Choi, J.-P., Bard, A.J.: Electrogenerated chemiluminescence (ECL) 79. Reductive-oxidation ECL of tris(2,2′-bipyridine)ruthenium(II) using hydrogen peroxide as a coreactant in pH 7.5 phosphate buffer solution. Anal. Chim. Acta **541**, 143–150 (2005)
83. Santa Cruz, T.D., Akins, D.L., Birke, R.L.: Chemiluminescence and energy transfer in systems of electrogenerated aromatic anions and benzoyl peroxide. J. Am. Chem. Soc. **98**, 1677–1682 (1976)
84. Zhou, P., Fu, W., Ding, L., Yan, Y., Guo, W., Su, B.: Toward mechanistic understanding of electrochemiluminescence generation by tris(2,2′-bipyridyl)ruthenium(II) and peroxydisulfate. Electrochim. Acta **439**, 141716 (2023)
85. Khan, P., Idrees, D., Moxley, M.A., Corbett, J.A., Ahmad, F., von Figura, G., Sly, W.S., Waheed, A., Hassan, I.: Luminol-based chemiluminescent signals: clinical and non-clinical application and future uses. Appl. Biochem. Biotechnol. **173**, 333–355 (2014)
86. Hiramoto, K., Villani, E., Iwama, T., Komatsu, K., Inagi, S., Inoue, K.Y., Nashimoto, Y., Ino, K., Shiku, H.: Recent advances in electrochemiluminescence-based systems for mammalian cell analysis. Micromachines **11**, 530 (2020)
87. Irkham, R.R.R., Ivandini, T.A., Fiorani, A., Einaga, Y.: Electrogenerated chemiluminescence of luminol mediated by carbonate electrochemical oxidation at a boron-doped diamond. Anal. Chem. **93**, 2336–2341 (2021)
88. Rahmawati, I., Einaga, Y., Ivandini, T.A., Fiorani, A.: Enzymatic biosensors with electrochemiluminescence transduction. ChemElectroChem **9**, e202200175 (2022)
89. Rahmawati, I., Fiorani, A., Irkham, W.W.T., Wahyuono, R.A., Einaga, Y., Ivandini, T.A.: Comparison between electrochemiluminescence of luminol and electrocatalysis by Prussian blue for the detection of hydrogen peroxide. Anal. Methods **17**, 1790–1796 (2025)
90. Rahmawati, I., Fiorani, A., Sanjaya, A.R., Du Irkham, J., Saepudin, E., Einaga, Y., Ivandini, T.A.: Modification of boron-doped diamond electrode with polyaniline and gold particles to enhance the electrochemiluminescence of luminol for the detection of reactive oxygen species (hydrogen peroxide and hypochlorite). Diam. Relat. Mater. **144**, 110956 (2024)
91. Syukur, J.M., Rahmawati, I., Sanjaya, A.R., Ridwan, M., Fiorani, A., Einaga, A., Ivandini, T.A.: Modification of screen-printed carbon electrodes with gold particles to enhance luminol electrochemiluminescence for hydrogen peroxide detection. Sens. Mater. **35**, 1785–1799 (2023)

92. Rahmawati, I., Saepudin, E., Fiorani, A., Einaga, Y., Ivandini, T.A.: Electrogenerated chemiluminescence of luminol at boron-doped diamond electrode for the detection of hypochlorite. Analyst **147**, 2696–2702 (2022)
93. Lippi, G., Henry, B.M., Adeli, K.: Diagnostic performance of the fully automated Roche Elecsys SARS-CoV-2 antigen electrochemiluminescence immunoassay: a pooled analysis. Clin. Chem. Lab. Med. **60**, 655–661 (2022)
94. Liang, M., Klakamp, S.L., Funelas, C., Lu, H., Lam, B., Herl, C., Umble, A., Drake, A.W., Pak, M., Ageyeva, N., Pasumarthi, R., Roskos, L.K.: Detection of high- and low-affinity antibodies against a human monoclonal antibody using various technology platforms. Assay Drug Dev. Technol. **5**, 655–662 (2007)
95. Knezevic, S., Bouffier, L., Liu, B., Jiang, D., Sojic, N.: Electrochemiluminescence microscopy: from single objects to living cells. Curr. Opin. Electrochem. **35**, 101096 (2022)
96. Descamps, J., Colin, C., Tessier, G., Arbault, S., Sojic, N.: Ultrasensitive imaging of cells and sub-cellular entities by electrochemiluminescence. Angew. Chem. Int. Ed. **62**, e202218574 (2023)
97. Zhao, W., Chena, H.-Y., Xu, J.-J.: Electrogenerated chemiluminescence detection of single entities. Chem. Sci. **12**, 5720–5736 (2021)

The Role of Electrode Materials in Electrogenerated Chemiluminescence

Andrea Fiorani and Yasuaki Einaga

Abstract In electrogenerated chemiluminescence (ECL), electrode materials affect the mechanism of signal generation, and therefore the emission intensity. This is particularly important in ECL imaging, where not only the total light emission, but its local state is crucial for visualizing the emission layer and exhibit an accurate and precise ECL image. The careful selection of electrode material is pivotal in ECL imaging because it has to be paired with the ECL system under investigation for the best results. This chapter reports an accurate selection of the most important electrode materials and their impact on ECL generation, with a special focus on the tris(2,2′-bipyridine)ruthenium(II)/tri-*n*-propylamine ECL system owing its extensive use in research and commercial clinical analyzers.

Keywords Electrogenerated chemiluminescence · Imaging · Electrode materials · Platinum · Carbon · Boron-doped diamond

1 Introduction

The choice of electrode materials for ECL do not differ from the general knowledge applied in other electroanalytical techniques. Nevertheless, particular attention has to be taken during the selection of the materials as it affects the ECL mechanisms, therefore the signal generation and emission intensity, to assure the best match with the ECL system of interest.

A convenient distinction can be based on the annihilation and coreactant ECL mechanisms. The requirement of organic solvents in annihilation ECL, mostly for investigation of new luminophores [1–3], does not dictate forcibly the use of a material over the others, except for the reversibility of the electrochemical reaction or

A. Fiorani · Y. Einaga (✉)
Department of Chemistry, Keio University, Yokohama, Japan
e-mail: einaga@keio.jp

A. Fiorani
e-mail: andrea.fiorani@keio.jp

F. Paolucci et al. (eds.), *Electrogenerated Chemiluminescence Microscopy and Imaging*,
https://doi.org/10.1007/978-3-032-14602-1_2

adsorption of molecules (e.g., sulfur compounds are better investigated with a glassy carbon electrode than a gold electrode) [4]. On the other hand, coreactant ECL, which is conducted in aqueous solvent for analytical applications, is highly affected by the electrode interaction with water and its response during potential application. Moreover, not only the material, but also the chemical and physical state of the electrode surface, which affect its electrochemical response, result in large variations of the ECL emission.

Nowadays, ECL is a powerful transduction technique in analytical chemistry, with important applications in clinical analyzers based on the coreactant ECL system tris(2,2′-bipyridine)ruthenium(II), ($Ru(bpy)_3^{2+}$)/tri-*n*-propylamine (TPrA). Therefore, the discussion on electrode materials will mainly focus on this luminophore/coreactant couple for two main reasons: its utmost importance in commercialized instrumentations, and the large amount of acquired knowledge compared to other ECL systems.

The choice of the electrode material requires the knowledge of its electrochemical response to find the best combination with the ECL system of interest [5].

2 Metals: Platinum and Gold

Platinum (Pt) and gold (Au) are even now widely used electrode materials in electrochemistry, and so in ECL, especially for the fast kinetics of electron transfer reactions [6]. This is particularly useful in the characterization of new luminophores, where organic solvents and aprotic condition enable a potential window of about 6 V, enough to explore a wide range of annihilation mechanisms [7]. Metals can have high purity (up to 99.99%), the shape is highly reproducible and the ductility allows the fabrication of electrodes with different size (from cm to μm). Another advantage of metal electrodes, in combination with organic solvents, is their superior resistance to fouling when compared to glassy carbon electrodes.

However, some drawbacks arise when metal electrodes are employed in aqueous solutions, a general environment for ECL analytical applications.

First, the narrow potential window that is limited by the small overpotential for the hydrogen evolution and the oxygen evolution reactions, which compete with luminophore or coreactant electrochemical reactions. In addition, bubbles formation at the electrode surface hampers the reactants diffusion perturbing the ECL generation.

Second, if we refer to the coreactant TPrA, its oxidation (0.88 V vs. SCE) [8] corresponds with the simultaneous generation of an oxide layer on the electrode surface (0.2 ~ 1.4 V vs. SCE) that prevents almost completely the heterogeneous electron transfer reaction of TPrA oxidation. This phenomenon is called electrode passivation that results is a serious quenching of the ECL signal and impairment of the technique sensitivity.

In aqueous solution, Pt can easily form oxide layers on its surface, while Au oxide formation is approximately 400 mV more positive than Pt oxide that gives Au 10 times higher ECL emission compared to Pt (Fig. 1) [9]. This delayed passivation for Au respect to Pt improves TPrA oxidation.

To prevent the oxide formation, or at least reduce its negative impact on the TPrA oxidation, the addition of surfactants, both ionic [10] and non-ionic [11, 12], and also alkanethiols in case of Au [11], in the electrolyte could alleviate this problem (Fig. 2). The formation of a surfactant layer on the electrode surface reduces the generation of the oxide. The effect of the surfactant is not limited at preventing the oxide formation, furthermore its hydrophobic nature promotes the TPrA oxidation

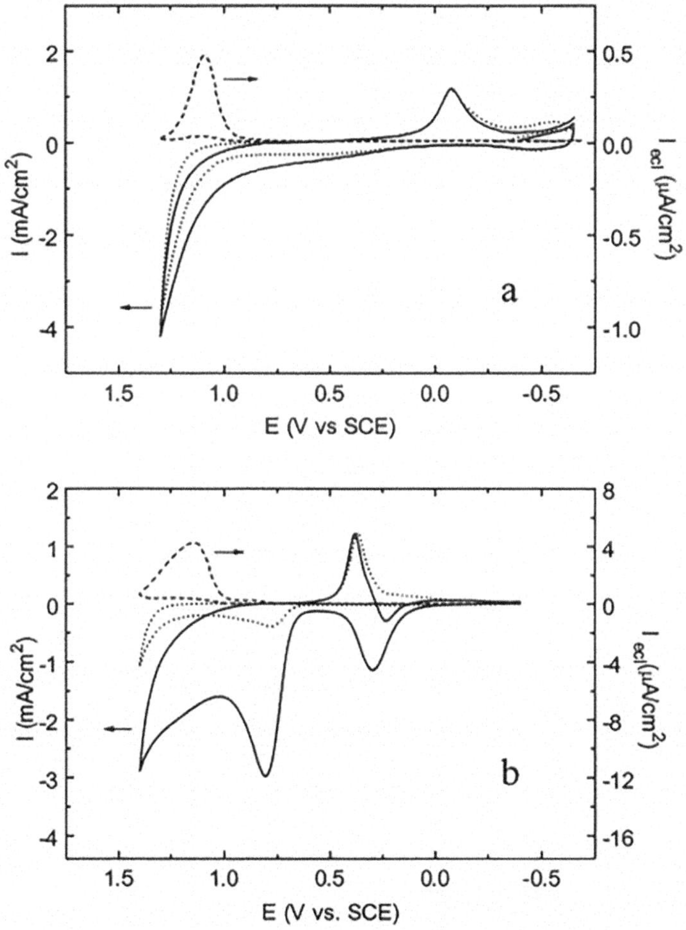

Fig. 1 Cyclic voltammogram and ECL curve at **a** platinum and **b** gold electrodes in 0.15 M phosphate buffer solutions (pH 7.5) containing 100 mM TPrA and 1 μM $Ru(bpy)_3^{2+}$. The dotted line represents data in the absence of both $Ru(bpy)_3^{2+}$ and TPrA. Potential scan rate, 0.1 V s^{-1}. Adapted with permission from [9], Copyright (2000) American Chemical Society

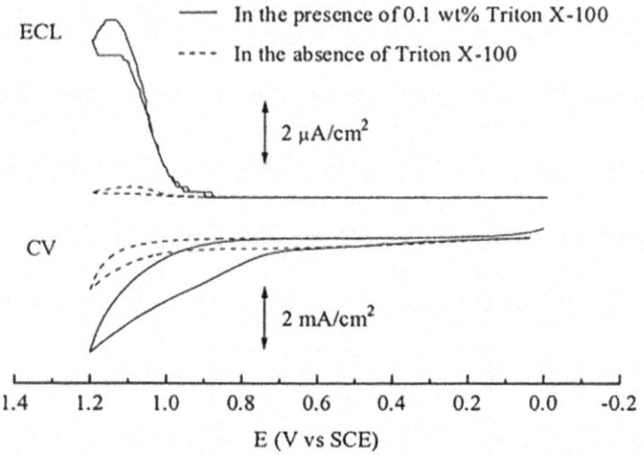

Fig. 2 Cyclic voltammograms and ECL curves at a Pt electrode in 0.15 M phosphate buffer solution (pH 7.5) containing 100 mM TPrA and 1 μM $Ru(bpy)_3{}^{2+}$ in the presence and absence of 0.1 wt% Triton X-100. Potential scan rate, 0.1 V s^{-1}. Adapted with permission from [11], Copyright (2001) American Chemical Society

rate enhancing the ECL emission [11, 13, 14]. Among the surfactants, the non-ionic shows the best result.

After ECL generation, metal electrodes can adsorb the coreactant by-products on their surface that might cause a poisoning of the electrode, modifying the electrochemical response [15, 16]. Therefore, either surfactants are used or not, the surface of metal electrodes requires a cleaning procedure (mechanical and/or electrochemical) after the ECL measurement to restore its initial state and assure signal reproducibility.

Example of metal electrodes in ECL imaging includes the use of Au electrode for the visualization of cell spheroids respiratory and secretory activities (Fig. 3). In presence of cell spheroids, the emission of ECL from luminol around the spheroids is suppressed due to O_2 consumption by the spheroids [17–19].

Historically, Au electrodes were initially employed in the first commercially available ECL instrument (Origen I analyzer) introduced by IGEN International in 1994 [20, 21]. However, in contemporary technology for electrochemiluminescence immunoassay (Fig. 4), now owned by Roche Diagnostics, Pt is utilized instead of Au [22, 23].

3 Carbon-Based Electrodes

Carbonaceous materials used in the fabrication of carbon electrodes consist of different types, namely carbon black, carbon nanotubes, graphene, graphite, and glassy carbon (GC), therefore mainly sp^2 carbon hybridization, although the production process affects the sp^2/sp^3 ratio. Moreover, diamond electrodes are also

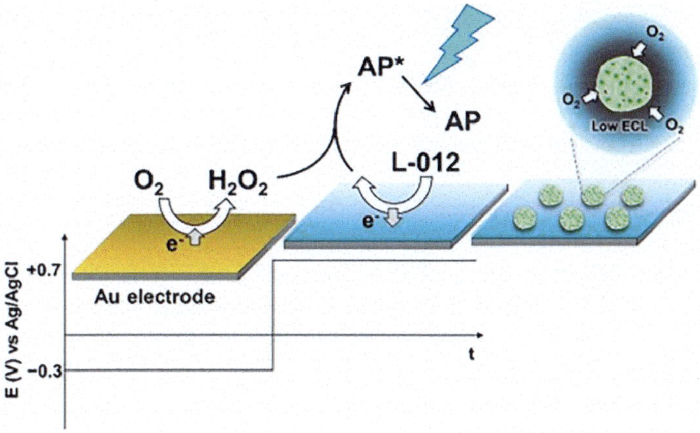

Fig. 3 Schematic illustration of electrochemiluminescence imaging using sequential potential steps. The generation of H_2O_2 from the electrochemical reduction of O_2 triggers the luminol ECL. L-012: luminol derivative; AP: 3-aminophthalate; AP*: excited state of AP. Adapted with permission from [17], Copyright (2021) Elsevier

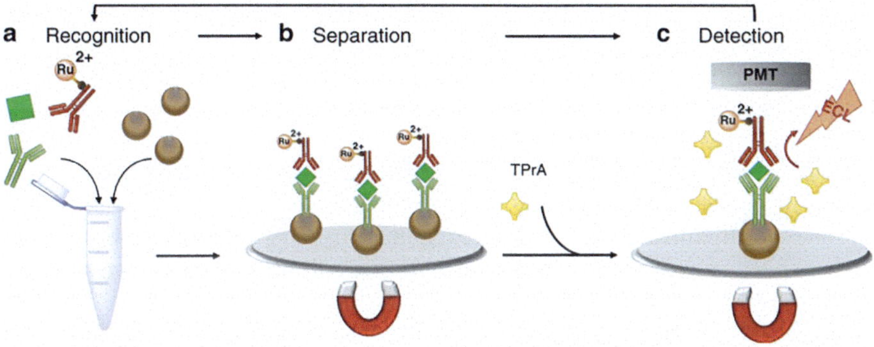

Fig. 4 **a** Recognition: a specific antigen (green square) is mixed together with an antigen-specific biotinylated monoclonal antibody (in green) and an antigen-specific monoclonal antibody labeled with tris(2,2′-bipyridine)ruthenium(II) (Ru^{2+}, in red) to form a sandwich immunoassay which binds to streptavidin-coated microbeads (brown spheres); **b** separation: microbeads are magnetically captured onto the surface of the electrode and unspecific substances are removed; **c** detection: a voltage is applied to the electrode inducing ECL emission, which is measured by a photomultiplier tube (PMT). TPrA: tri-*n*-propylamine. Adapted with permission from [24], Copyright (2020) Springer Nature

employed in electrochemistry, and ECL as well, which will be discussed in conclusion of this paragraph.

The general advantages of carbon electrodes over metal electrodes include low cost, wide potential window, and electrocatalytic activity for a variety of redox reactions [25].

Compared to metal, carbon electrodes show fast kinetics of TPrA heterogeneous electron transfer reaction that results in higher ECL emission, therefore better sensitivity. This is combined with a sluggish kinetics for both oxygen evolution and hydrogen evolution as well, which minimize competing reaction and bubbles generation at the electrode surface, thus making carbon electrodes preferentially used in aqueous electrolytes.

For example, direct comparison of GC (Fig. 5) to Au and Pt electrodes results in a ECL intensity 10 times and 100 times higher, respectively [9]. In case of GC, it has been used in combination of ECL imaging to decipher the mechanism of heterogeneous ECL mimicking the bead-based immunoassay [26].

In analogy with metal electrodes, glassy carbon requires a cleaning procedure (mechanical and/or electrochemical) after the ECL measurement to restore its initial state, because anodic oxidation during ECL generation [27], and adsorption of intermediates [15] lead to signal decrease.

This drawback has been alleviated by introducing a reduction step in the electrochemical procedure that trigger the ECL at GC electrode that is effective to sustain the ECL signal during ECL imaging [28, 29]. The cathodic regenerative treatment of the electrode surface allows to restore the initial TPrA oxidation current therefore to record a sequence of ECL image of cells (Fig. 6).

In addition, defects in the carbon structure or functional groups on the surface could cause generation of unspecific signal that increases the background [30].

Carbon is extremely useful in the fabrication of screen-printed electrode (SPE), which can be made transparent, or with high degree of transmittivity by using carbon nanotubes and graphene (see paragraph 4).

In case of SPE, the quality of carbon is extremely important to dictate the ECL signal (Fig. 7), while the incorporation of nanomaterials in the screen-printed electrode does not significantly enhance the ECL intensity [31], although nanostructured carbon can act as scaffold to bind the recognition unit of biosensors, for example, antibodies [32] or enzymes [33, 34].

Like Pt, also carbon finds application in clinical analyzers, in particular for its ability to generate intense ECL in aqueous electrolyte, although this approach is restricted to disposable platforms. The Multiarray Technology of Meso Scale Discovery uses screen-printed carbon ink both as an electrode and as a platform for the assay (Fig. 8). In fact, it resembles the enzyme linked immunosorbent assay (ELISA), where the recognition element is bound on the bottom of a well microtiter plates, and a sandwich immunoassay is formed by a secondary antibody labeled with the luminophore [35, 36].

Included in carbon-based electrodes is diamond, an intrinsic semiconductor with an indirect bandgap of 5.47 eV that was introduced into ECL relatively recently [37], compared to other carbon electrodes, with the first report in 2003 [38]. The metallic conductivity necessary for electrochemistry is generally assured by boron doping at concentration around or higher than 10^{20} [B] cm^{-3} [39]. Well-known characteristics include wide potential window in aqueous electrolyte (low catalytic activity for oxygen and hydrogen evolution reactions), low capacitive current, and physical and chemical stability.

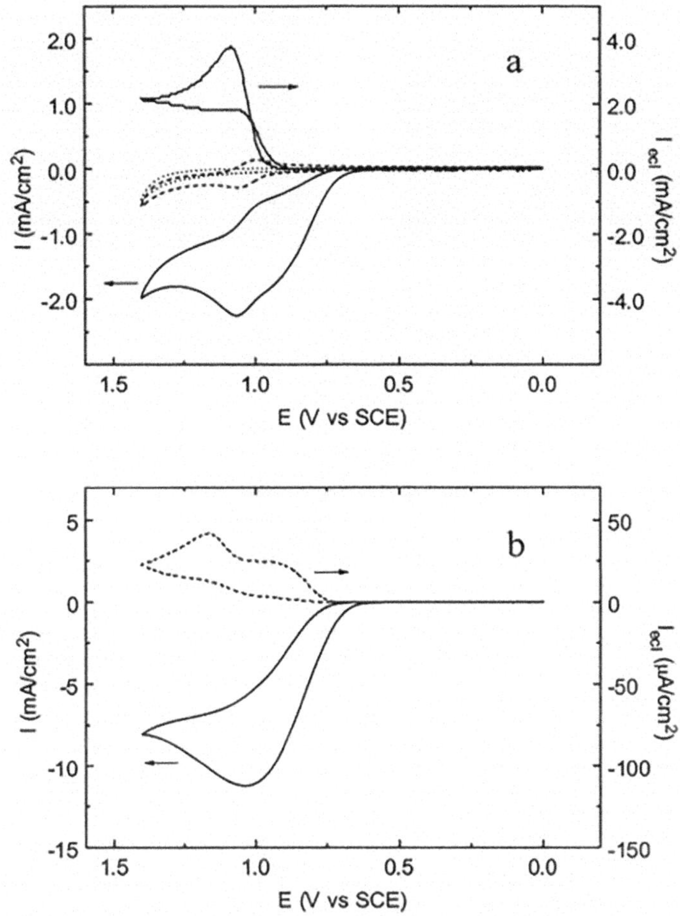

Fig. 5 Cyclic voltammograms and ECL curve of 1 mM $Ru(bpy)_3^{2+}$ at a glassy carbon electrode in 0.15 M phosphate buffer solution (pH 7.5) in the presence (solid line) and absence (dashed line) of 10 mM TPrA. The dotted line represents data in the absence of both $Ru(bpy)_3^{2+}$ and TPrA (**a**). Cyclic voltammogram and ECL curve at a glassy carbon electrode in 0.15 M phosphate buffer solution (pH 7.5) containing 100 mM TPrA and 1 μM $Ru(bpy)_3^{2+}$ (**b**). Potential scan rate, 0.1 V s^{-1}. Adapted with permission from [9], Copyright (2000) American Chemical Society

For the $Ru(bpy)_3^{2+}$/TPrA couple, BDD showed higher emission in the heterogeneous system with microbeads compared to conditions mimicking the commercial immunoassay with Pt electrode (Fig. 9) [40].

A special feature of BDD is the ability to produce directly in-situ the coreactant for the $Ru(bpy)_3^{2+}$ from sulfate or carbonate oxidation which generates $S_2O_8^{2-}$ or H_2O_2, respectively, for the "reductive-oxidation" ECL mechanism [41, 42]. The generation of the H_2O_2 coreactant directly in-situ from carbonate oxidation was also reported for luminol ECL [43].

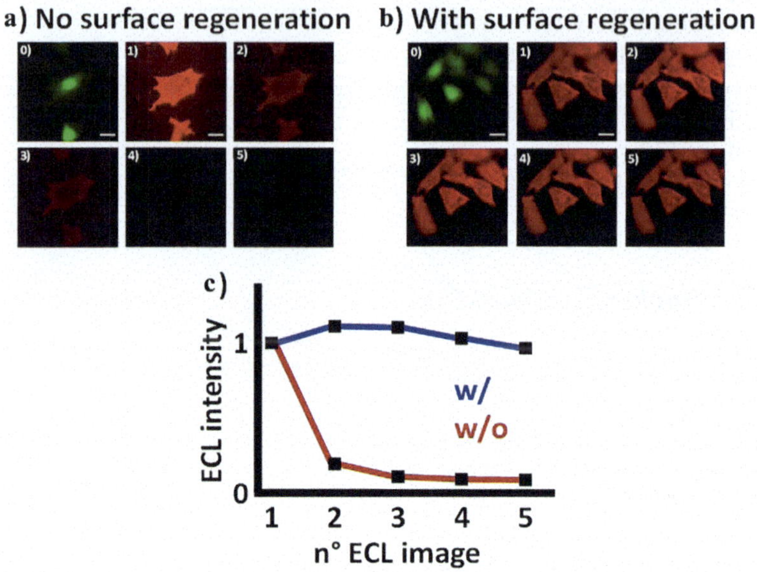

Fig. 6 Images of mammalian Chinese Hamster Ovary cells, (0) PL (green) and (1–5) ECL (red), grown on a GC electrode: without (**a**), and with (**b**) surface regeneration. Normalized ECL intensity (**c**) as function of ECL image sequence: without (red), and with (blue) surface regeneration. Regeneration treatment (30 pulses of − 1.8 V for 1 s) before the ECL imaging (1.45 V for 8 s). PBS solution containing 100 mM TPrA (pH 7.4). Scale bar: 20 μm. Reference electrode: Ag/AgCl/KCl (3 M). Adapted with permission from [28], Copyright (2021) American Chemical Society

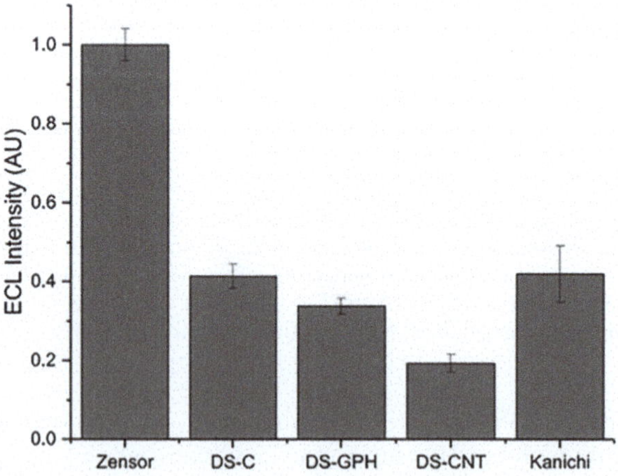

Fig. 7 Relative ECL intensities from various carbon-based SPEs for bead-based DNA assay. From DropSens (http://www.dropsens.com/): unmodified carbon (DS-C), carbon nanotubes (DS-CNT), graphene (DS-GPH). From eDAQ (http://www.edaq.com/): Zensor and Kanichi varieties. Adapted with permission from [31], Copyright (2021) Kerr, Alexander, Francis, Guijt, Barbante and Doeven

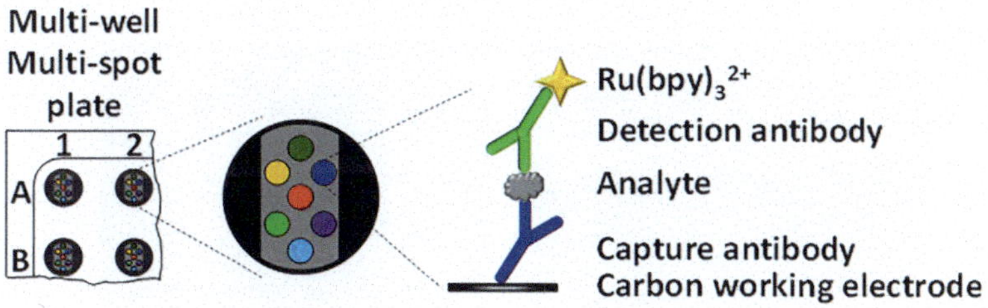

Fig. 8 Depiction of Meso Scale Discovery multiplexing sandwich immunoassay [36]

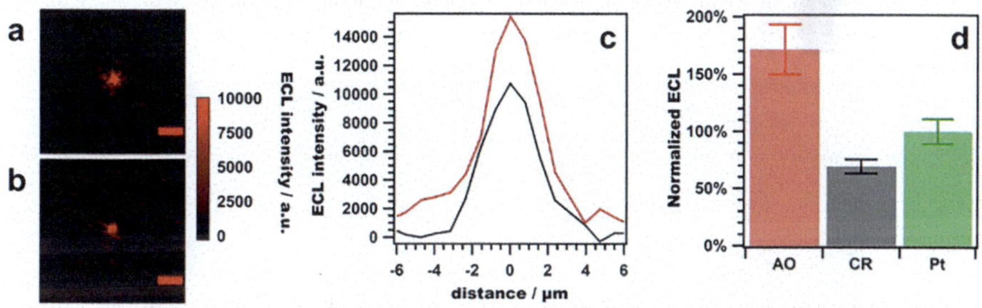

Fig. 9 ECL imaging of 2.8 µm bead labeled with $Ru(bpy)_3^{2+}$ for **a** oxidated BDD and **b** reduced BDD (scale bar: 10 µm). Comparison of the bead profile lines **c** oxidated BDD (red) and reduced BDD (black). ECL of 2.8 µm bead labeled with $Ru(bpy)_3^{2+}$ for oxidated BDD (AO, red), reduced BDD (CR, black), and platinum (Pt, green). Adapted with permission from [40], Copyright (2022) American Chemical Society

ECL imaging of BBD surface enabled to visualize the heterogeneity of the electrochemical activity of polycrystalline diamond [44], for (100), (111), and randomly oriented crystals (Fig. 10).

The (100)-oriented microcrystal shows the lowest ECL intensity, while the ECL intensities for the (111) and the randomly oriented BDD microcrystals are similar.

For the (100) the ECL increases linearly with the potential, indicating that more overpotential is required to drive the ECL reaction. This suggests a remarkably lower conductivity of (100)-oriented microcrystals compared to (111) which can be ascribed to microcrystal heterogeneity of boron concentration because heavily doped polycrystalline diamond contains microcrystals with different boron doping levels. This results in semiconductor and semimetallic diamond microcrystals leading to electrochemical reactivity heterogeneity at the heavily doped polycrystalline BDD.

The highest ECL peaks at the edges of the (100) and (111) facets are ascribed to diffusion effects. In fact, the ridges of microcrystals form pointed-edge lines toward

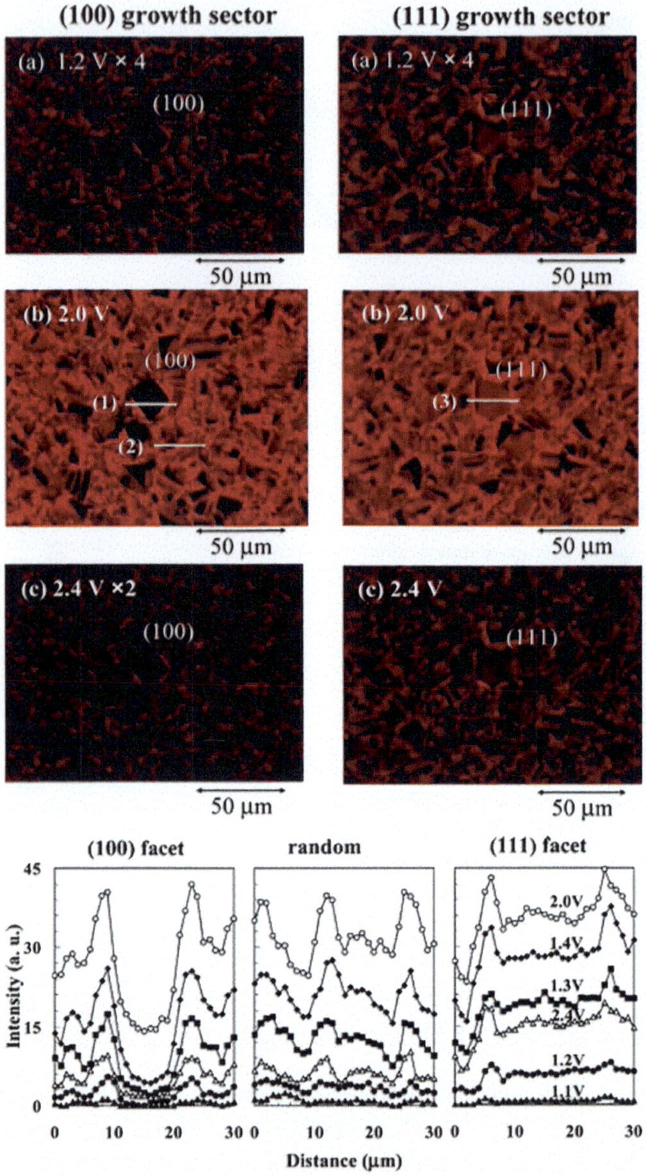

Fig. 10 ECL images (top) of polycrystalline BDD electrode during chronoamperometric measurements at **a** 1.2, **b** 2.0, and **c** 2.4 V (vs. Ag/AgCl) for 5 s, in 200 mM phosphate buffer solution containing 300 μM $Ru(bpy)_3^{2+}$ and 100 mM TPrA. "×4", e.g., refers to the number of magnification (4 times) of the light intensity. ECL intensity profiles (bottom) as a function of applied potential for three selected regions. The three selected regions are presented in the top Figure as horizontal bars: 1 for (100) facet, 2 for random, and 3 for (111) facet. Adapted with permission from [44], Copyright (2004) American Chemical Society

the solution and the reactant molecules can be supplied by semispherical diffusion, therefore enhancing the ECL intensity.

ECL imaging for analytical purposes was employed to investigate a nanoelectrode and ultramicroelectrode array platform fabricated on BDD by polycarbonate layer deposition and following e-beam lithography. In this case, the ECL imaging allows mapping and comparing simultaneously the spatial distribution of the ECL-emitting regions with a single image as a function of the geometrical features of the array as well as of the concentration of the TPrA coreactant (Fig. 11) [45].

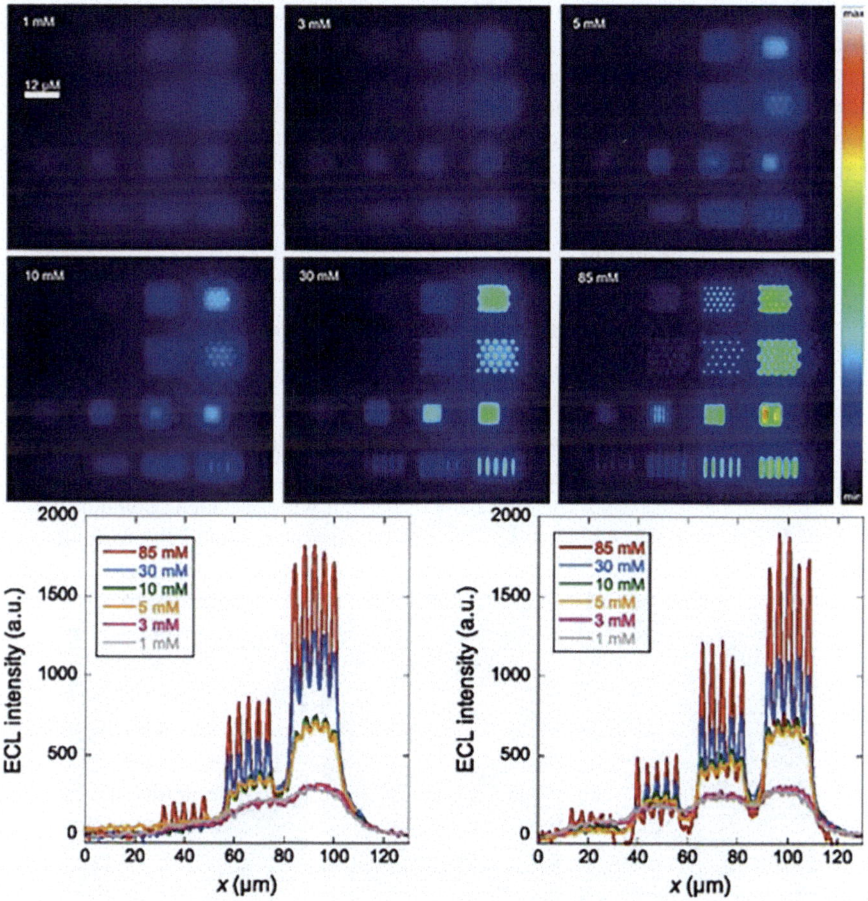

Fig. 11 Top: ECL images of BDD electrode array in PBS solution (pH 7.4) containing both 1 mM $Ru(bpy)_3^{2+}$ and increasing concentrations of TPrA. Images were recorded in the dark with a × 50 objective when applying a constant potential of 1.2 V versus Ag/AgCl/KCl. Bottom: effect of the TPrA concentration on the ECL intensity profiles. Adapted with permission from [45], Copyright (2016) Springer Nature

4 Transparent Electrodes

The ECL microscopy has benefitted particularly by the introduction of transparent electrode, where the light can be detected in inverted mode configuration directly through the electrode. This can bring easy visualization of processes limited on the electrode surface proximity by avoiding refraction of thick layers of electrolyte that can decrease the image resolution. In particular, for imaging of biological samples such as cells (Fig. 12) [46–48], single biomolecules [49], single nanoparticles [50], and as a light source [51].

Generally, transparent electrodes for ECL applications can be classified into two main groups: metal oxides on glass, such as indium-tin oxide (ITO) and fluorine-doped tin oxide (FTO), and carbon-based electrodes on glass or polymer support, such as graphene [52, 53] or nanotubes [54]. The main difference lies in the electrochemistry, especially the electron transfer constant of TPrA oxidation, for example, on a carbon nanotubes is about 30 times faster than on ITO, that is 2.6×10^{-2} cm s^{-1}

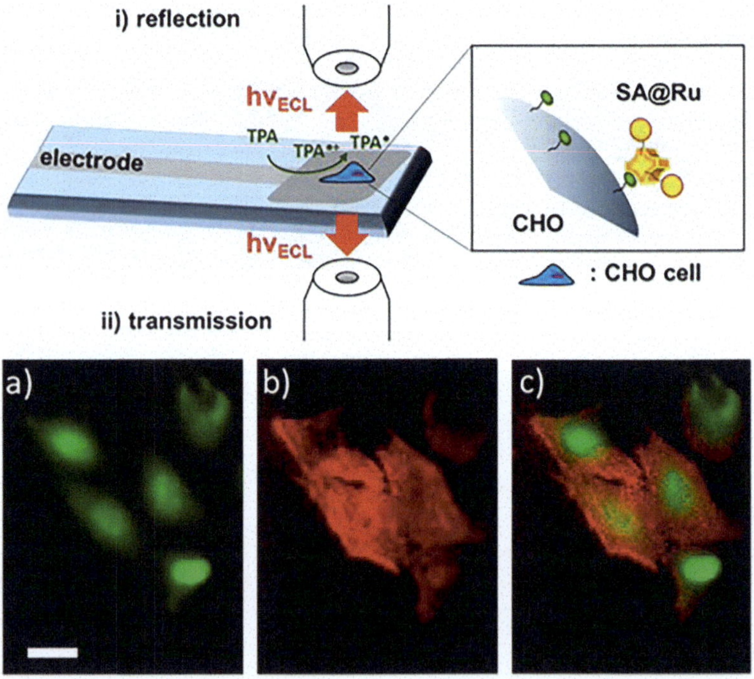

Fig. 12 Top: depiction of ECL microscopy for imaging cells labeled with streptavidin-ruthenium complex (SA@Ru) in reflection (i), and transmission (ii) configurations. Bottom: PL (**a**), ECL (**b**) and overlay of both luminescence images (**c**) of mammalian Chinese Hamster Ovary cells in the transmission configuration through the carbon nanotubes inkjet-printed electrode in the ECL focal plane (scale bar: 20 μm). Adapted with permission from [46], Copyright (2018) American Chemical Society

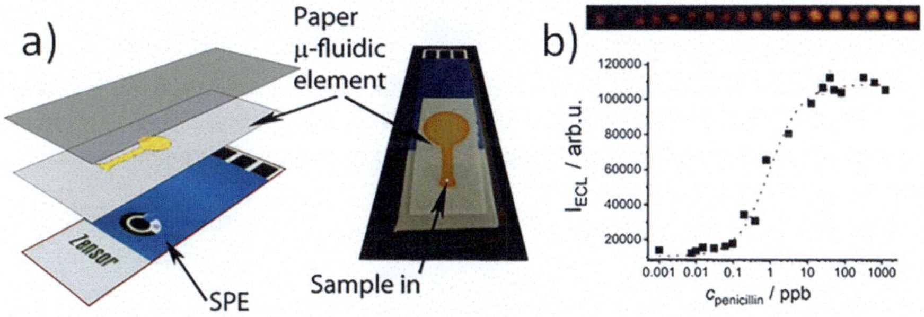

Fig. 13 Coupling of paper-based microfluidic with a screen-printed carbon electrode (**a**). Adapted with permission from [56], Copyright (2011) American Chemical Society. ECL signal from penicillin detection in milk diluted with phosphate buffer (1:3) on a screen-printed gold electrode coupled with a glass fiber paper strip and imaged by a smartphone (**b**). Adapted with permission from [57], Copyright (2021) Wiley

and 8×10^{-4} cm s^{-1}, respectively [54] resulting in higher ECL emission on carbon nanotubes.

5 Paper-Based Microfluidic and Screen-Printed Electrodes

Inkjet printing technology can be employed to fabricate paper microfluidic substrates, or alternatively using strip-based rapid test, that in combination with screen-printed electrodes create simple, easy to use, and disposable sensors [55]. Microfluidic paper-based analytical devices (mPADs) represent a platform that can be successfully coupled with smartphone camera to image and quantify the ECL signal, and develop portable devices (Fig. 13) [56, 57].

6 Conclusion

As an electrochemical technique, ECL is intrinsically affected by the heterogeneous electron transfer, thus highly dependent from the electrode material and its surface state. Upon the electrode material nature, the ECL intensity can results in large variations that translate into different sensitivity levels. This is particularly evident in the sensor applications with the coreactant TPrA oxidation reaction, which is an inner-sphere electron transfer.

We reported the most common electrode materials used in ECL and ECL imaging that can be categorized into two main groups: metal and carbon-based. While their use in organic and aprotic solvent could be nearly interchangeable, although differences exist, a remarkable dissimilarity arises in aqueous electrolyte. The surface generation of metal oxide, which hinders the heterogeneous electrochemical reactions, requires some form of protection or mitigation (e.g., surfactants), conversely to carbon-based that can be used as such.

On the other hand, carbon-based electrodes are more dependent from the fabrication process and the carbon form (e.g., carbon black, glassy carbon, graphene, nanotubes), and electrochemical pretreatments can be beneficial to enhance the ECL signal.

Within this context, the judicious design of the ECL system requires the pairing of the electrode material with the experimental parameters, such as coreactant, solvent, and analyte to develop an optimal ECL signal.

References

1. Fiorani, A., Valenti, G., Iurlo, M., Marcaccio, M., Paolucci, F.: Electrogenerated chemiluminescence: a molecular electrochemistry point of view. Curr. Opin. Electrochem. **8**, 31–38 (2018)
2. Villani, E., Sakanoue, K., Einaga, Y., Inagi, S., Fiorani, A.: Photophysics and electrochemistry of ruthenium complexes for electrogenerated chemiluminescence. J. Electroanal. Chem. **921**, 116677 (2022)
3. Haghighatbin, M.A., Laird, S.E., Hogan, C.F.: Electrochemiluminescence of cyclometalated iridium (III) complexes. Curr. Opin. Electrochem. **7**, 216–223 (2018)
4. Valenti, G., Fiorani, A., Di Motta, S., Bergamini, G., Gingras, M., Ceroni, P., Negri, F., Paolucci, F., Marcaccio, M.: Molecular size and electronic structure combined effects on the electrogenerated chemiluminescence of sulfurated pyrene-cored dendrimers. Chem. Eur. J. **21**, 2936–2947 (2015)
5. Valenti, G., Fiorani, A., Li, H., Sojic, N., Paolucci, F.: Essential role of electrode materials in electrochemiluminescence applications. ChemElectroChem **3**, 1990–1997 (2016)
6. Forster, R.J., Loughman, P., Keyes, T.E.: Effect of electrode density of states on the heterogeneous electron-transfer dynamics of osmium-containing monolayers. J. Am. Chem. Soc. **122**, 11948–11955 (2000)
7. Zoski, C.G. (eds.): Handbook of Electrochemistry. Elsevier B.V (2007)
8. Kanoufi, F., Zu, Y., Bard, A.J.: Homogeneous oxidation of trialkylamines by metal complexes and its impact on electrogenerated chemiluminescence in the trialkylamine/$Ru(bpy)_3{}^{2+}$ system. J. Phys. Chem. B **105**, 210–216 (2001)
9. Zu, Y., Bard, A.J.: Electrogenerated chemiluminescence. 66. The role of direct coreactant oxidation in the ruthenium tris(2,2′)bipyridyl/tripropylamine system and the effect of halide ions on the emission intensity. Anal. Chem. **72**, 3223–3232 (2000)
10. Xu, G., Pang, H.-L., Xu, B., Dong, S., Wong, K.Y.: Enhancing the electrochemiluminescence of tris(2,2′-bipyridyl)ruthenium(II) by ionic surfactants. Analyst **130**, 541–544 (2005)
11. Zu, Y., Bard, A.J.: Electrogenerated chemiluminescence. 67. Dependence of light emission of the tris(2,2′)bipyridylruthenium(II)/tripropylamine system on electrode surface hydrophobicity. Anal. Chem. **73**, 3960–3964 (2001)
12. Kirschbaum-Harriman, S., Duerkop, A., Baeumner, A.J.: Improving ruthenium-based ECL through nonionic surfactants and tertiary amines. Analyst **142**, 2648–2653 (2017)
13. Li, F., Zu, Y.: Effect of nonionic fluorosurfactant on the electrogenerated chemiluminescence of the tris(2,2′-bipyridine)ruthenium(II)/Tri-*n*-propylamine system: lower oxidation potential and higher emission intensity. Anal. Chem. **76**, 1768–1772 (2004)
14. Zu, Y., Li, F.: Characterization of the low-oxidation-potential electrogenerated chemiluminescence of tris(2,2′-bipyridine)ruthenium(II) with tri-*n*-propylamine as coreactant. Anal. Chim. Acta **550**, 47–52 (2005)

15. Adenier, A., Chehimi, M.M., Gallardo, I., Pinson, J., Vilà, N.: Electrochemical oxidation of aliphatic amines and their attachment to carbon and metal surfaces. Langmuir **20**, 8243–8253 (2004)
16. Aufray, M., Roche, A.A.: Is gold always chemically passive? Study and comparison of the epoxy-amine/metals interphases. Appl. Surf. Sci. **254**, 1936–1941 (2008)
17. Hiramoto, K., Ino, K., Komatsu, K., Nashimoto, Y., Shiku, H.: Electrochemiluminescence imaging of respiratory activity of cellular spheroids using sequential potential steps. Biosens. Bioelectron. **181**, 113123 (2021)
18. Hiramoto, K., Komatsu, K., Shikuwa, R., Konno, A., Sato, Y., Hirano-Iwata, A., Ino, K., Shiku, H.: Evaluation of respiratory and secretory activities of multicellular spheroids *via* electrochemiluminescence imaging. Electrochim. Acta **458**, 142507 (2023)
19. Hiramoto, K., Konno, A., Nashimoto, Y., Hirano-Iwata, A., Ino, K., Shiku, H.: Osteogenic differentiation of mesenchymal stem cell spheroids: a microfluidic device and electrochemiluminescence imaging study. Electrochim. Acta **491**, 144291 (2024)
20. Leland, J.K., Powell, M.J.: Electrogenerated chemiluminescence: an oxidative-reduction type ECL reaction sequence using tripropyl amine. J. Electrochem. Soc. **137**, 3127–3131 (1990)
21. Yang, H., Leland, J.K., Yost, D., Massey, R.J.: Electrochemiluminescence: a new diagnostic and research tool. ECL detection technology promises scientists new "yardsticks" for quantification. Nat. Biotechnol. **12**, 193–194 (1994)
22. Erler, K.: Elecsys immunoassay systems using electrochemiluminescence detection. Wien. Klin. Wochenschr. Suppl. **3**, 5–10 (1998)
23. Imai, K., Watari, S., Sakazume, T., Mitsuyama, S.: Clinical chemistry and immunoassay testing supporting the individual healthy life. Hitachi Rev **57**, 1–7 (2008)
24. Zanut, A., Fiorani, A., Canola, S., Saito, T., Ziebart, N., Rapino, S., Rebeccani, S., Barbon, A., Irie, T., Josel, H.-P., Negri, F., Marcaccio, M., Windfuhr, M., Imai, K., Valenti, G., Paolucci, F.: Insights into the mechanism of coreactant electrochemiluminescence facilitating enhanced bioanalytical performance. Nat. Commun. **11**, 2668 (2020)
25. McCreery, R.L.: Advanced carbon electrode materials for molecular electrochemistry. Chem. Rev. **108**, 2646–2687 (2008)
26. Sentic, M., Milutinovic, M., Kanoufi, F., Manojlovic, D., Arbault, S., Sojic, N.: Mapping electrogenerated chemiluminescence reactivity in space: mechanistic insight into model systems used in immunoassays. Chem. Sci. **5**, 2568–2572 (2014)
27. Chen, Z., Zu, Y.: Electrogenerated chemiluminescence of the tris(2,2′-bipyridine) ruthenium(II)/tri-n-propylamine (TPrA) system: crucial role of the long lifetime of TPrA + cation radicals suggested by electrode surface effects. J. Phys. Chem. C **112**, 16663–16667 (2008)
28. Han, D., Goudeau, B., Jiang, D., Fang, D., Sojic, N.: Electrochemiluminescence microscopy of cells: essential role of surface regeneration. Anal. Chem. **93**, 1652–1657 (2021)
29. Dutta, P., Han, D., Goudeau, B., Jiang, D., Fang, D., Sojic, N.: Reactivity mapping of luminescence in space: insights into heterogeneous electrochemiluminescence bioassays. Biosens. Bioelectron. **165**, 112372 (2020)
30. Wu, A.-H., Su, X.-L., Fang, Y.-M., Sun, J.-J., Chen, G.-N.: Electrogenerated chemiluminescence at bare glassy carbon electrode in basic media. Electrochem. Commun. **10**, 1344–1346 (2008)
31. Kerr, E., Alexander, R., Francis, P.S., Guijt, R.M., Barbante, G.J., Doeven, E.H.: A comparison of commercially available screen-printed electrodes for electrogenerated chemiluminescence applications. Front. Chem. **8**, 628483 (2021)
32. Zamolo, V.A., Valenti, G., Venturelli, E., Chaloin, O., Marcaccio, M., Boscolo, S., Castagnola, V., Sosa, S., Berti, F., Fontanive, G., Poli, M., Tubaro, A., Bianco, A., Paolucci, F., Prato, M.: Highly sensitive electrochemiluminescent nanobiosensor for the detection of palytoxin. ACS Nano **6**, 7989–7997 (2012)
33. Rahmawati, I., Einaga, Y., Ivandini, T.A., Fiorani, A.: Enzymatic biosensors with electrochemiluminescence transduction. ChemElectroChem **9**, e202200175 (2022)
34. Fiorani, A., Merino, J.P., Zanut, A., Criado, A., Valenti, V., Prato, M., Paolucci, F.: Advanced carbon nanomaterials for electrochemiluminescent biosensor applications. Curr. Opin. Electrochem. **16**, 66–67 (2019)

35. Miao, W.: Electrogenerated chemiluminescence and its biorelated applications. Chem. Rev. **108**, 2506–2553 (2008)
36. Meso Scale Discovery (2024). https://www.mesoscale.com/technical_resources/our_technology/our_immunoassays. Accessed 09 Sep 2024
37. Fiorani, A., Valenti, G., Paolucci, F., Einaga, Y.: Electrogenerated chemiluminescence at boron-doped diamond electrodes. Chem. Commun. **59**, 7900–7910 (2023)
38. Honda, K., Yoshimura, M., Rao, T.N., Fujishima, A.: Electrogenerated chemiluminescence of the ruthenium tris(2,2′)bipyridyl/amines system on a boron-doped diamond electrode. Phys. Chem. B **107**, 1653–1663 (2003)
39. Lagrange, J.P., Deneuville, A., Gheeraert, E.: Activation energy in low compensated homoepitaxial boron-doped diamond films. Diam. Relat. Mater. **7**, 1390–1393 (1998)
40. Sakanoue, K., Fiorani, A., Santo, C.I., Irkham, V.G., Paolucci, F., Einaga, Y.: Boron-doped diamond electrode outperforms the state-of-the-art electrochemiluminescence from microbeads immunoassay. ACS Sens. **7**, 1145–1155 (2022)
41. Irkham, W.T., Fiorani, A., Valenti, G., Paolucci, F., Einaga, Y.: Co-reactant-on-demand ECL: electrogenerated chemiluminescence by the in situ production of $S_2O_8{}^{2-}$ at boron-doped diamond electrodes. J. Am. Chem. Soc. **138**, 15636–15641 (2016)
42. Irkham, F.A., Valenti, G., Kamoshida, N., Paolucci, F., Einaga, Y.: Electrogenerated chemiluminescence by in situ production of coreactant hydrogen peroxide in carbonate aqueous solution at a boron-doped diamond electrode. J. Am. Chem. Soc. **142**, 1518–1525 (2020)
43. Irkham, R.R.R., Ivandini, T.A., Fiorani, A., Einaga, Y.: Electrogenerated chemiluminescence of luminol mediated by carbonate electrochemical oxidation at a boron-doped diamond. Anal. Chem. **93**, 2336–2341 (2021)
44. Honda, K., Noda, T., Yoshimura, M., Nakagawa, K., Fujishima, A.: Microstructural heterogeneity for electrochemical activity in polycrystalline diamond thin films observed by electrogenerated chemiluminescence imaging. J. Phys. Chem. B **108**, 16117–16127 (2004)
45. Sentic, M., Virgilio, F., Zanut, A., Manojlovic, D., Arbault, S., Tormen, M., Sojic, N., Ugo, P.: Microscopic imaging and tuning of electrogenerated chemiluminescence with boron-doped diamond nanoelectrode arrays. Anal. Bioanal. Chem. **408**, 7085–7094 (2016)
46. Voci, S., Goudeau, B., Valenti, G., Lesch, A., Jović, M., Rapino, S., Paolucci, F., Arbault, S., Sojic, N.: Surface-confined electrochemiluminescence microscopy of cell membranes. J. Am. Chem. Soc. **140**, 14753–14760 (2018)
47. Oba, K., Ino, K., Utagawa, Y., Abe, H., Shiku, H.: Comprehensive cell adhesion analysis using electrochemiluminescence imaging and electrochemical impedance spectroscopy. Electrochemistry **92**, 022009 (2024)
48. Ino, K., Komatsu, K., Hiramoto, K., Utagawa, Y., Nashimoto, Y., Shiku, H.: Electrochemiluminescence imaging of cellular adhesion in vascular endothelial cells during tube formation on hydrogel scaffolds. Electrochim. Acta **415**, 140240 (2022)
49. Liu, Y., Zhang, H., Li, B., Liu, J., Jiang, D., Liu, D., Sojic, N.: Single biomolecule imaging by electrochemiluminescence. J. Am. Chem. Soc. **143**, 17910–17914 (2021)
50. Zhu, M.-J., Pan, J.-B., Wu, Z.-Q., Gao, X.-Y., Zhao, W., Xia, X.-H., Xu, J.-J., Chen, H.-J.: Electrogenerated chemiluminescence imaging of electrocatalysis at a single Au-Pt Janus nanoparticle. Angew. Chem. Int. Ed. **57**, 4010–4014 (2018)
51. Maus, R.G., Wightman, R.M.: Microscopic imaging with electrogenerated chemiluminescence. Anal. Chem. **73**, 3993–3998 (2001)
52. Watanabe, T., Ishikawa, R., Hara, N., Iwasaki, T., Miyachi, M., Shiigi, Y., Takahashi, M., Kuroki, D., Koh, S.: Single-layer graphene as a transparent electrode for electrogenerated chemiluminescence biosensing. Electrochem. Commun. **138**, 107290 (2022)
53. Cristarella, T.C., Chinderle, A.J., Hui, J., Rodríguez-López, J.: Single-layer graphene as a stable and transparent electrode for nonaqueous radical annihilation electrogenerated chemiluminescence. Langmuir **31**, 3999–4007 (2015)
54. Valenti, G., Zangheri, M., Sansaloni, S.E., Mirasoli, M., Penicaud, A., Roda, A., Paolucci, F.: Transparent carbon nanotube network for efficient electrochemiluminescence devices. Chem. Eur. J. **21**, 12640–12645 (2015)

55. Gross, E.M., Durant, H.E., Hipp, K.N., Lai, R.Y.: Electrochemiluminescence detection in paper-based and other inexpensive microfluidic devices. ChemElectroChem **4**, 159–1603 (2017)
56. Delaney, J.L., Hogan, C.F., Tian, J., Shen, W.: Electrogenerated chemiluminescence detection in paper-based microfluidic sensors. Anal. Chem. **83**, 1300–1306 (2011)
57. Climent, E., Rurack, K.: Combining electrochemiluminescence detection with aptamer-gated indicator releasing mesoporous nanoparticles enables PPT sensitivity for strip-based rapid tests. Angew. Chem. Int. Ed. **60**, 26287–26297 (2021)

Electrochemiluminescence Microscopy Instrumentation

Lurong Ding, Wenxuan Fu, Yafeng Wang, and Bin Su

Abstract Electrochemiluminescence (ECL) is a luminous phenomenon triggered by electrochemical reactions. Owing to its high sensitivity, low background interference, and excellent spatiotemporal control, ECL has become as a powerful analytical technique across diverse fields. In the past two decades, the integration of ECL with microscopy has emerged as a significant advancement, enabling single-molecule/entity imaging, high-throughput analysis, and electrocatalysis studies. This chapter provides essential knowledge for conducting ECL microscopy experiments. The core apparatus is firstly introduced, typically a custom setup integrating an optical microscope, detector, electrochemical cell, and workstation. Then, critical components that determine the experimental performance are summarized, including working electrode materials and types, ECL reagents, and electrochemical methods. Furthermore, ECL imaging modes and their corresponding application scenarios are discussed. Finally, the chapter addresses the challenges and future perspectives in ECL imaging.

Keywords Electrochemiluminescence · Microscope · Instrumentation · Working electrode · Imaging mode

1 Introduction

Electrochemiluminescence (ECL) is a light emission phenomenon triggered by electrochemical reactions [1–3]. Thanks to its high sensitivity, low background, and excellent spatiotemporal controllability, ECL has manifested itself as a powerful tool in diverse fields. In particular, the combination of ECL and microscopy has opened a new venue for imaging study of single molecules and single entities, as well as high-throughput analysis and electrocatalysis. However, in comparison

L. Ding · W. Fu · B. Su (✉)
Department of Chemistry, Zhejiang University, Hangzhou, China
e-mail: subin@zju.edu.cn

Y. Wang
Department of Clinical Laboratory, Sir Run Run Shaw Hospital, School of Medicine, Zhejiang University, Hangzhou, China

F. Paolucci et al. (eds.), *Electrogenerated Chemiluminescence Microscopy and Imaging*,
https://doi.org/10.1007/978-3-032-14602-1_3

with traditional light intensity-based measurements, ECL microscopy is relatively complex, requiring specialized instrumentation and expertise in both electrochemistry and optics. This chapter is aimed at providing the readers with key knowledges regarding experimentation of ECL microscopy. We shall introduce the apparatus of ECL microscope, which is usually customized in research laboratories and basically include optical microscope, optical detector, electrochemical cell, and electrochemical workstation. The working electrode materials, types of working electrodes, ECL reagents, and electrochemical methods will also be summarized, as they are closely related to the performance of ECL microscope. Furthermore, we shall discuss the modes of ECL imaging and their relevant application scenarios. The chapter will end up with a discussion of new challenges and future perspectives of ECL imaging from the aspects of luminophores, working electrodes, and spatial resolution.

2 The Apparatus of ECL Microscope

In this section, the necessary equipment for ECL microscopy is outlined. Unlike other imaging techniques, ECL microscope does not require an external excitation light, thus featuring a simplified apparatus that can be readily adapted for different experiments. An ECL microscope is primarily composed of four key units: optical microscope, optical detector, electrochemical cell, and electrochemical workstation. This section will provide a detailed overview of these units. In addition, a dark environment is always needed to avoid interference from ambient light, since ECL is a kind of dim light.

2.1 Optical Microscope

A standard commercial microscope is usually used to build an ECL microscope. By incorporating additional accessories, it can be integrated with bright-field, dark-field, or fluorescence microscope to facilitate and complement ECL imaging. According to the relative position of objective lens and the sample stage, ECL microscope can be generally categorized into three configurations: upright, inverted, and side-view microscopes. As illustrated in Fig. 1, upright and inverted microscopes position the objective above and below the sample stage, respectively. While in the side-view microscope, the objective is positioned in the lateral direction of sample stage.

The upright microscope is the most versatile one among three configurations, because it is compatible with various types of electrodes, including metal oxide electrode, metal electrode, and carbon electrode (Fig. 1a), no matter they are transparent or not. This configuration is often used to study aqueous ECL systems. The spatial resolution and light collection efficiency can be significantly improved by using a water-immersion objective. The inverted microscope is suitable for both aqueous and organic ECL systems, which offers higher spatial resolution and light collection

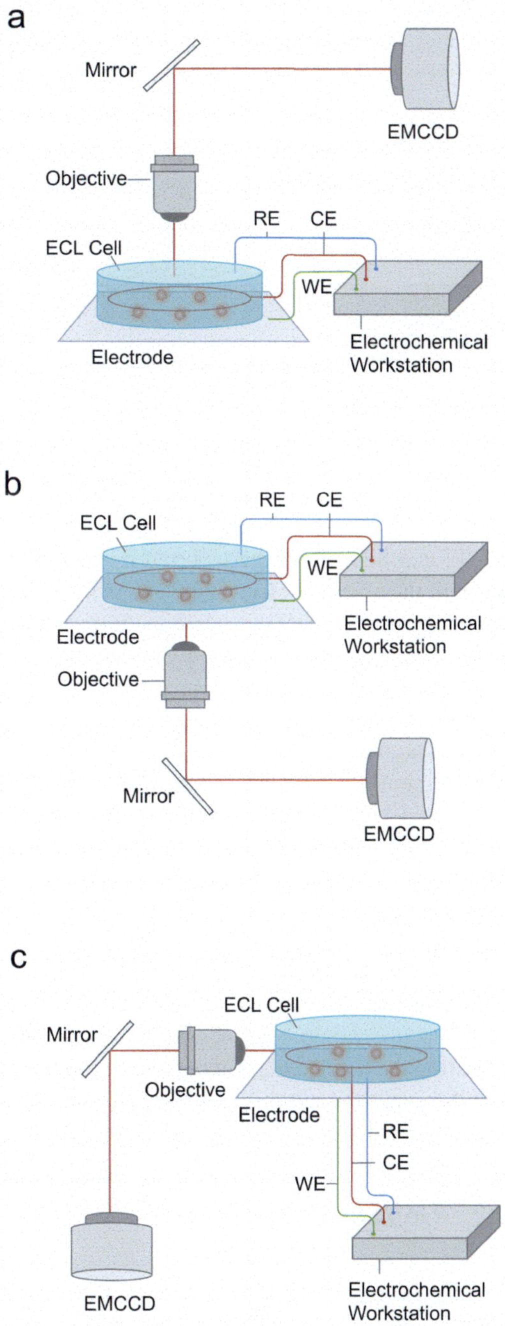

Fig. 1 Schematics of ECL microscopes: **a** upright microscope, **b** inverted microscope, and **c** side-view microscope. WE: working electrode, RE: reference electrode, CE: counter electrode

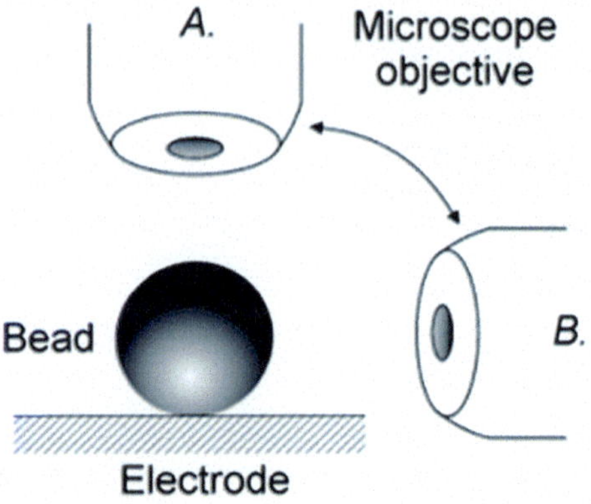

Fig. 2 Illustration of integrating top-view (**a**) and orthogonal side-view (**b**) configurations to image the functionalized bead. Reproduced with permission from Ref. [4] Copyright 2014, Royal Society of Chemistry

efficiency for ECL imaging, since it can be equipped with an oil-immersion objective (details regarding the effect of objective on ECL imaging is provided in the next section). However, only transparent working electrodes, such as indium tin oxide (ITO), can be used in the inverted microscope, because ECL signals are collected beneath the working electrode (Fig. 1b). However, a sluggish electrochemical reaction kinetic of coreactants, such as tertiary amines, on the mentioned ITO working electrodes severely limits the ECL intensity and thus an enhanced strategy is often needed for the application of inverted microscope in ECL imaging analysis.

The side-view microscope is meticulously designed to record the distribution of ECL signals in the vertical direction of electrode surface, which can efficiently complement top-view ECL images obtained by upright microscope. The integration of top-view and orthogonal side-view configurations enables spatially resolved mapping of ECL reactivity of the luminophores labeled on microbeads (Fig. 2) [4]. This combined strategy helps elucidate ECL reaction mechanisms of the bead-based bioassays. Moreover, it also facilitates the deep understanding of reaction mechanisms of redox-mediated ECL amplification strategy in bead-based bioassays [5]. However, it should be noted that in the side-view microscope the position of objective relative to the microbead is crucial for acquiring an accurate image. In addition, confocal microscope can also provide the vertical distribution of ECL signals by adjusting the focal plane, but the resolution of reconstructed images is usually limited and the experimentation is even more complicated.

2.2 *Objective*

As aforementioned, the objective is a critical hardware of an optical microscope, which significantly influences the performance of microscope. There are many parameters to define the properties of objective, here we mainly discuss the magnification, immersion medium, numerical aperture (NA), working distance, and depth of field (DOF), as they largely determine the quality of captured ECL images.

Magnification is the parameter by which the objective enlarges the image of sample, describing the increase in linear dimensions rather than areas. Immersion medium is the medium between the objective and the sample, mainly including air, water, or oil. NA is a dimensionless parameter representing the angular range of light that an objective can collect. An objective with a higher value of NA indicates a better light collection efficiency and a higher spatial resolution, which is the preferential choice of high-resolution ECL imaging. The value of NA for an objective is given by the following equation,

$$\mathrm{NA} = n \times \sin\alpha \tag{1}$$

where n is the refractive index of immersion medium (typically 1.00, 1.33, and 1.52 for air, pure water, and immersion oil, respectively) and α is the half-angle of the cone that light can enter the objective. For a given objective, the value of α is invariable. Therefore, increasing the refractive index of immersion medium turns to be an efficient way to increase the value of NA, leading to the development of water-immersion and oil-immersion objectives.

In an upright microscope, both dry and water-immersion objectives are commonly used. Due to their better light collection efficiency, water-immersion objectives are more preferred for ECL microscopy. Furthermore, water-immersion objectives are also compatible well with aqueous ECL systems. While inverted microscopes can be equipped with all the three types of objectives, because the immersion medium and the electrolyte solution are physically separated from each other. To be noted that, among three, oil-immersion objectives can usually offer the highest resolution when the magnifications of objectives are the same. As for the side-view microscope, dry objectives are usually used because of the structural limitation.

Working distance is the distance between the sample and the front lens of objective when the microscope is in focus, which generally decreases with the increase of magnification and NA. In this case, when using an inverted microscope equipped with an oil-immersion objective for ECL imaging, the transparent electrode should be thin enough to ensure that the electrode surface can be precisely focused. Similar to working distance, DOF also decreases with the increase of magnification and NA. It refers to the vertical distance along the optical axis within which the sample remains in focus. The value of DOF is usually very small, especially for water-immersion and oil-immersion objectives with high magnification, which is typically smaller than 1 μm. Therefore, when conducting ECL imaging experiment, a relatively smooth electrode surface is required. Meanwhile, the focal plane should be carefully selected.

2.3 *Optical Detector*

Optical detector is a significant unit of ECL microscope, which records ECL signals, converts them into electrical signals and transmits them to a computer for data processing. The commonly used optical detectors include charge-coupled device (CCD), electron-multiplying CCD (EMCCD), and complementary metal–oxide–semiconductor (CMOS) cameras. The detector plays a vital role in ensuring accurate and high-quality microscopic imaging.

A CCD camera primarily consists of imaging area, memory area, serial readout register, and output amplifier. The imaging area is composed of a certain number of independent lithography units, namely pixels, each of which can generate and store photoelectrons when exposed to the light. Upon stimulation by specific pulse sequences, these stored photoelectrons are transferred directionally to the output amplifier, where they are converted into electrical signals and sent to a computer for processing. Therefore, CCD camera capable of rapidly capturing light signals over a large area is widely used in optical imaging. Based on the spectral type, CCD camera can be categorized as color or monochrome. A color CCD, with its inherent color selectivity, is suitable for studying potential- or color-resolved multiplex ECL systems [6, 7]. However, this function is at the cost of spatial resolution. In contrast, a monochrome CCD can provide a higher spatial resolution. However, as the size and concentration of target analytes decrease, CCD cameras face challenges due to their insufficient sensitivity, making them less effective for obtaining high-quality ECL images. The emergence of EMCCD technology effectively addresses the limitations of conventional CCD cameras. EMCCD incorporates hundreds of gain registers between serial readout register and output amplifier, enabling remarkable signal amplification. This design makes EMCCD ideally suitable for ECL imaging, as they enhance the signal-to-noise ratio (S/N) through gain register amplification without increasing the readout noise. In contrast, traditional CCD camera mainly improves the S/N by increasing the exposure time. So far, EMCCD camera represents the most sensitive imaging technology available, with the sensitivity approaching the single-photon level [8, 9]. They are frequently used in micro- and nanoscale ECL imaging, enabling the visualization of samples with size from micrometers to nanometers, such as liposomes [10], single cells [11, 12], organelles [13, 14], and nanoparticles [15].

Unlike CCD camera, CMOS camera has a separate analog-to-digital converter (ADC) for each row and the voltage signal can be output directly from each pixel rather than as photoelectrons. This mode enables CMOS camera to achieve a faster readout rate compared with CCD camera [16]. But traditional CMOS camera suffers from a high noise level and a low quantum efficiency. To cope with these limitations, scientific CMOS (sCMOS) camera has been developed, achieving reduced noise, high frame rates, increased dynamic range, improved quantum efficiency and resolution. Another advantage of CMOS camera is that it can be readily integrated into a miniaturized device. For example, a CMOS chip can be used as a built-in detector in ECL biosensor, demonstrating its potentiality for portable and efficient diagnostic applications [17]. Nevertheless, for ECL systems with ultraweak intensity or applications requiring single-photon sensitivity, EMCCD camera is still the superior choice due to its unique ultrahigh sensitivity.

2.4 *Electrochemical Cell*

The electrochemical cell is the key unit of ECL microscope as ECL is light emission generated by electrochemical reactions. Depending on the experimental conditions, the electrochemical cell is usually customized in laboratories. A three-electrode configuration consisting of reference, counter, and working electrodes is often used. For aqueous ECL systems, silver/silver chloride (Ag/AgCl) electrode and saturated calomel electrode can serve as the reference electrode. For non-aqueous ECL systems, a silver wire is usually used as the quasi-reference electrode. In most cases, gold, platinum, or carbon electrode can function as the counter electrode. As for the working electrode, different materials and geometries can be used. Accordingly, the electrochemical cells are usually designed in terms of the geometrical shape of working electrode. Therefore, we mainly discuss about the working electrode, design of electrochemical cell, and ECL reagents in this section.

2.4.1 Materials of Working Electrode

Noble metals, such as gold and platinum, are frequently used as the working electrode materials in ECL imaging due to their good electrochemical properties, e.g., fast electron transfer kinetics and catalytic function toward ECL reactions [18]. Additionally, the surface of gold electrode can be easily modified with biomolecules, such as DNA, aptamers, proteins, and enzymes, through the gold-thiol bonds [19–22]. However, a significant drawback of noble metal electrodes is their susceptibility to surface oxidation at high positive potentials during anodic ECL reactions, which will passivate the electrode surface and thus deactivate the electrochemical properties of electrode. The strategies to mitigate this issue include modifying the electrode surface with surfactants or alkanethiols, coating the electrode with polymers, or adding a small number of halides in the electrolyte to compete with surface oxidation and to reduce the fouling caused by surface oxidation [18, 23]. Very recently, Sivakumar et al. proposed a novel approach to eliminate the formation of passivating oxide on the gold electrode by selectively deactivating the gold active sites and knocking out the metal surface asperities using Cu-Fenton reagents [24]. The ECL intensity of tris(2,2′bipyridyl)ruthenium ($Ru(bpy)_3^{2+}$)/tri-*n*-propylamine (TPrA) system can be enhanced by 2.7-fold, with an ECL efficiency of 163%.

Carbon-based electrodes have been widely used in ECL imaging, among which glassy carbon electrode (GCE) is most frequently considered. In the classic $Ru(bpy)_3^{2+}$/TPrA coreactant system, GCE exhibits significantly faster kinetics for TPrA oxidation in comparison with other electrode materials [25]. As a result, the ECL intensity of GCE is 10 times and 100 times higher than that of gold and platinum electrodes, respectively [23]. Moreover, the wide potential window of GCE makes it suitable for both anodic and cathodic ECL systems. However, a key disadvantage of GCE is its complicated pretreatment, such as polishing, before use. Apart from

GCE, other carbon-based materials, such as carbon nanotubes (CNTs) and graphene, are gaining attention as promising alternatives [26, 27].

In comparison with noble metal or carbon-based electrodes, the primary advantage of metal oxide electrodes is their high transmittance in the visible and near-infrared region, which covers the emission wavelength of most ECL luminophores. The transparency of metal oxide electrodes also makes them compatible with both upright and inverted microscopes. ITO and fluorine-doped tin oxide (FTO) electrodes are the most commonly used metal oxide electrodes, particularly the former. After cleaning with appropriate solutions or plasma, the amount of hydroxyl groups on the surface of ITO increases, enabling further a diverse range of surface derivatizations. For instance, after modifying with glutaraldehyde or 3-aminopropyltriethoxysilane (APTES), ITO can be readily immobilized with biomolecules such as DNA and antibodies, thus allowing the construction of various ECL biosensors [28–30]. However, ITO is unstable at very negative potentials due to the reduction of indium and tin oxides, which somehow limits its applications in cathodic ECL reactions [31]. In contrast, FTO behaves well at negative potentials [32]. Besides the cathodic instability, the relatively poor electrochemical properties of metal oxide electrodes lead to the sluggish oxidation kinetics of coreactants, such as TPrA, which also restricts the practical applications of electrodes. This can be addressed by proper modification of the electrode surface. For example, the electrochemical oxidation of TPrA on ITO electrode can be remarkably promoted by modifying the ITO electrode surface with an ultrathin hydrophobic polydimethylsiloxane (PDMS) layer or nanoclusters [33–35].

2.4.2 Types of Working Electrodes

There are various types of working electrodes, including planar electrodes, microelectrodes, bipolar electrodes (BPEs) and electrode arrays, used in ECL imaging.

Planar electrodes exhibit remarkable compatibility with all the three types of microscopes, which are thus most widely used. The planar structure allows the efficient loading of diverse research objects, such as fingerprints [36], cells [11] and microbeads [37, 38]. This versatility is realized by choosing planar electrodes with suitable size and designing properly the electrochemical cell. Consequently, the planar electrodes demonstrate excellent adaptability across a wide range of scenarios.

In addition to conventional electrodes, several new types of microelectrodes have emerged, such as microtube, microwire, capillary and microbead electrodes, to realize specific ECL imaging experiments. Microtube electrodes, fabricated by electroless deposition of gold onto the inner walls of microchannels, are ideal for imaging and measuring the thickness of ECL layer from top or bottom directions, since the ECL signals are confined to the electrode surface and restricted in the circular region of the microtube electrode [39]. Measuring the thickness of ECL layer provides valuable insights into the mechanisms of different ECL systems, because the thickness correlates with the dominant ECL reaction pathway [40]. For instance, the mechanism for ECL generation of $Ru(bpy)_3^{2+}/S_2O_8^{2-}$ system can be unraveled by measuring the thickness of ECL layer using bismuth microwire electrode [41].

Capillaries modified with conductive layers, luminophores, and other molecules can also be used as microelectrodes. Thanks to the micro-scaled capillary tips, the electrode can easily penetrate into single cells to visualize intracellular molecules. For example, modification of oxidase on the electrode enabled the imaging of intracellular hydrogen peroxide (H_2O_2) at the single cell level [42]. In addition, gold-coated microbeads can act as spherical microelectrodes to enhance ECL emission by narrowing the distance between coreactant radicals and ECL luminophores [43].

BPEs are electrically conductive materials immersed in the electrolyte solution, which do not require direct ohmic contact. A bipolar electrochemical system typically consists of two driving electrodes and a wirelessly connected BPE. When a sufficient driving voltage is applied across the two driving electrodes in an electrolyte solution, a potential difference forms between solution and BPE. This potential difference simultaneously drives cathodic and anodic reactions at the two poles of BPE (Fig. 3a) [44]. BPEs can be generally classified into open and closed configurations [45]. In an open configuration, both cathode and anode are immersed in the same electrolyte solution (Fig. 3b). Due to the high solution resistance, the external voltage applied to the two driving electrodes is usually quite high [46]. Combining BPEs with ECL imaging, small molecules and DNA hybridization can be detected [47, 48]. Furthermore, a large-scale microarray consisting of 1000 individual BPEs has been developed for multiplex and quantitative analysis [49]. In addition to the design shown in Fig. 3b, nanopipettes with platinum deposited on their inner surfaces can also function as open BPEs. Thanks to their small tip size, the nanopipettes fulfill the requirements of intracellular electroanalysis [50]. To be mentioned that, in a closed configuration, the cathode and anode of BPE are immersed in two different electrolyte solutions (Fig. 3c). The physical separation of oxidation and reduction reactions minimizes the mutual interference of anodic and cathodic poles, enhancing the reliability of experimental results. This configuration also allows to simultaneously study aqueous and non-aqueous ECL systems at the two poles, which is particularly advantageous for multicolor ECL system in which different luminophores are dissolved in different solvents [51, 52]. In addition, compared with open BPEs, the driving voltage required to initiate the redox reactions at two poles of a closed BPE is significantly lower [45].

Electrode arrays enable multiplex and high-throughput analysis. There are several ways to fabricate electrode array, with the microwell electrode array being the most frequently used (Fig. 4a) [53]. One method to prepare microwell electrode involves spatially spotting of Sepharose beads onto a substrate and subsequently coating it with graphite and PDMS layer. After peeling off the PDMS-graphite polymer, the microwell array electrode is prepared, with the beads trapped in the microwells [54]. Another approach uses a silicon wafer mold with micropillars to fabricate a PDMS slide containing microwells by reversing the mold. An electrode array can be then created by coating a conductive layer on the PDMS slide [55]. Besides, the microwell electrode array can be fabricated through photolithography. In a typical process, an ITO substrate is sequentially spin-coated with a layer of photoresist, covered with a mask and exposed to light illumination. After removing the unpolymerized

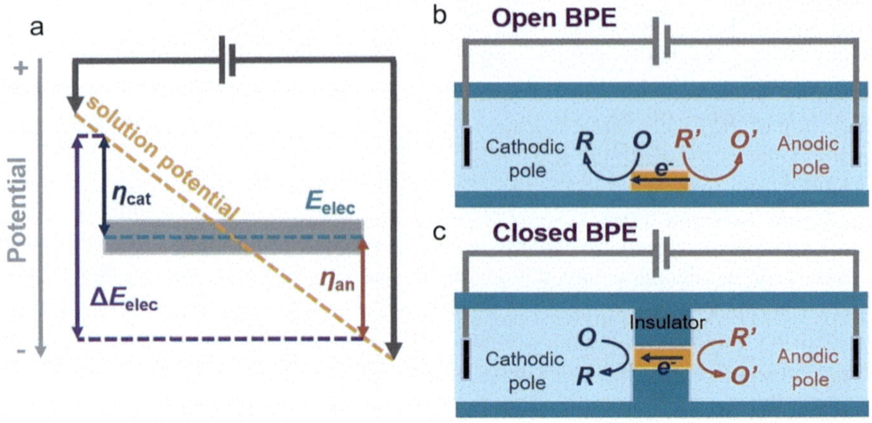

Fig. 3 **a** Interfacial potential difference of BPE. **b–c** Schematic diagram of open (**b**) and closed (**c**) BPEs. Reproduced with permission from Ref. [45] Copyright 2022, Wiley–VCH

photoresist through development, an ITO microwell electrode array is prepared [56, 57]. Sputtering and etching are also often used to fabricate electrode arrays [58]. The advantage of molding and lithography is the ability to precisely control the shape of electrode array by designing different molds or masks [59]. Alternatively, electrode arrays can be fabricated by pasting a porous sticker with a well array onto the electrode, which is simple and efficient [22, 60, 61].

As shown in Fig. 4b, BPE arrays can be prepared by increasing the number of conductive materials and spatially organizing them at a regular distance [62]. Due to their unique structure, gold electrodeposited porous anodic aluminum oxide (AAO) membranes can be used as nano-BPE arrays, which do not need a complex fabrication process (Fig. 4c) [63]. Similarly, a uniform bipolar ultramicroelectrode (UME) array consisting of over 146,000 carbon electrodes embedded in a Parylene C membrane has also been prepared, allowing the simultaneous study of electrochemical reactions on more than one hundred thousand of electrodes in the millisecond timescale [64].

2.4.3 Design of Electrochemical Cell

The electrochemical cell, as mentioned above, is usually customized in terms of the geometrical shape of working electrode and specific ECL experiments. For planar working electrodes, the electrochemical cell is relatively simple, which can be prepared by punching a hole in inert materials such as PDMS and attaching it to the surface of working electrode (Fig. 5a). Before ECL imaging, the electrolyte solution is added into the cell, followed by immersing reference and counter electrodes in the solution. For microelectrodes, electrode arrays and BPEs prepared on the planar substrate, the electrochemical cell is similar. For cylindrical or rod-shaped working electrodes, such as classical GCE, the electrochemical cell is often home-made to

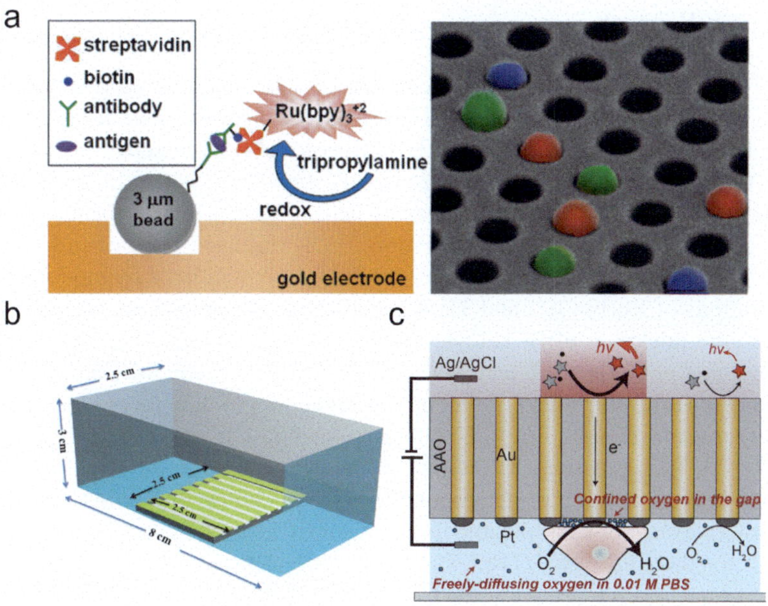

Fig. 4 **a** Illustration of a sandwich ECL immunoassay based on microwell electrode array. Reproduced with permission from Ref. [53] Copyright 2009, American Chemical Society. **b** Schematic illustration of a BPE array. Reproduced with permission from Ref. [62] Copyright 2015, American Chemical Society. **c** Schematic illustration for a closed bipolar nanoelectrode array based on AAO membrane. Reproduced with permission from Ref. [63] Copyright 2021, Wiley–VCH

fit the specific structure of the working electrode (Fig. 5b). The cell should have an opening to accommodate the working electrode, allowing it to be inversely inserted into the cell. This cell is only suitable for upright microscope.

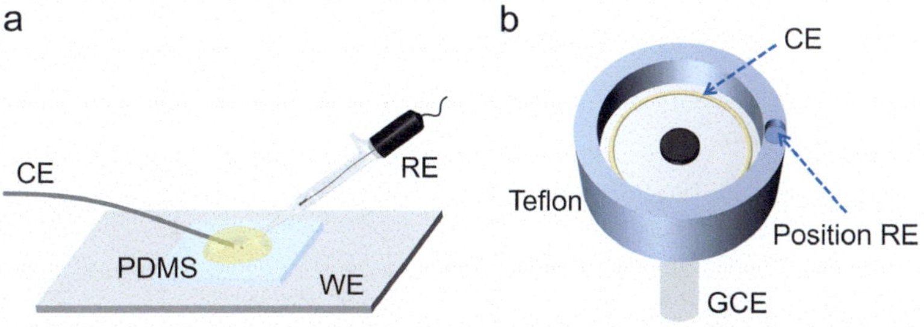

Fig. 5 Schematic illustrations of electrochemical cells used for planar (**a**) and cylindrical (**b**) electrodes

2.4.4 ECL Reagents

A variety of ECL luminophores and coreactants have been developed so far. For ECL imaging, $Ru(bpy)_3^{2+}$ and luminol (its analogues) are the most frequently used luminophores, with tertiary amines and H_2O_2/oxygen as the coreactants, respectively. The $Ru(bpy)_3^{2+}$/TPrA system is the most widely studied coreactant ECL system. For example, it has been used for imaging single cells, organelles, microbeads, nanoparticles, nanorods, nanoplates, etc. [9, 65, 66]. ECL reactions are electrode surface-confined and the thickness of ECL layer can be readily regulated by modulating the concentrations of $Ru(bpy)_3^{2+}$ and TPrA according to the objects studied. Generally speaking, the thickness of ECL layer increases and decreases with increasing the concentrations of $Ru(bpy)_3^{2+}$ and TPrA, respectively, because the lifetime of electrogenerated $Ru(bpy)_3^{3+}$ (tens of seconds) is much longer than that of TPrA radical cations (sub-milliseconds) [67]. When the concentration of $Ru(bpy)_3^{2+}$ is higher than 500 μM, ECL reactions are dominated by the catalytic route (Fig. 6b) instead of the oxidative-reduction route (Fig. 6a) [23, 68], leading to a remarkable extension of ECL layer. For example, the thickness of ECL layer was ca. 886 nm and ca. 6000 nm for the $Ru(bpy)_3^{2+}$/TPrA system with concentrations of 50 μM/100 mM and 500 μM/1 mM, respectively [38]. In addition, the thickness of ECL layer can also be changed by using tertiary amines with different chemical structures, since the lifetime of tertiary amine radical cations is quite different [38]. Other commonly used tertiary amines include 2-(dibutylamino) ethanol (DBAE), bis-(2-hydroxyethyl) amino-tris(hydroxymethyl) methane (BIS–TRIS), *N*-(2-hydroxyethyl)piperazine-*N*′-2-ethanesulfonic acid (HEPES), 3,3′-(piperazine-1,4-diyl)bis(2-hydroxypropane-1-sulfonic acid) (POPSO) [39, 69].

Choosing the appropriate coreactant and controlling the concentrations of both luminophore and coreactant are essential for effective ECL imaging. To probe object in the close proximity of electrode surface, a coreactant that can generate short-lived intermediates or a low concentration of luminophore is preferred. For object with

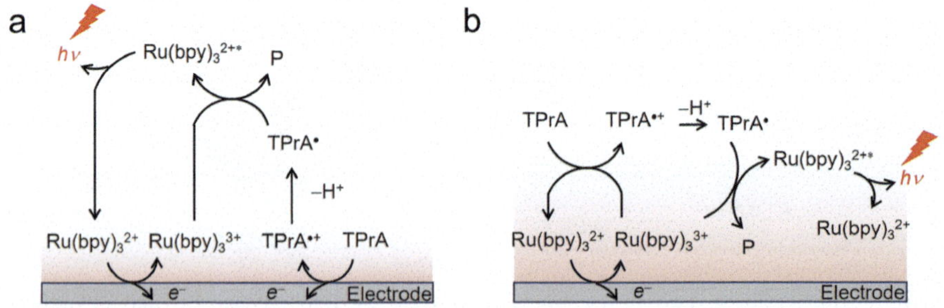

Fig. 6 Schematic illustrations of the oxidative-reduction route (**a**) and catalytic route (**b**) in ECL reactions. When increasing the concentration of $Ru(bpy)_3^{2+}$, the ECL layer changes from confined (**a**) to extended (**b**)

a thickness (height) of several micrometers, using a coreactant that can yield long-lived intermediates or increasing the concentration of luminophore is suggested. By adjusting the relative concentrations of luminophore and coreactant, the thickness of ECL layer can be tailored to match the spatial location of different cell junctions. For instance, when the concentration of luminophore is low, ECL generation is confined to the electrode surface, thus ECL imaging only reveals cell–matrix adhesions at the basal cell membrane. Upon increasing the concentration of luminophore, the ECL layer can be significantly extended to enable the imaging of cell–cell junctions near the apical surface of cells [70].

Compared with $Ru(bpy)_3{}^{2+}$, luminol and its analogues require only a relatively low potential to initiate ECL reactions, which is beneficial to live cells and other biological samples, as high potentials may affect or even damage the samples [71, 72]. Another advantage of luminol and its analogues arises from the use of H_2O_2 as coreactant, which enables the ECL imaging of intracellular H_2O_2 or H_2O_2 released from single cells [42, 73]. Furthermore, since many enzymatic reactions can produce H_2O_2, luminol and its analogues can be also used for imaging detection of cholesterol, glucose, and sphingomyelinase [50, 56, 74]. It should be noted that the ECL signal of luminol is relatively strong in alkaline solutions but weak in neutral solutions. However, the solutions in bioanalysis are usually neutral. This incompatibility can be solved by using an analogue of luminol, L012 (8-amino-5-chloro-7-phenylpyrido[3,4-d]pyridazine-1,4(2H, 3H)-dione), which is highly emissive in a neutral solution. Very recently, luminol and its analogues are used for studying oxygen reduction reaction (ORR) and oxygen evolution reaction (OER), since the reactive oxygen species (ROS) generated during ORR or OER can serve as coreactants. This strategy has enabled the study of electrocatalytic activity of single-atom nanocatalysts [75], single nanoplates [76], and single graphene sheets [77] with a high spatiotemporal resolution.

2.5 Electrochemical Workstation

ECL reactions are typically initiated by a direct current (DC) potential, commonly applied using an electrochemical workstation. Cyclic voltammetry (CV) and chronoamperometry are the two most widely used electrochemical techniques. The electrochemical workstation can be synchronized with the optical detector via TTL. Compared with chronoamperometry, CV can provide valuable insights into the potential-related behavior of ECL systems, such as ECL onset potential.

When a constant positive potential is applied to the working electrode, the electrode undergoes passivation due to surface oxidation and fouling from the oxidation products of coreactants, resulting in an irreversible decrease of ECL intensity. This issue can be addressed by applying a negative potential to renew the working electrode surface, thus restoring the ECL signal [78]. It is then not a surprise that the stability of ECL signal can be improved by applying an alternating positive and negative pulse potentials [74]. However, it should be noticed that the exposure time of optical detector should match with the duration of applied potential.

3 ECL Microscopy Imaging

ECL imaging can be generally categorized into labeled imaging and label-free imaging.

3.1 Labeled Imaging

In labeled imaging, ECL luminophores are labeled on the target objects. ECL signals are generated mainly through the low oxidation potential (LOP) route (Fig. 7a), in which only the coreactants are electrochemically oxidized on the electrode surface and the generated coreactant radicals diffuse and react with labeled luminophores to emit light.

Luminophore used for ECL imaging can be generally divided into molecular luminophores and nanoprobes. Among molecular luminophores, $Ru(bpy)_3^{2+}$ is the most commonly used one. Dolci et al. recorded ECL images of $Ru(bpy)_3^{2+}$ labeled polystyrene (PS) beads with a diameter of 8 μm [37]. The images showed that the ECL signals were only generated on PS beads that loaded on electrode surface, achieving a spatial resolution of 400 nm. This work demonstrated the potential of ECL imaging as a powerful tool for single-cell analysis. In 2017, Valenti et al. reported the ECL imaging of single cells for the first time by labeling $Ru(bpy)_3^{2+}$ on cell membrane (Fig. 7b). Due to the short lifetime and hindered diffusion of coreactant radical cations, only cell borders were imaged eventually [11]. After permeabilizing the cell membrane with surfactant, coreactant radical cations could freely diffuse into cells, enabling the ECL imaging of entire basal membrane of single cells [12].

ECL nanoprobes are typically prepared using porous silica or polymer as the physical carrier. As shown in Fig. 7c, Liu's group realized single protein molecule imaging by using $Ru(bpy)_3^{2+}$-doped silica/gold nanoparticles (RuDSNs/AuNPs) as ECL nanoprobe [30]. Each RuDSNs/AuNP carried more than ten thousand $Ru(bpy)_3^{2+}$ molecules, which remarkably enhanced the ECL signal and enabled the imaging of single protein molecules. Mesoporous silica nanoparticle, metal–organic frameworks, and zeolite have also been employed as carriers to load $Ru(bpy)_3^{2+}$. Thanks to the nanoconfinement effect of mesoporous nanoarchitectures on ECL reactions, visualization of single molecule movement at the cellular membrane and ultrasensitive ECL analysis have been realized [29, 79, 80]. Furthermore, through surface modification of nanoprobe with stimuli-responsive DNA polymer hydrogel shell, targeted ECL imaging of intracellular ATP distribution has also been conducted,

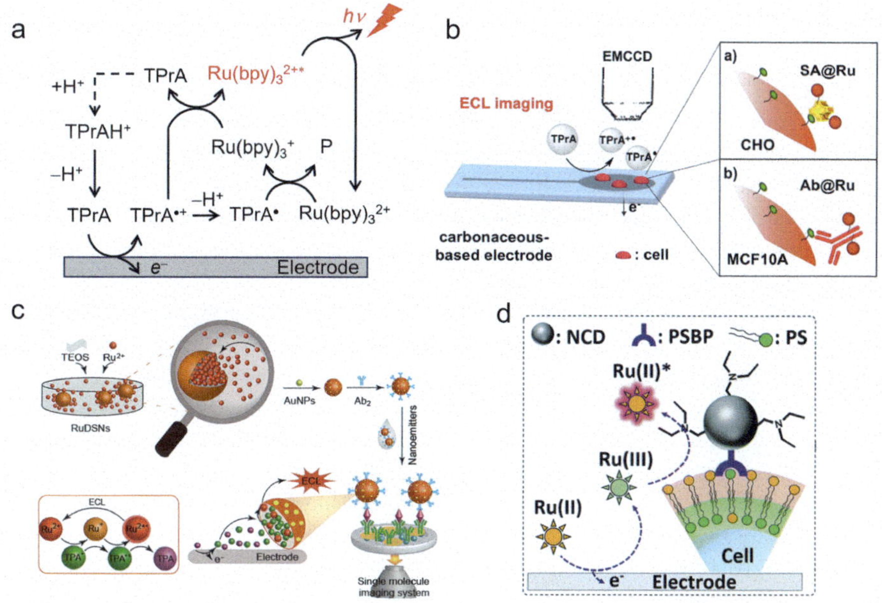

Fig. 7 **a** Schematic of the LOP route involved in the $Ru(bpy)_3{}^{2+}$/TPrA system. **b** Schematic illustration of imaging single cells labeled with $Ru(bpy)_3{}^{2+}$. Reproduced with permission from Ref. [11] Copyright 2017, American Chemical Society. **c** Schematic of ECL imaging of single protein labeled with RuDSNs/AuNP. Reproduced with permission from Ref. [30] Copyright 2021, American Chemical Society. **d** Schematic illustration of nitrogen-doped carbon dots (NCD) as label for ECL imaging of upper cell membrane. Reproduced with permission from Ref. [84] Copyright 2021, Royal Society of Chemistry

offering a strategy for selectively imaging of intracellular species [81]. Nanoparticles like semiconductor quantum dots (QDs) [82], polymer dots [83], and noble metal clusters [26], can also be used as nanoprobes for ECL imaging.

The labeled imaging can be also performed in a reverse way, in which coreactant conjugated nanoprobes instead of luminophores are used as the labels. For example, Ma et al. prepared nitrogen-doped carbon dots (NCD) as the label. The prepared nanocoreactants labels can generate strong ECL signals through the so-called catalytic route, in which freely diffusive $Ru(bpy)_3{}^{2+}$ in solution mediates the ECL reaction and eventually emits to realize the imaging. In this way, labeling the externalized phosphatidylserine with the nano dots, in situ ECL imaging of single-cell membrane was achieved. In the catalytic route, the thickness of ECL layer is principally determined by the vertical diffusion distance of $Ru(bpy)_3{}^{3+}$. Due to the long lifetime of $Ru(bpy)_3{}^{3+}$, the ECL patterns of the whole cell membrane were imaged by adjusting the focal plane to different heights of the cell (Fig. 7d) [84].

3.2 Label-Free Imaging

In label-free ECL imaging, both luminophores and coreactants diffuse freely in the solution. The ECL intensity will change upon the presence of target object on the electrode surface. For instance, when the electrode surface is coated with an electrochemically inert object, the corresponding region appears as a relatively dark area against the bright background, due to the hindered ECL reactions and the restricted diffusion of ECL reagents. Based on this idea, Ding et al. mapped cell–matrix adhesions of live cells that adhere to a silica nanochannel membrane (SNM) coated ITO electrode. Thanks to the enhancement of ECL intensity by SNM, cell–matrix adhesions were imaged with a high visual contrast and spatial resolution (Fig. 8a) [85]. Cell–matrix adhesion structures, such as spikes, podosomes, lamellae, ruffles, and focal or fibrillar adhesions, were clearly identified in ECL images. Furthermore, T-cell antigen recognition, dynamic variation of cell–matrix adhesions, cell morphology and collective migration of cells under extracellular stimulations were also imaged and analyzed [86–89]. In addition, small organelles like living mitochondria can be also studied by ECL imaging [13, 14].

Interestingly, intracellular biomolecules can also serve as coreactants for ECL imaging. In this context, ECL signal is generated primarily via the so-called catalytic route, namely electrogenerated $Ru(bpy)_3^{3+}$ molecules diffuse into the cell and react with biomolecules to emit light. For example, Ma et al. used guanine as the coreactant to image intracellular structure. Guanine-rich intracellular structures, such as nucleus and endoplasmic reticulum, were visualized without the requirement of multiple labeling (Fig. 8b). Time-resolved ECL snapshot sequences further revealed the dynamic transport process of $Ru(bpy)_3^{3+}$, uncovering heterogeneous diffusion behavior within cells [72].

Like other optical imaging techniques, the spatial resolution of ECL imaging is also restricted by the optical diffraction limit (~ 200 nm). Dong et al. proposed a super-resolution ECL imaging strategy, realizing spatial resolutions of 22 nm and 150 nm for imaging nanoelectrodes and cell–matrix adhesions, respectively (Fig. 8c) [90]. The strategy was based on separating ECL reactions in space and time to ensure that each pixel of EMCCD camera only captured single photon emitted by individual ECL reactions during a snapshot. The super-resolution ECL images were subsequently reconstructed by single-molecule localization.

In contrast to electrochemically inert objects, when the target objects are electrochemically active, such as gold nanoparticles, ECL images will show bright regions against a relatively dark background. Zhu et al. synthesized gold-platinum Janus nanoparticles (Au-Pt JNPs) and used ECL imaging to evaluate their electrocatalytic performance [15]. The results revealed that ECL generated on individual Janus nanoparticles was brighter and more stable. This enhanced performance was attributed to the unique structure of Janus nanoparticles, which could not only enhance the mass transport but also decrease the formation of oxidation layer at the particles.

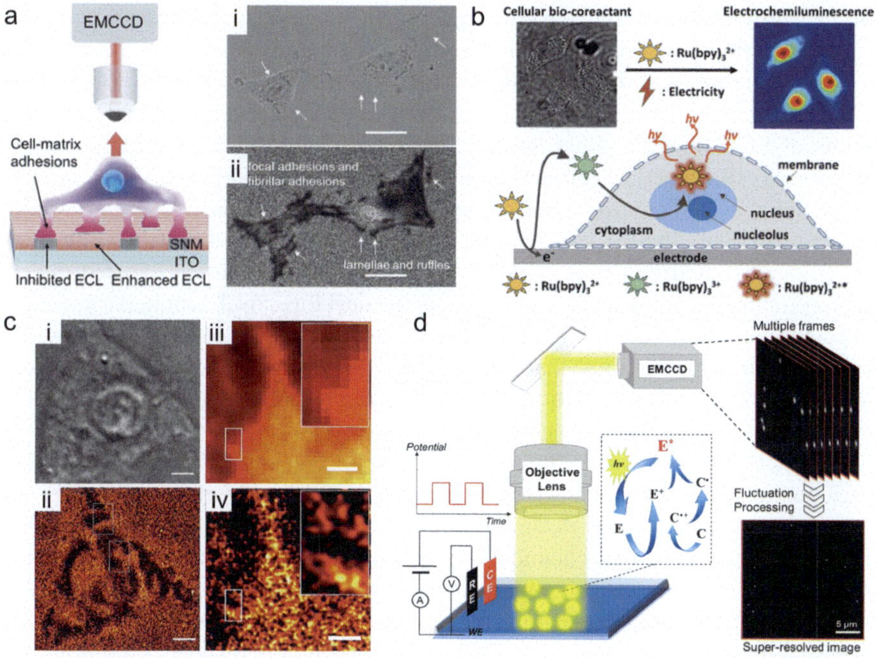

Fig. 8 **a** Illustration of ECL imaging of cell–matrix adhesions: (i) Bright-field and (ii) ECL images of single cell. Reproduced with permission from Ref. [85] Copyright 2020, Wiley–VCH. **b** Principle of imaging intracellular biomolecules by ECL. Reproduced with permission from Ref. [72] Copyright 2021 Wiley–VCH. **c** Super-resolved images of single cell: (i) Bright-field and (ii) super-resolved ECL images of single cell. (iii) Diffraction-limited and (iv) super-resolved ECL images of a local adhesion region. Reproduced with permission from Ref. [90] Copyright 2021, Nature. **d** Schematic illustration of super-resolution ECL imaging by combining ECL with SRRF. Reproduced with permission from Ref. [91] Copyright 2021, American Chemical Society

ECL imaging can be also combined with the super-resolution radial fluctuation (SRRF) algorithm to improve the spatial resolution to ca. 100 nm, as the generation of photons during ECL reactions obey Poisson statistics [91]. This approach was used to study the spatial distribution of catalytic activities at single gold nanorod and nanoplate, unraveling the facet- and defect-dependent surface activity and the temporal fluctuation of reactivity within individual nanoparticles (Fig. 8d).

4 Conclusion

In recent years, significant advancements have been achieved in ECL imaging and related instruments, with the spatial resolution improving from micrometers to tens of nanometers and the remarkably enhanced detection sensitivity to image single photons. Three types of ECL imaging systems incorporating upright, inverted, and side-view microscope have been developed to address a variety of research requirements. Selecting appropriate configurations and objectives enables the study of diverse ECL systems and samples with different sizes.

Nevertheless, several important issues need to be addressed to extend the applications of ECL imaging. First, to date, the luminophores used for ECL imaging are primarily $Ru(bpy)_3^{2+}$, luminol and their derivatives. However, their emission wavelengths are difficult to be tuned. Another limitation is their relatively low ECL efficiency/intensity, which restricts the spatiotemporal resolution of ECL imaging. One strategy to address these issues is to increase the amount of labeled molecular luminophores by using nanocarriers or to develop novel luminophores, such as quantum dots, with high ECL efficiency and readily tuned wavelengths. An alternative strategy is to enhance the ECL efficiency of available luminophores, primarily through the elucidation of ECL reaction mechanisms and the development of novel coreactants. Second, the kinds of working electrode that can be used in ECL imaging is still limited, especially in the inverted microscopes. The development of working electrodes with excellent electrochemical properties and high transmittance is highly desirable, probably through advanced thin-film fabrication techniques, such as chemical vapor deposition. Third, although the emergence of super-resolution ECL microscopy has significantly improved the spatial resolution of ECL imaging, it remains insufficient compared with the Ångström-resolution fluorescence microscopy [92]. In recently reported super-resolution ECL microscopy, luminophores are dissolved in the solution and a diffusional blurring exists. Labeling luminophores provides a potential way to mitigate this issue, but the resolution remains constrained by the diffraction limit. In this context, combining ECL with the state-of-the-art super-resolution microscopy techniques, such as stimulated emission depletion microscopy (STED), holds promise for overcoming the limitation.

References

1. Richter, M.M.: Electrochemiluminescence (ECL). Chem. Rev. **104**, 3003–3036 (2004)
2. Miao, W.: Electrogenerated chemiluminescence and its biorelated applications. Chem. Rev. **108**, 2506–2553 (2008)
3. Hu, L., Xu, G.: Applications and trends in electrochemiluminescence. Chem. Soc. Rev. **39**, 3275–3304 (2010)
4. Sentic, M., Milutinovic, M., Kanoufi, F., Manojlovic, D., Arbault, S., Sojic, N.: Mapping electrogenerated chemiluminescence reactivity in space: mechanistic insight into model systems used in immunoassays. Chem. Sci. **5**, 2568–2572 (2014)
5. Kerr, E., Knezevic, S., Francis, P.S., Hogan, C.F., Valenti, G., Paolucci, F., Kanoufi, F., Sojic, N.: Electrochemiluminescence amplification in bead-based assays induced by a freely diffusing iridium(III) complex. ACS Sens. **8**, 933–939 (2023)
6. Lu, H.J., Zhao, W., Xu, J.J., Chen, H.Y.: Visual electrochemiluminescence ratiometry on bipolar electrode for bioanalysis. Biosens. Bioelectron. **102**, 624–630 (2018)
7. Wang, N., Chen, L., Chen, W., Ju, H.: Potential- and color-resolved electrochemiluminescence of polymer dots for array imaging of multiplex microRNAs. Anal. Chem. **93**, 5327–5333 (2021)

8. Liu, Y., Guo, W., Su, B.: Recent advances in electrochemiluminescence imaging analysis based on nanomaterials and micro-/nanostructures. Chin. Chem. Lett. **30**, 1593–1599 (2019)
9. Gou, X., Xing, Z., Ma, C., Zhu, J.J.: A close look at mechanism, application, and opportunities of electrochemiluminescence microscopy. Chem. Biomed. Imaging **1**, 414–433 (2023)
10. Ben Trad, F., Wieczny, V., Delacotte, J., Morel, M., Guille-Collignon, M., Arbault, S., Lemaitre, F., Sojic, N., Labbe, E., Buriez, O.: Dynamic electrochemiluminescence imaging of single giant liposome opening at polarized electrodes. Anal. Chem. **94**, 1686–1696 (2022)
11. Valenti, G., Scarabino, S., Goudeau, B., Lesch, A., Jovic, M., Villani, E., Sentic, M., Rapino, S., Arbault, S., Paolucci, F., Sojic, N.: Single cell electrochemiluminescence imaging: from the proof-of-concept to disposable device-based analysis. J. Am. Chem. Soc. **139**, 16830–16837 (2017)
12. Voci, S., Goudeau, B., Valenti, G., Lesch, A., Jovic, M., Rapino, S., Paolucci, F., Arbault, S., Sojic, N.: Surface-confined electrochemiluminescence microscopy of cell membranes. J. Am. Chem. Soc. **140**, 14753–14760 (2018)
13. Ma, Y., Colin, C., Descamps, J., Arbault, S., Sojic, N.: Shadow electrochemiluminescence microscopy of single mitochondria. Angew. Chem. Int. Ed. **60**, 18742–18749 (2021)
14. Descamps, J., Colin, C., Tessier, G., Arbault, S., Sojic, N.: Ultrasensitive imaging of cells and sub-cellular entities by electrochemiluminescence. Angew. Chem. Int. Ed. **62**, e202218574 (2023)
15. Zhu, M.J., Pan, J.B., Wu, Z.Q., Gao, X.Y., Zhao, W., Xia, X.H., Xu, J.J., Chen, H.Y.: Electrogenerated chemiluminescence imaging of electrocatalysis at a single Au-Pt janus nanoparticle. Angew. Chem. Int. Ed. **57**, 4010–4014 (2018)
16. Dong, J., Feng, J.: Electrochemiluminescence from single molecule to imaging. Anal. Chem. **95**, 374–387 (2023)
17. Abbasi, R., Liu, J., Suarasan, S., Wachsmann-Hogiu, S.: SE-ECL on CMOS: a miniaturized electrochemiluminescence biosensor. Lab Chip **22**, 994–1005 (2022)
18. Valenti, G., Fiorani, A., Li, H., Sojic, N., Paolucci, F.: Essential role of electrode materials in electrochemiluminescence applications. ChemElectroChem **3**, 1990–1997 (2016)
19. Liu, X., Dong, M., Qi, H., Gao, Q., Zhang, C.: Electrogenerated chemiluminescence bioassay of two protein kinases incorporating peptide phosphorylation and versatile probe. Anal. Chem. **88**, 8720–8727 (2016)
20. Yang, X., Zhao, Y., Sun, L., Qi, H., Gao, Q., Zhang, C.: Electrogenerated chemiluminescence biosensor array for the detection of multiple AMI biomarkers. Sens. Actuators B Chem. **257**, 60–67 (2018)
21. Liu, G., Wang, Z.Q., Lei, C.C., Wang, F.: An electrochemiluminescence imaging sensor for the analysis of lactate in foods via a single gold microsphere. J. Food Compost. Anal. **119**, 105270 (2023)
22. Li, L.L., Chen, W.W., Hu, X.F., Tang, Z.W., Wang, C., Ju, H.X.: Coupled poly(ethylenimine) coreactant to enhance electrochemiluminescence of polymer dots for array imaging of protein biomarkers. Anal. Chem. **96**, 4308–4313 (2024)
23. Zu, Y., Bard, A.J.: Electrogenerated chemiluminescence. 66. the role of direct coreactant oxidation in the ruthenium tris(2,2′)bipyridyl/tripropylamine system and the effect of halide ions on the emission intensity. Anal. Chem. **72**, 3223–3232 (2000)
24. Leka Kottaiveedu Sivakumar, I., Bouffier, L., Sojic, N., Senthil Kumar, S.: Enhanced electrochemiluminescence by knocking out gold active sites. Angew. Chem. Int. Ed. **64**, e202421185 (2025)
25. Ding, J., Zhou, P., Su, B.: Quantum efficiency of electrochemiluminescence generation by tris(2,2′-bipyridine)ruthenium(II) and tri-*n*-propylamine revisited from a kinetic reaction model. ChemElectroChem **9**, e202200236 (2022)
26. Liu, G., Jin, B.K., Ma, C., Chen, Z., Zhu, J.J.: Potential-resolved electrochemiluminescence nanoprobes for visual apoptosis evaluation at single-cell level. Anal. Chem. **91**, 6363–6370 (2019)
27. Wang, J., Shang, P., Zhong, J., Lin, S., Chi, Y.: A multilayer-graphene nanosheet film deposited on a ceramic substrate without a catalyst for constructing an electrochemiluminescence imaging platform. Nanoscale **11**, 12132–12138 (2019)

28. Cao, J.T., Wang, Y.L., Zhang, J.J., Dong, Y.X., Liu, F.R., Ren, S.W., Liu, Y.M.: Immunoelectrochemiluminescent imaging of a single cell based on functional nanoprobes of heterogeneous $Ru(bpy)_3{}^{2+}$@SiO_2/Au nanoparticles. Anal. Chem. **90**, 10334–10339 (2018)
29. Lu, Y., Ning, Y., Li, B., Liu, B.: Dual-signal imaging mode based on fluorescence and electrochemiluminescence for ultrasensitive visualization of SARS-CoV-2 spike protein. Anal. Chem. **96**, 463–470 (2024)
30. Liu, Y., Zhang, H., Li, B., Liu, J., Jiang, D., Liu, B., Sojic, N.: Single biomolecule imaging by electrochemiluminescence. J. Am. Chem. Soc. **143**, 17910–17914 (2021)
31. Benck, J.D., Pinaud, B.A., Gorlin, Y., Jaramillo, T.F.: Substrate selection for fundamental studies of electrocatalysts and photoelectrodes: inert potential windows in acidic, neutral, and basic electrolyte. PLoS ONE **9**, e107942 (2014)
32. Zhang, H., Liu, Y., Yao, M., Han, W., Zhang, S.: Cathodic electrochemiluminesence microscopy for imaging of single carbon nanotube and nucleolin at single tumor cell. Anal. Chem. **95**, 570–574 (2023)
33. Lee, S.B., Ju, Y., Kim, Y., Koo, C.M., Kim, J.: Electrooxidative grafting of amine-terminated dendrimers encapsulating nanoparticles for spatially controlled surface functionalization of indium tin oxide. Chem. Commun. **49**, 8913–8915 (2013)
34. Kim, Y., Kim, J.: Modification of indium tin oxide with dendrimer-encapsulated nanoparticles to provide enhanced stable electrochemiluminescence of $Ru(bpy)_3{}^{2+}$/tripropylamine while preserving optical transparency of indium tin oxide for sensitive electrochemiluminescence-based analyses. Anal. Chem. **86**, 1654–1660 (2014)
35. Li, T., Ding, J., Wang, Y., Su, B.: Regulating the work function and surface hydrophobicity of an indium tin oxide electrode for enhanced electrochemiluminescence analysis. Chem. Commun. **60**, 15007–15010 (2024)
36. Xu, L.R., Li, Y., Wu, S.Z., Liu, X.H., Su, B.: Imaging latent fingerprints by electrochemiluminescence. Angew. Chem. Int. Ed. **51**, 8068–8072 (2012)
37. Dolci, L.S., Zanarini, S., Della Ciana, L., Paolucci, F., Roda, A.: Development of a new device for ultrasensitive electrochemiluminescence microscopy imaging. Anal. Chem. **81**, 6234–6241 (2009)
38. Wang, Y., Ding, J., Zhou, P., Liu, J., Qiao, Z., Yu, K., Jiang, J., Su, B.: Electrochemiluminescence distance and reactivity of coreactants determine the sensitivity of bead-based immunoassays. Angew. Chem. Int. Ed. **62**, e202216525 (2023)
39. Guo, W., Zhou, P., Sun, L., Ding, H., Su, B.: Microtube electrodes for imaging the electrochemiluminescence layer and deciphering the reaction mechanism. Angew. Chem. Int. Ed. **60**, 2089–2093 (2021)
40. Zhou, P., Hu, S., Guo, W., Su, B.: Deciphering electrochemiluminescence generation from luminol and hydrogen peroxide by imaging light emitting layer. Fundam. Res. **2**, 682–687 (2022)
41. Zhou, P., Fu, W., Ding, L., Yan, Y., Guo, W., Su, B.: Toward mechanistic understanding of electrochemiluminescence generation by tris(2,2′-bipyridyl)ruthenium(II) and peroxydisulfate. Electrochim. Acta **439**, 141716 (2023)
42. He, R., Tang, H., Jiang, D., Chen, H.Y.: Electrochemical visualization of intracellular hydrogen peroxide at single cells. Anal. Chem. **88**, 2006–2009 (2016)
43. Yang, X.R., Hang, J.M., Qu, W.Y., Wang, Y.L., Wang, L., Zhou, P., Ding, H., Su, B., Lei, J.P., Guo, W.L., Dai, Z.H.: Gold microbeads enabled proximity electrochemiluminescence for highly sensitive and size-encoded multiplex immunoassays. J. Am. Chem. Soc. **145**, 16026–16036 (2023)
44. Fosdick, S.E., Knust, K.N., Scida, K., Crooks, R.M.: Bipolar electrochemistry. Angew. Chem. Int. Ed. **52**, 10438–10456 (2013)
45. Qin, X., Gao, J., Jin, H.J., Li, Z.Q., Xia, X.H.: Closed bipolar electrode array for optical reporting reaction-coupled electrochemical sensing and imaging. Chem. Eur. J. **29**, e202202687 (2023)
46. Mavré, F., Anand, R.K., Laws, D.R., Chow, K.F., Chang, B.Y., Crooks, J.A., Crooks, R.M.: Bipolar electrodes: a useful tool for concentration, separation, and detection of analytes in microelectrochemical systems. Anal. Chem. **82**, 8766–8774 (2010)

47. Zhan, W., Alvarez, J., Crooks, R.M.: Electrochemical sensing in microfluldic systems using electrogenerated chemiluminescence as a photonic reporter of redox reactions. J. Am. Chem. Soc. **124**, 13265–13270 (2002)
48. Chow, K.F., Mavré, F., Crooks, R.M.: Wireless electrochemical DNA microarray sensor. J. Am. Chem. Soc. **130**, 7544–7545 (2008)
49. Chow, K.F., Mavré, F., Crooks, J.A., Chang, B.Y., Crooks, R.M.: A large-scale, wireless electrochemical bipolar electrode microarray. J. Am. Chem. Soc. **131**, 8364–8365 (2009)
50. Wang, Y., Jin, R., Sojic, N., Jiang, D., Chen, H.Y.: Intracellular wireless analysis of single cells by bipolar electrochemiluminescence confined in a nanopipette. Angew. Chem. Int. Ed. **59**, 10416–10420 (2020)
51. Wang, Y.Z., Xu, C.H., Zhao, W., Guan, Q.Y., Chen, H.Y., Xu, J.J.: Bipolar electrode based multicolor electrochemiluminescence biosensor. Anal. Chem. **89**, 8050–8056 (2017)
52. Wu, S., Zhou, Z., Xu, L., Su, B., Fang, Q.: Integrating bipolar electrochemistry and electrochemiluminescence imaging with microdroplets for chemical analysis. Biosens. Bioelectron. **53**, 148–153 (2014)
53. Deiss, F., LaFratta, C.N., Symer, M., Blicharz, T.M., Sojic, N., Walt, D.R.: Multiplexed sandwich immunoassays using electrochemiluminescence imaging resolved at the single bead level. J. Am. Chem. Soc. **131**, 6088–6089 (2009)
54. Marquette, C.A., Blum, L.c.J.: Conducting elastomer surface texturing: a path to electrode spotting: application to the biochip production. Biosens. Bioelectron. **20**, 197–203 (2004)
55. Xia, J., Zhou, J., Zhang, R., Jiang, D., Jiang, D.: Gold-coated polydimethylsiloxane microwells for high-throughput electrochemiluminescence analysis of intracellular glucose at single cells. Anal. Bioanal. Chem. **410**, 4787–4792 (2018)
56. Xu, J., Huang, P., Qin, Y., Jiang, D., Chen, H.-Y.: Analysis of intracellular glucose at single cells using electrochemiluminescence imaging. Anal. Chem. **88**, 4609–4612 (2016)
57. Xu, J.J., Jiang, D.P., Qin, Y.L., Xia, J., Jiang, D.C., Chen, H.Y.: C_3N_4 nanosheet modified microwell array with enhanced electrochemiluminescence for total analysis of cholesterol at single cells. Anal. Chem. **89**, 2216–2220 (2017)
58. Szunerits, S., Tam, J.M., Thouin, L., Amatore, C., Walt, D.R.: Spatially resolved electrochemiluminescence on an array of electrode tips. Anal. Chem. **75**, 4382–4388 (2003)
59. Wang, N.N., Ao, H., Xiao, W.C., Chen, W.W., Li, G.M., Wu, J., Ju, H.X.: Confined electrochemiluminescence imaging microarray for high-throughput biosensing of single cell-released dopamine. Biosens. Bioelectron. **201**, 113959 (2022)
60. Wang, N., Feng, Y., Wang, Y., Ju, H., Yan, F.: Electrochemiluminescent imaging for multi-immunoassay sensitized by dual dna amplification of polymer dot signal. Anal. Chem. **90**, 7708–7714 (2018)
61. Wang, N., Wang, Z., Chen, L., Chen, W., Quan, Y., Cheng, Y., Ju, H.: Dual resonance energy transfer in triple-component polymer dots to enhance electrochemiluminescence for highly sensitive bioanalysis. Chem. Sci. **10**, 6815–6820 (2019)
62. Khoshfetrat, S.M., Ranjbari, M., Shayan, M., Mehrgardi, M.A., Kiani, A.: Wireless electrochemiluminescence bipolar electrode array for visualized genotyping of single nucleotide polymorphism. Anal. Chem. **87**, 8123–8131 (2015)
63. Qin, X., Jin, H.J., Li, X.X., Li, J., Pan, J.B., Wang, K., Liu, S.Q., Xu, J.J., Xia, X.H.: Label-free electrochemiluminescence imaging of single-cell adhesions by using bipolar nanoelectrode array. Chem. Eur. J. **28**, e202103964 (2022)
64. Anderson, T.J., Defnet, P.A., Zhang, B.: Electrochemiluminescence (ECL)-based electrochemical imaging using a massive array of bipolar ultramicroelectrodes. Anal. Chem. **92**, 6748–6755 (2020)
65. Zanut, A., Fiorani, A., Rebeccani, S., Kesarkar, S., Valenti, G.: Electrochemiluminescence as emerging microscopy techniques. Anal. Bioanal. Chem. **411**, 4375–4382 (2019)
66. Sojic, N., Knezevic, S., Han, D., Liu, B., Jiang, D.: Electrochemiluminescence microscopy. Angew. Chem. Int. Ed. **63**, e202407588 (2024)
67. Fu, W., Qi, M., Rong, Y., Lin, C., Guo, W., Su, B.: Remote on-paper electrochemiluminescence-based high-safety and multilevel information encryption. Angew. Chem. Int. Ed. **64**, e202420184 (2025)

68. Chovin, A., Garrigue, P., Vinatier, P., Sojic, N.: Development of an ordered array of opto-electrochemical individually readable sensors with submicrometer dimensions: application to remote electrochemiluminescence imaging. Anal. Chem. **76**, 357–364 (2004)
69. Yuan, Y., Han, S., Hu, L., Parveen, S., Xu, G.: Coreactants of tris(2,2′-bipyridyl)ruthenium(II) electrogenerated chemiluminescence. Electrochim. Acta **82**, 484–492 (2012)
70. Ding, H., Zhou, P., Fu, W., Ding, L., Guo, W., Su, B.: Spatially selective imaging of cell-matrix and cell-cell junctions by electrochemiluminescence. Angew. Chem. Int. Ed. **60**, 11769–11773 (2021)
71. Batista Napotnik, T., Polajžer, T., Miklavčič, D.: Cell death due to electroporation—a review. Bioelectrochem. **141**, 107871 (2021)
72. Ma, C., Wu, S., Zhou, Y., Wei, H.F., Zhang, J., Chen, Z., Zhu, J.J., Lin, Y., Zhu, W.: Bio-coreactant-enhanced electrochemiluminescence microscopy of intracellular structure and transport. Angew. Chem. Int. Ed. **60**, 4907–4914 (2021)
73. Ding, H., Guo, W., Ding, L., Su, B.: Confined electrochemiluminescence at microtube electrode ensembles for local sensing of single cells. Chin. J. Chem. **39**, 2911–2916 (2021)
74. Zhou, J., Ma, G., Chen, Y., Fang, D., Jiang, D., Chen, H.Y.: Electrochemiluminescence imaging for parallel single-cell analysis of active membrane cholesterol. Anal. Chem. **87**, 8138–8143 (2015)
75. Kong, Y.C., Ye, D., Xu, C.H., Ma, Z., Zhao, H., Zhao, W.: Electrogenerated chemiluminescence imaging of single-atom nanocatalysts. Angew. Chem. Int. Ed. **63**, e202318748 (2024)
76. Zhao, X., Li, Y., Cui, Y., Saqib, M., Zhang, X., Hao, R., Zheng, Z.: Spatiotemporally and chemically resolved imaging of electrocatalytic oxygen evolution on single nanoplates of cobalt-layered hydroxide. J. Am. Chem. Soc. **145**, 20897–20906 (2023)
77. Zhu, H., Jiang, D., Zhu, J.J.: High-resolution imaging of catalytic activity of a single graphene sheet using electrochemiluminescence microscopy. Chem. Sci. **12**, 4794–4799 (2021)
78. Dutta, P., Han, D.N., Goudeau, B., Jiang, D.C., Fang, D.J., Sojic, N.: Reactivity mapping of luminescence in space: insights into heterogeneous electrochemiluminescence bioassays. Biosens. Bioelectron. **165**, 112372 (2020)
79. Huang, X., Li, B., Lu, Y., Liu, Y., Wang, S., Sojic, N., Jiang, D., Liu, B.: Direct visualization of nanoconfinement effect on nanoreactor via electrochemiluminescence microscopy. Angew. Chem. Int. Ed. **62**, e202215078 (2022)
80. Li, B., Huang, X., Lu, Y., Fan, Z., Li, B., Jiang, D., Sojic, N., Liu, B.: High electrochemiluminescence from $Ru(bpy)_3{}^{2+}$ embedded metal–organic frameworks to visualize single molecule movement at the cellular membrane. Adv. Sci. **9**, 2204715 (2022)
81. Li, B., Lu, Y., Huang, X., Sojic, N., Jiang, D., Liu, B.: Stimuli-responsive DNA nanomachines for intracellular targeted electrochemiluminescence imaging in single cells. Angew. Chem. Int. Ed. **64**, e202421658 (2025)
82. Wang, L., Wang, Z.G., Ning, D., Hu, Y., Liu, S.L., Pang, D.W.: Real-time monitoring of biomolecular dynamics on cell membranes by quantum dot-based multicolor electrochemiluminescence. Nano Today **50**, 101855 (2023)
83. Wang, N., Gao, H., Li, Y., Li, G., Chen, W., Jin, Z., Lei, J., Wei, Q., Ju, H.: Dual intramolecular electron transfer for in situ coreactant-embedded electrochemiluminescence microimaging of membrane protein. Angew. Chem. Int. Ed. **60**, 197–201 (2021)
84. Ma, C., Wang, M.X., Wei, H.F., Wu, S., Zhang, J.R., Zhu, J.J., Chen, Z.: Catalytic route electrochemiluminescence microscopy of cell membranes with nitrogen-doped carbon dots as nano-coreactants. Chem. Commun. **57**, 2168–2171 (2021)
85. Ding, H., Guo, W., Su, B.: Imaging cell-matrix adhesions and collective migration of living cells by electrochemiluminescence microscopy. Angew. Chem. Int. Ed. **59**, 449–456 (2020)
86. Yan, Y., Zhou, P., Ding, L., Hu, W., Chen, W., Su, B.: T cell antigen recognition and discrimination by electrochemiluminescence imaging. Angew. Chem. Int. Ed. **62**, e202314588 (2023)
87. Gao, H., Han, W., Qi, H., Gao, Q., Zhang, C.: Electrochemiluminescence imaging for the morphological and quantitative analysis of living cells under external stimulation. Anal. Chem. **92**, 8278–8284 (2020)

88. Ding, L., Ding, H., Zhou, P., Xi, L., Su, B.: Surface-sensitive imaging analysis of cell-microenvironment interactions by electrochemiluminescence microscopy. Anal. Chem. **94**, 10885–10892 (2022)
89. Ding, L., Zhou, P., Yan, Y., Su, B.: Electrochemiluminescence imaging of cellular contact guidance on microfabricated substrates. Chem. Biomed. Imaging **1**, 558–565 (2023)
90. Dong, J., Lu, Y., Xu, Y., Chen, F., Yang, J., Chen, Y., Feng, J.: Direct imaging of single-molecule electrochemical reactions in solution. Nature **596**, 244–249 (2021)
91. Chen, M.M., Xu, C.H., Zhao, W., Chen, H.Y., Xu, J.J.: Super-resolution electrogenerated chemiluminescence microscopy for single-nanocatalyst imaging. J. Am. Chem. Soc. **143**, 18511–18518 (2021)
92. Reinhardt, S.C.M., Masullo, L.A., Baudrexel, I., Steen, P.R., Kowalewski, R., Eklund, A.S., Strauss, S., Unterauer, E.M., Schlichthaerle, T., Strauss, M.T., Klein, C., Jungmann, R.: Ångström-resolution fluorescence microscopy. Nature **617**, 711–716 (2023)

Electrochemiluminescence Instrumentation Toward Single Events

Tianyu Wei, Zhenzhong Cai, Jessica Elizabeth Melissa Winslade, Emma Mae Lord, Jian-bin Pan, and Zhifeng Ding

Abstract This chapter summarizes the development and refinement of electrochemiluminescence (ECL) instrumentation, with a focus on achieving single-event detection. It provides an in-depth discussion of advanced instrument design, including high-sensitivity photodetectors, synchronized electrochemical control systems, and optimized signal acquisition techniques. This chapter also presents experimental examples that illustrate the capability of instrumentation to perform in situ ECL spectroscopy, determine absolute quantum efficiency, and, more importantly, capture single ECL events at the nanosecond scale. This discussion underscores the critical role of precision instrumentation in expanding the possibilities of ECL research.

Keywords Electrochemiluminescence (ECL) instrumentation · Photodetector · ECL spectroscopy · Absolute quantum efficiency · Single ECL events

1 Introduction

1.1 ECL Instrumentation Concepts and Applications

Electrochemiluminescence (ECL) experiments typically involve an electrochemical workstation, often a potentiostat or bipotentiostat, paired with a light detection system. Initially, silicon photodiodes (PDs) were used to capture light [1]. Since the late 1980s, photomultiplier tubes have largely taken their place [2, 3]. Some

Authors Tianyu Wei, Zhenzhong Cai are equally contributed authors

T. Wei · Z. Cai · J. E. M. Winslade · E. M. Lord · Z. Ding (✉)
Department of Chemistry, The University of Western Ontario, London, Canada
e-mail: zfding@uwo.ca

J. Pan
State Key Laboratory of Analytical Chemistry for Life Science, School of Chemistry and Chemical Engineering, Nanjing University, Nanjing, China

F. Paolucci et al. (eds.), *Electrogenerated Chemiluminescence Microscopy and Imaging*,
https://doi.org/10.1007/978-3-032-14602-1_4

studies, however, have preferred charge-coupled device (CCD) cameras for their spectroscopic capabilities [4]. Notably, commercial ECL systems for immunoassays have been developed by companies like Roche [5], while in China, ECL analyzer systems created by the Changchun Institute of Applied Chemistry and produced by Xi'an Remax Electronic Co. Ltd. are widely used [2]. Xi'an Heyongzhong Electronic Technology Co., Ltd. was established in 2015 [6] while the State Key Laboratory of Analytical Chemistry for Life Science at Nanjing University has been producing ECL analyzers since 2016.

In addition to these commercial systems, many researchers opt for custom-built/home-made ECL setups. These require a data acquisition system that can record and analyze applied potential, electrochemical current, and ECL intensity as photocurrent. These systems often incorporate external tools like National Instruments (NI) data acquisition boards and software such as LabVIEW [3]. Several computer-controlled potentiostats, including models such as Princeton Applied Research 273A, 263A (AMETEK, Inc., Berwyn, PA), CHI 660, 760 (CH Instruments, Austin, TX), Autolab PGSTAT302N, PGSTAT30 (Metrohm, Herisau, Switzerland) listed a few, offer auxiliary signal channels to capture both electrochemical and ECL data simultaneously.

1.2 Major Routes of ECL Realized by Instrumentation

Initially, cyclic voltammetry is utilized to acquire the electrochemical characteristics of the sample, identifying oxidation potentials in the anodic region and reduction potentials in the cathodic region for generating radical cations and anions, which react to generate excited states emitting light. The ECL intensity, both in the anodic and cathodic regions, is influenced by the reactivity and the time gap between the formation of these radicals. If this delay exceeds the radicals' lifetime, their interactions and subsequent formation of the excited species are inefficient. Instruments can be used to adjust the applied potential profile from scanning to pulsing to enhance ECL emission [7]. While the ECL signal is recorded through the auxiliary channel of the potentiostat, this channel may not be fast enough to follow rapid potential pulsing. Therefore, a high-speed data acquisition system may be required.

Using the same instrumentation, ECL intensity can be further enhanced by introducing a coreactant. Unlike the annihilation mechanism, the coreactant pathway enables ECL generation by applying a potential in a single direction. This approach involves the breakdown of the coreactant, leading to the formation of a highly reactive intermediate [8]. This intermediate then reacts with the radical ion of the luminophore to generate its excited states giving off ECL. A typical example of a highly efficient ECL system is the combination of $Ru(bpy)_3^{2+}$ and tripropylamine (TPrA). The mechanistic pathway between $Ru(bpy)_3^{2+}$ and TPrA is thoroughly explained in the 2002 article by Miao et al. [9]. This system has become the foundation for commercial systems used in immunoassays and DNA analysis, demonstrating its broad applicability and potential in various analytical techniques. Applied potential pulsing can be further used to generate enhanced ECL signal for coreactant systems.

1.3 Challenges from Absolute ECL Quantum Efficiency

The development of absolute ECL quantum efficiency (ϕ_{ECL}) has significantly enhanced the standardization of ECL measurements, yielding more reliable and impactful results for a wide range of researchers. In 1973, the Bard research group utilized a rotating ring-disk electrode (RRDE) paired with a PD to determine the ECL efficiency of 5–6% for $Ru(bpy)_3^{2+}$ in acetonitrile solution [1]. While this early ECL efficiency was useful for relative quantum efficiency calculations, inconsistencies in solutions, concentrations, and testing conditions led to significant errors and unreliable comparisons between luminophores. In 2021, the Ding group introduced a strategy for determining absolute ECL quantum efficiency, combining an electrochemical workstation, a calibrated PMT, a picoammeter, and a spectrometer [10]. This approach incorporates a series of calibrations and self-absorption corrections to ensure the accuracy and consistency of measurements.

1.4 Spectroscopy

Over recent years, spooling ECL spectroscopy and accumulated ECL spectra have been used to analyze the ECL generated through the annihilation and coreactant pathways, as well as to determine the ϕ_{ECL}. The experimental setup involves a spectrograph paired with a cooled charge-coupled device (CCD) camera and an electrochemical workstation. Synchronization between the electrochemical process and spectral acquisition can be achieved using a trigger generated at the beginning of an electrochemical procedure. When measuring ECL spectroscopy, it is important to consider self-absorption effects in concentrated solutions, which can affect the measured ϕ_{ECL}. Additionally, spooling ECL spectroscopy can be used to explore the behavior behind light emission, including the pathways of electron transfer, intermediate species formation, and the generation of excited states. By studying the efficiency of light emission and quantum yields, ECL spectroscopy aids in designing better luminophores and optimizing conditions for maximum luminescence intensity [11].

1.5 Single ECL Events by Means of UMEs

Ultramicroelectrodes (UMEs) enable the detection of single ECL events due to their minimized non-faradaic currents, enhanced mass transport through radial diffusion, and reduced double-layer capacitance. These properties enable UMEs to operate efficiently at high scan rates and pulsing frequencies, making them ideal for ECL generation [12–19].

In 1994, Collinson and Wightman utilized an UME with a 1 μm radius to generate ECL at high pulsing frequencies. By comparing the experimental results with simulated 1D data, they determined the ion-annihilation rate of DPA, DMA, and $Ru(bpy)_3^{2+}$ [19]. A year later, they improved the temporal resolution to 5 ns and reduced the DPA concentration to 15 μM, successfully detecting single-event ECL at UMEs [12].

More recently, Ding's research group has further advanced single-event ECL detection at UMEs, using 2D simulations to confirm the ion-annihilation rate constant for $Ru(bpy)_3^{2+}$, consistent with Wightman's earlier findings [19, 20]. More are anticipated.

1.6 A Short Review of Previous Books and Reviews on ECL Instrumentation

Over the years, ECL has seen significant advancements, and numerous reviews have recorded its evolution by means of advanced instrumentation, offering valuable insights into its diverse applications and underlying mechanisms. In 1973, Bard et al. discussed the use of a rotating ring-disk electrode (RRDE) system [21]. In this method, an electrode, consisting of a planar disk electrode and a concentric ring electrode separated by an insulating gap, is rotated. The solution is drawn upwards toward the electrode and radially outwards on the electrode surface. This causes the produced species on the disk to flow into the ring-produced species, resulting in a light-producing reaction. This system was able to produce several minutes of steady light emission. In 1977, Faulkner and Bard discussed the use of a silicon PD to determine the efficiency of light collection [22]. The ECL efficiency was measured by using a fixed distance from the electrode to the PD, where the total intensity of emission is equal to $2D\pi r^2$ of the RRDE. This was able to provide an uncorrected ECL efficiency, to which corrections would have to factor in the reflectivity of the electrode and the self-absorbance by the compound. In 2004, Fan discussed the use of a bare PMT combined with an electrometer to detect ECL [23]. This was a significant development over the years as the PMT was known to have a high sensitivity in comparison to the previously used PDs and to be used in a photocurrent or photon-counting mode, offering a higher level of versatility. Finally, in 2007 and 2008, Miao discussed the use of commercially available ECL instruments [2, 8]. Miao highlighted the development of the Sector HTS Imager, which contained screen-printed carbon ink electrodes placed within multiwell plates. This allowed for several binding domains to react with specific targets, acting as an efficient high-throughput immunoassay. Significantly, this device also utilized a CCD camera, allowing for efficient detection. These reviews and chapters outlined the advancements in ECL instrumentation that have not only refined detection sensitivity and efficiency, but also paved the way for its application in high-throughput analysis, solidifying ECL as a powerful tool in clinical diagnostics, biomedical applications, optoelectronics, and many other fields.

2 Electrochemistry and Electrochemiluminescence

2.1 ECL in Annihilation and Coreactant Routes

Electrode Preparation and Electrochemical Cells. Here, we describe a typical example of ECL experiments along with electrochemistry. Cyclic voltammograms (CVs), differential pulse voltammograms (DPVs), chronoamperograms (CAs), and ECL experiments can be conducted using a three-electrode system. The working electrode (WE) consisted of a 2 mm diameter platinum disk sealed in a glass tube, while two platinum wire coils served as the counter electrode (CE) and quasi-reference electrode (RE), Fig. 1 [11]. Experimental potentials were referenced to the Fc/Fc^+ redox couple, corresponding to 0.342 V versus the saturated calomel electrode (SCE) [24]. Before each experiment, the WE was polished to achieve a mirror-like surface. This was done using a polishing pad (Buehler Ltd., Lake Bluff, IL) and a 0.05 μm alumina slurry mixed with water, followed by rinsing with acetone and drying under argon gas. Electrochemical polishing of the WE was performed by cycling the potential between −0.9 V and 0.75 V at a scan rate of 0.1 V/s for 400 cycles in a 0.1 M H_2SO_4 solution. The CE and RE were cleaned by sequential ultrasonication in acetone, isopropyl alcohol, and ultrapure water for 5 min each. After cleaning, they were dried at 120 °C for 10 min and cooled before use. ECL experiments were conducted in a custom-designed cylindrical glass cell with a flat Pyrex window at its bottom, allowing the transmission of emitted light. In order to hold an ECL cell on top of a spectrograph and shield it from environmental light, a light-tight cell holder is required, as shown in Fig. 1 [11].

To minimize the oxygen and moisture, the ECL cell was assembled inside a nitrogen-filled glovebox (Model Nexus I, Vacuum Atmospheres Company, Hawthorne, CA). The electrodes were installed, and the cell was sealed with an air-tight cap equipped with electrical connection leads before being removed from the glovebox for the experiments. The electrolyte solution used in the ECL experiments consisted of a 1:1 mixture of acetonitrile: benzene, chosen for its wide potential window and high solubility for the luminophore.

General ECL Instrumentation. A routine ECL instrumentation setup is illustrated in Fig. 2 in photon counting mode [7], while ECL setup in photocurrent

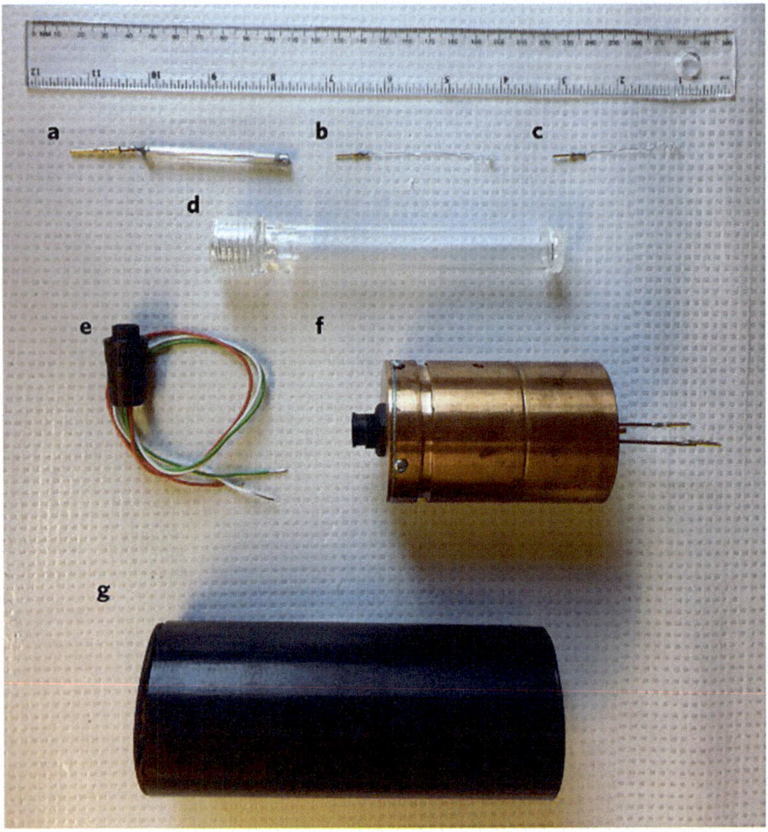

Fig. 1 Required components to assemble a light-tight ECL cell. **a** A Pt disk electrode as WE. **b**, **c** Two coiled Pt Wires as CE and RE, respectively. **d** A typical glass cylinder ECL cell with a flat quartz or Pyrex window at the bottom. **e** A nex ternal connector with electrode cables to attach to ECL cell cap [11]. **f** ECL cell cap for shielding external light and connecting the three electrodes internally. **g** A plastic cylinder to shield light and hold the ECL cell at the spectrograph entrance [11]

mode can be found in our previous publication [25]. A potentiostat (Autolab, PGSTAT302N) can be utilized for performing CVs, DPVs, and CAs. The ECL signal was recorded by a photodetector which is a side-on photon-counting head (PCH), which is an all-in-one unit incorporating a photomultiplier tube (PMT) and high-speed amplifier/discriminator/high-voltage power supply circuits in a compact housing (model H6420-02, Hamamatsu, Japan). This unit operates with a +5 Vdc power input, which was supplied via the voltage delivery pins of a standard USB Type-A connection. Before being recorded, the output signal from the PCH was routed through an attenuator to lower the pulse magnitude to approximately + 200 mV. The conditioned signal was then fed into an SR430 multichannel scaler/ averager (Stanford Research Systems, Sunnyvale, CA) for data acquisition. The

SR430 organizes incoming voltage pulses into sequential "bins," with a discriminator level set at +100 mV; this value is exactly one-half of the attenuated signal to improve the signal-to-noise ratio. The bin width (T_{bin}), defined as the time interval for one bin, is the key factor affecting temporal resolution. The bin per record (N) specifies the number of bins in each measurement. Together, these parameters define the total acquisition time of a single measurement, as expressed in Eq. 1:

$$T = (N \times T_{bin}) + (N \times 250\,\text{ns}) + 150\,\mu\text{s} \quad (1)$$

The constant values of 250 ns and 150 μs correspond to the required accumulation time per bin, and the instrument processing overhead, respectively [26]. For example, if 1024 bins per record (N) with a bin width of 5.120 μs (T_{bin}) are used, the total acquisition time per record (T) is:

$$T = (1024 \times 5.12\,\mu\text{s}) + (1024 \times 0.25\,\mu\text{s}) + 0.15\,\text{ms} = 5.65\,\text{ms}$$

Additionally, the records per scan determine the number of repeated measurements (or records) in a single experiment (or scan). The results from repeated measurements are dynamically summed, allowing the statistical accumulation of individual

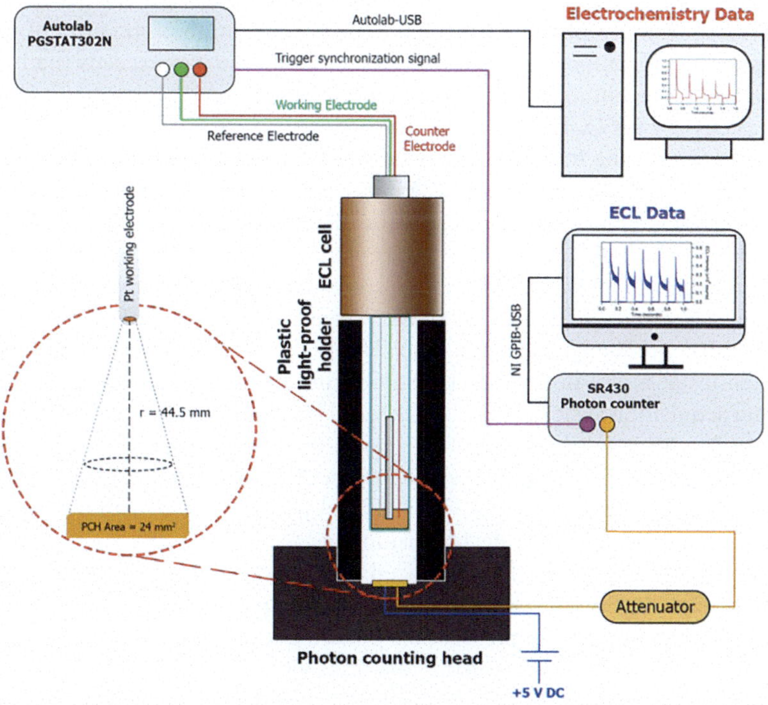

Fig. 2 Illustrative instrumentation of the electrochemistry and ECL experiment [7]

photons generated from ECL reactions. To ensure synchronized data collection, an external voltage trigger from the Autolab workstation initiated each recording, aligning the electrochemical current measurement (from the PGSTAT302N) with the corresponding ECL signal, recorded as photon counts by the PCH.

Typical examples. By using the "gold standard" sample tris(2,2′-bipyridine) ruthenium (II) hexafluorophosphate ($Ru(bpy)_3(PF_6)_2$) in ECL as an example, CV and CA profiles of $Ru(bpy)_3^{2+}$ corresponding to its ECL voltage curve were generated from the instrumentation set in Fig. 2. Figure 3a shows the CV (red) and the corresponding ECL-voltage (blue) curve of 1 mM $Ru(bpy)_3^{2+}$ in anhydrous acetonitrile using a 2 mm diameter platinum disk electrode at a scan rate of 100 mV/s [7]. The cathodic scan shows three successive reductions of $Ru(bpy)_3^{2+}$ at formal potentials of −1.31 V, −1.58 V, and −1.78 V, while the anodic scan shows a single oxidation at a formal potential of +1.30 V. In the anodic region, $Ru(bpy)_3^{2+}$ has a sharp ECL peak with an intensity of approximately 300 counts at the first oxidation peak, occurring at the potential of +1.30 V. A broader ECL peak with a lower intensity of about 10 counts is observed in the cathodic region, spanning from the first reduction peak to the third reduction peak. The observed ECL is from the excited states generated from interactions between the equivalent radical cation and radical anion of $Ru(bpy)_3^{2+}$, $Ru(bpy)_3^{3+}$, and $Ru(bpy)_3^{1+}$, respectively. The lower ECL intensity on the cathodic side compared to the anodic side indicates reduced stability of the radical cations. Additionally, the relatively low ECL intensity on both anodic and cathodic sides is primarily due to the significant time gap (~33 s) between the formation of radical anions and radical cations during the CV scan. This time gap exceeds the lifetime of the radicals, preventing efficient interaction and subsequent formation of excited species necessary for ECL generation.

The issue about the long-time gap between the formation of radical anions and radical cations in scanning CV can be addressed by applying a potential profile that rapidly alternates between the anodic and cathodic regions. As shown in Fig. 3b, potentials of −1.36 V and +1.35 V were applied to the working electrode at a pulsing frequency of 5 Hz [7]. This range corresponds to the completion of the first oxidation peak and the first reduction peak. The current (red) and ECL (blue) signals were recorded simultaneously over time. The ECL recording an auxiliary channel of a potentiostat might not be fast enough during the applied potential pulsing. A rapid data acquisition system might be needed. The maximum ECL intensity reached more than 7×10^5 counts, which is more than 2300 times stronger than that observed during CV scanning. This setup enabled the anodic and cathodic radical species of $Ru(bpy)_3^{2+}$ to be electrogenerated within a brief period of 20 ms, allowing them to undergo ECL reactions before decomposition.

ECL intensity can be further enhanced in coreactant systems using the same instrumentation. In Fig. 4a, the CV (red) and ECL intensity (blue) curves for the $Ru(bpy)_3^{2+}$/TPrA system are presented [7]. During an anodic potential sweep, two consecutive peaks were observed at 0.82 and 1.27 V corresponding to the oxidation of tripropylamine and $Ru(bpy)_3^{2+}$, respectively. A minor, but non-negligible ECL signal is present as early as 0.70 V, followed by a second prominent ECL peak at an onset voltage of 1.10 V, the latter which increased steadily to a maximum of approximately

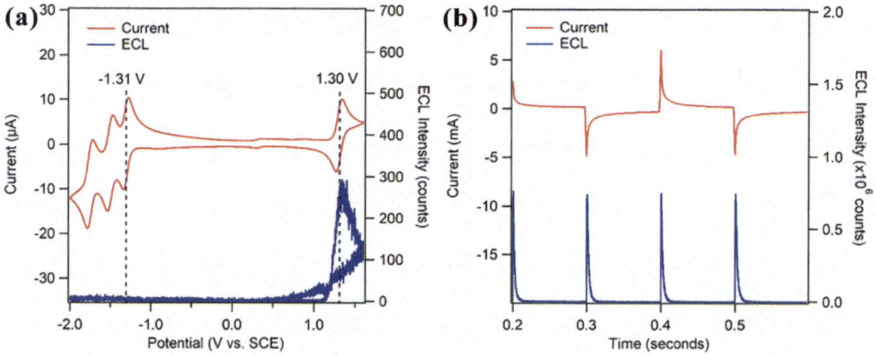

Fig. 3 **a** CV (red) and ECL–voltage curve (blue) of 1 mM $Ru(bpy)_3(PF_6)_2$ in acetonitrile with 0.1 M $TBAPF_6$ as the supporting electrolyte. Dotted lines highlight the potentials for the first reduction and oxidation. The scan rate for the cyclic voltammogram was 100 mV/s. Bin width = 10.48 ms, number of bins = 14,336, trigger level = 0.3 V, records per scan = 1. **b** Potential stepping of $Ru(bpy)_3^{2+}$ between 1.35 and −1.36 V at 5 Hz (100 ms pulse width). Current is shown in red, ECL photon counts are shown in blue, and ECL is measured by the H6420-02 photon counting head. Bin width = 40.96 ms, number of bins = 13,312, trigger level = −1.10 V, records per scan = 1 [7]

2 million counts, showing an enhancement of approximately 6600 times compared to the annihilation pathway. The first and second ECL signals concur well with the scheme proposed by Miao, Choi, and Bard in 2002 [9]. The TPrA radical ($TPrA^{\bullet}$) can participate in the reduction of $Ru(bpy)_3^{2+}$ to form $Ru(bpy)_3^{+}$, thereby enabling the generation of the excited state $Ru(bpy)_3^{2+*}$ at potentials lower than that required for the direct electro-oxidation of $Ru(bpy)_3^{2+}$ due to the interaction between $TPrA^{\bullet+}$ and $Ru(bpy)_3^{+}$ (the first ECL peak). The second ECL peak is enhanced because of the interactions between $TPrA^{\bullet}$ and $Ru(bpy)_3^{3+}$, as well as between $Ru(bpy)_3^{1+}$ and $Ru(bpy)_3^{3+}$ [9].

Figure 4b shows the ECL-time (blue) and current–time (red) responses of the $Ru(bpy)_3^{2+}$/TPrA system under a pulsed potential (black dash) alternating between 0 V and +1.5 V at a frequency of 5 Hz [7]. The ECL intensity achieved by the $Ru(bpy)_3^{2+}$/TPrA potential pulsing system reaches 1 million counts, representing a remarkable 3000-fold enhancement compared to the annihilation CV scanning pathway. Upon the application of an anodic potential step, there is a sharp increase in the ECL signal, followed by an exponential decay as the electrogenerated radicals are gradually depleted. The overall profile of the ECL peak is also much different from the one observed in annihilation, as it does not fully decay to the baseline level during the anodic steps. TPrA was selected as the coreactant due to its ability to form a radical cation ($TPrA^{\bullet+}$) in the anodic region which contributes to the persistence of the ECL signal since both $TPrA^{\bullet+}$ and $Ru(bpy)_3^{\bullet+}$ radical anions are being continuously generated over the duration of the anodic pulse.

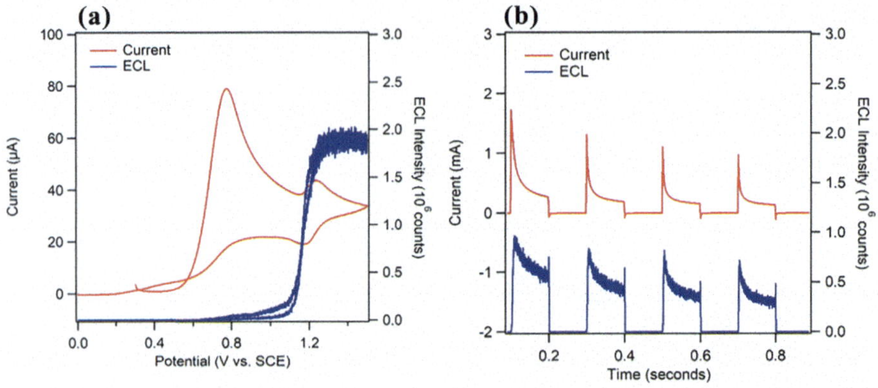

Fig. 4 **a** CV (red) and ECL–voltage curve (blue) of 1 mM $Ru(bpy)_3{}^{2+}$ in acetonitrile with 10 mM TPrA. The scan rate for the cyclic voltammogram was 100 mV/s. Bin width = 2.62 ms, number of bins = 11,264, trigger level = 0.30 V, records per scan = 1. **b** Potential step experiment of 1 mM $Ru(bpy)_3{}^{2+}$ with 10 mM TPrA, between 0 and 1.50 V at 10 Hz (100 ms pulse width). Current is shown in red, and ECL intensity (as photon counts) is shown in blue. Bin width = 81.92 μs, number of bins = 13,312, trigger level = −1.40 V, records per scan = 1 [7]

2.2 ECL Spectroscopy

To further analyze the ECL generated from the annihilation and coreactant pathways and to determine the absolute ΦECL, spooling ECL spectroscopy and accumulated ECL spectra have been investigated the intensity of light produced from the reactions at each wavelength. The ECL cell setup, electrode preparation, and reagents used are the same as those mentioned in Sect. 2.1: **ECL in the annihilation and coreactant routes**.

ECL spectra were recorded using a spectrograph (Model SP2300i, Princeton Instruments, Trenton, NJ) coupled with a Charge-Coupled Device (CCD) camera (Andor DU401-BR-DD-352, Oxford Instruments, UK) cooled to −65 °C, in conjunction with an electrochemical workstation (Model 610A, CH Instruments Inc., Austin, TX), as shown in Fig. 5 [27]. Wavelength calibration was accomplished with a mercury lamp (HG-1, Ocean Optics, Dunedin, FL). Accumulated ECL spectra were captured either as a single spectrum over an entire CV scanning or during a 10-s CA at a frequency of 5 Hz. Spooling ECL spectra were obtained throughout the CV scanning, with a spectrum recorded every second. Synchronization between the electrochemical driving process and ECL spectral acquisition was achieved using the 5 V TTL voltage output from the CHI analyzer, which served as an external trigger for the CCD camera. This synchronization was implemented through programming. ECL spectra were acquired using AndorSolis software (Andor, Oxford Instruments, UK). The wavelengths of the spectrograph-CCD camera setup were calibrated with a mercury argon lamp (HG-1 lamp, Ocean Insight, Orlando, FL) using a center wavelength of 546.07 nm. More detailed equipment setup is provided in our previous work [11].

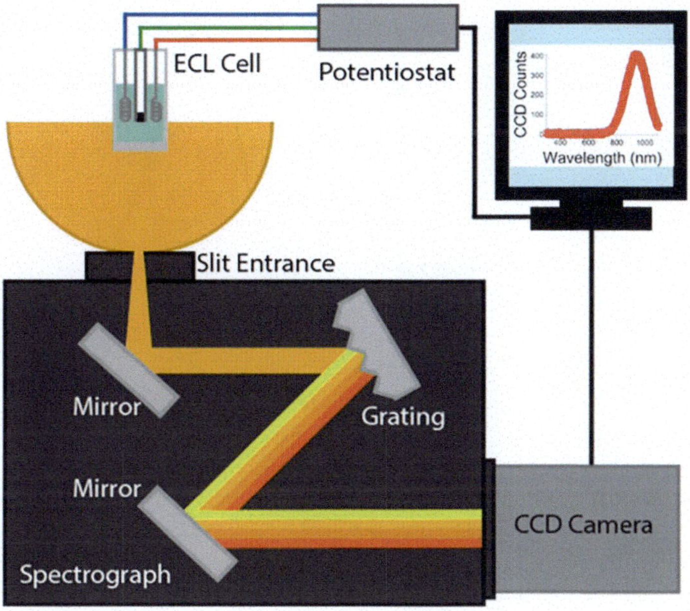

Fig. 5 Illustrative instrumentation of the accumulated ECL spectra and spooling ECL spectroscopy [27]

Figure 6 shows an example of the accumulated ECL spectra generated from the instrument setup from Fig. 5 during the annihilation pathway, performed between the first oxidation peak and the first reduction peak at a frequency of 5 Hz [10]. These voltage ranges were selected based on prior studies confirming their ability to achieve optimal ECL performance. The black trace shows the ECL signal captured by the CCD camera, resulting in an emission spectrum, where a peak wavelength of 620 nm is observed. For the annihilation pathway, spooling ECL spectroscopy and accumulated ECL spectra during CV scanning were challenging to obtain due to the relatively low ECL intensity during the CV scan. The charge-coupled device (CCD) camera requires an ECL intensity of at least approximately 20 nA on a PMT to detect photons as counts and generate ECL spectra. Consequently, only accumulated ECL spectra obtained during CA were used to study the ECL behavior of $Ru(bpy)_3^{2+}$ under the annihilation pathway. Additionally, the red trace in Fig. 6 corresponds to the absorption spectrum of the same $Ru(bpy)_3^{2+}$ solution used in this study. The overlap between the black and red traces provides strong evidence of self-absorption, a phenomenon commonly observed in concentrated solutions. In contrast to photoluminescence (PL) studies, where luminophore concentrations are much lower, ECL experiments often require significantly higher concentrations, leading to intensely colored solutions. If self-absorption is not considered, the loss of higher-energy emitted light can substantially affect the accuracy of the measured ΦECL [10].

Fig. 6 Absorption spectrum (red) and background-subtracted and normalized ECL spectrum (black) of the same $Ru(bpy)_3{}^{2+}$ electrolyte solution [10]

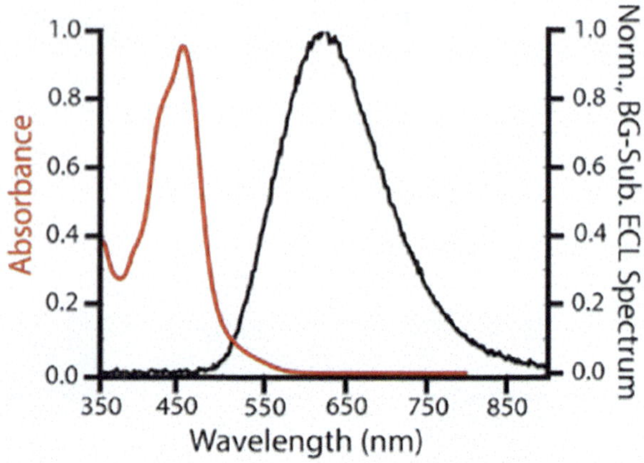

For the $Ru(bpy)_3{}^{2+}$/TPrA coreactant system, accumulated ECL spectra from both CV scanning and CA were used to investigate the ECL behavior of $Ru(bpy)_3{}^{2+}$ under the coreactant pathway [7]. Figure 7b shows the accumulated ECL spectra during CA between 0 V and 1.5 V at a frequency of 5 Hz in the coreactant pathway, overlapping with the accumulated ECL spectra from the annihilation pathway. In this mode, the peak wavelength appears at 620 nm—identical to the peak wavelength observed in the annihilation pathway (Fig. 6) [7].

Spooling ECL spectroscopy, performed using the instrument setup in Fig. 5, is a technique coupled with electrochemistry. It provides time or potential-resolved (up

Fig. 7 Background-corrected and normalized intensity ECL emission spectra of $Ru(bpy)_3{}^{2+}$ annihilation and TPrA coreactant systems [7]

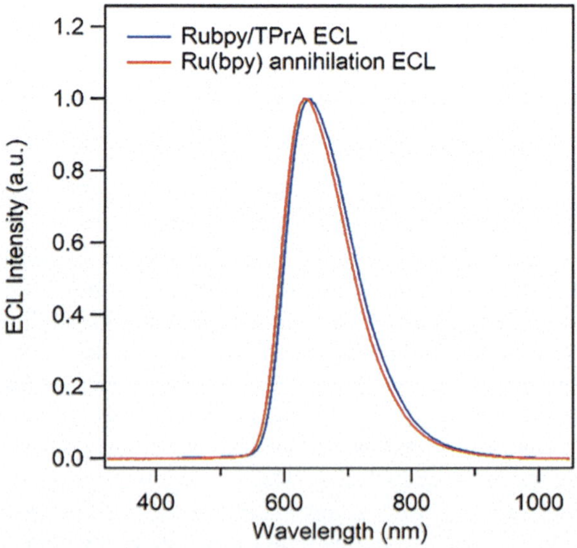

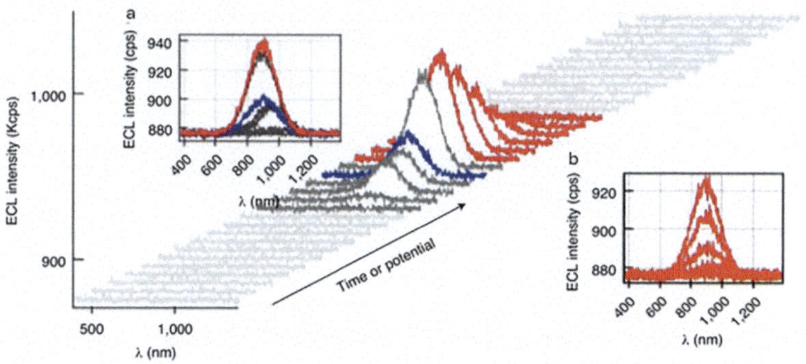

Fig. 8 Spooling ECL spectra of an Au_{25}^{1-}/ TPrA system in acetonitrile: benzene (1:1) mixture containing 0.1 M TBAP from –0.33 to 1.33 back to 0.33 V at scan rate of 100 mV/s to accumulate 34 spectra at 1 s intervals. **a** stacked spectra of evolution direction (forward scan), and **b** stacked spooling spectra of devolution direction (backward scan). The dark gray, blue, and red ECL spectra highlight the spectra with wavelength change during spooling ECL spectroscopy experiment [11]

to 1 mV) detail for each individual excited state that emits light during an electrochemical process [11]. Note that potential range and scan rate connect the applied potential to time. Our group used this approach to investigate the ECL mechanisms of atomically precise gold nanoclusters showing their near-infrared (NIR) light emissions. These probes were then applied to image vascular complexity to a depth of more than 4 mm in living mice [11].

Figure 8 demonstrates ECL spooling spectra of 0.1 mM Au_{25}^{1-} clusters in the presence of 5 mM TPrA [11]. It is obvious that the spectrum intensity (ECL) on spectra displayed in blue is slightly higher than the ones in gray. However, in the next spectrum, the intensity is significantly enhanced. Another option for plotting data is the stacked approach, in which the peak position and intensity changes of a set of spectra can be easily identified (Fig. 8). For example, in Fig. 8a (stacked spectra of evolution direction) one can easily determine that the spectrum peak wavelength shifted from 950 nm (gray spectra) to 890 nm (blue spectrum) within an interval, followed by an obvious peak intensity enhancement while the peak wavelength position remained at 890 nm [11].

2.3 *Absolute ΦECL Determination*

The absolute ECL quantum efficiency (ΦECL) was determined using Eq. 1 [7, 10], where the numerator represents the total number of photons, and the denominator represents the total number of electrons. The total number of photons was determined from the ECL intensity diagrams obtained from the instrument setup in Fig. 2 or Fig. 9 and the ECL spectra from the setup in Fig. 5 through a series of calibrations, along with a MATLAB program [28]. The instrument setup shown in Fig. 9 is similar to

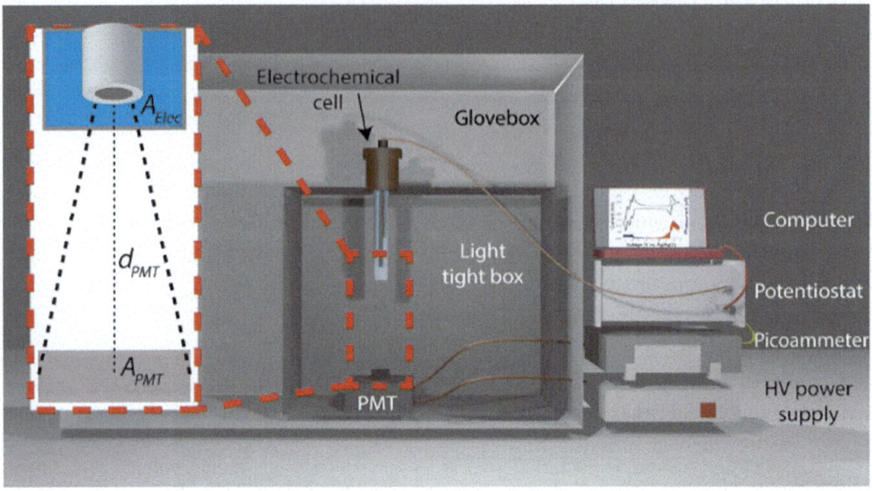

Fig. 9 Example instrumentation for QE measurements with an inset of solid angle descriptions [10]

the one in Fig. 2 but uses a different type of light detector. Instead of using a photon-counting PMT as the light detector, the setup in Fig. 9 employs a PMT (R3896, Hamamatsu, Japan) powered by a high-voltage power supply set at −750 V. This PMT measures ECL intensity as photocurrent with a corresponding gain of 4.78×10^5, which is defined as the ratio of the anode output current to the photoelectric current from the photocathode. A picoammeter (model 6487, Keithley, Cleveland, OH) was used to convert the photocurrent from the PMT into a voltage signal [10].

$$\Phi ECL = \frac{\text{total photons}}{\text{total electrons}} \times 100\% \tag{2}$$

A series of calibrations/corrections must be involved during the absolute ΦECL determination. These calibrations include adjustments for the detection area factor and the reflectivity of the platinum electrode, as the light emission from the ECL reaction follows a non-Lambertian emission profile. The next calibration accounts for hardware- and wavelength-specific factors due to variations in hardware sensitivity to incoming light. Additionally, self-absorption correction must be considered by applying Gaussian curve fitting, as some light generated by the reaction is absorbed by the dark-colored solution, leading to a red shift in the wavelength window as mentioned in the previous section. More details on these calibrations are available in our previous work [10, 27], and were also reviewed by Paolucci, Valenti and Sojic et al. in 2023 [29]. The total number of electrons was derived from the total charge, which was calculated by integrating the electrochemical current over the duration of the experiment. This determination process also involves a calibration to correct for significant non-Faradaic current that does not directly contribute to ECL emission.

More detailed equations to determine the total number of photons and calibration methods are provided in our previous work [10].

The absolute ΦECL of $Ru(bpy)_3^{2+}$ was determined in our previous work under both annihilation and coreactant pathways at different concentrations and frequencies. In 2021, our group published a paper that determined the absolute ΦECL of a 1 mM $Ru(bpy)_3^{2+}$ solution during CV scanning and CA using a calibrated PMT in photocurrent mode [10]. During CV scanning, the absolute ΦECL was measured at **0.0019%**. This low ΦECL is attributed to two factors: the large time gap between radical generation events and the unnecessary reduction of $Ru(bpy)_3^{2+}$. At a scan rate of 0.1 V/s, as shown in Fig. 3a, the minimum time between radical generation events is 33 s, which is 23 s longer than the 0.1 Hz pulsing frequency. This long-time gap between the formation of radical anions and cations significantly lowers the quantum efficiency due to radical quenching. During CA, the absolute ΦECL of $Ru(bpy)_3^{2+}$ was measured at different frequencies, including 0.1 Hz, 1 Hz, 10 Hz, 100 Hz, and 1000 Hz. The ΦECL increased from **0.09%** at 0.1 Hz to **2.43%** at 100 Hz, as higher frequencies facilitate the generation of more $Ru(bpy)_3^{2+}$ radical cations and anions, leading to a greater number of excited $Ru(bpy)_3^{2+}$ species and, consequently, higher photon emission. However, at 1000 Hz, the absolute ΦECL dropped to **0.28%**, likely because the potential was switching too rapidly for sufficient redox reactions to occur, causing a significant portion of the current to contribute to electrical double-layer formation instead [10].

For the $Ru(bpy)_3^{2+}$/TPrA system, our group published another paper in 2021 that determined the absolute ΦECL of the $Ru(bpy)_3^{2+}$/TPrA system in photon counting mode during CV scanning and CA at different TPrA concentrations, including 5 mM, 10 mM, 20 mM, and 50 mM [7]. The highest ECL efficiency for this system was observed at a TPrA concentration of 10 mM, reaching **3.1 ± 0.7%** during CV scanning and **10.0 ± 1.1%** during CA [7]. When the TPrA concentration exceeds 10 mM, the calculated ECL efficiencies begin to decline. This decrease is likely due to an imbalance in the relative concentrations of the electrogenerated radicals of $Ru(bpy)_3^{2+}$ and TPrA. The results indicat that an optimal ratio between these two species is necessary to maximize successful electron transfers between the luminophore and the coreactant. This optimal balance ensures the highest production of ECL photons while minimizing unnecessary oxidation or reduction of ECL reagents. However, when TPrA concentrations surpass this critical threshold, excessive oxidation of TPrA molecules contributes more significantly to the electrochemical current than to ECL photon generation, ultimately reducing the overall ECL quantum efficiency [7].

Our group also published a paper in 2022 that determined the absolute ΦECL in the annihilation pathway and in the Au_{25} cluster/TPrA system using a spectroscopy charge-coupled device (CCD) camera at various TPrA concentrations [27]. This method allows for the direct measurement of ECL emission across different wavelengths, enabling accurate and instrument-independent absolute ΦECL assessments. In the annihilation pathway, the absolute ΦECL for 10 Hz pulsing was determined to be **0.64%**, a value very close to previous studies [10], confirming the accuracy of the CCD-based absolute ΦECL technique. In the coreactant pathway, the same phenomenon observed in a previous study was confirmed [7], where a significant decrease in absolute ΦECL was recorded, from **3.2 x 10^{-3}%** at **6.25 mM** TPrA to **6.5 x 10^{-6}%** at **200 mM** TPrA during CV scanning. This decrease is attributed to increased quenching of the excited state at higher TPrA concentrations, as previously explained [27].

3 Fast Data Acquisition and Kinetics Modeling Developed by Bard's Group

In 2010, Bard's group published a paper exploring the electrochemistry and ECL properties of a dithienylbenzothiadiazole-based fluorophore using CV, CA, and ECL measurements [30]. The molecule consists of a benzothiadiazole core (an electron acceptor) flanked by thiophene-based electron donors, leading to distinct oxidation and reduction behaviors. Their electrochemical analysis showed two sequential oxidations at +1.01 V and +1.24 V and a single reduction at −1.18 V. Their experimental ECL transients were obtained through chronoamperometry, using an Autolab potentiostat 100 connected to a photomultiplier tube to generate ECL intensity-time curves. Compared to routine potentiostats, such as the early version CH Instruments 660, which can only provide a minimum time interval of 10 ms during CA, the Autolab potentiostat enables faster data acquisition with a time interval on the submicrosecond scale, allowing for a higher time-resolution ECL-time or ECL-transient profile. With this advantage of the Autolab potentiostat, an ECL transient study was performed by comparing experimental ECL transients to simulated ones under the annihilation pathway [30].

By performing digital simulations of CV and comparing them to the experimental CVs, they determined that the heterogeneous electron-transfer rate constant for oxidation is greater than 0.1 cm/s, which is significantly higher than that for reduction, which is approximately 1.5×10^{-3} cm/s.

Simulations were performed using a 1D planar diffusion model with COMSOL Multiphysics software v.3.2, as the working electrode used in the experiment was a 2 mm diameter Pt disk electrode. By comparing the experimental and simulated ECL transients, Bard's group reported a $\mathbf{k_{ann}}$ value of **1 x $10^9$$M^{-1}s^{-1}$** for this donor–acceptor dithienylbenzothiadiazole derivative.

Using the 1D COMSOL code kindly provided by Bard's group, we conducted a study on the kinetic behavior of $Ru(bpy)_3^{2+}$ using a 1D simulation model in COMSOL Multiphysics software v.3.5 with a 2 mm platinum electrode. Later, we developed a 2D simulation model in COMSOL Multiphysics software v.5.3 to further investigate the kinetic information of $Ru(bpy)_3^{2+}$. This model enables digital simulations of ultramicroelectrodes (UMEs), which are commonly defined as electrodes with a critical diameter of less than 25 μm [31]. In UMEs, the diffusive flux at the electrode surface follows a hemispherical geometry. With the growing popularity of micro- and nanoelectrode electrochemistry and ECL studies—due to minimal ohmic drop, reduced interference from double-layer capacitance, and extremely fast diffusion kinetics—for probing single-molecule reactions, it is essential to develop simulations that can reliably model these environments.

4 Single Electrochemiluminescence Events with Ultramicroelectrodes

Ultramicroelectrodes (UMEs), defined as electrodes with diameters below 25 μm, exhibit unique properties due to their small size, including reduced ohmic drop, enhanced mass transport, and minimized double-layer capacitance [16, 18, 32]. They also respond more quickly to voltage changes because of their low time constant, representing the delay between voltage application and the full electrochemical response [12, 16, 17, 32]. In fact, the time constant is the product of uncompensated solution resistance and double-layer capacitance, which is much lower for UMEs. As a result, higher frequency techniques, especially potential pulsing, are expected to be effectively conducted on UMEs to investigate ECL kinetics while reducing side reactions that lower ECL efficiency [12, 14, 15, 19].

To achieve single-event detection of ECL, two key factors must be addressed. Firstly, the amount of reagents should be significantly reduced, which is usually achieved by lowering the reagent concentration and reducing reagent volume [32]. Reagent volume can be reduced by utilizing UMEs and decreasing diffusion layer thickness, typically achieved by applying high pulsing frequency. Secondly, the time interval (bin width) for ECL data acquisition must be minimized to ensure that no more than one photon is detected within each interval.

Wightman's group previously detected individual photons from ECL reactions of 9,10-diphenylanthracene (DPA) using a UME with 5.4 μm radius, a concentration of 15 μM, and a time interval (bin width) of 5 ns [32]. In the resulting ECL transient profile, (Fig. 10e &1f) each bin contains no more than one photon, thereby confirming the single-event detection of ECL.

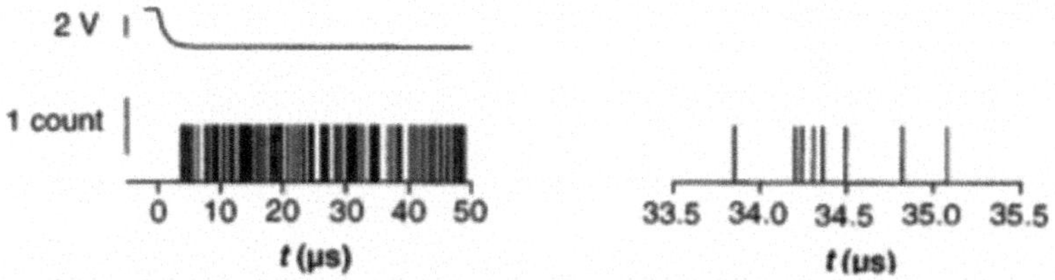

Fig. 10 ECL transients from 15 μM DPA in acetonitrile containing 0.1 M $TBAPF_6$. The electrode was pulsed from 1.7 to −2.1 V at 550 μs intervals. Data was collected with a bin width of 5 ns [32]

4.1 Instruments for Single ECL Event Detection

Instruments in Ding's lab. Figure 11 illustrates a schematic overview of the instrumental setup used to collect individual photons from ECL reactions from our research group. CH instrument 660F (CHI) potentiostats is used to controls the potential on electrodes and collects electrochemistry data. As described in Sect. 2.1 of this chapter, the ECL signal is recorded by a photon counting head (PCH), an all-in-one unit that includes a photomultiplier tube (PMT) and high-speed amplification/discrimination circuits. The output signal from the PCH is passed through an attenuator to lower the pulse magnitude before being fed into an SR430 multichannel scaler. The SR430 multichannel scaler is synchronized with the CH Instrument 660F potentiostat, which controls the potential on the working electrode and collects electrochemical data. Once triggered by the CHI, the SR430 records photons and sorts them into bins, providing the basis for single-event detection of ECL.

To minimize reagent volume and thus achieve single-event detection, a UME is used as working electrode and higher pulsing frequency is applied.

Instruments in Wightman's Lab. Figure 12 illustrates a simplified view of the instrument setup for ECL single photon counting used in Wightman's group in the 1990 s [12]. Instead of a conventional electrochemical cell, they incorporated an ECL flow cell and a flow-injection analysis system. The flow cell is positioned in a light-tight box, around 4 cm from photocathode of a Hamamatsu 4632 PMT, which detects the ECL signal. This PMT was selected for its low dark counts (50–100 counts per second), minimizing the need for external cooling system. A high-voltage power supply (Bertan Series 230) applied −800 V to the PMT. The PMT was connected to a fast preamplifier (EGG Ortec VT120A, 150 MHz bandpass, 200 gain), which amplify the signal from PMT. The output of the preamplifier was directed to the discriminator of a multichannel scaler (EGG Ortec T-914), where a discriminator was

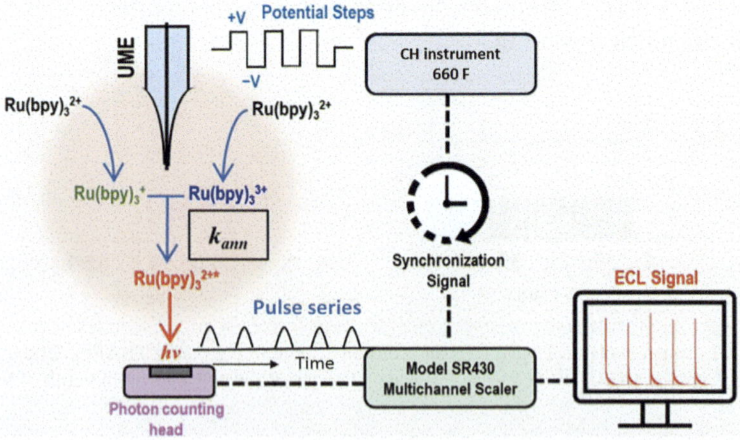

Fig. 11 Schematic overview of the instrumental setup for single-event detection in Ding's Lab [33]

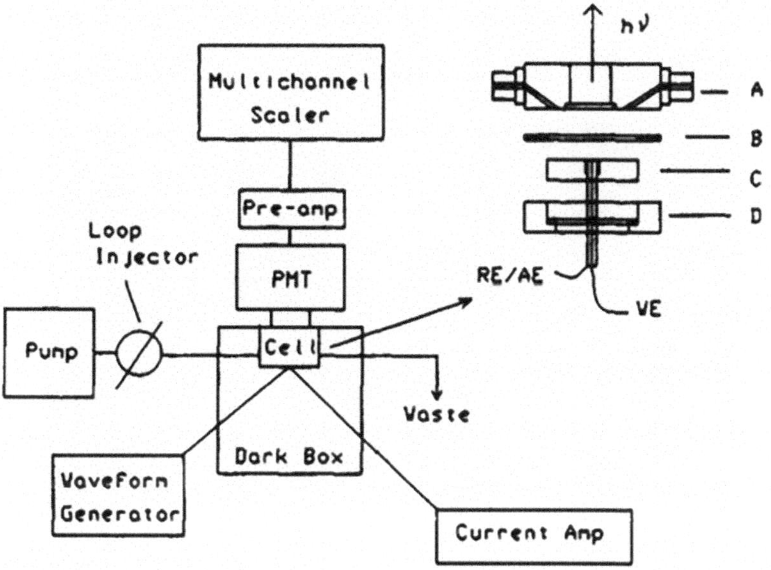

Fig. 12 Simplified view of the instrumental setup for single-event detection in Wightman's Lab [12]

set at −600 mV. Instead of a potentiostat, a Wavetek Model 143 function generator was used to apply a symmetric square wave of electrode potential, and a Nicolet 320 digital oscilloscope was used to monitor the waveform.

4.2 Electrode and Electrochemical Cell Preparation

A three-electrode system was commonly used for ECL experiments. As illustrated in Fig. 1 of Sect. 2.1, two platinum coils serve as the counter and reference electrodes. All electrochemical analyses were performed in a glass cylindrical cell with a flat quartz window at its bottom to allow for the detection of ECL. The electrochemical cell is assembled within a glovebox to minimize oxygen and moisture. The copper head is securely fitted on top of the glass cylindrical cell to prevent gas exchange and sits stably on the black cell holder, ensuring a light-tight environment.

The working electrode is an UME with a diameter of 25 μm. Before use, UMEs are hand polished using FibrMet abrasive disks adhered with 0.3 μm alumina (Fig. 13). Its electrochemical performance is tested using a solution of 0.9 mM ferrocenemethanol (FcMeOH) and 0.1 M KCl to verify the presence of characteristic steady-state oxidation current [31]. The choice of solvent and supporting electrolytes depends on the lumiphores used. For example, in our lab, electrochemical grade of

Fig. 13 Video micrograph of a 25 μm diameter electrode after polished with alumina [34]

anhydrous acetonitrile and 0.1 M of $TBAPF_6$ are usually used for $Ru(bpy)_3(PF_6)_2$ [7, 10, 33].

4.3 Electrochemistry and ECL of $Ru(bpy)_3^{2+}$ with a UME

Figure 14a shows the cyclic voltammetry (CV) of 0.1 mM $Ru(bpy)_3^{2+}$ at a scan rate of 0.1 V/s. Three reduction reactions occur at formal potential of −1.25, −1.45, and −1.65 V along with one oxidation reaction occuring at formal potential of +1.40 V. The steady-state current responses observed are characteristics of electrochemical behavior at UMEs. In this single ECL event experiment, higher frequency potential pulsing is commonly employed for higher quantum yield due to effective generation of radical ions and reduced reagent volume [7, 10, 25]. For a $Ru(bpy)_3^{2+}$ study, only the first reduction and oxidation were selected for potential pulsing, thereby restricting the production of ECL to the simplest energy sufficient annihilation route only involving the interaction of $Ru(bpy)_3^{+\bullet}$ and $Ru(bpy)_3^{3+\bullet}$ to generate the excited state $Ru(bpy)_3^{2+*}$.

Figure 14b depicts the potential profile and corresponding ECL transient signal, obtained using the instruments introduced in Fig. 11. A sharp ECL rise occurs at the onset of each potential pulsing step, indicating the rapid reaction between present-step-generated radical ions and previous-step-generated ions, which generates excited radicals to emit photons. As the previous-step-generated ions are consumed, the ECL decays accordingly. For this experiment, a pulsing frequency of 5 Hz was applied with a bin width of 40.96 μs.

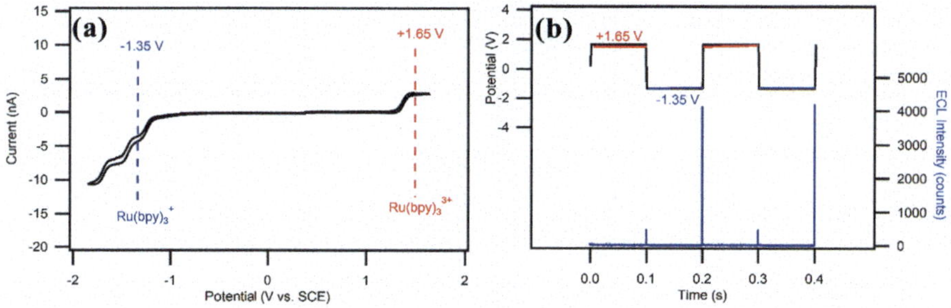

Fig. 14 **a** Cyclic voltammogram of 0.1 mM $Ru(bpy)_3(PF_6)_2$ in acetonitrile with 0.1 M $TBAPF_6$ as supporting electrolyte at a Pt UME with a diameter of 25 μm at a scan rate of 0.1 V/s. **b** Potential stepping versus time program (black) and corresponding ECL transient profile (blue). The bin width was 40.96 μs and the bins per record was 10,240. A total of 1000 records were collected [33]

Figure 15 demonstrates the ECL transient profile of $Ru(bpy)_3(PF_6)_2$ under conditions optimized for single-event detection, achieved by applying a higher pulsing frequency of 50 Hz and using a narrower bin width of 20.48 μs. In this experiment, a total of 5000 records are collected, but the ECL signal in the sharpest region remains below 2000, corresponding to an average of less than 0.4 counts per bin, thus meeting the criteria of single-event detection of ECL.

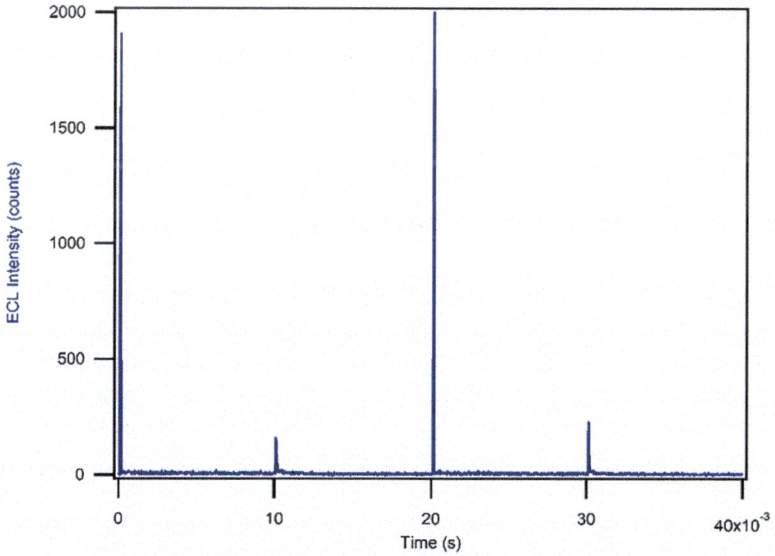

Fig.15 ECL transient profile of 0.1 mM $Ru(bpy)_3(PF_6)_2$ in acetonitrile with 0.1 M $TBAPF_6$ as supporting electrolyte at a platinum ultramicroelectrode with a diameter of 25 μm. The pulsing frequency, bin width, and records per scan are 50 Hz, 20.48 μs, and 5000, respectively [33]

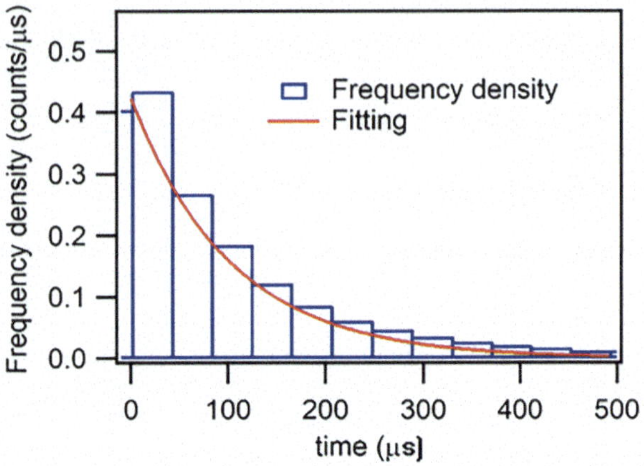

Fig. 16 Poisson description of single reaction events obtained in an acetonitrile solution containing 0.1 mM $Ru(bpy)_3^{2+}$ [33]

To further justify the single-event nature of this ECL process at the UME, a frequency density vs. time interval histogram generated using ECL transient at a pulse frequency of 5 Hz was constructed, as shown in Fig. 16, which follows an exponential decay, as expressed in Eq. 3, where frequency (λ) gives the mean rate of the event and A represents the fitting amplitude. This exponential decay is characteristic of random single reaction events [35], as Whiteman and Collinson described for DPA ECL system in the annihilation pathway [32] and our group reported for a Au38 cluster single ECL event system [36].

$$f(t) = A \cdot \exp(-\lambda\tau) \tag{3}$$

4.4 Comparing Experimental Results with Simulated Ones

Numerical modeling of the ECL reaction at UMEs offers an important link between experimental observations and theoretical insights. By employing finite element analysis software, such as COMSOL Multiphysics, the spatial and temporal dynamics of the electrochemical species involved in ECL reactions can be accurately modeled, and researchers can explore complex reaction kinetics and elucidate key mechanistic details that are difficult to isolate experimentally. Moreover, simulation also holds the potential to improve and optimize ECL quantum efficiency by predicting how changes in electrochemical properties, stability, and applied potential profiles might enhance light emission. Expanding research efforts into the numerical simulation of ECL could thus advance both the fundamental understanding and practical applications of ECL.

Shen, Bard, and co-workers elegantly demonstrated ECL numerical simulations by using 1-dimensional approach to model the planar diffusion from the electrode via COMSOL Multiphysics software v.3.2 [30]. Similarly, the Wightman group, along with Pastore, employed the finite difference Hopscotch algorithm in 1 dimension to model the high-frequency ECL at microelectrodes [12, 14, 15]. By comparing experimental data with a theoretical model, Wightman group determined the bimolecular annihilation rate constant for various compounds, including $Ru(bpy)_3^{2+}$ [19]. However, most of these simulation models were developed in 1 dimension to accurately reproduce the planar diffusion characteristics of macroelectrode, which becomes unreliable for UMEs due to their hemispherical diffusion layer, requiring a 2 or 3 dimensional model for accurate simulations. Valenti, Paolucci, and colleagues reported numerical simulation of a doped-silica nanoparticle/coreactant ECL system in 2 dimensions [37], and the Su group conducted numerous studies exploring scanning electrochemical microscopy using 2D simulations [38, 39].

In our lab, we have demonstrated ECL simulations at UMEs using COMSOL 5.3 and compared the simulated results with experimental results to determine the annihilation bimolecular rate constants (k_{ann}) [33]. Simulating ECL transients at varying annihilation rate constant and matching them with experimental ECL ones allow us to determine the annihilation rate constants [19, 33]. Figure 17 shows the comparison between the experimental ECL transients (at 0.5 Hz with a concentration of 0.1 mM and a 2 mm diameter platinum electrode) to the simulated $Ru(bpy)_3^{2+}$ annihilation ECL transient with a k_{ann} value of $(9 \pm 1) \times 10^9\ M^{-1}S^{-1}$. The strong agreement between the simulated and experimental transients in both anodic and cathodic pulse regions confirms the reliability of this k_{ann} estimation.

5 Concluding Remarks

The aim of this chapter has been to present recent design and refinement of ECL instrumentation with a focus on single-event detection. The development in high-sensitivity photodetectors, synchronized electrochemical control system, and optimized signal acquisition techniques has been shown to accelerate data acquisition and enhance detection sensitivity, temporal resolution, which enhance overall analytical performance. By combining these developments, modern ECL instruments are now capable of performing spooling ECL spectroscopy, determining absolute ECL quantum efficiency, and detecting single ECL events with nanosecond precision. These developments not only extend the analytical capabilities of ECL-based systems but also reveal the potential for improving spatial and temporal resolution through continued advances in instrumentation design and experimental techniques.

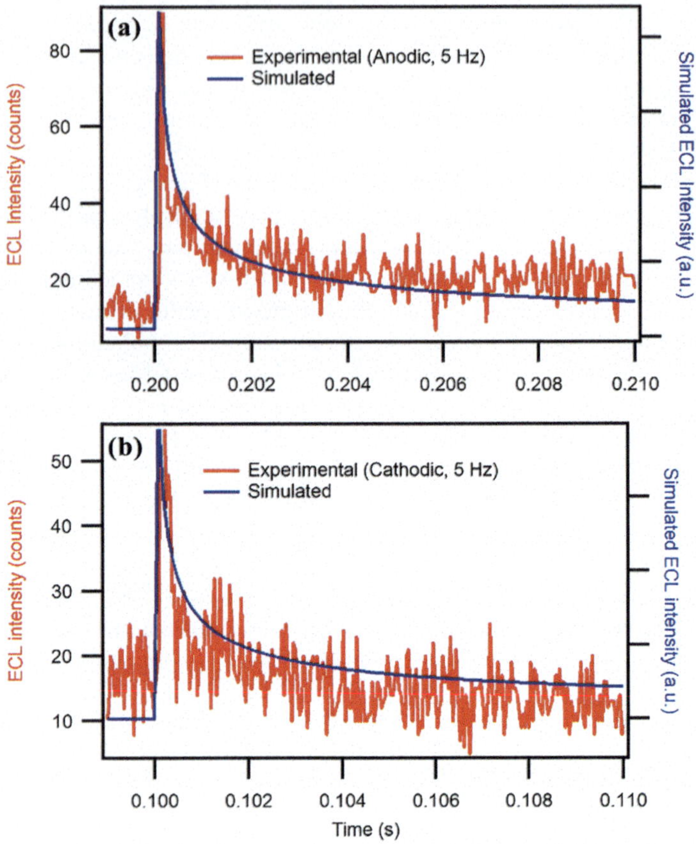

Fig. 17 Comparison of experimental (red traces) and simulated (blue traces) ECL transients of 0.5 mM $Ru(bpy)_3^{2+}$ via the annihilation pathway. **a** 5 Hz, anodic ECL peak, **b** 5 Hz, cathodic ECL peak. For the experimental trace, the bin width was 40.96 μs and the bins per record was 10,240, a total of 1000 records were collected [33]

Acknowledgements We appreciate the great support to this research from the Natural Sciences and Engineering Research Council of Canada (NSERC, DG RGPIN-2013-201697, DG RGPIN-2018-06556, DG RGPIN-2023-05337 and SPG STPGP-2016-493924), New Frontiers in Research Fund (NFRFR-2021-00272), Canada Foundation of Innovation/Ontario Innovation Trust (CFI/OIT, 9040), and the University of Western Ontario. We are grateful to the quality service from Electronic Shop in the Department of Chemistry and Science Stores at Western University.

References

1. Tokel-Takvoryan, N.E., Hemingway, R.E., Bard, A.J.: Electrogenerated chemiluminescence. Xiii. Electrochemical and electrogenerated chemiluminescence studies of ruthenium chelates. J. Am. Chem. Soc. **95**, 6582–6589 (1973)
2. Miao, W.: Electrogenerated chemiluminescence and its biorelated applications. Chem. Rev. **108**, 2506–2553 (2008)
3. Booker, C., Wang, X., Haroun, S., Zhou, J., Jennings, M., Pagenkopf, B.L., Ding, Z.: Tuning of electrogenerated silole chemiluminescence. Angew. Chem. Int. Ed. **47**, 7731–7735 (2008)
4. Chen, S., Ma, H., Padelford, J.W., Qinchen, W., Yu, W., Wang, S., Zhu, M., Wang, G.: Near infrared electrochemiluminescence of rod-shape 25-Atom Auag nanoclusters that is hundreds-fold stronger than that of Ru(Bpy)3 standard. J. Am. Chem. Soc. **141**, 9603–9609 (2019)
5. Roche Diagnostics Corp., www.Roche.Com
6. Xi'an Heyongzhong Electronic Technology Co., Ltd. http://www.xianhyz.com/
7. Chu, K., Adsetts, J.R., Ma, J., Zhang, C., Hesari, M., Yang, L., Ding, Z.: Physical strategy to determine absolute electrochemiluminescence quantum efficiencies of coreactant systems using a photon-counting photomultiplier device. J. Phys. Chem. C **125**, 22274–22282 (2021)
8. Miao, W.: 13 - Electrogenerated chemiluminescence. In: Zoski, C.G. (ed.) Handbook of electrochemistry, pp. 541–590. Elsevier, Amsterdam (2007)
9. Miao, W., Choi, J.-P., Bard, A.J.: Electrogenerated chemiluminescence 69: the Tris(2,2'-Bipyridine)Ruthenium(Ii), (Ru(Bpy)32+)/Tri-N-Propylamine (Tpra) system revisiteda new route involving Tpra•+ Cation radicals. J. Am. Chem. Soc. **124**, 14478–14485 (2002)
10. Adsetts, J.R., Chu, K., Hesari, M., Ma, J., Ding, Z.: Absolute electrochemiluminescence efficiency quantification strategy exemplified with $Ru(Bpy)_3{}^{2+}$ in the annihilation pathway. Anal. Chem. **93**, 11626–11633 (2021)
11. Hesari, M., Ding, Z.: Spooling electrochemiluminescence spectroscopy: development, application and beyond. Nat. Protoc. **16**, 2109–2130 (2021)
12. Collinson, M.M., Wightman, R.M.: High-Frequency generation of electrochemiluminescence at microelectrodes. Anal. Chem. **65**, 2576–2582 (1993)
13. Collinson, M.M., Wightman, R.M.: Observation of individual chemical reactions in solution. Science (Washington, D. C.) **268**, 1883–1885 (1995)
14. Collinson, M.M., Pastore, P., Maness, K.M., Wightman, R.M.: Electrochemiluminescence interferometry at microelectrodes. J. Am. Chem. Soc. **116**, 4095–4096 (1994)
15. Pastore, P., Magno, F., Collinson, M.M., Wightman, R.M.: Simulation of electrochemiluminescence produced by a high-frequency square-wave potential waveform. effects of ohmic drop and cell time constant on electrode potential, current and light emission. J. Electroanal. Chem. **397**, 19–26 (1995)
16. Wightman, R.M., Ritchie, E.: Detection of individual chemical reactions by electrogenerated chemiluminescence. Abstracts Papers Am. Chem. Soc.**,** **213**, 179-Phys (1997)
17. Wightman, R.M., Curtis, C.L., Flowers, P.A., Maus, R.G., McDonald, E.M.: Imaging microelectrodes with high-frequency electrogenerated chemiluminescence. J. Phys. Chem. B **102**, 9991–9996 (1985)
18. Gross, E.M., Pastore, P., Wightman, R.M.: High-Frequency electrochemiluminescent investigation of the reaction pathway between Tris(2,2'-Bipyridyl)Ruthenium(Ii) and tripropylamine using carbon fiber microelectrodes. J. Phys. Chem. B **105**, 8732–8738 (2001)
19. Collinson, M.M., Wightman, R.M., Pastore, P.: Evaluation of ion-annihilation reaction kinetics using high-frequency generation of electrochemiluminescence. J. Phys. Chem. **98**, 11942–11947 (1994)
20. Wei, T., Cai, Z., Chu, K., Lee, I., Winslade, J.E.M., Lord, E.M., Zhan, Z., Zhang, C., Adsetts, J., Ding, Z.: Determining biomolecular rate constants for annihilation reactions of electrogenerated $Ru(Bpy)3^{2+}$ radical ions by means of electrochemiluminescence at a conventional electrode and an ultramicroelectrode. Anal. Chem. **97**, 19028–19037 (2025)

21. Bard, A.J., Keszthelyi, C.P., Tachikawa, H., Tokel, N.E.: On the efficiency of electrogenerated chemiluminescence. In: Cormier, M.J., Hercules, D.M., Lee, J. (eds.) Chemiluminescence and Bioluminescence, pp. 193–208. Springer Science+Business Media, LlC, New York (1973)
22. Faulkner, L.R., Bard, A.J.: Techniques of electrogenerated chemiluminescence. In: Bard, A.J. (ed.) Etectroanalytical Chemistry, vol. 10, pp. 1–95. Marcel Dekker, Inc., New York (1977).
23. Bard, A.J.: Electrogenerated chemiluminescence. Marcel Dekker, New York (2004)
24. Nepomnyashchii, A.B., Bröring, M., Ahrens, J., Bard, A.J.: Synthesis, photophysical, electrochemical, and electrogenerated chemiluminescence studies. multiple sequential electron transfers in bodipy monomers, dimers, trimers, and polymer. J. Am. Chem. Soc. **133**, 8633–8645 (2011).
25. Zhan, Z., Qin, X., Chu, K., Sun, X., Ding, Z.: Absolute quantum efficiencies for electrochemiluminescence and chemiluminescence of protoporphyrin Ix dimethyl ester**. J. Electroanal. Chem. **932**, 117220 (2023)
26. Myung, N., Ding, Z., Bard, A.: Electrogenerated chemiluminescence of cdse nanocrystals. Nano Lett. **2**, 1315–1319 (2002)
27. Adsetts, J.R., Chu, K., Hesari, M., Whitworth, Z., Qin, X., Zhan, Z., Ding, Z.: Absolute electrochemiluminescence quantum efficiency of Au nanoclusters by means of a spectroscopy charge-coupled device camera. J. Phys. Chem. C **126**, 20155–20162 (2022)
28. Adsetts, J.R., Chu, K., Hesari, M.A.-O., Ma, J., Ding, Z.A.-O.: Absolute electrochemiluminescence efficiency quantification strategy exemplified with Ru(Bpy)(3)(2+) in the annihilation pathway
29. Fracassa, A., Mariani, C., Marcaccio, M., Xu, G., Sojic, N., Valenti, G., Paolucci, F.: Methods and strategies for robust electrochemiluminescence signal quantification. Curr. Opin. Electrochem. **41**, 101375 (2023)
30. Shen, M., Rodríguez-López, J., Huang, J., Liu, Q., Zhu, X.-H., Bard, A.J.: Electrochemistry and electrogenerated chemiluminescence of dithienylbenzothiadiazole derivative. differential reactivity of donor and acceptor groups and simulations of radical cation−anion and dication−radical anion annihilations. J. Am. Chem. Soc. **132**, 13453–13461 (2010).
31. Ying, Y.-L., Ding, Z., Zhan, D., Long, Y.-T.: Advanced electroanalytical chemistry at nanoelectrodes. Chem. Sci. **8**, 3338–3348 (2017)
32. Collinson, M.M., Wightman, R.M.: Observation of individual chemical reactions in solution. Science **268**, 1883–1885 (1995)
33. Wei, T., Cai, Z., Chu, K., Lee, I., Winslade, J.E.M., Lord, E.M., Zhan, Z., Zhang, C., Adsetts, J., Ding, Z.: Determining bimolecular rate constants for annihilation reactions of electrogenerated Ru(Bpy)32+ radical ions by means of electrochemiluminescence at a conventional electrode and an ultramicroelectrode submitted
34. Ding, Z., Quinn, B., Bard, A.: Kinetics of heterogeneous electron transfer at liquid/liquid interfaces as studied by Secm. J. Phys. Chem. B **105**, 6367–6374 (2001)
35. Van Kampen, N.G.: Chapter I—stochastic variables. In: Van Kampen, N.G. (ed.) Stochastic Processes in Physics and Chemistry (Third Edition), pp. 1–29. Elsevier, Amsterdam (2007)
36. Hesari, M., Ma, H., Ding, Z.: Monitoring single Au_{38} nanocluster reactions via electrochemiluminescence. Chem. Sci. **12**, 14540–14545 (2021)
37. Imai, K., Valenti, G., Villani, E., Rapino, S., Rampazzo, E., Marcaccio, M., Prodi, L., Paolucci, F.: Numerical simulation of doped silica nanoparticle electrochemiluminescence. J. Phys. Chem. C **119**, 26111–26118 (2015)
38. Zhu, R., Qin, Z., Noel, J.J., Shoesmith, D.W., Ding, Z.: Analyzing the influence of alloying elements and impurities on the localized reactivity of titanium grade-7 by scanning electrochemical microscopy. Anal. Chem. **80**, 1437–1447 (2008)
39. Zhao, X., Zhu, R., Anikovskiy, M., Wu, Q., Ding, Z.: Profiling H2o2 from single cos-7 cells by means of scanning electrochemical microscopy. Biosens. Bioelectron. **227**, 115123 (2023)

Insights into Reaction Mechanisms by ECLM

Claudio Ignazio Santo, Alessandro Fracassa, Chiara Mariani, and Alessandra Zanut

Abstract Electrochemiluminescence (ECL) is a surface-confined process, where light emission arises from short-lived intermediates generated near the electrode after an electrochemical stimulus. The spatial extent of the emitting region, known as the thickness of the ECL layer (TEL), is primarily governed by the diffusion behavior of these intermediates. In this chapter, the use of Electrochemiluminescence Microscopy (ECLM) is explored as a powerful technique for visualizing and understanding TEL formation, providing spatially resolved insight into ECL mechanisms. A variety of strategies to probe and control TEL are presented, including the use of advanced electrode architectures and tailored luminophores and coreactants. Modeling approaches, especially finite element simulations, are also discussed as essential tools to interpret complex spatial and kinetic behaviors in ECL systems. Through this framework, ECLM is shown not only as an imaging tool, but as a mechanistic probe that offers a deeper understanding of light generation processes and supports the development of improved ECL-based sensing platforms.

Keywords Electrochemiluminescence Microscopy (ECLM) · ECL layer thickness (TEL) · Mechanistic imaging · Electrode architecture · Finite element simulations

1 Introduction

Electrochemiluminescence (ECL) emission is a surface-confined phenomenon, meaning the light emission occurs primarily near the electrode surface. The thickness of the ECL layer (TEL) is largely determined by the diffusion characteristics of the reactive intermediates involved in the process. In annihilation-based ECL processes,

C. I. Santo · A. Fracassa · C. Mariani
Department of Chemistry "Giacomo Ciamician", Alma Mater Studiorum—University of Bologna, Via. P. Gobetti 85, 40129 Bologna, Italy

A. Zanut (✉)
Department of Chemical Sciences, University of Padova, Via Francesco Marzolo 1, 35122 Padova, Italy
e-mail: alessandra.zanut@unipd.it

F. Paolucci et al. (eds.), *Electrogenerated Chemiluminescence Microscopy and Imaging*,
https://doi.org/10.1007/978-3-032-14602-1_5

the TEL is limited due to the short diffusion lengths of electrogenerated species produced at a single electrode. In contrast, for coreactant-based ECL, the TEL varies depending on the specific mechanism involved. In the oxidative-reduction and low oxidation potential pathways, the TEL is largely determined by the diffusion range of the $TPrA^{+\cdot}$ intermediate. Meanwhile, in the catalytic pathway, it is controlled by the diffusion length of $[Ru(bpy)_3]^{3+}$.

This chapter explores Electrochemiluminescence Microscopy (ECLM) as a powerful tool for investigating ECL mechanisms, combining spatially resolved imaging with electrochemical control to study the emitting layer with high sensitivity. In this chapter we will introduce key tools and techniques used to probe the emitting layer via ECLM. Next, we examine the role of ECL reagents, luminophores, and coreactants in elucidating reaction pathways and how advanced materials and electrode architectures enhance the mechanistic insight attainable through ECLM by improving spatial control and detection sensitivity. Finally, the final section delves into modeling approaches, particularly the use of finite element simulations, to describe and predict the behavior of the ECL-emitting layer.

This includes models tailored for bead-based systems and alternative computational strategies aimed at simulating complex ECL mechanisms. Through this comprehensive framework, the chapter aims to elucidate how ECLM serves not only as a visualization tool but also as a mechanistic probe in the study of ECL systems.

1.1 Tools for Studying the Emitting Layer Through ECLM

Measuring the TEL using Electrochemiluminescence Microscopy (ECLM) is essential for unraveling the underlying ECL mechanisms and fine-tuning the dominant coreactant pathways [1]. Modern ECLM systems feature a comprehensive setup that includes a bright-field microscope, an electrochemical workstation (comprising an electrochemical cell and a potentiostat), and a charge-coupled device (CCD or EMCCD) camera to detect ECL signals (Fig. 1). Researchers have used ECLM to investigate how precisely regulating the ECL process through variations in reactant concentrations can influence the outcome [2–6]. This approach has enabled high-resolution imaging of structures such as cell–matrix adhesions and cell–cell junctions within the ECL-emitting layer, extending several micrometers from the electrode surface [4], primarily under static conditions [6], but also in dynamic environments [7, 8]. In addition to visualizing the spatial distribution of excited luminophores, further studies have focused on monitoring coreactant radicals near the electrode interface [9], addressing issues such as coreactant lifetimes and limited mass transport. ECLM has thus proven to be a powerful tool for probing TEL and refining coreactant-based ECL pathways. However, one notable limitation is its inability to resolve TEL values below approximately 200 nm due to the constraints of optical diffraction [10]. In contrast, scanning electrochemical microscopy (SECM) offers precise distance control (less than 100 nm) between the probe and the substrate, enabling detailed investigation of coreactant species with extremely short lifetimes

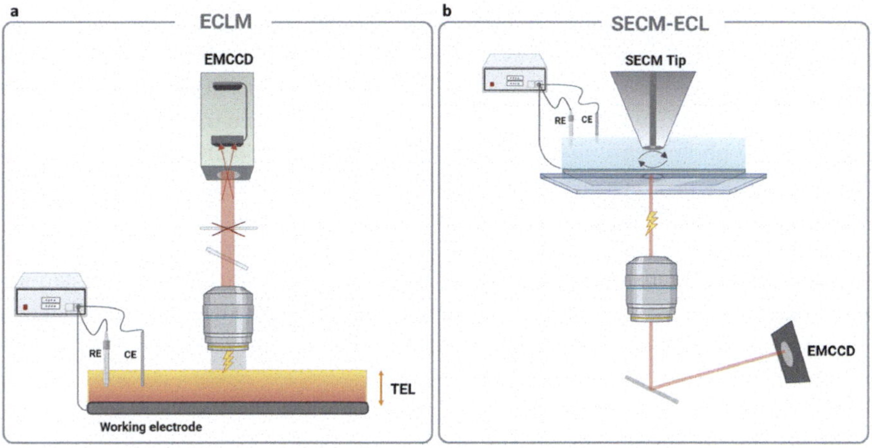

Fig. 1 (Left) Schematic representation of ECLM including an electrochemical cell, a microscope objective, and a charge-coupled device (EMCCD) to capture the ECL signal. (Right) Schematic representation of SECM-ECL including a small-scale probe (SECM-tip), an inverted microscope objective, and an EMCCD to capture the ECL signal. Reproduced with permission from Ref. [14]

[1]. Significant advancements in the understanding of ECL mechanisms have been made by integrating SECM with ECLM. This integration was achieved by incorporating a photomultiplier tube (PMT) into a SECM setup mounted on an inverted microscope, enabling simultaneous visualization and electrochemical analysis.

Using this combined SECM-ECL system, Miao et al. validated the low oxidation potential mechanism of the $[Ru(bpy)_3]^{2+}$/TPrA system [11]. Through digital simulations, they were able to estimate both the diffusion length and the half-life of the $TPrA^{+\cdot}$ radical. By carefully controlling the gap between a hemispherical microelectrode and the substrate with a micromanipulator, they mapped the ECL distribution within ~ 1 μm from the electrode surface.

Building on this work, Zanut et al. explored ECL efficiency at sub-micrometer distances (≤ 1 μm). They used a transparent indium tin oxide (ITO) electrode functionalized with a $[Ru(bpy)_3]^{2+}$ monolayer as the light-emitting surface, combined with hemispherical platinum microelectrodes (1.5 mm or 0.5 mm in diameter) to oxidize TPrA. This system, integrated with an EMCCD camera and a micropositioner for precise control of tip-substrate spacing, enabled us to investigate ECL enhancement in close proximity. The use of labeled microbeads further revealed a previously overlooked ECL pathway, producing signals up to ten times stronger than those observed at greater distances (> 1 μm) [9].

In a separate study, Lopez et al. investigated the distance-dependence of ECL emission using $[Ru(bpy)_3]^{2+}$ and rubrene as model luminophores [12].

By adjusting the distance between the SECM tip (cathode) and a transparent microsubstrate (anode), the researchers investigated the steady-state generation of radical ions. They observed that ECL emission increased as the tip neared the substrate, enabling the creation of light-based approach curves for precise tip positioning. Notably, light emission was detected even before electrochemical reduction began due to inverse photoemission, in which radical cations at the substrate transferred holes to the tip, while electrons from an external source were injected into the tip, leading to photon emission. More recently, Zhou et al. demonstrated the ECL-enhancing effect of silica nanochannel membranes (SNMs) on the $[Ru(bpy)_3]^{2+}$/TPrA system using SECM [13]. They attributed this enhancement to the SNM's negatively charged, ultrasmall nanochannels, which enrich and attract charged reactants. By combining SECM approach curve analysis with finite element simulations, they quantified the mass transfer rate of $[Ru(bpy)_3]^{2+}$ through the nanochannels (~ 0.3 mm/s), which was significantly faster than the diffusion rate observed in bulk solution.

2 ECL Reagents and Their Role in Mechanism Elucidation

2.1 *Mechanistic Insights Through Luminophores*

Although ECL provides remarkable spatial resolution, a major challenge is extending the thin ECL-emitting layer beyond the narrow region adjacent to the electrode surface. This limitation is primarily due to the short-lived nature of electrogenerated radicals, which restricts the vertical reach of the emission zone. In this context, the role of the luminophore is more critical than ever. Traditionally, ECL emitters have been classified into inorganic and organic probes, based on the nature of their functional components. However, beyond their composition, the methods used to position and confine luminophores have become increasingly important for enhancing signal control and interpretability. Techniques such as electrodeposition, surface grafting onto objects or microstructures, and spatial confinement strategies are now widely employed to precisely localize luminophores and extend the active ECL region. These innovations not only improve emission intensity and resolution but also enable a deeper understanding of the underlying electrochemical pathways.

Recently, the easy functionalization and favorable biocompatibility have rendered organic molecules, like luminol, promising candidates for ECL applications. Hu et al. selectively electropolymerized luminol on ITO electrode after applying a sebaceous fingerprint. The fingerprint's inert components reduced the electrochemical activity in specific electrode regions. Cyclic voltammetry (CV) allowed the electropolymerization, resulting in a polyluminol film whose thickness affected current peaks. After the 3rd CV cycle, current signals decreased due to hindrance in mass transport and charge transfer at the ITO electrode caused by the higher polyluminol film thickness. The reduction in current was attributed to redox transformations of polyluminol involving ion transport through the film. This process produced recognizable negative fingerprint images captured using a CCD camera under a constant 1.2 V potential with a 100-s exposure time [15].

In contrast to other visualization technologies, this method stands out for its simplicity and minimal reliance on complex instruments. However, the unstable ECL emission of luminol, caused by the self-decomposition of H_2O_2, has led to the exploration of alternative approaches. In this context, dissolved dioxygen (O_2) has emerged as a promising substitute for H_2O_2 due to its low toxicity and inherent stability [16]. Nevertheless, the luminol-O_2 system suffers from an ultra-low ECL signal due to the inefficient production of reactive oxygen species (ROS). Despite efforts to enhance ECL signals through nanomaterials [17] or ultrasound [18], the limitations of poor solubility of organic molecules and their reliance on annihilation reactions have limited their application in ECLM [19].

Differently to organic molecules, inorganic ECL probes, including coordination complexes such as $[Ru(bpy)_3]^{2+}$ and its derivatives, as well as Ir(III) complexes [20] and multicolor complex luminophores [21] are widely employed in chemical analysis. These probes offer advantages such as tunable wavelengths, high efficiency, and excellent stability in aqueous solutions [22]. Frequently, instead of directly incorporating them into the ECL system's detection solution, they are commonly immobilized on the surface of an electrode and coupled with sacrificial coreactants [23]. However, a substantial drawback in this scenario lies in the notable reduction in ECL intensity over time, which stems from the diffusion of coreactant radicals away from the electrode surface.

Incorporating luminophore molecules into a nanoemitter has been recognized as an effective and straightforward strategy to enhance the ECL signal. For instance, dye-doped silica nanoparticles (DDSNs), well-known ECL materials capable of concentrating high amounts of ECL-active molecules, have been successfully employed as single tags in ECL imaging, achieving a high signal-to-noise ratio.

In this context, the integration of ECL with nanotechnology has gained significant attention, enabling the development of efficient ECL-active systems. The earliest report of nanomaterial-generated ECL dates back to 2002, when Bard and colleagues utilized semiconductor silicon nanocrystals (NCs) to generate ECL via both annihilation and coreactant mechanisms in acetonitrile solutions [24]. Building on this approach, DDSNs have been widely used to investigate and optimize ECL emission by systematically tuning their physicochemical characteristics. Studies have demonstrated that incorporating luminophores into DDSNs can significantly improve their chemical and electrochemical stability. Moreover, the efficiency of ECL emission has been shown to depend on multiple factors, including nanoparticle size, type and level of doping, surface charge [25], supporting electrolyte, electrode material, and buffer composition [26]. These findings highlight the importance of nanoparticle design in maximizing ECL performance. A later study showed that the ECL performance of covalently doped $[Ru(bpy)_3]^{2+}$ core–shell silica–PEG nanoparticles is primarily governed by the doping level. The authors identified two distinct mechanisms contributing to ECL generation, both influenced by the dye concentration and the overall nanoparticle charge which affects the spatial distribution of charged species involved in the ECL process within and around the nanoparticles [27].

This challenge primarily stems from limitations inherent in the synthetic strategies used for DDSNs. To overcome these limitations, a new class of DDSNs was developed using a reverse microemulsion synthesis method. These nanoparticles, referred to as bio-Triton@RuNPs, were specifically designed to enhance performance through improved structural and functional properties. This approach enabled a high dye-loading capacity—approximately 4800 Ru complexes per nanoparticle—without inducing a positive surface charge, thereby supporting strong ECL emission [28]. When tested with TPrA as a coreactant, these DDSNs exhibited a significant enhancement in ECL signal compared to conventional immunoassay conditions, which typically employ antibodies labeled with only six dye molecules.

To elucidate the enhancement mechanisms of ruthenium-doped silica nanoparticles in ECL, advancements in the detection of weak optical signals have played a crucial role. The development of high-sensitivity micro-imaging techniques has enabled the performance of single-particle and even single-molecule ECL imaging [7, 29]. These approaches allow for spatial resolution of individual emitters, offering significantly more detailed information compared to traditional bulk ECL detection methods. This has opened new avenues for studying and optimizing the performance of nanoparticle-based ECL systems. In one recent study, it was observed that the presence of free $[Ru(bpy)_3]^{2+}$ in the electrolyte significantly amplifies the ECL signal, up to 478-fold, by enabling a diffusion-accelerated charge-transfer mechanism involving electro-generated $[Ru(bpy)_3]^{3+}$. This mechanism, faster than internal electron hopping within the nanoparticle, facilitates efficient oxidation of the doped luminophore. Leveraging this insight, a real-time ECL imaging approach was developed to monitor dynamic biological processes, such as surfactant-induced changes in cell permeability, offering new opportunities for digital ECL analysis and high-resolution nano-imaging [30].

An alternative enhancement strategy was introduced by Liu et al., who integrated gold nanoparticles with Ru-doped silica nanoparticles to form RuDSNs/AuNPs nanoemitters [31]. In this configuration, ECL emission is localized at the surface of the RuDSNs, resulting in a two-fold increase in signal intensity. When used with TPrA as the sacrificial coreactant, these nanoemitters enabled effective visualization of individual protein molecules on the electrode surface as well as membrane proteins on cells. Specifically, RuDSNs/AuNPs conjugated with monoclonal antibodies against cytokeratin 19 (CK19) were used to form sandwich immunocomplexes on an ITO electrode. Unlike wide-field fluorescence microscopy, ECL imaging provided near-zero background, leading to a high signal-to-noise (S/N) ratio for protein detection. Nonetheless, a key limitation emerged: the restricted accessibility of reaction intermediates to $[Ru(bpy)_3]^{2+}$ within the RuDSNs/AuNPs led to substantial dye leakage, causing a decline in luminescence during continuous imaging.

More recently (2022), Han et al. introduced $[Ru(bpy)_3]^{2+}$-functionalized microgels (Ru-MGs) to decorate polystyrene beads (Ru-MG@bead) as a strategy to extend the size of the ECL-emitting layer. Ru-MGs were synthesized by an aqueous free-radical precipitation polymerization with the $[Ru(bpy)_3]^{2+}$ complex successfully grafted to the poly(*N*-isopropylacrylamide) hydrogel matrix (pNIPAM). Efficient labeling and stable emission of the Ru-MG@bead were imaged by both top and side-view configurations. Remarkably, the ECL intensity of Ru-MGs@beads showed a ninefold enhancement compared to beads functionalized by a sandwich immunoassay or an amide bond. This improvement was attributed to the higher density of ruthenium sites on the beads located in close proximity to the electrode surface. These ruthenium sites underwent direct oxidation, which contributed more effectively to the overall ECL signals [32]. In addition to traditional molecular luminophores, alternative materials such as quantum dots (QDs) have emerged as promising ECL emitters due to their tunable optical properties, high brightness, and stability [33–35]. The development of wavelength-tunable ECL systems using such nanomaterials offers a powerful strategy for real-time, dynamic visualization of biomolecular interactions. For example, QD-based ECL nanoemitters functionalized with glucose oxidase (GOx) were employed to track the movement of wheat germ agglutinin (WGA) on cell membranes. The ECL signal decreased over time due to H_2O_2 consumption, while a control system maintained stable emission. By applying 2D Gaussian fitting, individual spots were precisely localized and their trajectories reconstructed. This dual-color single-particle tracking (SPT) approach enabled detailed investigation of WGA-cholesterol interactions, highlighting the potential of multicolor ECL imaging for probing dynamic biological processes at the nanoscale [36].

2.2 *Role of Coreactants in Reaction Pathways*

In parallel with the ongoing development of novel coreactants and strategies to enhance ECL efficiency, ECL microscopy has emerged as a powerful analytical technique for elucidating the mechanistic pathways of coreactant-driven luminescence at electrochemical interfaces. The oxidation of TPrA is a key step in the heterogeneous mechanism with $[Ru(bpy)_3]^{2+}$, and tuning it has been shown to improve sensitivity. Zanut et al., demonstrated that the integration of ECL with high-resolution microscopy, combined with the use of labeled microbeads as spatial probes, allows precise mapping of ECL generation within ≤ 1 µm of the electrode surface, thereby revealing a previously unrecognized auxiliary pathway contributing to the overall ECL emission where a dipropylamine radical is generated by C–N bond cleavage. Building on these findings, a new branched amine coreactant, *N*-dipropyl isobutyl amine (DPIBA), was proposed, offering a more stable carbocation. DPIBA, when mixed with TPrA, maintained high efficiency at long distances and enhanced short-distance contribution, leading to a significant increase in ECL intensity and sensitivity in Elecsys assays on the Roche Diagnostics Cobas e801 immunoassay analyzer [9]. The additive positively affected ECL efficiency, increasing the signal by 66% compared to TPrA alone. Successively, Ding and coworkers employed single-photon counting to demonstrate that triethanolamine (TEOA) enhances the ECL of $[Ru(bpy)_3]^{2+}$/TPrA by 10.8% and expands the ECL layer thickness [37]. This technique, which allows the study of ECL reaction rates tracking single luminescence events, revealed increased photon counts and event frequency, indicating that

$TEOA^{+*}$ accelerates TPrA oxidation via a chemical pathway, thereby boosting ECL efficiency and expanding the emitting layer.

Alternative studies have explored the use of iridium-based redox mediators as additives to enhance ECL performance. For instance, Fracassa et al. developed a theoretical framework for ECL activation through molecular electrocatalysis, demonstrating that a redox-active Ir(III) complex could electrocatalytically enhance coreactant ECL in the $[Ru(bpy)_3]^{2+}$/TPrA homogeneous system [26]. Their findings revealed previously uncharted mechanisms that enabled Ru-based ECL at lower electrode potentials, bypassing direct oxidation of $[Ru(bpy)_3]^{2+}$ or TPrA, and highlighted the dual role of the Ir(III) complex as both a redox mediator and luminophore. Implementing high-resolution ECL microscopy, they combined freely diffusing water-soluble Ir(III) complexes with $[Ru(bpy)_3]^{2+}$-labeled microbeads system, showing that ECL emission could be finely tuned, with changes in emission linked to the redox potentials of the mediators [38]. Specifically, more facile reduction leads to scavenging of the critical radical species derived from the coreactant, resulting in quenching of the ECL of $[Ru(bpy)_3]^{2+}$ (− 75%). In contrast, the generation of the oxidized Ir(IV) species can enhance the ECL signal (+ 108%), particularly when the direct reduction route is not thermodynamically favorable. This electrocatalytic pathway not only promotes efficient radical generation but also results in an increased TEL, as the longer-lived oxidized species from the Ir complex can diffuse farther from the electrode surface compared to the short-lived coreactant-derived radicals. Further studies confirmed these hypotheses by integrating ECL with self-interference spectroscopy, demonstrating that the redox mediators also enable manipulation of the distance that the reactive intermediates can diffuse from the working electrode surface (TEL) [39]. Building on these findings, additional ECLM studies focused on overcoming the slow oxidation of TPrA on boron-doped diamond (BDD) electrodes by incorporating an Ir(III) complex as a redox mediator, significantly enhancing the ECL signal by up to 20.7-fold in cyclic voltammetry and 46.3-fold in chronoamperometry [40]. This approach improves the viability of BDD as a high-performance platform for commercial bioanalysis, demonstrating the potential of Ir(III)-based mediators to optimize ECL in biosensing and analytical applications. Recently Wang et al. investigated alternative five tertiary amines coreactants for the ECL generation of $[Ru(bpy)_3]^{2+}$, focusing on the relationship between coreactant reactivity, radical cation lifetime and ECL distance. Using ECL self-interference spectroscopy (ECLIS) and statistical analysis of single-microbead ECL images, they demonstrated that both ECL distance and coreactant reactivity jointly determine emission intensity and immunoassay sensitivity. ECLIS measures the distance between ECL emitters and the electrode by analyzing interference between emitted and reflected light in a multilayer electrode setup, with detection via fiber-coupled spectrometry. Notably, 2,2-bis(hydroxymethyl)-2,2′,2″-nitrilotriethanol (BIS–TRIS) optimally balanced reactivity and ECL distance, enhancing sensitivity by 236% compared to TPrA in a bead-based immunoassay for carcinoembryonic antigen (CEA) detection [41]. Other researchers have leveraged the *heterogeneous catalytic ECL mechanism*, where the Ru complex is dissolved in solution and the coreactant is encapsulated within micro- and nanomaterials, to enhance the efficiency and sensitivity of ECL systems. Ma et al. exploited the ECLM technique for imaging upper cell membranes labeled with amine-rich nitrogen-doped carbon dots (NCDs) as nano-coreactants using $[Ru(bpy)_3]^{2+}$ as the emitter. This approach extends the ECL-emitting layer due to the prolonged lifetime of electro-oxidized $[Ru(bpy)_3]^{3+}$, enabling remote ECL imaging away from the electrode surface. The amine-rich NCDs significantly enhance ECL intensity (by 245 times) and serve as efficient tags for visualizing externalized phosphatidylserine on apoptotic cell membranes [6]. By performing *z*-axis scanning, they achieved three-dimensional ECL imaging of single beads and upper cell membranes (Fig. 2), enabling the extension of the TEL and allowing detailed visualization of ECL distribution in the vertical dimension, thus demonstrating the potential of this method for high-resolution topographic studies in biological systems.

In the same years, Wang et al. designed a dual intramolecular electron transfer strategy and tertiary amine conjugated polymer dots (TEA-Pdots) as a coreactant-embedded ECL mechanism and microimaging system. To improve the performance of coreactant-embedded Pdots, two types of anodic coreactants were coupled to the backbone of poly[(9,9-dioctylfuor-enyl-2,7-diyl)-alt-co-(1,4-benzo-{2,1,3}-thiadiazole)] (PFBT), resulting in the creation of diethylamine-conjugated PFBT (TEA-PFBT) and dipropylamine-conjugated PFBT (TPrA-PFBT). Using a similar set-up, with homemade ECLM setup implemented by integrating an inverted microscope with an EM-CCD camera, TEA-Pdots showed an ECL efficiency 132 and 45 times stronger than those from the mixture of Pdots with TEA at equivalent and 62.5 times higher amounts, respectively. The ECL efficiency was even higher than that of a typical $[Ru(bpy)_3]^{2+}$ system. This strong ECL emission was attributed to the fast intramolecular electron transfer, which significantly enhances the ECL emission by shortening the electron transfer pathway and reducing the complexity of transporting radical intermediates in intermolecular electron transfer-based ECL processes [42]. This enhancement, along with the negligible cytotoxicity of TEA-Pdots, led to interesting applications of the microimaging system in the visual analysis of membrane proteins on single living cells and the in situ evaluation of their expression change upon drug treatment. Adopting a more biocompatible coreactant for ECL microscopy, or even employing coreactant-free approaches, offers several advantages. It reduces toxicity, allowing for the study of biological samples under physiological conditions while preserving cellular integrity, and enhances the signal-to-noise ratio and sensitivity for real-time analysis in complex environments. Ding et al. developed a technique for imaging cell adhesions by employing a silica nanochannel membrane as the substrate in conjunction with a $[Ru(bpy)_3]^{3+}$/HEPES (*N*-(2-hydroxyethyl)piperazine-*N*′-2-ethanesulfonic acid) oxidative-reduction ECL system. In this setup, HEPES acts as a biocompatible coreactant with an ECL mechanism analogous to that of TPrA. The silica nanochannel membrane was used to modify the ITO electrode, thereby protecting living cells from direct electrical stimulation [43]. This approach, combining direct ECL microscopy and post-acquisition image analysis using MATLAB, enabled the visualization of collective cell migration in a wound healing assay and revealed ECL from extended and concealed structures that were undetectable under bright-field imaging. In addition, some

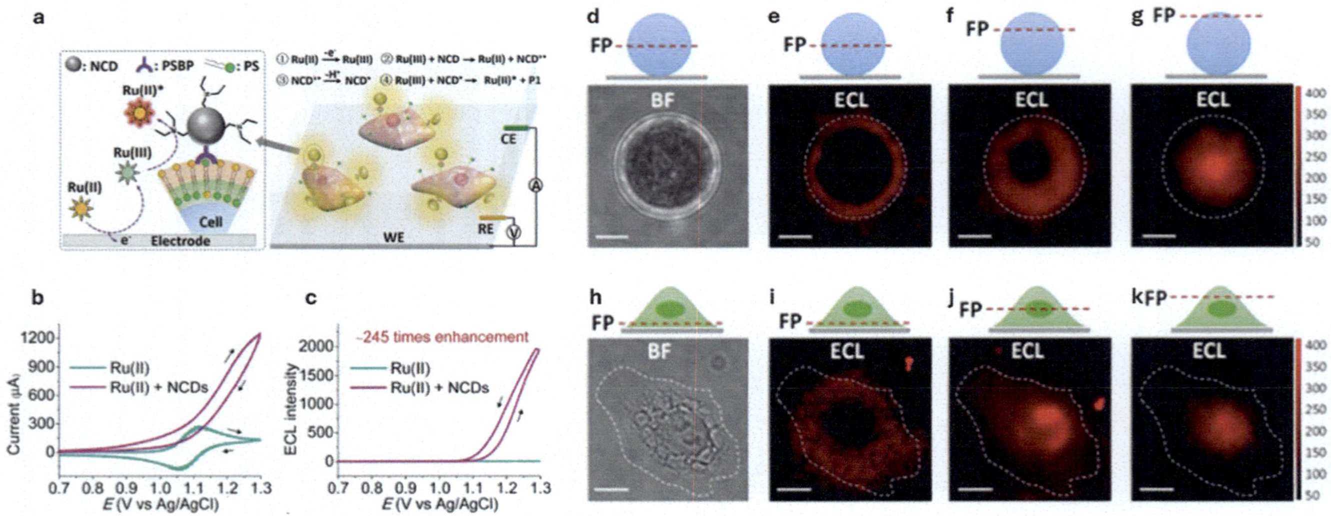

Fig. 2 **a** Schematic representation of the catalytic route ECL microscopy of upper cell membranes with $[Ru(bpy)_3]^{2+}$ as the luminophore and NCDs as the nano-coreactants. Ru(II), Ru(III), Ru(II)* and P_1 represent $[Ru(bpy)_3]^{2+}$, $[Ru(bpy)_3]^{3+}$, $[Ru(bpy)_3]^{2+*}$ and the decomposition product of NCD*, respectively. Cyclic voltammetry curves (**b**) and ECL–potential curves (**c**) in 10 mM PBS electrolyte (pH 7.4) containing 10 mM $[Ru(bpy)_3]^{2+}$ and also NCDs (1 mg mL^{-1}). Scan rate is 0.1 V s^{-1}. Working electrode is ITO. **d–g** Bright field (BF) image and corresponding ECL images of a single bead (10 μm) at different focal planes (FP). The beads were modified with NCDs via an amido bond. Scale bar is 3 μm. **h–k** BF image and corresponding ECL images of a single apoptotic cell at different FPs. The apoptotic cells (induced by lipopolysaccharide) were fixed with paraformaldehyde for 30 min and then incubated with PSBP@NCDs, which recognize the externalized PS on apoptotic cell membranes. Scale bar is 10 μm. Dashed lines represent the outlines of the bead and the cell. The working electrode is ITO. Electrolyte is 10 mM PBS electrolyte (pH 7.4) containing 10 mM $[Ru(bpy)_3]^{2+}$. Applied voltage is 1.3 V pulse (1 s). Exposure time is 1 s. Reproduced and adapted with permission from Ref. [6]

biomolecules with amine moieties such as DNA and RNA can act as the coreactants of $[Ru(bpy)_3]^{2+}$. To generate ECL, these intracellular biomolecules react with electrogenerated $[Ru(bpy)_3]^{3+}$ to drive the ECL reactions. Accordingly, intracellular structures that contain most intracellular DNA and RNA, including nucleolus and endoplasmic reticulum, were identified. This bio-coreactant enhanced ECL microscopy provides a new method for studying intracellular biomolecules and structures. The Rusling group first demonstrated that guanine in DNA can be oxidized twice by $[Ru(bpy)_3]^{3+}$, leading to an excited state $[Ru(bpy)_3]^{2+*}$ and resultant ECL [44]. According to this principle, Chen et al. designed a biocompatible and highly efficient ECL coreactant, Hi-AuNF@G-ssDNA-Apt, for single-cell imaging. This nanoprobe integrates high-index faceted gold nanoflowers (Hi-AuNF) with guanine-rich single-stranded DNA (G-ssDNA) and a CEA aptamer, enabling both signal amplification and selective recognition of CEA-overexpressing cancer cells. Using Hi-AuNF@G-ssDNA-Apt nanoprobes to label the cell surface and a homemade ECL microscope equipped with a 40 × water immersion objective, they studied the catalytic properties of Hi-AuNF and the "catalytic route" reaction between guanine and $[Ru(bpy)_3]^{2+}$.The system significantly enhances ECL vertical resolution and overall intensity, thanks to the longer lifetime of the oxidized luminophore, while maintaining low biotoxicity [45]. This work demonstrates the first ECL imaging of CEA on cytomembranes, offering a powerful platform for bioanalysis and single-cell studies. Other researchers employed direct ECLM for cell visualization and monitoring. Liu and coworkers developed a permeable chitosan film deposited on the FTO/TiO_2/CS electrode before cell immobilization, and then, a cell stimulating agent was added which triggered cells with hydrogen peroxide (H_2O_2) production [46]. H_2O_2 is one of the reactive oxygen species (ROS) associated with the regulation of cellular processes in vivo and acts as a coreactant for ECL generation in the presence of luminol, allowing single-cell direct ECL imaging, also thanks to its longer lifetime in aqueous solution compared to the short-lived radicals of commonly used coreactants. Then, Cui et al. demonstrated steady-state bright ECL at single titanium dioxide (TiO_2) nanoparticle substrate for local electrochemical sensing of H_2O_2 efflux from single living cells [47]. By leveraging oxygen vacancies on TiO_2 for sustained hydrogen peroxide adsorption, the researchers achieve continuous ECL from freely diffusing luminol under physiological conditions, enabling subcellular electroanalysis and real-time cellular activity monitoring at the single-nanoparticle level. A particularly promising and valuable domain for sensoristic applications is also forensic science, where fingerprint analysis is paramount. Several investigations have been effectively executed to detect latent fingerprints using ECL microscopy and to identify external substances present on fingerprint residues. Li et al. recently introduced an integrated smartphone-based imaging system that combines fingerprint mapping with on-site biochemical sensing. Their study demonstrated the benefits of using a point-of-care device to detect exogenous compounds in fingerprint residues: nicotine, for example, amplifies the ECL signal due to its tertiary amine character (akin to TPrA), whereas trinitrotoluene (TNT) quenched the ECL output by acting as an electron acceptor that interferes with signal generation [48]. Moreover, fine details such as lakes, bifurcations, terminations, and crossovers are critical for personal identification. Based on these insights, a method was devised to visualize latent fingerprints by selectively inducing the electropolymerization of luminol on an ITO electrode without resorting to toxic reagents; here, the fingerprint residue serves as a mask, ensuring that polyluminol is formed only on the unmasked electrode regions, thereby producing a negative ECL fingerprint pattern [15]. The continuous discovery of novel coreactants through ECL microscopy and coreactant-free ECL strategies not only expands the understanding of ECL mechanisms but also paves the way for the development of more efficient and versatile electrochemiluminescent systems, with each choice tailored to the specific requirements of the system under study. This progress is further supported by the increase in emitting layer thickness, which enables ECL emission at greater distances from the working electrode, thereby enhancing spatial resolution for imaging of cells and biological species.

3 Advanced Materials and Electrode Architectures in ECLM Studies

Heterogeneous electron transfer is the fundamental step in ECL and is profoundly influenced by the electrode's properties [49]; thus, the efficiency of the ECL process is intrinsically linked to the choice of electrode material. This section explores the interplay between electrode materials, miniaturization strategies, and advanced configurations like bipolar electrodes (BPEs) in optimizing ECL mechanisms.

The choice of the right electrode material is essential for enhanced signal generation and precise control of the emitting layer thickness, particularly with the classical ECL system, tris(2,2-bipyridyl) ruthenium (II) ($Ru(bpy)_3^{2+}$)/tri-*n*-propylamine (TPrA). For instance, the surface properties of the electrode affect the lifetime of $TPrA^{+\cdot}$ and their subsequent deprotonation, leading to increased radical generation near the electrode and enhancing oxidative consumption. Therefore, the investigation of diverse electrode materials is crucial for optimizing key parameters such as the operational potential range [50] and electron transfer kinetics [51, 52]. Metallic electrodes, like Pt and Au, are widely used in electrochemistry and ECL, especially in organic solvents and aprotic media, as well as for the characterization of new luminophores. However, they show weaker ECL emission compared to carbon-based materials, due to the formation of surface oxides that significantly reduce the direct oxidation of TPrA [53–56]. On the other hand, carbon-based electrodes, like GC and CNTs, represent excellent materials for ECL applications as their surface exhibits enhanced affinity toward TPrA, improving its oxidation rate, which is widely considered one of the rate-determining steps for signal generation [57–60].

A diverse array of nanomaterials has been identified as effective platforms for enhancing ECL signals. Rebeccani et al. introduced Ru-functionalized carbon nanotubes (CNT-Ru) as promising materials to improve the spatial distribution of the ECL-emitting layer [61]. These carbon nanotubes form a conductive shell around the beads, thereby extending the ECL-active region and enabling direct oxidation of the luminophore on this new conductive surface, an effect not observed in conventional bead systems, where direct luminophore oxidation is largely absent. To assess the amplification capability of CNTs, the authors functionalized magnetic microbeads with CNT-Ru using a robust biotin-streptavidin interaction (beads@CNT-Ru) and evaluated their performance on platinum electrodes. The ECL output was compared with that of standard 2.8 μm microbeads functionalized with a biotinylated antibody labeled with the $[Ru(bpy)_3]^{2+}$ complex (beads@Ru), revealing a notable fourfold enhancement. This result highlights the strategic role of CNTs in significantly boosting the ECL response.

A distinctive carbon allotrope successfully applied to ECL is boron-doped diamond, thanks to its wide potential window in aqueous electrolytes, low capacitive current, and excellent physical and chemical stability. ECLM studies on BDD electrodes have shown that, for the $[Ru(bpy)_3]^{2+}$/TPrA system, emission was higher in heterogeneous setups with microbeads than under conditions simulating commercial immunoassays with Pt electrodes [62].

Transparent electrodes are particularly useful for visualizing the ECL process, as they allow optical access to the electrode surface. They are often combined with microscopy to image surface-confined phenomena, especially in biological samples. Common types include glass or polymer-supported electrodes coated with indium-doped tin oxide. In this context, recently Meg et al. introduced an innovative method that leveraged temperature to shift between different ECL pathways. They engineered a temperature-responsive ECL layer by integrating a digital heating element onto the non-conductive side of an ITO electrode, allowing them to observe the formation of shadow regions from adherent cells. Both finite element simulations and experimental data revealed that the TEL could be reversibly modulated by adjusting the electrode temperature (Te), enabling the imaging of single-cell topography at varying heights [5].

Beyond material selection, nanoconfinement and miniaturization offer a promising avenue for studying and extending ECL emission. By confining reactants within nanoscale regions, collision efficiency is increased, leading to stronger ECL emission through interactions between luminescent molecule radicals and coreactants [38, 63]. Advances in high-resolution ECL imaging, coupled with micro/nano devices and bipolar electrochemistry, could address challenges related to coreactant lifetime and mass transport [64].

3.1 Micro- and Nanoelectrodes for Mechanistic Resolution

Micro and nanoelectrodes can be defined as electrodes that have at least one dimension (called the "critical dimension") on the order of micrometers or nanometers, respectively. This accelerates mass transport, reducing diffusion limitations and solution resistance, while providing low capacitive currents and high current densities.

These properties enhance analytical sensitivity and lower detection limits, enabling analyses with minimal sample volumes, ideal for in vivo studies. However, low current outputs necessitate micro/nanoelectrode arrays (MEA/NEA) to amplify signals, preserving the benefits of individual miniaturized electrodes [65]. ECL imaging on Pt microelectrode arrays was first reported in the 1980s [66]. Following this, due to advancements in imaging technology, the optical resolution of ECL imaging reached the micron level [67] and optical fibers, with their excellent light transmission capabilities, emerged as promising substrates for micro- and nanoelectrode array fabrication [64]. In 2003, Walt et al. pioneered this approach by creating a tapered Au microelectrode array on an optical fiber bundle using chemical wet etching [68]. This innovative design enabled the capture of spatially resolved ECL images from the microelectrode array. Furthermore, through subsequent post-treatment, the tapered Au microelectrodes were further refined into a nanoelectrode array, enabling higher resolution imaging due to reduced overlap between the diffusion layers of adjacent electrodes. Building upon these advancements, Ugo et al. constructed more complex nanoelectrode arrays to systematically investigate the influence of electrode shape and distribution on ECL imaging quality [69]. A layer of polycarbonate (PC) was spin-coated onto a BDD substrate and subsequently patterned using electron-beam lithography to define the desired nanoelectrode array geometry. This approach enabled precise control over the size, shape, and location of individual nanoelectrodes. ECL imaging studies revealed a strong correlation between electrode size and inter-electrode distance and the observed ECL intensity (Fig. 3a). Specifically, larger electrodes and closer inter-electrode spacing resulted in higher ECL intensities, likely due to the overlap of diffusion layers from neighboring electrodes. Furthermore, the concentration of the coreactant, namely TPrA, significantly influenced the spatial resolution of ECL imaging. As the TPrA concentration increased, the ECL emission zone around each nanoelectrode contracted, leading to sharper and more defined ECL images, consistent with molecular diffusion theory. Very recently, the measurement of the TEL using microtube electrode ensembles has been reported, showing that the combined use of microelectrode ensembles and ECL imaging technique is indeed a powerful tool for deciphering ECL reaction mechanisms [70, 71]. Their findings revealed that increasing the luminophore concentration led to an increase in the TEL from approximately 3.1 μm to over 4.5 μm (Fig. 3b). They attributed this change to the growing influence of a catalytic pathway, which caused a shift in the ECL pattern from a ring shape to a spot. In a successive work, they gold-coated UMEA to visualize ECL reactions involving $[Ru(bpy)_3]^{2+}$ and TPrA at the single microwell using ECLM. In their work, the whole surface of the microwell array is continuously coated with an ultrathin layer of gold, meaning that its top/bottom surface and vertical sidewall are all capable of triggering ECL reactions. Regardless of luminophore concentration, the strongest emission is distributed at the junction between the sidewall and the bottom of UMEA [72].

Using a similar strategy, they visualized and quantified the thickness of the ECL layer for the luminol/H_2O_2 system [71]. The ECL pattern evolved from a spot to a ring with increasing pH, indicating a decrease in TEL from > 9.1 μm to ~ 4.3 μm.

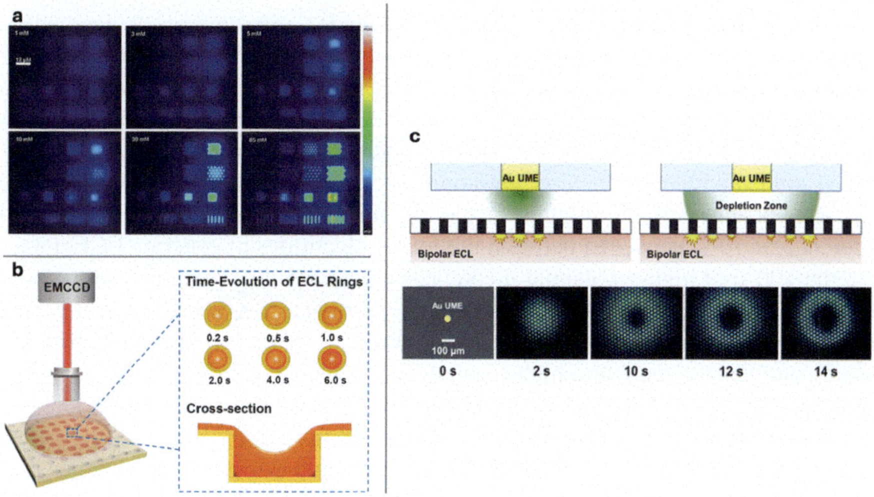

Fig. 3 **a** ECL images of the MNEAP obtained in a PBS solution (pH 7.4) containing both 1 mM $[Ru(bpy)_3]^{2+}$ and increasing concentrations of TPrA (indicated in top-left corner of each box). Images were recorded in the dark with a × 50 objective when applying a constant potential of 1.2 V versus Ag/AgCl/KCl. All images were coded according to the same false color scale (right). Reproduced with permission from Ref. [69]. **b** Microtube electrodes for imaging the electrochemiluminescence layer and deciphering the reaction mechanism. Reproduced with permission from Ref. [70]. **c** Up: schematic showing the formation of a depletion zone in a generator-collector setup. $Fe(CN)_6^{4-}$ is oxidized on a 50 μm Au UME generating $Fe(CN)_6^{3-}$. The incident $Fe(CN)_6^{3-}$ on the biased array (+ 2.0 V) is visually detected by anodic ECL. A dark redox depletion zone is formed due to the consumption of $Fe(CN)_6^{4-}$ in the ~ 100 μm microgap between the UME and array. Down: Progression of the ECL response during the depletion time lapse. Reproduced with permission from Ref. [73]

This suggests a pH-dependent shift in reaction mechanism, from long-lived neutral intermediates at low pH to short-lived radical species at high pH.

3.2 Bipolar Electrodes in Spatially Resolved ECLM

Bipolar electrodes provide a distinct advantage in creating miniaturized ECL systems that operate wirelessly [22]. For instance, a BPE acts as an electronically conductive element in contact with an ionically conductive phase. Under a sufficiently high electric field, faradaic reactions occur at the BPE ends without a direct connection to an external power source. This wireless configuration, though unconventional, maintains control over interfacial potential differences and has opened new possibilities for ECL applications. For instance, in 2006, Crooks et al. fabricated a micro-sized bipolar electrode array comprising 1000 individual Au BPEs, each measuring 500 μm in length and 50 μm in width [74]. ECL imaging revealed distinct and independent emission from each BPE, demonstrating the potential of these arrays for high-efficiency ECL sensing. However, achieving micrometric or nanometric objects with BPEs presents challenges due to the high electric fields required for polarization and the limited sensitivity associated with small sample volumes. To overcome these challenges, researchers have developed innovative strategies. For instance, Voci et al. designed a solid-state micropore decorated with a rhombus-shaped gold microelectrode, concentrating the electric field within the micropore to enable localized ECL emission. Due to the micropore features, only a potential difference of 10 V is needed between the feeder electrodes to generate ECL, rather than the 100-fold higher potential theoretically required in a configuration without the micropore [75]. Similarly, Anderson et al. fabricated a massive array of closed bipolar ultramicroelectrodes (UMEs) in a hexagonal pattern to optimize electrode spacing and density, enabling high-resolution imaging of dynamic redox processes (Fig. 3c) [73]. The resulting device was used to study the generation, diffusion, and depletion of $Fe(CN)_6^{3-}$ on an Au UME. By applying + 0.8 V oxidizing potential, they produced $Fe(CN)_6^{3-}$ through $Fe(CN)_6^{4-}$ oxidation, resulting in a larger illuminated zone expanding to approximately 400 μm within 2 s. However, after 3 s, the center's intensity decreased, forming a dark depletion zone under the Au UME position. At the outset, the spatial resolution was found to be limited by the size of the bipolar UMEs and their spacing. This limitation could potentially be addressed by reducing electrode dimensions and interelectrode spacing. Conversely, the temporal resolution was limited only by the data transfer rate of the camera. These advancements underscore the importance of electrode geometry and material selection in enhancing ECL performance.

The synergistic combination of microfabricated BPEs and ECL imaging has opened new possibilities for high-performance sensing. For example, Iwama and coworkers used a closed BPE array to image the respiratory activity of MCF-7 spheroids with high spatiotemporal resolution, offering 2D insights into cell respiration. An evaluation of the oxygen consumption of MCF-7 was carried out using bipolar ECLM (BEM), which revealed a noticeable reduction in the surrounding ECL spheroids as a result of oxygen consumption. This BEM methodology offered improved spatial resolution, providing 2D insights into cell respiration compared to traditional methods [76]. However, a common element in both recent studies is the incorporation of an initial assessment to evaluate the consistency of the potential and ECL response. This step is essential because the cell geometry significantly influences the distribution and strength of the potential gradient generated on the surface of the BPE, and monitoring this process is challenging due to the wireless nature of the BPE. Interestingly, Villani et al. mapped potential gradients in bipolar electrochemical cells with different geometries, revealing how ECL emission patterns vary with applied voltage and electrode configuration. They exploited the stable ECL signals generated through luminol/H_2O_2. In a linear bipolar arrangement, the potential of the BPE varies uniformly across its surface, allowing the expansion of the ECL area by increasing the applied total voltage (E_{tot}). When configured as a cylinder, the anodic potential gradient on the BPE's surface transitioned from a spot to a ring-like shape with rising values of E_{tot}. Utilizing a single BPE in a U-type electrolytic cell primarily generated ECL emission on the cathode side. As E_{tot} increased, the

ECL region expanded toward the center until reaching a shielding wall. However, employing a pair of BPEs resulted in an asymmetric ECL emission pattern, accentuated at higher E_{tot} values as a result of the floating potential of the BPE adjusting to the solution potential. Conversely, electrically connecting the two BPEs using an ammeter transformed the asymmetrical ECL emission into a single, uniform ECL emission region, aligning with the pattern observed with a single BPE [77].

In addition to the uniformity of the ECL response, another critical aspect involves the choice of the electrode and the deliberate control of the electric field's focus and shape. This constitutes a notable and effective strategy for improving the efficiency of bipolar electrochemical and ECL cells. Subsequently, Douman et al. created a wireless ECL system based on 3D-printed titanium arrays as feeder electrodes and gold microparticles (45 μm diameter), functionalized with 11-mercaptoundecanoic acid, as microemitters. These titanium arrays, created using selective laser melting, produced intense electric fields whose distribution depends on the geometry of the array, enabling conductive microparticles as individual bipolar electrodes and localized ECL. Using $[Ru(bpy)_3]^{3+}$ and TPrA, Au microparticles generate ECL at a minimum electric field strength, and the intensity increases until it levels out as the field strength increased. This behavior aligns with the influence of heterogeneous electron transfer across the titanium oxide layer on ECL generation. Furthermore, the use of charged microparticles as a massively parallel array of bipolar electrodes enhanced ECL through electrokinetic transport, facilitating the entry of more particles into the ECL generation zone per unit of time [78].

In conclusion, the integration of advanced electrode materials, miniaturization strategies, and bipolar electrode technology has significantly advanced ECL systems. From biosensing to high-resolution imaging, these innovations provide valuable insights into the fundamental processes of ECL and pave the way for next-generation analytical tools. By pushing the boundaries of ECL to the micro- and nanometer regime, researchers continue to unlock exciting possibilities for understanding and harnessing electroluminescence.

4 Modeling the ECL-Emitting Layer for Mechanistic Understanding

4.1 Fundamentals of Finite Element Simulation in ECLM

As discussed throughout the chapter, ECL arises from cascade electron transfers occurring within the Marcus inverted region, naturally leading to the excitation of a given luminophore, L^*. Since the subsequent radiative relaxation enables optical readout in the visible region, ECLM stands out as a powerful tool for mapping L^* distribution. However, excitation represents only the final step of a complex sequence of chemical reactions among electrogenerated radicals. Relying on the experimental L^* distribution as the sole descriptor to speculate about the ECL mechanism risks overlooking crucial kinetic factors. Therefore, formulating a comprehensive mechanistic hypothesis ideally requires a complementary tool capable of describing the mass transport and chemical reactivity of ECL reaction intermediates.

The physics governing any ECL system is fully described by the Nernst–Planck–Poisson partial differential equations (PDEs), where electrochemical perturbations are defined through boundary conditions. Consequently, the behavior of ECL-active species can be unraveled by solving such PDEs within an appropriate geometry and timescale. Since Feldberg's seminal work in the 1960s [79], the steady advancement of computing power has progressively enabled computational electrochemistry to handle increasingly complex problems, including ECL. In this context, simulations are widely employed to study the time-dependent behavior of transient species involved in the reaction model under specific initial conditions. Typically, mechanistic interpretations are validated by comparing experimental and computed ECL distribution until convergence is achieved. Once the thermodynamic and kinetic pictures are correctly defined, the resulting concentration profiles of short-lived intermediates offer valuable mechanistic insights, assisting in a more comprehensive understanding of the reaction mechanism.

From a kinetic perspective, a general ECL mechanism could be modeled with a sequence of extremely fast (i.e., close to diffusion limit of ~ 10^{10} M^{-1} s^{-1}) second-order reactions. This process results in a traveling narrow reaction front, where rapid first-order radiative decay restricts L^* concentration to very small values. Such a scenario, combined with the usually non-uniform current distribution, introduces severe mathematical complications in the PDEs describing the space and time evolution of any variables in the ECL model. Although analytical solutions to nonlinear PDEs can sometimes be derived [80], they often rely on key assumptions such as linear semi-infinite diffusion and the steady-state approximation. These formulations offer a rough a priori estimate of the ECL behavior, but they only apply to a few ideal scenarios and fail to describe the time-dependent evolution of the system, demanding alternative mathematical methods to obtain physically meaningful results.

One of the most popular methods is the finite element method (FEM). This numerical approach involves breaking down the whole domain (in this case, the electrolyte) into a collection of non-overlapping subdomains, known as finite elements, connected to each other at nodal points to form a mesh. PDEs are discretized over this mesh applying the Galerkin method. First, the equations are transformed into an integral form, or weak formulation, allowing them to be conveniently manipulated, similarly to vectors [81, 82]. The approximate solution is then expressed as a linear combination of polynomial trial functions and substituted into the weak formulation, resulting in a system of equations where the unknowns represent variable values at specific points. This system is ultimately formulated as a matrix equation, which is solved by incorporating original boundary conditions and initial values to determine the approximate solution. Since the integrals in the discretized equations involve products of trial functions, the approximate solution allows non-zero values only where basis functions overlap at neighboring nodes. Consequently, the solution is explicitly determined at the nodes only, while values within each element are computed through interpolation. This process of determining a solution defines a close interplay between mesh accuracy and result accuracy. Finer elements fit more precisely the model geometry and capture smoother gradients, providing a more accurate representation of the physical problem. Conversely, coarser elements, due to their larger area, are more prone to distort the flux in proximity of angles and edges and may misinterpret concentration distribution when facing steep gradients, as the interpolation accuracy is inherently reduced. However, a finer mesh comes at a significantly higher computational cost.

The most widely employed FEM-based software is COMSOL Multiphysics, a nonspecialized multipurpose simulation platform applied across various fields, including electromagnetics, acoustics, and structural mechanics [83]. With specialized modules for chemistry, COMSOL can solve electrochemical problems involving both heterogeneous and homogeneous electron transfer processes, becoming a leading tool in computational electrochemistry.

Over the past 30 years, Amatore and colleagues have been highly prolific in enhancing the accuracy of computing general electrochemical problems, with a particular focus on translating this knowledge to ECL simulations [67, 84–89]. To reduce the complexity of mesh modeling for less experienced users, in 2011 they introduced a mathematical framework, implemented in the KISSA software, to optimize the simulated ECL accuracy while simultaneously minimizing the need for user intervention [90].

Among their works, Amatore et al. conducted extensive studies on the mapping accuracy of nano- and microtubes [91]. Specifically, when the radius of these structures is larger than molecular size and the thickness of the layer separating two adjacent pores is significant (i.e., comparable or even larger than the pore radius), strong flux inhomogeneities arise in the close vicinity of pore entrance, as observed for recessed microdisks. To precisely address mass transport at the interface between the tube edges and the bulk solution, they formulated a multi-step coordinate transformation to achieve a closed rectangular space. Visualization of homogeneous rectangular meshing in the transformed space after back-transformation highlights how the coordinate change was particularly adequate to suppress the critical challenges introduced by the discontinuities and boundary angles introduced by the thick separating layer.

Su and colleagues combined the same electrode geometry with ECLM to determine the ECL-emitting layer thickness of $[Ru(bpy)_3]^{2+}$/TPrA and Luminol/H_2O_2 under various experimental conditions [71, 92]. Although the results of these studies have been discussed throughout the chapter, it is noteworthy from a computational electrochemistry perspective that each work was accompanied with finite element simulations conducted on COMSOL to support the proposed mechanistic hypothesis on the basis of observed ECL patterns.

Despite the challenges of mapping the microtube geometry with a grid that precisely captures flux curvatures at the edges and steep concentration gradients, the authors successfully generated a locally refined mesh. As a result, the simulations achieved convergence while maintaining smooth L^* profiles.

Notably, they observed that the experimental ECL thickness was significantly affected by $[Ru(bpy)_3]^{2+}$ concentration for $[Ru(bpy)_3]^{2+}$/TPrA and by solution pH for luminol/H_2O_2. On the basis of these findings, they hypothesized the existence of two distinct reaction pathways for each system, whose dominance varies with experimental conditions. In both works, the simulated L^* emission shows either confinement or extension along the radial direction depending on the initial reaction model. These models resulted in ECL thicknesses of ca. 1.5 µm and > 4.5 µm for $[Ru(bpy)_3]^{2+}$/TPrA at 10 and 500 µM of $[Ru(bpy)_3]^{2+}$, respectively, while the simulated thickness of luminol/H_2O_2 was esteemed to be > 9.1 µm at pH < 9 and ca. 4.1 µm at pH 10–12 (Fig. 4a). The simulations closely match the shrinkage and expansion of experimental ECL profiles (i.e., 3.1 µm and > 4.5 µm for $[Ru(bpy)_3]^{2+}$/TPrA; > 9.1 µm and ca. 4.3 µm for luminol/H_2O_2), proving that ECL-emitting layer variations are associated to different reaction routes. In a more recent study, Mariani et al. successfully simulated the ECL spatial extension over recessed microdisks, 2.5 µm in diameter, made of different materials, specifically gold and glassy carbon. By adjusting the oxidation rate constant of TPrA according to the nature of the electrode surface, they accurately reproduced the experimentally observed increase in ECL spot thickness from glassy carbon to gold. Their computational results demonstrate that higher concentrations of electrogenerated TPrA radicals confine $[Ru(bpy)_3]^{3+}$ closer to the electrode surface: on glassy carbon, faster oxidation rates reduce the thickness (~ 4 µm) of the emitting layer by enhancing the oxidative-reduction mechanism, while on gold, slower oxidation promotes catalytic oxidation of TPrA by $[Ru(bpy)_3]^{3+}$, thereby extending the emitting layer (~ 7 µm) [93].

When meticulous care is taken in configuring the electrochemical problem, these scenarios highlight COMSOL effectiveness in simulating general ECL processes. Despite challenges related to electrode geometry and mass transport, COMSOL allows users to explicitly control mesh size and element-to-element expansion rates at any point, boundary, or volume within the model, simplifying the construction of a mesh that conforms to small geometric features such as faces and edges. Additionally, the well-established understanding of conventional ECL reaction mechanisms, along with the growing body of literature on finite element simulations of ECL systems, provides valuable guidance for designing reliable ECL simulations, even in the absence of precise a priori chemical knowledge to refine the grid around traveling concentration gradients.

4.2 Finite Element Modeling of Bead-Based ECL Systems

While these features alone do not provide a significant advantage over KISSA, it should be noted that the software developed by Amatore and his coworkers is limited to simulate ECL emission on specific microelectrode geometries. Currently, most studies involving ECL focus on its application as transduction technique for bioanalysis that heavily rely on bead-based immunoassays. This setup represents a non-standard geometry that KISSA cannot accommodate. As a versatile Multiphysics platform, COMSOL thrives in such cases, featuring a wide range of geometric primitives, operations, and modeling tools to handle fancy configurations.

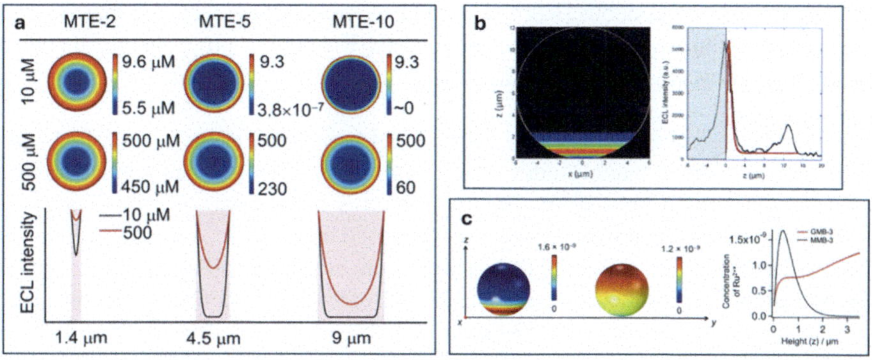

Fig. 4 **a** Concentration distribution of $[Ru(bpy)_3]^{2+*}$ in a top-view configuration simulated at microtubes with diameters of 1.4 (left), 4.5 (middle), and 9 μm (right), respectively. The upper section shows the simulated [Ru(bpy)3]2+* intensity profiles along the radial direction, where the pink region represents the microtubes inner region. The ECL behavior of each system was simulated for 10 and 500 μM of $[Ru(bpy)_3]^{2+*}$. Used with permission of Wiley, from Angew. Chem. Int. Ed., 2021, 60, 2089–2093; permission conveyed through Copyright Clearance Center, Inc. **b** Left: concentration distribution of $[Ru(bpy)_3]^{2+*}$ in a side-view configuration simulated at the surface of a 12 μm bead. Right: comparison between the experimental (black line) and simulated (red line) ECL profiles at the level of a single bead. The ECL profile was simulated considering a value of 2920 s^{-1} for the $TPrA^{\cdot+}$ deprotonation rate constant. The hatched zone represents the reflection of the ECL light on the electrode surface. Used with permission of Royal Society of Chemistry, from Chem. Sci., 2014, 5, 2568–2572; permission conveyed through Copyright Clearance Center, Inc. **c** Left: concentration distribution of $[Ru(bpy)_3]^{2+*}$ in a side-view configuration simulated at the surface of Ru@PMB (left) and Ru@GMB (right). Right: Dependence of the simulated concentrations of $[Ru(bpy)_3]^{2+*}$ at Ru@PMB (black line) and Ru@GMB (red line) on the distance from the substrate electrode surface. Reprinted (adapted) with permission from J. Am. Chem. Soc. 2023, 145, 29, 16,026–16,036. Copyright 2023 American Chemical Society. For interpretation of the references to color in these figures' legends, the reader is referred to the Web version of this article

In particular, bead-based systems are typically modeled as 2D axisymmetric hemispherical surfaces positioned on top of a planar electrode, where mesh density is dramatically increased at both the electrode and bead boundaries. The behavior of every chemical species within this system is governed by two key physical interfaces (pre-built packages of physical equations with associated boundary conditions): the Transport of Concentrated Species interface to express the flux and reactivity of freely diffusing species in solution, and the Surface Reactions interface to describe the distribution and reactivity of bound luminophores with radicals approaching the bead surface. For the described interfaces, time-dependent problems are solved by integration of the PDEs over time according to an implicit backward differentiated formula (BDF) method [94]. Even though the rate constant values may differ from the actual ones, until reasonable relative magnitudes between the kinetics of competing reactions are kept, good agreement with experimental results should still be achieved. For example, Danis et al. leveraged spectroelectrochemistry to monitor the consumption of $[Ru(bpy)_3]^{2+}$ at the electrode surface in presence of TPrA during an anodic potential sweep. By parametrically analyzing individual homogeneous rate constants included in the homogeneous $[Ru(bpy)_3]^{2+}$/TPrA model, they observed that the best quality fit between simulated and experimental $d[Ru(bpy)_3{}^{2+}]/dt$ plots occurs when the reaction between $TPrA^{\cdot}$ and $[Ru(bpy)_3]^{3+}$ is 100 times faster than the other competing reactions [95].

Despite beads being proposed as commercial bioanalytical tools as early as late 1990s [96, 97], fundamental understanding of the reaction mechanism driving the ECL emission of $[Ru(bpy)_3]^{2+}$ on the microspheres surface was lacking. Not much later, in their cornerstone work, Miao et al. proposed the so-called remote mechanism where diffusion of $TPrA^{\cdot+}$ and $TPrA^{\cdot}$ governs the emission as Ru(II) labels cannot directly interact with electrode as basically all of them are confined further away than the tunneling distance. In particular, the bottleneck of this mechanism was identified with the fast deprotonation of $TPrA^{\cdot+}$ ($\tau_{1/2}$ ~ 0.2 ms) that would confine the ECL-emitting layer close to the electrode surface. Since the presence of the bead is supposed to deviate the radicals mass transport from the ideal planar diffusion, this discovery raised a parallel debate regarding the optimal bead dimension to involve as much $[Ru(bpy)_3]^{2+}$ as possible in the remote mechanism.

This question remained unanswered until a decade later, when Sentic et al. imaged Ru(II)-coated 12 μm single beads using side-view ECL microscopy to reconstruct the extension of the ECL emitting region. This layer experimentally extends along the normal to the electrode surface as far as 3 μm. Finite element simulation was eventually employed to rationalize the results as a function of the concentration profiles of coreactant radicals by considering the remote mechanism reaction scheme. Coherently with the mechanistic interpretation of Bard and his coworkers, the authors found out the key kinetic parameter is the rate constant for the decomposition of $TPrA^{\cdot+}$ as it controls the position of the maximum ECL intensity and the thickness of the emitting layer: more durable diffusion allowed by slower deprotonation stretches the emitting region which, on the other hand, gets progressively more confined upon increasing the rate constant. Best fitting of the experimental ECL profile along the z-axis was obtained with a rate constant of 2920 s^{-1} ($\tau_{1/2}$ ~ 0.24 ms) that closely agrees with the formerly proposed one (Fig. 4b). Noteworthy, smaller radical concentrations are observed on the bead surface compared to bulk, particularly for the radical cation. This behavior arises from the hindrance of mass transfer by the bead, simultaneously allowing for an expansion of the concentration profiles and, consequently, of the ECL-active region compared to planar diffusion.

The latter observation is supported by a recent work from Feng et al. [98], who performed similar finite element simulations, but modeling a series of Ru(II)-functionalized beads with increasing diameter, from 5 to 18 μm. In particular, while they agree on the smaller concentration of radicals at the bead surface and on the key role of the deprotonation constant, they reported that the $TPrA^{\cdot+}$ concentration at the bead surface drops to 0.1% of its maximum value at 2.7, 3.4, and 4 μm on 5, 10, and 18 μm beads, respectively. This trend translates into progressively thicker emitting layer from 3 to 3.5, to 4 μm. Accordingly, they hypothesized that, in order to be detected by the microscope objective, most, if not all, the ECL emission must travel around the bead. In this case it is reasonable to assume that optics could play a pivotal role in shaping the imaged ECL patterns from microbeads as already hinted by the focusing effect of the bead that produces photonic nanojets at the top despite lacking any chemical reactivity.

In this regard, Han et al. observed major discrepancies between the experimental ECL profile of 12 μm beads recorded in top-view configuration and the simulated counterpart considering only the remote mechanism, without accounting for any optical effects [99]. Despite in both profiles a stronger ECL is observed at the bead edges than in the center, the experimental profile reports a 40% drop in the middle while it is completely absent in the simulated profile. Additionally, the ECL patterns are deeply different as the simulated model predicts a narrower profile and only 2 peaks instead of the 4 ones actually observed. Thus, they introduced a discrete dipole approximation (DDA) in the simulation to model the optical wave propagation as the ECL emission from beads can be reasonably assimilated to a radiating dipole [100]. The resulting simulated profile very closely matches the experimental one as it shows two couples of symmetric peaks and quasi-plateau in the middle that is 64% of the normalized ECL maximum. The generalizability of this approach was further demonstrated by achieving excellent correlation between the experimental top-view ECL profile of a 2.8 μm bead with the profile simulated by means of complementary chemical and optical modeling. Their work demonstrated that electromagnetic scattering governs the ECL images recorded with a top-view microscope configuration.

All mechanistic studies on the remote ECL mechanism agree that its efficiency, governed by the coreactant, is hindered by the TPrA slow oxidation rate and the $TPrA^{\cdot+}$ instability. Recently, Kerr et al. proposed to introduce a free-to-diffuse Ir(III)-based redox mediator, namely $[Ir(sppy)_3]^{3-}$, in a TPrA solution to boost the ECL signal of $[Ru(bpy)_3]^{2+}$ both in homogeneous phase and on microbeads surface [23, 101]. In the subsequent work [38] within a comprehensive series [23, 38, 101–105], Fracassa et al. established a set of thermodynamic reactions for a heterogeneous Ru(II)/TPrA/Ir(III) system through a comparative study involving three Ir(III) complexes with different electrochemical properties. While this work mainly focused on thermodynamics, a follow-up investigation by Knežević et al. explored the kinetics of this redox-mediated system in homogeneous phase [103]. Notably, both works were supported by finite element simulations. Despite including different geometrical models, such as 1D box and 2D axisymmetric hemisphere, perhaps influencing the distribution of electroactive species, mechanistic interpretations consistently converge, highlighting the complementarity of these simulations in systematically studying the role of the Ir(III) mediator within the coreactant ECL mechanism.

First, finite element simulations match experimental data, proving the correct thermodynamic understanding of the system where the ECL enhancement simultaneously relies on (i) the electrocatalytic homogeneous oxidation of TPrA (k_{cat}) and (ii) the redox-mediated oxidation of $[Ru(bpy)_3]^+$ (k_{IrRu}), both mediated by oxidized $[Ir(sppy)_3]^{2-}$. To evaluate their impact on the $[Ru(bpy)_3]^{2+}$ ECL emission, the respective rate constants were systematically varied. Intuitively, increasing k_{IrRu} progressively enhances $[Ru(bpy)_3]^{2+}$ emission by introducing an alternative excitation pathway. Similarly, increasing k_{cat} boosts the ECL intensity up to $3 \times 10^3\ M^{-1}\ s^{-1}$, a coherent value with the experimental one. However, beyond this k_{cat}, any further increase accelerates $[Ir(sppy)_3]^{2-}$ depletion, shifting the ECL mechanism from being controlled by the oxidized redox mediator to the weaker $TPrA^{\cdot+}$-governed process. The depletion rate of $[Ir(sppy)_3]^{2-}$ significantly affects also the emitting layer thickness: even though faster k_{cat} extends the $TPrA^{\cdot}$ diffusion layer, it simultaneously shrinks that of $[Ir(sppy)_3]^{2-}$, resulting in a thinner $[Ru(bpy)_3]^{2+}$ ECL-emitting layer.

Beyond exploring the reactivity of conventional and revised ECL mechanisms, COMSOL versatility makes it an ideal tool to simulate even more complex scenarios. Recently, Yang et al. proposed replacing conventional polystyrene microbeads (PMBs) with gold microbeads (GMBs) as platforms for binding $[Ru(bpy)_3]^{2+}$ units [106]. The remarkable ECL enhancement was attributed to GMBs acting as spherical ultramicroelectrodes, directly oxidizing TPrA on their surface. This system was successfully modeled in COMSOL by reshaping classical boundaries, extending the electrode surface to the bead boundary while incorporating the conventional remote ECL reaction scheme with standard rate constants (Fig. 4c). Concerning radicals propagation at the microsphere surface, the simulated concentration profiles of $TPrA^{\cdot+}$ share similar tendencies, despite being constantly higher at GMBs than at PMBs. Remarkably, at GMBs, $TPrA^{\cdot+}$ concentration remains constant beyond 1.5 μm, whereas at PMBs, it drops to zero aligning with the radical cation short half-life. The concentration profiles of $TPrA^{\cdot}$ follow similar patterns on both platforms but, contrarily to what observed for $TPrA^{\cdot+}$, $TPrA^{\cdot}$ concentration at GMBs is smaller throughout the whole surface than that simulated at PMBs. Nonetheless, the simulated concentrations of $[Ru(bpy)_3]^{2+*}$ are significantly greater on GMBs than on PMBs, and present two antithetical trends. Despite both profiles rising with the distance from the substrate electrode up to 0.4 μm, beyond this point the $[Ru(bpy)_3]^{2+*}$ distribution on GMBs steadily increases reaching a maximum at the top, while it progressively declines on PMBs and vanishes above 2.5 μm due to the depletion of $TPrA^{\cdot+}$ far away from the electrode surface. Results of digital simulations provide unprecedented mechanistic insights into a novel ECL system, demonstrating the potential of conductive platforms to efficiently extend the ECL layer thickness across their entire surface.

4.3 Alternative Computational Approaches to Simulate ECL Mechanisms

Although COMSOL has repeatedly proven to be a valuable tool to simulate ECL systems, the FEM relies on continuum approximation, a mathematical idealization used to model the collective behavior of discrete systems. This approach works well in relatively concentrated environments but falls short in mesoscopic or stochastic conditions, where only a small number of particles are involved, and fluctuations of their density or the stochasticity of reactions become significant.

To address this limitation, Mathwig developed a two-dimensional random walk algorithm in Python to simulate diffusion, electrochemical and chemical reactions, and photon generation [107]. Since coding in Python presents significant challenges compared to the ease of using the user-friendly COMSOL, the author leveraged a large language model like ChatGPT o1 through an iterative Q&A process, circumventing the lack of advanced coding experience. This model simulates freely diffusing $[Ru(bpy)_3]^{2+}$ in a nanofluidic channel with a 100 nm thickness and 20 μm length, where independent electrodes embedded in the channel ceiling and floor enable the simulation of annihilation ECL. This channel, containing approximately 3600 ions at 1 μM concentration, operates within a stochastic regime, necessitating a random walk simulation to accurately model phenomena such as the collision radius. The resulting ECL profiles, simulated at various photon emission rates, were compared to finite element results with matching $[Ru(bpy)_3]^{2+}$ concentrations. Overall, the concentration profiles of $[Ru(bpy)_3]^{2+}$, $[Ru(bpy)_3]^{3+}$, $[Ru(bpy)_3]^{+}$, and $[Ru(bpy)_3]^{2+*}$ match very well. Additionally, unlike COMSOL, stochastic simulations enable noise analysis on time traces, such as determining power spectral density in the frequency domain. This analysis was applied to both particles and photons, revealing a characteristic noise pattern associated to Brownian motion in nanochannels and the random nature of excited state decay, respectively.

5 Conclusion and Future Perspectives

Electrochemiluminescence Microscopy has emerged as a powerful tool for elucidating the fundamental mechanisms underlying ECL generation. In particular, measuring the thickness of the ECL layer provides crucial insights into the spatial dynamics of coreactant diffusion and the dominance of specific reaction pathways. Advances such as scanning electrochemical microscopy have enabled sub-200 nm resolution in TEL analysis, offering precise control over the electrode interface and enhancing mechanistic understanding.

Efforts to optimize TEL through the design of new luminophores, coreactants, and nanostructured materials, such as dual-doped silica nanoparticles and carbon-based nanoemitters, have significantly improved ECL efficiency and imaging quality. However, challenges remain in overcoming the short lifetimes of reactive intermediates and in adapting ECLM for intracellular applications.

The integration of ECL with nanotechnology, microfabricated devices, and bipolar electrochemistry has further expanded the scope of ECLM, enabling high-resolution, multiplexed, and spatially controlled studies. Despite technical hurdles, particularly in fabricating dense bipolar electrode arrays, continued innovation is expected to simplify device design and broaden their analytical applications.

Ultimately, TEL measurement using ECLM not only advances our understanding of ECL mechanisms across different systems but also offers a promising strategy for screening and optimizing bioanalytical and immunoassay platforms. As the field continues to evolve, ECLM stands at the forefront of mechanistic electrochemical imaging with growing impact in both fundamental research and applied biosensing.

References

1. Wang, Y., Su, B.: Deciphering the mechanisms of electrochemiluminescence by spatially resolved measurements. Anal. Sens. **1**(4), 148–155 (2021). https://doi.org/10.1002/anse.202100037
2. Zhang, Z., Ma, C., Xu, Q., Zhu, J.J.: Recent progress in electrochemiluminescence microscopy analysis of single cells. Analyst **147**(13), 2884–2894 (2022). https://doi.org/10.1039/D2AN00709F
3. Ding, H., Zhou, P., Fu, W., Ding, L., Guo, W., Su, B.: Spatially selective imaging of cell-matrix and cell-cell junctions by electrochemiluminescence. Angew. Chem. Int. Ed. **60**(21), 11769–11773 (2021). https://doi.org/10.1002/ANIE.202101467
4. Ma, C., Wu, S., Zhou, Y., Wei, H.F., Zhang, J., Chen, Z., Zhu, J.J., Lin, Y., Zhu, W.: Bio-coreactant-enhanced electrochemiluminescence microscopy of intracellular structure and transport. Angew. Chem. Int. Ed. **60**(9), 4907–4914 (2021). https://doi.org/10.1002/ANIE.202012171
5. Ma, C., Xing, Z., Gou, X., Jiang, L.P., Zhu, J.J.: A temperature-tuned electrochemiluminescence layer for reversibly imaging cell topography. Chem. Sci. **13**(46), 13938–13947 (2022). https://doi.org/10.1039/D2SC04944A
6. Ma, C., Wang, M.X., Wei, H.F., Wu, S., Zhang, J.R., Zhu, J.J., Chen, Z.: Catalytic route electrochemiluminescence microscopy of cell membranes with nitrogen-doped carbon dots as nano-coreactants. Chem. Commun. **57**(17), 2168–2171 (2021). https://doi.org/10.1039/D0CC08223F
7. Dong, J., Lu, Y., Xu, Y., Chen, F., Yang, J., Chen, Y., Feng, J.: Direct imaging of single-molecule electrochemical reactions in solution. Nature **596**(7871), 244–249 (2021). https://doi.org/10.1038/s41586-021-03715-9
8. Han, D., Fang, D., Valenti, G., Paolucci, F., Kanoufi, F., Jiang, D., Sojic, N.: Dynamic mapping of electrochemiluminescence reactivity in space: application to bead-based assays. Anal. Chem. **95**(42), 15700–15706 (2023). https://doi.org/10.1021/ACS.ANALCHEM.3C02960
9. Zanut, A., Fiorani, A., Canola, S., Saito, T., Ziebart, N., Rapino, S., Rebeccani, S., Barbon, A., Irie, T., Josel, H.P., Negri, F., Marcaccio, M., Windfuhr, M., Imai, K., Valenti, G., Paolucci, F.: Insights into the mechanism of coreactant electrochemiluminescence facilitating enhanced bioanalytical performance. Nat. Commun. **11**(1), 1–9 (2020). https://doi.org/10.1038/s41467-020-16476-2
10. Chen, M.-M., Xu, C.-H., Zhao, W., Chen, H.-Y., Xu, J.-J.: Super-resolution electrogenerated chemiluminescence microscopy for single-nanocatalyst imaging. J. Am. Chem. Soc. **143**(44), 18511–18518 (2021). https://doi.org/10.1021/jacs.1c07827
11. Miao, W., Choi, J.-P., Bard, A.J.: Electrogenerated chemiluminescence 69: the tris(2,2′-bipyridine)ruthenium(II), ($Ru(Bpy)_3{}^{2+}$)/Tri-n-Propylamine (TPrA) system revisited a new route involving $TPrA^{\cdot+}$ cation radicals. J. Am. Chem. Soc. **124**(48), 14478–14485 (2002). https://doi.org/10.1021/ja027532v
12. Rodríguez-López, J., Shen, M., Nepomnyashchii, A.B., Bard, A.J.: Scanning electrochemical microscopy study of ion annihilation electrogenerated chemiluminescence of rubrene and $[Ru(Bpy)_3]^{2+}$. J. Am. Chem. Soc. **134**(22), 9240–9250 (2012). https://doi.org/10.1021/ja301016n
13. Zhou, P., Su, B.: Enhanced electrochemiluminescence at silica nanochannel membrane studied by scanning electrochemical microscopy. J. Electroanal. Chem. **904**, 115943 (2022). https://doi.org/10.1016/j.jelechem.2021.115943

14. Mariani, C., Bogialli, S., Paolucci, F., Pastore, P., Zanut, A., Valenti, G.: Enhancing electrochemiluminescence intensity through emission layer control. Electrochim. Acta **489**, 144256 (2024). https://doi.org/10.1016/J.ELECTACTA.2024.144256
15. Hu, S., Cao, Z., Zhou, L., Ma, R., Su, B.: Electrochemiluminescence imaging of latent fingerprints by electropolymerized luminol. J. Electroanal. Chem. **870**, 114238 (2020). https://doi.org/10.1016/j.jelechem.2020.114238
16. Jiang, X., Wang, Z., Wang, H., Zhuo, Y., Yuan, R., Chai, Y.: A novel metal-organic framework loaded with abundant *N*-(aminobutyl)-*N*-(ethylisoluminol) as a high-efficiency electrochemiluminescence indicator for sensitive detection of mucin1 on cancer cells. Chem. Commun. **53**(70), 9705–9708 (2017). https://doi.org/10.1039/C7CC05495E
17. Wu, F.-F., Zhou, Y., Zhang, H., Yuan, R., Chai, Y.-Q.: Electrochemiluminescence peptide-based biosensor with hetero-nanostructures as coreaction accelerator for the ultrasensitive determination of tryptase. Anal. Chem. **90**(3), 2263–2270 (2018). https://doi.org/10.1021/acs.analchem.7b04631
18. Zhang, H., Du, L., Wei, Z., Wang, X., Sojic, N., Zhou, X., Wang, Z.: Boosting the electrochemiluminescence of luminol-O_2 system by high-intensity focused ultrasound. Anal. Bioanal. Chem. **414**(29), 8309–8315 (2022). https://doi.org/10.1007/s00216-022-04365-0
19. Wu, K., Zheng, Y., Chen, R., Zhou, Z., Liu, S., Shen, Y., Zhang, Y.: Advances in electrochemiluminescence luminophores based on small organic molecules for biosensing. Biosens. Bioelectron. **223**, 115031 (2023). https://doi.org/10.1016/j.bios.2022.115031
20. Swanick, K.N., Ladouceur, S., Zysman-Colman, E., Ding, Z.: self-enhanced electrochemiluminescence of an Iridium(III) complex: mechanistic insight. Angew. Chem. Int. Ed. **51**(44), 11079–11082 (2012). https://doi.org/10.1002/anie.201206074
21. Guo, W., Ding, H., Gu, C., Liu, Y., Jiang, X., Su, B., Shao, Y.: Potential-resolved multicolor electrochemiluminescence for multiplex immunoassay in a single sample. J. Am. Chem. Soc. **140**(46), 15904–15915 (2018). https://doi.org/10.1021/jacs.8b09422
22. Meng, C., Knežević, S., Du, F., Guan, Y., Kanoufi, F., Sojic, N., Xu, G.: Recent advances in electrochemiluminescence imaging analysis. eScience **2**(6), 591–605 (2022). https://doi.org/10.1016/j.esci.2022.10.004
23. Kerr, E., Knezevic, S., Francis, P.S., Hogan, C.F., Valenti, G., Paolucci, F., Kanoufi, F., Sojic, N.: Electrochemiluminescence amplification in bead-based assays induced by a freely diffusing Iridium(III) complex. ACS Sens. **8**(2), 933–939 (2023). https://doi.org/10.1021/acssensors.2c02697
24. Myung, N., Ding, Z., Bard, A.J.: Electrogenerated chemiluminescence of CdSe nanocrystals. Nano Lett. **2**(11), 1315–1319 (2002). https://doi.org/10.1021/NL0257824
25. Alemu, Y.A., Difonzo, M., Genovese, D., Paolucci, F., Prodi, L., Rampazzo, E., Valenti, G.: Dye self-organization in doped silica nanoparticles increases the electrochemiluminescence emission in magnetic bead-based assays. Chem. Commun. **60**(83), 11936–11939 (2024). https://doi.org/10.1039/D4CC03488K
26. Fiorani, A., Han, D., Jiang, D., Fang, D., Paolucci, F., Sojic, N., Valenti, G.: Spatially resolved electrochemiluminescence through a chemical lens. Chem. Sci. **11**(38), 10496–10500 (2020). https://doi.org/10.1039/D0SC04210B
27. Valenti, G., Rampazzo, E., Bonacchi, S., Petrizza, L., Marcaccio, M., Montalti, M., Prodi, L., Paolucci, F.: Variable doping induces mechanism swapping in electrogenerated chemiluminescence of $Ru(Bpy)_3{}^{2+}$ core-shell silica nanoparticles. J. Am. Chem. Soc. **138**(49), 15935–15942 (2016). https://doi.org/10.1021/JACS.6B08239
28. Zanut, A., Palomba, F., Rossi Scota, M., Rebeccani, S., Marcaccio, M., Genovese, D., Rampazzo, E., Valenti, G., Paolucci, F., Prodi, L.: Dye-doped silica nanoparticles for enhanced ECL-based immunoassay analytical performance. Angew. Chem. Int. Ed. **59**(49), 21858–21863 (2020). https://doi.org/10.1002/ANIE.202009544
29. Zhu, M.J., Pan, J.B., Wu, Z.Q., Gao, X.Y., Zhao, W., Xia, X.H., Xu, J.J., Chen, H.Y.: Electrogenerated chemiluminescence imaging of electrocatalysis at a single Au-Pt janus nanoparticle. Angew. Chem. Int. Ed. **57**(15), 4010–4014 (2018). https://doi.org/10.1002/ANIE.201800706

30. Hu, X., Yu, S., Wang, C., Zhang, X., Pan, J., Ju, H.: Electrochemiluminescence imaging at a single nanoparticle scale to elucidate diffusion-accelerated charge transfer and monitor cell permeability. Anal. Chem. **95**(9), 4496–4502 (2023). https://doi.org/10.1021/ACS.ANALCHEM.2C05250
31. Liu, Y., Zhang, H., Li, B., Liu, J., Jiang, D., Liu, B., Sojic, N.: Single biomolecule imaging by electrochemiluminescence. J. Am. Chem. Soc. **143**(43), 17910–17914 (2021). https://doi.org/10.1021/JACS.1C06673
32. Han, D., Goudeau, B., Lapeyre, V., Ravaine, V., Jiang, D., Fang, D., Sojic, N.: Enhanced electrochemiluminescence at microgel-functionalized beads. Biosens. Bioelectron. **216** (2022). https://doi.org/10.1016/j.bios.2022.114640
33. Yang, L., De-Jager, C.R., Adsetts, J.R., Chu, K., Liu, K., Zhang, C., Ding, Z.: Analyzing near-infrared electrochemiluminescence of graphene quantum dots in aqueous media. Anal. Chem. **93**(36), 12409–12416 (2021). https://doi.org/10.1021/ACS.ANALCHEM.1C02441
34. Mehta, J., Dilbaghi, N., Kumar Singhal, N., Marrazza, G., Kaushik, A., Kumar, S.: Electrochemiluminescent quantum dots as emerging next generation sensing platforms. Chem. Eng. J. **477**, 146958 (2023). https://doi.org/10.1016/J.CEJ.2023.146958
35. O'Connor, S., Dennany, L., O'Reilly, E.: Evolution of nanomaterial electrochemiluminescence luminophores towards biocompatible materials. Bioelectrochemistry **149**, 108286 (2023). https://doi.org/10.1016/J.BIOELECHEM.2022.108286
36. Wang, L., Wang, Z.-G., Ning, D., Hu, Y., Liu, S.-L., Pang, D.-W.: Real-time monitoring of biomolecular dynamics on cell membranes by quantum dot-based multicolor electrochemiluminescence. Nano Today **50**, 101855 (2023). https://doi.org/10.1016/j.nantod.2023.101855
37. Ding, J., Su, B.: Dual-coreactants enhanced electrochemiluminescence. Chem. Eur. J. **31**(7), e202403804 (2025). https://doi.org/10.1002/CHEM.202403804
38. Fracassa, A., Santo, C.I., Kerr, E., Knežević, S., Hayne, D.J., Francis, P.S., Kanoufi, F., Sojic, N., Paolucci, F., Valenti, G.: Redox-mediated electrochemiluminescence enhancement for bead-based immunoassay. Chem. Sci. **15**(3), 1150–1158 (2024). https://doi.org/10.1039/D3SC06357G
39. Adamson, N.S., Blom, S.J., Doeven, E.H., Connell, T.U., Hadden, C., Knežević, S., Sojic, N., Fracassa, A., Valenti, G., Paolucci, F., Wang, Y., Ding, J., Su, B., Hua, C., Francis, P.S.: Electrochemiluminescence enhanced by a non-emissive dual redox mediator. Angew. Chem. **136**(50), e202412097 (2024). https://doi.org/10.1002/ANGE.202412097
40. Fracassa, A., Mariani, C., Fiorani, A., Einaga, Y., Hogan, C.F., Paolucci, F., Sojic, N., Francis, P.S., Valenti, G.: Overcoming kinetic barriers of remote electrochemiluminescence on boron-doped diamond via catalytic coreactant oxidation. Chem. Commun. (2025). https://doi.org/10.1039/D5CC00270B
41. Wang, Y., Ding, J., Zhou, P., Liu, J., Qiao, Z., Yu, K., Jiang, J., Su, B.: Electrochemiluminescence distance and reactivity of coreactants determine the sensitivity of bead-based immunoassays. Angew. Chem. **135**(16), e202216525 (2023). https://doi.org/10.1002/ANGE.202216525
42. Wang, N., Gao, H., Li, Y., Li, G., Chen, W., Jin, Z., Lei, J., Wei, Q., Ju, H.: Dual intramolecular electron transfer for in situ coreactant-embedded electrochemiluminescence microimaging of membrane protein. Angew. Chem. Int. Ed. **60**(1), 197–201 (2021). https://doi.org/10.1002/ANIE.202011176
43. Ding, H., Guo, W., Su, B.: Imaging cell-matrix adhesions and collective migration of living cells by electrochemiluminescence microscopy. Angew. Chem. Int. Ed. **59**(1), 449–456 (2020). https://doi.org/10.1002/ANIE.201911190
44. Dennany, L., Forster, R.J., Rusling, J.F.: Simultaneous direct electrochemiluminescence and catalytic voltammetry detection of DNA in ultrathin films. J. Am. Chem. Soc. **125**(17), 5213–5218 (2003). https://doi.org/10.1021/JA0296529
45. Chen, Y., Gou, X., Ma, C., Jiang, D., Zhu, J.J.: A synergistic coreactant for single-cell electrochemiluminescence imaging: guanine-rich SsDNA-loaded high-index faceted gold nanoflowers. Anal. Chem. **93**(21), 7682–7689 (2021). https://doi.org/10.1021/ACS.ANALCHEM.1C00602

46. Liu, G., Ma, C., Jin, B.K., Chen, Z., Zhu, J.J.: Direct electrochemiluminescence imaging of a single cell on a chitosan film modified electrode. Anal. Chem. **90**(7), 4801–4806 (2018). https://doi.org/10.1021/ACS.ANALCHEM.8B00194
47. Cui, C., Chen, Y., Jiang, D., Chen, H.Y., Zhang, J., Zhu, J.J.: Steady-state electrochemiluminescence at single semiconductive titanium dioxide nanoparticles for local sensing of single cells. Anal. Chem. **91**(1), 1121–1125 (2019). https://doi.org/10.1021/ACS.ANALCHEM.8B04778
48. Li, S., Lu, Y., Liu, L., Low, S.S., Su, B., Wu, J., Zhu, L., Li, C., Liu, Q.: Fingerprints mapping and biochemical sensing on smartphone by electrochemiluminescence. Sens. Actuators B Chem. **285**, 34–41 (2019). https://doi.org/10.1016/J.SNB.2019.01.035
49. Wang, Y., Ding, J., Zhou, P., Liu, J., Qiao, Z., Yu, K., Jiang, J., Su, B.: Electrochemiluminescence distance and reactivity of coreactants determine the sensitivity of bead-based immunoassays. Angew. Chem. Int. Ed. **62**(16), e202216525 (2023). https://doi.org/10.1002/anie.202216525
50. Valenti, G., Fiorani, A., Li, H., Sojic, N., Paolucci, F.: Essential role of electrode materials in electrochemiluminescence applications. ChemElectroChem **3**(12), 1990–1997 (2016). https://doi.org/10.1002/celc.201600602
51. Han, S., Lee, H.J., Kim, T., Lim, S.Y., Kim, J.: Flexible and dynamic light-guided electrochemiluminescence for spatiotemporal imaging of photoelectrochemical processes on hematite. Anal. Chem. **96**(28), 11146–11154 (2024). https://doi.org/10.1021/acs.analchem.3c05097
52. Chen, Z., Zu, Y.: Electrogenerated chemiluminescence of the tris(2,2′-bipyridine)ruthenium(II)/tri-n-propylamine (TPrA) system: crucial role of the long lifetime of $TPrA^{\cdot+}$ cation radicals suggested by electrode surface effects. J. Phys. Chem. C **112**(42), 16663–16667 (2008). https://doi.org/10.1021/jp802873e
53. Sardesai, N.P., Barron, J.C., Rusling, J.F.: Carbon nanotube microwell array for sensitive electrochemiluminescent detection of cancer biomarker proteins. Anal. Chem. **83**(17), 6698–6703 (2011). https://doi.org/10.1021/ac201292q
54. Zu, Y., Bard, A.J.: Electrogenerated chemiluminescence. 66. The role of direct coreactant oxidation in the ruthenium tris(2,2′)bipyridyl/tripropylamine system and the effect of halide ions on the emission intensity. Anal. Chem. **72**(14), 3223–3232 (2000). https://doi.org/10.1021/ac000199y
55. Kitte, S.A., Wang, C., Li, S., Zholudov, Y., Qi, L., Li, J., Xu, G.: Electrogenerated chemiluminescence of tris(2,2′-bipyridine)ruthenium(II) using N-(3-aminopropyl)diethanolamine as coreactant. Anal. Bioanal. Chem. **408**(25), 7059–7065 (2016). https://doi.org/10.1007/s00216-016-9409-z
56. Aufray, M., Roche, A.A.: Is gold always chemically passive?: Study and comparison of the epoxy-amine/metals interphases. Appl. Surf. Sci. **254**(7), 1936–1941 (2008). https://doi.org/10.1016/j.apsusc.2007.07.192
57. Avouris, P.: Graphene: electronic and photonic properties and devices. Nano Lett. **10**(11), 4285–4294 (2010). https://doi.org/10.1021/nl102824h
58. Miao, W.: Electrogenerated chemiluminescence and its biorelated applications. Chem. Rev. **108**(7), 2506–2553 (2008). https://doi.org/10.1021/cr068083a
59. Yuan, Y., Han, S., Hu, L., Parveen, S., Xu, G.: Coreactants of tris(2,2′-bipyridyl)ruthenium(II) electrogenerated chemiluminescence. Electrochim. Acta **82**, 484–492 (2012). https://doi.org/10.1016/j.electacta.2012.03.156
60. Richter, M.M.: Electrochemiluminescence (ECL). Chem. Rev. **104**(6), 3003–3036 (2004). https://doi.org/10.1021/cr020373d
61. Rebeccani, S., Wetzl, C., Zamolo, V.A., Criado, A., Valenti, G., Paolucci, F., Prato, M.: Electrochemiluminescent immunoassay enhancement driven by carbon nanotubes. Chem. Commun. **57**(76), 9672–9675 (2021). https://doi.org/10.1039/D1CC03457J
62. Sakanoue, K., Fiorani, A., Santo, C.I., Irkham, Valenti, G., Paolucci, F., Einaga, Y.: Boron-doped diamond electrode outperforms the state-of-the-art electrochemiluminescence from microbeads immunoassay. ACS Sens. **7**(4), 1145–1155 (2022). https://doi.org/10.1021/ACSSENSORS.2C00156

63. Zu, Y., Bard, A.J.: Electrogenerated chemiluminescence. 67. Dependence of light emission of the tris(2,2′)bipyridylruthenium(II)/tripropylamine system on electrode surface hydrophobicity. Anal. Chem. **73**(16), 3960–3964 (2001). https://doi.org/10.1021/ac010230b
64. Liu, Y., Guo, W., Su, B.: Recent advances in electrochemiluminescence imaging analysis based on nanomaterials and micro-/nanostructures. Chin. Chem. Lett. **30**(9), 1593–1599 (2019). https://doi.org/10.1016/j.cclet.2019.05.038
65. Heinze, J.: Ultramicroelectrodes in electrochemistry. Angew. Chem. Int. Ed. Engl. **32**(9), 1268–1288 (1993). https://doi.org/10.1002/anie.199312681
66. Chidsey, C.E., Feldman, B.J., Lundgren, C., Murray, R.W.: Micrometer-spaced platinum interdigitated array electrode: fabrication, theory, and initial use. Anal. Chem. **58**(3), 601–607 (1986). https://doi.org/10.1021/ac00294a026
67. Amatore, C., Pebay, C., Servant, L., Sojic, N., Szunerits, S., Thouin, L.: Mapping electrochemiluminescence as generated at double-band microelectrodes by confocal microscopy under steady state. ChemPhysChem **7**(6), 1322–1327 (2006). https://doi.org/10.1002/cphc.200500626
68. Szunerits, S., Tam, J.M., Thouin, L., Amatore, C., Walt, D.R.: Spatially resolved electrochemiluminescence on an array of electrode tips. Anal. Chem. **75**(17), 4382–4388 (2003). https://doi.org/10.1021/ac034370s
69. Sentic, M., Milutinovic, M., Kanoufi, F., Manojlovic, D., Arbault, S., Sojic, N.: Mapping electrogenerated chemiluminescence reactivity in space: mechanistic insight into model systems used in immunoassays. Chem. Sci. **5**(6), 2568–2572 (2014). https://doi.org/10.1039/C4SC00312H
70. Guo, W., Zhou, P., Sun, L., Ding, H., Su, B., Guo, W., Zhou, P., Sun, L., Ding, H., Su, B.: Microtube electrodes for imaging the electrochemiluminescence layer and deciphering the reaction mechanism. Angew. Chem. Int. Ed. **60**(4), 2089–2093 (2021). https://doi.org/10.1002/ANIE.202012340
71. Zhou, P., Hu, S., Guo, W., Su, B.: Deciphering electrochemiluminescence generation from luminol and hydrogen peroxide by imaging light emitting layer. Fundam. Res. **2**(5), 682 (2021). https://doi.org/10.1016/J.FMRE.2021.11.018
72. Ding, J., Zhou, P., Guo, W., Su, B.: Confined electrochemiluminescence generation at ultra-high-density gold microwell electrodes. Front. Chem. **8**, 630246 (2021). https://doi.org/10.3389/FCHEM.2020.630246
73. Anderson, T.J., Defnet, P.A., Zhang, B.: Electrochemiluminescence (ECL)-based electrochemical imaging using a massive array of bipolar ultramicroelectrodes. Anal. Chem. **92**(9), 6748–6755 (2020). https://doi.org/10.1021/ACS.ANALCHEM.0C00921
74. Chow, K.-F., Mavré, F., Crooks, J.A., Chang, B.-Y., Crooks, R.M.: A large-scale, wireless electrochemical bipolar electrode microarray. J. Am. Chem. Soc. **131**(24), 8364–8365 (2009). https://doi.org/10.1021/ja902683f
75. Voci, S., Ismail, A., Pham, P., Yu, J., Maziz, A., Mesnilgrente, F., Reynaud, L., Livache, T., Mailley, P., Buhot, A., Thierry, L., Kuhn, A., Leroy, L., Spinelli, A., Sojic, N.: Wireless enhanced electrochemiluminescence at a bipolar microelectrode in a solid-state micropore. J. Electrochem. Soc. **167**, 137509 (2020). https://doi.org/10.1149/1945-7111/abbbc1
76. Iwama, T., Inoue, K.Y., Abe, H., Matsue, T., Shiku, H.: Bioimaging using bipolar electrochemical microscopy with improved spatial resolution. Analyst **145**(21), 6895–6900 (2020). https://doi.org/10.1039/D0AN00912A
77. Villani, E., Inagi, S.: Mapping the distribution of potential gradient in bipolar electrochemical systems through luminol electrochemiluminescence imaging. Anal. Chem. **93**(23), 8152–8160 (2021). https://doi.org/10.1021/acs.analchem.0c05397
78. Douman, S.F., Collins, D., Cumba, L.R., Beirne, S., Wallace, G.G., Yue, Z., Iwuoha, E.I., Melinato, F., Pellegrin, Y., Forster, R.J.: Wireless electrochemiluminescence at functionalised gold microparticles using 3D titanium electrode arrays. Chem. Commun. **57**(38), 4642–4645 (2021). https://doi.org/10.1039/D1CC01010G
79. Feldberg, S.W., Auerbach, C.: Model for current reversal chronopotentiometry with second-order kinetic complications. Anal. Chem. **36**(3), 505–509 (1964). https://doi.org/10.1021/ac60209a055

80. Laborda, E., López-Asanza, J., Moreno-Molina, J., Gonzalez, J., Molina, A.: Current-potential-time response of the second-order co-reactant mechanism: an analytical theoretical approach. ChemElectroChem **12**(5), e202400633 (2025). https://doi.org/10.1002/celc.202400633
81. Antonietti, P.F., Cangiani, A., Collis, J., Dong, Z., Georgoulis, E.H., Giani, S., Houston, P.: Review of discontinuous galerkin finite element methods for partial differential equations on complicated domains. Lect. Notes Comput. Sci. Eng. **114**, 279–308 (2016). https://doi.org/10.1007/978-3-319-41640-3_9
82. Barrenechea, G.R., Brezzi, F., Cangiani, A., Georgoulis, E.H.: Building bridges: connections and challenges in modern approaches to numerical partial differential equations. **114** (2016). https://doi.org/10.1007/978-3-319-41640-3
83. COMSOL—Software for Multiphysics Simulation. https://www.comsol.com/
84. Amatore, C., Fosset, B., Maness, K.M., Wightman, R.M.: Theory of electrochemical luminescence at double band electrodes. An examination of "steady-state" diffusion at ultramicroelectrodes. Anal. Chem. **65**(17), 2311–2316 (1993). https://doi.org/10.1021/ac00065a024
85. Svir, I.B., Oleinick, A.I., Klimenko, A.V.: "ECL-PACKAGE"—software for electrochemiluminescence simulation at microelectrodes. J. Electroanal. Chem. **513**(2), 119–125 (2001). https://doi.org/10.1016/S0022-0728(01)00608-8
86. Svir, I.B., Oleinick, A.I.: The electrogenerated chemiluminescence kinetics at a microdisc electrode. J. Electroanal. Chem. **499**(1), 30–38 (2001). https://doi.org/10.1016/S0022-0728(00)00446-0
87. Amatore, C., Oleinick, A., Svir, I.: Theory of transient and steady-state ECL generation at double-hemicylinder assemblies using conformal mapping and simulations. Electrochem. Commun. **5**(12), 989–994 (2003). https://doi.org/10.1016/j.elecom.2003.09.011
88. Amatore, C., Oleinick, A., Klymenko, O.V., Thouin, L., Servant, L., Svir, I.: Confocal microscopy imaging of electrochemiluminescence at double band microelectrode assemblies: numerical solution of the inverse optical problem. ChemPhysChem **8**(11), 1664–1676 (2007). https://doi.org/10.1002/cphc.200700216
89. Svir, I., Oleinick, A., Klymenko, O.V., Amatore, C.: Strong and unexpected effects of diffusion rates on the generation of electrochemiluminescence by amine/transition-metal(II) systems. ChemElectroChem **2**(6), 811–818 (2015). https://doi.org/10.1002/celc.201402460
90. Klymenko, O.V., Svir, I., Amatore, C.: A new approach for the simulation of electrochemiluminescence (ECL). ChemPhysChem **14**(10), 2237–2250 (2013). https://doi.org/10.1002/cphc.201300126
91. Amatore, C.: Theoretical trends of diffusion and reaction into tubular nano- and mesoporous structures: general physicochemical and physicomathematical modeling. Chem. Eur. J. **14**(18), 5449–5464 (2008). https://doi.org/10.1002/chem.200701635
92. Guo, W., Zhou, P., Sun, L., Ding, H., Su, B.: Microtube electrodes for imaging the electrochemiluminescence layer and deciphering the reaction mechanism. Angew. Chem. Int. Ed. **60**(4), 2089–2093 (2021). https://doi.org/10.1002/anie.202012340
93. Mariani, C., Fracassa, A., Pastore, P., Bogialli, S., Paolucci, F., Valenti, G., Zanut, A.: Singling out the electrochemiluminescence profile in microelectrode arrays. Chem. Biomed. Imaging (2025). https://doi.org/10.1021/CBMI.5C00022
94. Press, W.H., Teukolsky, S.A., Vetterling, W.T., Flannery, B.P.: Numerical Recipes Third Edition-Numerical Recipes: The Art of Scientific Computing, Third Edition.
95. Danis, A.S., Potts, K.P., Perry, S.C., Mauzeroll, J.: Combined spectroelectrochemical and simulated insights into the electrogenerated chemiluminescence coreactant mechanism. Anal. Chem. **90**(12), 7377–7382 (2018). https://doi.org/10.1021/acs.analchem.8b00773
96. Blackburn, G.F., Shah, H.P., Kenten, J.H., Leland, J., Kamin, R.A., Link, J., Peterman, J., Powell, M.J., Shah, A., Talley, D.B., Tyagi, S.K., Wilkins, E., Wu, T.G., Massey, R.J.: Electrochemiluminescence detection for development of immunoassays and DNA probe assays for clinical diagnostics. Clin. Chem. **37**(9), 1534–1539 (1991). https://doi.org/10.1093/clinchem/37.9.1534

97. Sánchez-Carbayo, M., Mauri, M., Alfayate, R., Miralles, C., Soria, F.: Analytical and clinical evaluation of TSH and thyroid hormones by electrochemiluminescent immunoassays. Clin. Biochem. **32**(6), 395–403 (1999). https://doi.org/10.1016/S0009-9120(99)00032-6
98. Feng, Y., Zhou, W., Wang, X., Zhang, J., Zou, M., Zhang, C., Qi, H.: Imaging and simulation of ruthenium derivative coating microbeads at the opaque electrode with electrogenerated chemiluminescence. Chem. Biomed. Imaging **1**(7), 648–658 (2023). https://doi.org/10.1021/cbmi.3c00042
99. Han, D., Jiang, D., Valenti, G., Paolucci, F., Kanoufi, F., Chaumet, P.C., Fang, D., Sojic, N.: Optics determines the electrochemiluminescence signal of bead-based immunoassays. ACS Sens. **8**(12), 4782–4791 (2023). https://doi.org/10.1021/acssensors.3c01878
100. Rahmani, A., Chaumet, P.C., de Fornel, F., Girard, C.: Field propagator of a dressed junction: fluorescence lifetime calculations in a confined geometry. Phys. Rev. A (Coll. Park) **56**(4), 3245 (1997). https://doi.org/10.1103/PhysRevA.56.3245
101. Kerr, E., Hayne, D.J., Soulsby, L.C., Bawden, J.C., Blom, S.J., Doeven, E.H., Henderson, L.C., Hogan, C.F., Francis, P.S.: A redox-mediator pathway for enhanced multi-colour electrochemiluminescence in aqueous solution. Chem. Sci. **13**(2), 469–477 (2022). https://doi.org/10.1039/d1sc05609c
102. Blom, S.J., Adamson, N.S., Kerr, E., Doeven, E.H., Wenger, O.S., Schaer, R.S., Hayne, D.J., Paolucci, F., Sojic, N., Valenti, G., Francis, P.S.: Redox mediated enhancement and quenching of co-reactant electrochemiluminescence by iridium(III) complexes. Electrochim. Acta **484**, 143957 (2024). https://doi.org/10.1016/j.electacta.2024.143957
103. Knežević, S., Kerr, E., Valenti, G., Paolucci, F., Francis, P.S., Hogan, C.F., Sojic, N., Kanoufi, F.: Electrocatalytic amplification of coreactant electrochemiluminescence using redox mediators. Electrochim. Acta **499**, 144677 (2024). https://doi.org/10.1016/j.electacta.2024.144677
104. Adamson, N.S., Blom, S.J., Doeven, E.H., Connell, T.U., Hadden, C., Knežević, S., Sojic, N., Fracassa, A., Valenti, G., Paolucci, F., Wang, Y., Ding, J., Su, B., Hua, C., Francis, P.S.: Electrochemiluminescence enhanced by a non-emissive dual redox mediator. Angew. Chem. Int. Ed. **136**(50), e202412097 (2024). https://doi.org/10.1002/anie.202412097
105. Fracassa, A., Mariani, C., Fiorani, A., Einaga, Y., Hogan, C.F., Paolucci, F., Sojic, N., Francis, P.S., Valenti, G.: Overcoming kinetic barriers of remote electrochemiluminescence on boron-doped diamond via catalytic coreactant oxidation. Chem. Commun. **61**(19), 3900–3903 (2025). https://doi.org/10.1039/d5cc00270b
106. Yang, X., Hang, J., Qu, W., Wang, Y., Wang, L., Zhou, P., Ding, H., Su, B., Lei, J., Guo, W., Dai, Z.: Gold microbeads enabled proximity electrochemiluminescence for highly sensitive and size-encoded multiplex immunoassays. J. Am. Chem. Soc. **145**(29), 16026–16036 (2023). https://doi.org/10.1021/jacs.3c04250
107. Mathwig, K.: AI-driven simulation of stochastic electrochemiluminescence. ACS Electrochem. **1**(3), 280–285 (2024). https://doi.org/10.1021/ACSELECTROCHEM.4C00166

Super-Resolution Electrochemiluminescence Microscopy

Wenxin Zhu, Pinlong Zhao, and Jiandong Feng

Abstract Super-resolution microscopy has played an indispensable role in observing nanoscale structures, advancing our understanding of the molecular interaction mechanism. Mainstream super-resolution microscopy is based on the photoluminescence to generate detectable photons. The use of laser beams for excitation may introduce a series of laser effects such as high background and photon bleaching. On the other hand, electrochemiluminescence (ECL) is excited by the electrochemical reactions, which can eliminate the use of laser beams. Therefore, it is beneficial for applications that are restricted by the laser effects of fluorescence. Compared to the super-resolution fluorescence microscopy, which has achieved great success, super-resolution ECL microscopy is still in its early development stages. Based on the principles of non-radiative relaxation processes that generate luminescence, super-resolution ECL imaging can adopt reconstruction principles from super-resolution fluorescence methods, such as single-molecule localization microscopy, signal fluctuation-based super-resolution fluorescence microscopy, and deep learning-based super-resolution microscopy. Here we review the basic principles, performance, and applications of super-resolution ECL imaging based on these methods. By integrating the principles of super-resolution microscopy with ECL imaging, current super-resolution ECL microscopy has greatly improved the spatial and temporal resolution of conventional ECL imaging. These advancements have revolutionized ECL's applications in single-cell imaging, immunoassays and the characterization of active sites in chemical reactions. We think that with the future development of higher-efficiency ECL imaging systems, the performance of super-resolution ECL imaging can be further improved, pushing imaging resolution from the nanoscale to the atomic-scale.

The original version of the chapter has been revised. A correction to this chapter can be found at https://doi.org/10.1007/978-3-032-14602-1_15

W. Zhu · J. Feng (✉)
Laboratory of Experimental Physical Biology, Department of Chemistry, Zhejiang University, Hangzhou, China
e-mail: jiandong.feng@zju.edu.cn

P. Zhao
School of Cyberspace, Hangzhou Dianzi University, Hangzhou, China

F. Paolucci et al. (eds.), *Electrogenerated Chemiluminescence Microscopy and Imaging*, https://doi.org/10.1007/978-3-032-14602-1_6

Keywords Electrochemiluminescence · Super-resolution · Deep learning

1 Introduction

Since optical microscopy started to unveil the microscopic world, the ability to resolve material and biological structures has undergone three major advances: (1) the early stage optical microscopy enabled microscale visualization of imaging targets, (2) probe-enhanced fluorescence microscopy greatly improved imaging contrast and achieved nanoscale visualization of specific molecules [1, 2], and (3) super-resolution fluorescence techniques are now visualizing dynamic molecular interactions down to the molecular, and even atomic-level [3–5]. Reference number 5 is missing. Each advancement in resolution represents not only a technological breakthrough but also a fundamental redefinition of our understanding of the structures and the molecular interactions within matter and life. Consequently, fluorescence microscopy has become the most widely used imaging technique in biology. However, since fluorescence microscopy relies on laser excitation, it introduces a series of laser effects such as high background noise, autofluorescence, photodamage, and photobleaching. These effects limit the further application of fluorescence imaging in certain areas such as single-molecule detection at ultralow concentrations. To meet these challenges, it is necessary to develop alternative optical imaging techniques that bypass the limitations of fluorescence.

Among the various optical imaging techniques, electrochemiluminescence (ECL) incorporates electrochemical reactions with optical output owing to its unique luminescence mechanism [6, 7]. Since it does not rely on external excitation lasers, the laser effects can be effectively avoided. Thus, ECL has played an indispensable role from bioanalytics such as immunoassay, single-cell analysis [8, 9] to chemical analysis such as tracking of reaction active sites [10] with advantages of zero-background, high signal to noise ratio (SNR) and spatiotemporal controllability. However, unlike fluorescence super-resolution imaging which has achieved atomic-scale resolution [11, 12], ECL super-resolution imaging is still in the early stages of development. This delayed progress primarily stems from two reasons. First, the complex multi-step coupled mechanisms involved in ECL reactions make them difficult to control and leverage [13]. Second, the low photon emission efficiency in ECL is hard to meet the photon number requirements of mainstem super-resolution imaging methods. Besides, chemical diffusion during ECL reactions also restricts further improvements in spatial resolution. Thus, the spatial resolution of conventional ECL is primarily limited to the microscale by accumulating tens of seconds, which makes it difficult to reveal structures at the nanoscale. There is an urgent need to develop ECL-based super-resolution imaging methods for visualizing microscopic information.

Although ECL and fluorescence differ in their excitation mechanisms, both the generation of luminescent signals relies on the relaxation of excited states, with each emission process exhibiting flexible spatiotemporal tunability. Consequently, the success of fluorescence super-resolution imaging provides valuable insights for the development of ECL super-resolution imaging.

Existing fluorescence super-resolution imaging techniques can be fundamentally categorized into four main classes according to their mechanisms: (1) illumination-pattern-modulated approaches, e.g., structured illumination microscopy (SIM) [14] and stimulated emission depletion (STED) [15], which break the diffraction limit through engineered excitation light. (2) Single-molecule localization microscopy (SMLM), represented by photoactivated localization microscopy (PALM) [16], stochastic optical reconstruction microscopy (STORM) [17] and DNA-based point accumulation for imaging in nanoscale topography (DNA-PAINT) [18]. These approaches exploit stochastic photoactivation and localization of individual fluorophores to achieve nanoscale resolution. (3) Fluctuation-based super-resolution reconstruction methods, such as super-resolution optical fluctuation imaging (SOFI) [19] and super-resolution radial fluctuations (SRRF) [20]. These methods realize super-solution by analyzing spatiotemporal correlations of individual molecules. In addition to the above physical feature-based super-resolution methods, (4) computational methods that integrate deep learning and artificial intelligence (AI) with super-resolution imaging have been proposed as the fourth framework. Neural networks such as generative adversarial networks (GANs) and convolutional neural networks (CNNs) are trained to reconstruct high-resolution images from diffraction-limited inputs that effectively reduce acquisition and processing times [21, 22]. Even with different basic principles, each of these four approaches effectively addresses the limitations of spatial resolution in conventional fluorescence microscopy.

Considering the basic principles in fluorescence super-resolution methods, there is the possibility to realize ECL super-resolution by adapting some of the above strategies. Firstly, photon emission in ECL can also be spatiotemporally isolated for sparse acquisition and localization, except that it is controlled by electrochemical reactions. By leveraging this controllability of ECL reactions, single-emitter localization and super-resolution imaging could be realized [23]. Besides, taking advantages of the stochasticity in electrochemical reactions, the photon emission rate and intensity fluctuate with time, which exactly fits the basic requirements not only SMLM but also SOFI and SRRF. By exploiting the ECL intensity fluctuations and stochastic blinking of molecules, SOFI-like correlation analysis of ECL signals has the potential to resolve nanoscale structures. Finally, deep learning approaches also hold great potential for ECL super-resolution since they are mainly dependent on the quality of input images rather than the optical designs and underlying luminescent mechanisms [24].

The realization of super-resolution ECL would revolutionize its established applications, including single-molecule detection [25], single-cell imaging [26], immunoassays [27], and characterization of chemically active sites [28]. By improving the spatial resolution and the sensitivity of ECL, we can gain deeper insights into molecule reaction mechanisms, leading to remarkable advances in biochemistry, clinical diagnostics, and materials science. This chapter offers a structured overview of super-resolution ECL imaging and deep-learning enhanced ECL imaging, including the fundamental principles, performances, and applications. We believe that it will help readers navigate both the fundamentals and future potential of super-resolution ECL and deep learning-enhanced ECL microscopy.

2 Principles of Super-Resolution Imaging

2.1 Principles of Mainstream Super-Resolution Fluorescence Imaging

Driven by the pursuit of deeper exploration of the microscopic world, optical microscopy has become an indispensable tool in scientific research, enabling the visualization of microscopic structures. Over the past decades, optical microscopy has undergone dramatic developments, from conventional microscale imaging to breakthrough nanoscale super-resolution fluorescence imaging and cutting-edge computational microscopy imaging. Especially after the 2014 Nobel Prize in Chemistry, super-resolution fluorescence microscopy has empowered us to visualize material structures with molecular-scale resolution. Currently, mature super-resolution fluorescence microscopy can be categorized into four major techniques according to their basic principles.

The first is based on the modulation of illumination patterns to realize super-resolution, with STED [29] and SIM [30] being the representative methods. There are two laser beams used in STED, including an excitation beam and a depletion beam with zero center intensity. The excitation beam is used for stimulating all probe molecules in the field of view. And the depletion beam is used for depleting the fluorescence in the surrounding area of the center light (Fig. 1a) [31]. The integration of these two laser beams realizes the compression of the point spread function (PSF), thus improving the spatial resolution. By spatially scanning the sample with these two lasers, STED achieves resolution beyond the diffraction limit. According to its principle, the spatial resolution is proportional to the intensity of the laser beam, which means that the resolution improvement is tunable. However, such scanning mode results in low temporal resolution of STED, which is not suitable for real-time, ensemble observation. Different with fluorescence spot scanning mode in STED, another illumination pattern modulation method–SIM is a wide-field imaging mode with high temporal resolution. SIM excites samples with structured patterns to obtain detectable low-frequency signals. Multiple coupled images are acquired by shifting and rotating the illumination patterns. A super-resolved image can be obtained by decoupling the acquired multiple frequency information, effectively doubling the spatial resolution of conventional microscopy (Fig. 1b) [31]. Based on this principle, SIM has lower illumination intensity requirements than STED, which makes it suitable for live cell imaging. In addition, the temporal resolution of SIM is relatively high as it requires only nine frames to reconstruct an image, thus it has wide application in observing dynamic processes in living cells [30].

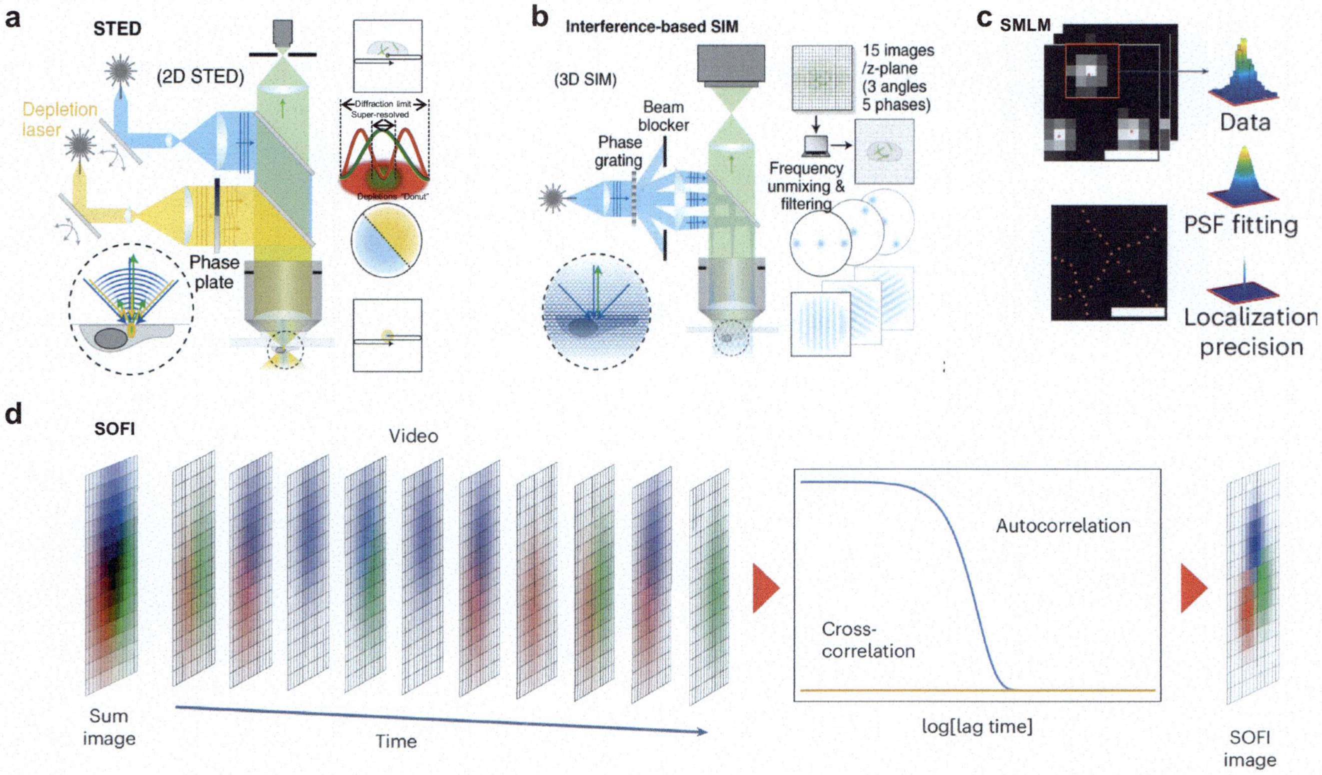

Fig. 1 Principles of the mainstream super-resolution fluorescence microscopy. **a** Illustration of the imaging setup and principle of STED [31]. **b** The generation of the excitation pattern and principle for SIM reconstruction [31]. (**a**, **b** are adapted with permission from Ref. [31]). Copyright © 2019, Springer Nature. **c** Principle of SMLM [32] Copyright © 2025, Springer Nature. **d** Image sequence acquisition and correlation analysis in SOFI [33]. Copyright © 2025, Springer Nature

The second is the SMLM techniques represented by PALM [16] and STORM [17], along with their derivative methods. SMLM mainly relies on the sparse activation and precise localization of individual molecules (Fig. 1c) [32]. The position of a single-molecule is calculated by Gaussian fitting, where the localization precision depends on the photon number. High-precision localization requires that there is no overlap between each molecule. Therefore, photo-switchable fluorescent probes are adopted in these two methods. The difference is that PALM uses a light-activated fluorescent protein as its probe but STORM adopts a light-switching fluorescent dye. PALM requires one laser beam to pre-activate the probe molecules and another beam to continuously irradiate the activated molecules to emit fluorescence. STORM controls the ON/OFF state of the probe molecules by alternating two laser beams. With these methods, molecules are sparsely activated in each imaging cycle. Through thousands of acquisitions, the center of each fluorophore is mapped. Then, a super-resolved image can be reconstructed by accumulating all localized positions. Based on this principle, SMLM can realize nanoscale localization precision, and the localization results can be used for single-molecule quantitative analysis. However, this method requires prolonged imaging times due to the multiple acquisitions, which is also not suitable for observing dynamic processes.

The third is fluctuation-based super-resolution imaging such as SOFI [19] and SRRF [20]. These two methods exploit stochastic blinking and intensity fluctuations of individual fluorescent probes over time. By applying temporal correlation analysis to pixel intensity fluctuations over thousands of image frames, spatial correlations of each molecule can be extracted (Fig. 1d) [33]. These correlations reveal sub-diffraction structural details, achieving nearly a twofold improvement in spatial resolution over conventional microscopy. This method is suitable for most fluorophores that have blinking or intensity fluctuation characteristics, and it does not need a complex optical design. Thus, it has wide applications in biology and chemical science.

The last is the deep learning-based super-resolution technique. CNN has been well-established in its application in image segmentation and recognition inrecent years. With high learning ability, CNN has great potential in the field of super-resolution imaging. By inputting raw data and ideal super-resolution imaging results into the CNN model and setting different evaluation indices, the model can continuously iterate to optimize various parameters for obtaining ideal imaging results. According to the different training data sources, the application of deep learning in super-resolution imaging includes accelerating reconstruction and extending imaging dimensions from two-dimensions to three-dimensions, which has all been demonstrated. The development of deep learning reduces the requirements of prior knowledge and programming ability of users while maintaining stability and robustness. Nevertheless, there also exist some limitations such as the need for a large amount of high-precision training datasets, which is a tedious task that greatly affects the effectiveness of the model. With the development of unsupervised deep learning, there are also methods based on a deep understanding of the luminescence mechanism and physical properties of the molecules that require smaller datasets as input.

To summarize, each of these four main super-resolution techniques has its own advantages and disadvantages concerning imaging spatial resolution, temporal resolution, and photodamage to samples. SMLM offers high spatial resolution but has a slower imaging speed and requires specific fluorescent molecules with blinking properties. SIM and other fluctuation-based methods, on the other hand, provide high temporal resolution but achieve limited improvements in spatial resolution. STED demonstrates a good balance between spatial and temporal resolution. However, it tends to cause severe photodamage to samples. Deep learning-enabled super-resolution can overcome the above limitations but it requires large datasets of high-quality inputs. Considering these advantages and disadvantages, the choice of these techniques depends on the requirements of the imaging application, balancing the trade-offs between spatiotemporal resolution and potential sample damage.

2.2 Luminescence Strategies Comparable to Fluorescence

Super-resolution fluorescence techniques have revolutionized our capability to visualize the nanoscale structures. They are still restricted by the laser effect and there is an urgent need to develop new imaging methods to provide an alternative to fluorescence. In principle, photoluminescence, chemiluminescence [34], and ECL are all categorized as cold-light phenomena, as their radiative processes occur without heat generation. Their respective luminescence is all formed by a non-radiative relaxation process, from the excited state to the ground state. In photoluminescence, photoemission originates from photoexcitation, and the material absorbs the excited photons to promote electrons to a higher energy state. Following the non-radiative relaxation process, the excited electrons return to the ground state and release energy through photon emission (Fig. 2a). This laser excitation–relaxation–emission process distinguishes photoluminescence as a photon-driven phenomenon. In contrast, chemiluminescence and ECL avoid photon excitation totally. Both rely on high-energy electron transfer reactions to generate excited states. In chemiluminescence, intermediates generated by redox-based chemical reactions produce higher energy states that emit photons directly following the radiative relaxation (Fig. 2b). Notably, ECL is a specialized chemiluminescence where the redox reactions are electrochemically triggered at the electrode surface (Fig. 2c). This introduces spatiotemporal control through applied voltages, enabling precise modulation of the intensity and duration of the luminescence. Based on their basic luminescence principles, these two other modalities hold promise for replacing flourescence imaging due to their zero background, high-sensitivity imaging. Compared with chemiluminescence, ECL is confined to the electrode, which reduces probe diffusion and allows high spatial resolution. Meanwhile, it is easy to control the luminescence with the electrode, distinguishing its great controllability, which is beneficial for super-resolution imaging. However, chemiluminescence emission follows the intrinsic kinetics of chemical reactions, offering low spatiotemporal controllability. Therefore, we focus here on ECL for super-resolution imaging.

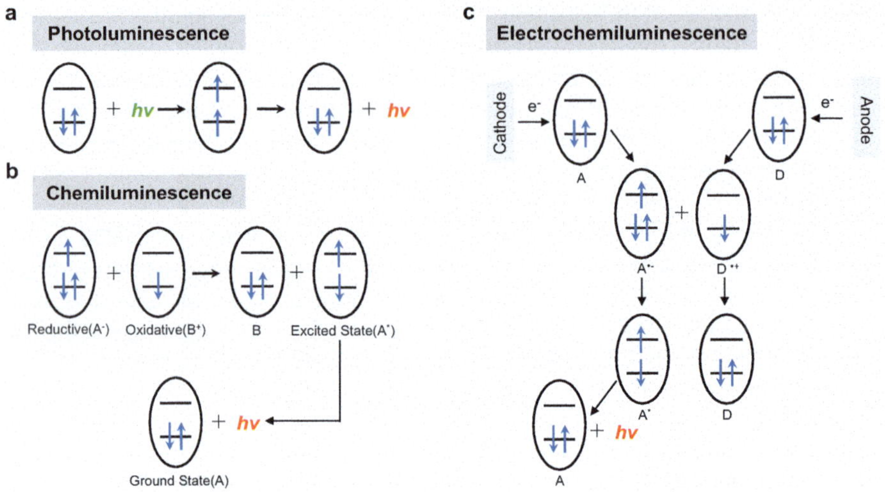

Fig. 2 General principles of photoluminescence, chemiluminescence, and electrochemiluminescence

2.3 *Advantages and Limitations of ECL Imaging*

Owing to the unique mechanisms of ECL imaging, it exhibits several advantages compared to other optical techniques, (1) High sensitivity. ECL demonstrates high sensitivity compared to photoluminescence due to its zero-background. The electrochemical reactions confine ECL within the diffusion layer, where the concentration of luminescent species is amplified. This spatial restriction minimizes non-specific signals and allows detection limits down to sub-femtomolar levels. (2) High SNR. ECL requires no external excitation. The absence of excitation lasers reduces optical noise, thereby improving the imaging SNR. (3) Spatiotemporal controllability. The emission rate and spatial distribution of ECL are strictly controlled by the applied voltage due to its unique excitation mechanism, which enables high flexibility in various applications. (4) Reasonable resolution. The spatial resolution of conventional ECL imaging is constricted at the microscale, as the luminescence is confined to the electrode surface with microscale diffusion distances of short-lived intermediates. With the advantages of spatiotemporal controllability, more precise spatial control of the luminescence position can also be achieved by patterning the electrode. The temporal resolution of ECL imaging is relatively low due to the low photon emission efficiency of ECL. Both the spatial and temporal resolution of ECL imaging need to be further improved to fit the nano-scale structure observation. In addition to the above features, ECL is also characterized by its simplicity in instrumentation and applicability to a broad range of analytes. With its unique voltage excitation mechanism and distinct advantages, ECL has great potential to provide an imaging method that is comparable to fluorescence imaging.

Despite the above advantages and wide application in analytical chemistry, unlike fluorescence microscopy that now even purchasing the super-duper resolution [4, 5], the spatial resolution of traditional ECL imaging is still limited by the optical diffraction until 2021 [35]. The main limitation is the low luminescence efficiency of ECL. Although it is characterized by zero background and high SNR, once ECL reaction emits only one photon, resulting in low photon utilization in many super-resolution techniques [23]. Taking example of SMLM, localization uncertainty is calculated as $\sigma/\sqrt{N}$, where σ is the diffraction limit and N represents photon counts emitted from a single molecule. The localization uncertainty is equal to the diffraction limit in single-molecule ECL imaging as calculated by this formula. In conventional ECL imaging, it usually extends the exposure time by tens of seconds to accumulate photon emissions for improving image contrast. While longer exposure times can improve the SNR by capturing more photons, they may also lead to molecular clustering. Thus, optimizing exposure time and sampling strategies is essential for super-resolution ECL imaging. Another strategy to improve the photon emission focuses on enhancing the quantum efficiency of ECL probes [36]. Recent advancement in developing ruthenium tris-bipyridine derivatives, nanomaterials such as quantum dots [37] and metal–organic frameworks [38] have greatly promoted the ECL emission, demonstrating higher radiative transition probabilities compared to conventional ruthenium complexes. These efforts hold promise for advancing super-resolution ECL imaging. However, critical material limitations persist. Many of these probes exhibit poor stability under long-time imaging, which may affect localization accuracy. Improving the ECL emission and effectively utilizing these emitted photons remains a major challenge in the field of super-resolution ECL imaging.

2.4 *Feasibility of Super-Resolution ECL Imaging*

Considering its photon emission and utilization principles, some super-resolved fluorescence imaging methods can be extended to ECL for super-resolution imaging. The integration of super-resolution fluorescence principles with ECL emission may reveal nanoscale structures in scenarios where fluorescence is limited by photobleaching or background excitation. Firstly, inspired by SMLM, it is possible to isolate and localize individual molecules by exploiting the stochastic blinking feature of the ECL reactions. Additionally, principles of SOFI and SRRF are also suitable for ECL systems. The blinking and intensity fluctuations of the ECL reaction fit well with these requirements. Finally, the achievements of deep learning in the field of super-resolution fluorescence imaging [39] are also expected to promote super-resolution ECL imaging. Especially, CNNs are effective in reconstructing high-resolution images from low-resolution inputs while significantly reducing the imaging time.

Some existing cases have validated the feasibility of the above-mentioned approaches. Single-molecule localization-based ECL microscopy has achieved a spatial resolution of up to 22 nm using $Ru(bpy)_3^{2+}$ emitters. It can also achieve super-resolved cell adhesion observations with a resolution of 150 nm [35]. SRRF-ECL has demonstrated a spatial resolvability as low as 100 nm for electrocatalytic active sites [40]. Deep learning enhanced ECL microscopy has improved the temporal resolution of conventional ECL imaging by nearly 100-fold [41]. Advanced photon harvesting and utilizing strategies have established ECL as a unique super-resolution imaging technique. Based on the principles for improving spatiotemporal resolution, we classify the above ECL imaging enhancement strategies into two main types: physical feature-based and computational-based ECL imaging. Specific principles and examples are presented in the following subsections.

3 Physical Feature-Based Super-Resolution ECL Imaging

The unique excitation mechanism of ECL makes it challenging to implement super-resolution techniques such as SIM or STED, which rely on precise modulation of the excitation light. ECL emission occurs near the electrode and greatly relies on the random electrochemical reactions, making it difficult to design specific excitation patterns. On the other hand, the principles of SMLM or fluctuation-based super-resolution imaging methods such as SOFI or SRRF provide a viable way to achieve ECL super-resolution.

3.1 Principles and Applications of SMLM

3.1.1 Principles and Applications of PALM

The basic principle of PALM has been discussed in the subsection 1. Betzig's group first proposed PALM in 2006, which used the photo-switching properties of photoactivatable green fluorescent proteins (PA-GFP) [16]. The PA-GFP does not emit fluorescence until it is activated. It can be sparsely activated by a 405 nm laser beam with low-energy pulses, and then switched to a 561 nm laser beam for continuous excitation. At this moment, only PA-GFP pre-activated with a 405 nm laser was distinguished as the ON state, while the other PA-GFP remained in the OFF state of non-emission. The positions of sparse activated single molecules are then acquired and localized until these single molecules are bleached. The next cycle repeats this activation–excitation–localization–bleaching process until all the PA-GFP were localized. A super-resolution image can be reconstructed by accumulating all the localizations (Fig. 3a) [42]. With this method, Betzig et al. imaged proteins in lysosomes and mitochondria, revealing their membrane widths to be about 10 nm and 20 nm, respectively. They also imaged proteins at the whole cell level, observing vinculin at focal adhesions, actin, and the retroviral protein at the plasma membrane with high resolution [16]. Moreover, Shroff et al. developed a live cell super-resolution imaging technique based on PALM, achieving the dynamic observation of tdEos-paxillin in a

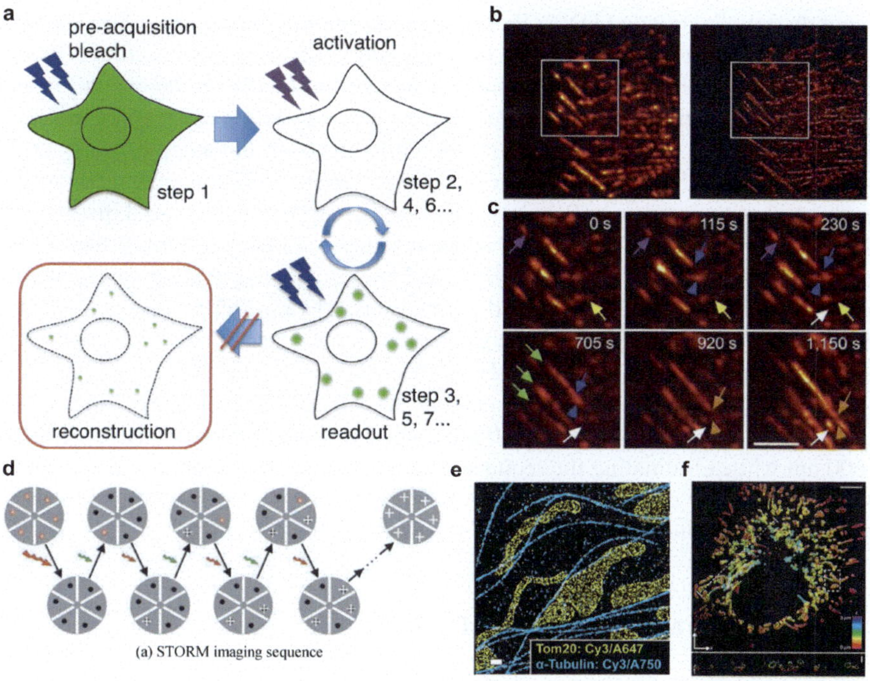

Fig. 3 Principles of PALM and STORM.**a** Principle illustration of PLAM [42]. Copyright © 2011, Wiley Periodicals, Inc. **b** Summed traditional TIRF (left) and PALM (right) images of the tdEos-paxillin in a CHO cell [43]. **c** Enlarged views of the white box marked PALM image in different times in '**b**' [43]. (**b**, **c** are adapted with permission from Ref. [43]). Copyright © 2008, Springer Nature. **d** Typical image sequence of a single Cy5 probe in STORM [17]. Copyright © 2006, Springer Nature. **e** Dual-color STORM imaging of microtubules (blue) and mitochondria (yellow) in a BS-C-1 cell [44]. Copyright © 2012 WILEY-VCH Verlag GmbH & Co. KGaA, Weinheim. **f** 3D STORM reconstruction of the mitochondrial network in a BS-C-1 cell [46]. Copyright © 2008, Springer Nature

CHO cell (Fig. 3b, c) [43]. These investigations confirmed that PALM is a powerful imaging technique that can reconstructed finer structures with nanoscale resolution in both organelles and the extracellular cell membrane.

3.1.2 Principles and Applications of STORM

Zhuang's group developed STORM in 2006 [17], which is similar in principle to PALM, except that it adopts different probe molecules (Fig. 3d). A molecule pair Cy3/Cy5 is typically used in STORM. Two laser beams with wavelengths of 633 nm and 532 nm are used to control the ON/OFF state of the Cy5 molecules. The Cy3 is used to accelerate the switching of Cy5 from the OFF to the ON state. With this method, Zhuang et al. achieve the visualization of two separate molecules with

46 nm on double-stranded DNA. Based on the excitation mechanism of STORM, it is possible to realize multi-color STORM imaging. The prerequisite is that different photo-switchable fluorescent probes are clearly differentiated in the excitation and emission wavelengths. And multi-color imaging can be achieved by using different fluorescent probes to label various target components of the sample. Bates et al. applied multicolor STORM to achieve dual-target super-resolution imaging of microtubules and mitochondria in BS-C-1 cells, clearly mapping their spatial distribution as shown in Fig. 3e [44]. Beyond 2D super-resolution, Zhuang's team further developed 3D STORM in 2008 [45]. A cylindrical lens is adopted in the imaging path to distinguish focal planes in the x- and y-directions. Then, different elliptical diffuse spots can be obtained at varying heights of the focal plane. The z-axis position is determined by Gaussian fitting to obtain the spot width, which is then correlated to a pre-calibrated curve. With this method, 3D STORM achieves precision with ~ 20–30 nm lateral resolution, ~ 50–60 nm axial resolution, and an imaging depth of ~ 600 nm while eliminating the requirements for mechanical scanning. Huang et al. reconstruct the 3D distribution of mitochondria in a BS-C-1 cell with this method, revealing the excellent performance of 3D-STORM (Fig. 3f) [46].

3.1.3 Principles and Applications of DNA-PAINT

As a derivative method of SMLM, DNA-PAINT was first proposed by Jungmann et al. in 2014 [18]. It uses DNA strand hybridization to generate molecule blinking in SMLM. In detail, one of the double helical DNA strands is labeled to the target molecule while the other is labeled with a fluorescent probe that diffuses in the solution. Since the DNA strands are highly specific to each other, these two DNA strands will spontaneously bind in solution, emitting fluorescence in the focal plane. By adjusting the binding strength and concentration of the imaging strands, transient binding can be generated repeatedly to obtain optimal blinking rates like those of the fluorescent dyes blinking in PALM and STORM. The final super-resolution image can be reconstructed by accumulating the obtained single-molecule localizations. Leveraging the unique binding mechanism of DNA-PAINT, it is easy to realize multiplexed super-resolution imaging by designing different pairs of double-helical DNA strands. Due to its high specificity, a high localization precision of 10 nm is achieved by Jungmann et al. (Fig. 4a–c) [18]. Moreover, they also introduce Exchange-PAINT for multiplexed imaging with only one dye, demonstrating ten-color in vitro super-resolution imaging and 3D super-resolution imaging in fixed cells (Fig. 4d–f) [18].

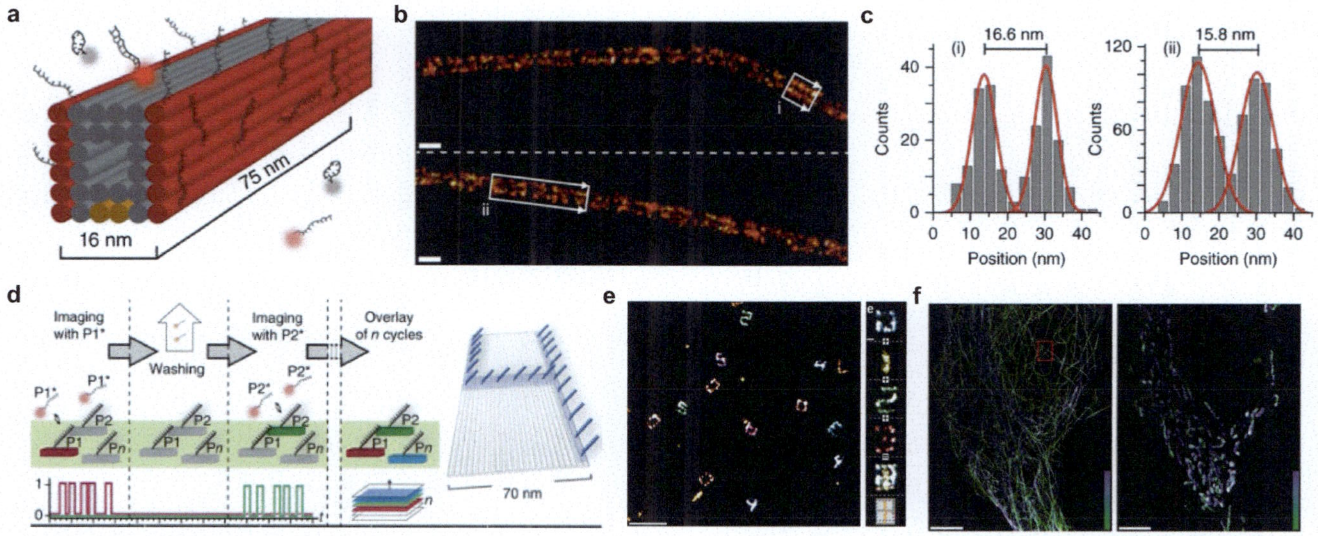

Fig. 4 Principle and application of DNA-PAINT [18]. **a** Illustration of DNA origami with a label space of 16 nm imaged by DNA-PAINT. **b** Corresponding super-resolved image of "**a**". **c** Cross-section histogram of white box in "**b**". **d** Principles of Exchange-PAINT. **e** Multiplexed super-resolution imaging result of ten Exchange-PAINT cycles. **f** 3D super-resolution imaging of microtubules and mitochondria. (**a**–**f** are all adapted with permission from Ref. [18]). Copyright © 2014, Springer Nature

3.1.4 Principles and Applications of MINFLUX

Despite the above three methods having achieved high spatial resolution as low as several nanometers. They still face critical limitations due to the reliance on time-consuming stochastic single-molecule activation, resulting in low temporal resolution. As a result, their applications in biology are greatly restricted to fixed cells or slow biological processes. Over the past decade, techniques like minimal photon fluxes (MINFLUX) have been developed to improve temporal resolution. Hell's group first introduced MINFLUX in 2017 [3], achieving a spatial resolution of at least 1 nm. Similar to STED, MINFLUX uses a donut-shaped laser beam for excitation rather than depletion. When the center of the laser beam aligns with a target molecule, no fluorescence is detected. However, any displacement between the center of the beam and the molecule will induce detectable fluorescence. The fluorescence intensity is directly related to the positional mismatch. By analyzing fluorescence intensity in relation to the beam position, the location of the molecule can be determined with nanometer precision (Fig. 5a). The development of MINSTED helps to reveal the structure of the nuclear pore, which distinguishes a better localization accuracy than STORM (Fig. 5b, c) [32]. Thanks to its high spatiotemporal resolution, Eilers et al. demonstrated rapid single-molecule tracking in 2018, achieving a localization precision of 2.4 nm and a temporal resolution of 400 μs [47].

3.2 SMLM-Inspired Super-Resolution ECL Imaging

The achievements of SMLM have also inspired ECL microscopy to break the diffraction limit. Instead of laser excitation, electrochemical excitation can be used to obtain stochastic blinking single-molecule luminescence signals since ECL reactions occur randomly and in isolation. Sparse activation can be realized by adjusting the reactant concentration and sampling frequency. Based on the similarities in principle with SMLM in fluorescence, the first super-resolution ECL microscopy was reported by Dong et al. in 2021 [35]. Taking advantage of the stochastic nature of ECL reactions, the photon emission of individual molecules is isolated both in time and space, with each reaction emitting a single photon. The isolated single-photon emissions are acquired through spatiotemporal acquisition and localized by single-molecule localization similar to fluorescence imaging. Although the localization precision of a single photon event is relatively low, enhanced localization accuracy can be obtained by spatiotemporally accumulating photon emissions from multiple ECL reactions occurring at the same site. Therefore, the first electrochemical reaction-driven super-resolution microscopy has been achieved by accumulating thousands of single-molecule localization results, resulting in an improved resolution of nearly 10 times that of conventional ECL imaging (Fig. 6a–c) [35]. With enhanced spatial resolution, Dong et al. also achieved the first super-resolution cell adhesion imaging, which pushes super-resolution ECL imaging into the biological regime (Fig. 6d, e) [35].

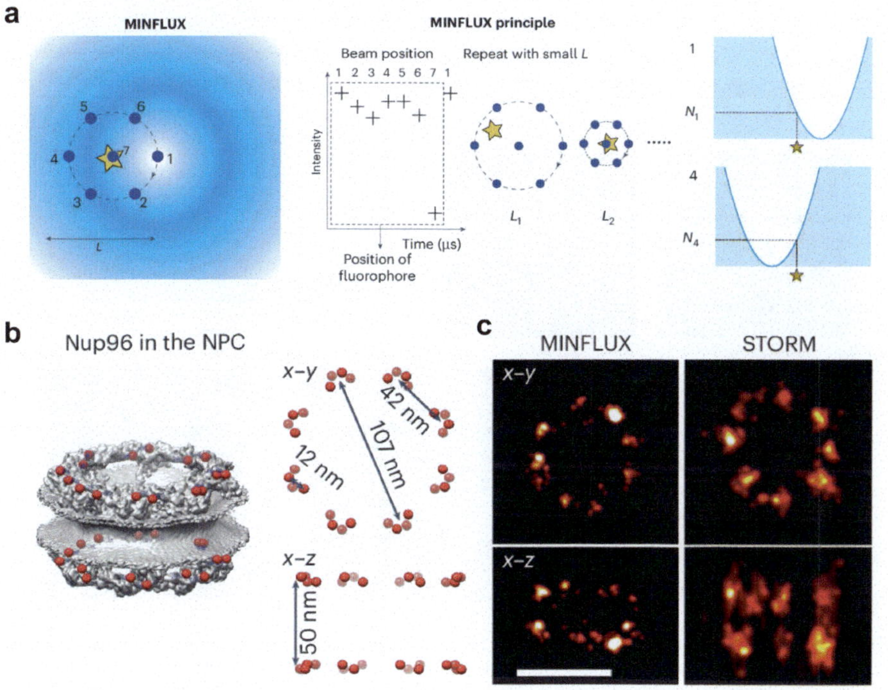

Fig. 5 Principles and performance of MINFLUX [32]. **a** Illustration of super-resolution principles of MINFLUX. **b** Detailed structures of nuclear pores. **c** Comparison of the reconstructed super-resolved images of nuclear pores obtained by MINFLUX and STORM. (**a**–**c** are all adapted with permission from Ref. [32]). Copyright © 2025, Springer Nature

3.3 *Fluctuation-Based Super-Resolution Fluorescence Imaging*

Although the development of SMLM-based ECL imaging has achieved spatial resolutions from 22 to 150 nm and open up applications in both biology and chemistry. It still faces the limitations of low temporal resolution due to the requirements for time-consuming stochastic single-molecule blinking acquisition. Meanwhile, overcoming the challenge of low temporal resolution imaging has driven the development of other super-resolution fluorescence methods. Among these, fluctuation-based super-resolution imaging such as SOFI and SRRF, greatly improves temporal resolution with reduced image sequences. These methods do not require specialized optical and probe designs and only rely on the characteristics of fluorescence signal fluctuations. Therefore, such methods not only feature fast imaging speeds but also have low phototoxicity, which is advantageous for applications in subcellular-scale live-cell imaging. The principles and applications of SOFI and SRRF will be analyzed in detail.

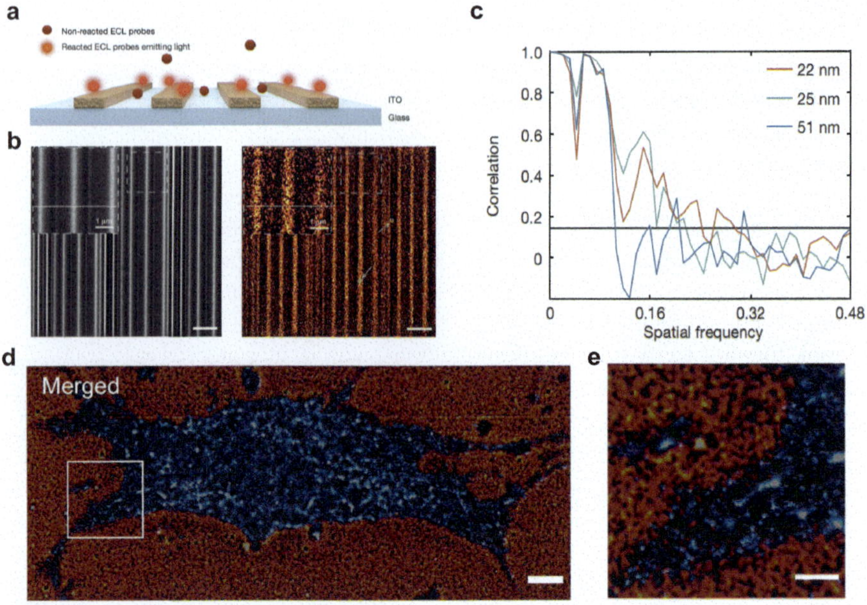

Fig. 6 Localization-based super-resolution ECL imaging [35]. **a** Representative single-molecule collision on the electrode surface. **b** Correlation of super-resolution ECL reconstruction and scanning electron microscopy. **c** Fourier ring correlation analysis of selected region of the interest in "b". **d** Merged super-resolution ECL reconstruction and STORM image of cell adhesion. **e** Enlarged white box region in "d". (**a**–**e** are all adapted with permission from Ref. [35]). Copyright © 2021, Springer Nature

3.3.1 Principles and Application of SOFI

As the earliest fluctuation-based super-resolution technique, SOFI was proposed by Enderlein's group in 2009 [19]. SOFI analyses the higher-order temporal autocorrelation of intensity fluctuation of individual molecules by acquiring a series of image sequences to extract the sub-diffraction structures. The enhancement in spatial resolution depends on the correlation orders. The higher the order, the higher the resolution. Although infinite resolution can be theoretically achieved using higher-order correlation analysis, both the fluorescence distribution and the PSF change during this analysis, resulting in clear artifacts when the higher orders are adopted. To address this problem, Lasser's group proposed the balanced fluorescence fluctuation super-resolution (bSOFI) technique in 2012 [48]. This method extracts molecular properties such as molecular lifetimes, fluorophore concentrations, and brightness distributions by analyzing different orders of temporal autocorrelation. These molecular properties can be used to balance the image contrast, effectively linearizing the luminescence intensity to reduce the artifacts. This correlation analysis-driven approach effectively resolves spatially overlapping signals based on their distinct temporal signatures and can achieve ~ 100 nm spatial resolution within minutes. Since SOFI is based on the

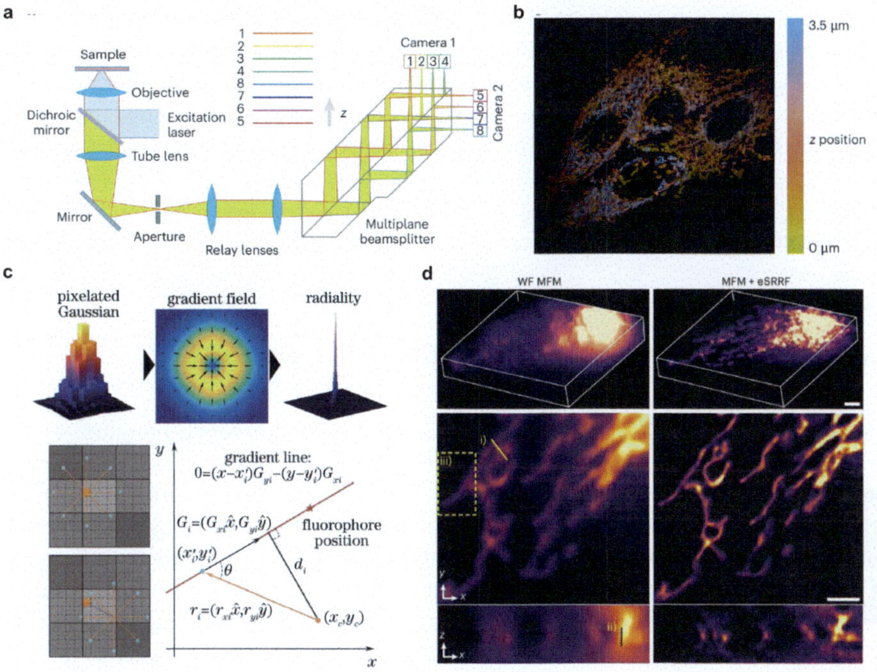

Fig. 7 SOFI and SRRF super-resolution imaging. **a** Illustration of imaging configuration in multi-color, wide-field SOFI imaging [33]. **b** 3D mitochondria imaging with SOFI [33]. (**a**, **b** are adapted with permission from Ref. [33]). Copyright © 2025, Springer Nature. **c** Principles of SRRF applied to wide-field microscopy [49]. Copyright © 2016, Springer Nature. **d** 3D-live-cell volumetric imaging of mitochondria in U2OS cells with eSFFR [20]. Copyright © 2023, Springer Nature

analysis of intensity correlation, it can resolve super-resolution information from the wide-field images. By designing the optical systems, it can realize the multi-color imaging simultaneously without complex excitation lasers (Fig. 7a) [33]. Moreover, SOFI is also capable of processing 3D imaging data thanks to its basic calculation principle (Fig. 7b).

3.3.2 Principles and Application of SRRF

Except for SOFI, there are also many other super-resolution imaging methods that have been developed by analyzing the correlation of spatiotemporal fluctuations of luminescent molecules such as SRRF and eSRRF, which further reduced the imaging time to within seconds. Henriques's group first introduced the SRRF method in 2016 [49]. This method achieves super-resolution imaging by analyzing the radial gradient changes of fluorescence fluctuation signals (Fig. 7c). It reconstructs super-resolution images without specialized optical design and has low requirements for probe molecules, providing a computable alternative to other super-resolution techniques such as STORM or MINFLUX. In 2023, eSRRF was proposed, which redefined several fundamental principles used for estimating the radial degree and temporal analysis [20]. The improvements of eSRRF minimize artifacts and non-linearities, thereby enhancing the overall fidelity of images and further improving the performance of super-resolution imaging. Henriques et al. also extended eSRRF to 3D imaging by combining it with multi-focus microscopy. Since eSRRF does not require high luminescence intensity, a low energy excitation laser beam was adopted, enabling super-resolution imaging of live cells (Fig. 7d) [20].

3.4 SRRF-Inspired Super-Resolution ECL Imaging

Based on the success of SRRF in super-resolution fluorescence imaging, recent studies have extended its application to ECL microscopy, as they share the same signal fluctuation characteristics, both rely on stochastic photon emission. In addition, the low requirement for absolute luminescence intensity in SRRF fits well with the low photon emission of ECL. Taking advantage of the zero background of ECL, the combination of SRRF and ECL makes it ideal for observing photosensitive bioprocesses or for long-term electrochemical dynamic imaging with high SNR. Chen et al. first applied the SRRF method to the ECL image sequence in 2021 (Fig. 8a) [40]. SRRF-ECL was used to investigate the active site of Au nano-catalysts. Au nano-catalysts can catalyze free ECL probes in solution to produce pronounced ECL emission on the catalyst surface. By capturing independent ECL events with high acquisition speed and analyzing them with SRRF, the catalytic activities of Au nanorods and nanoplates can be clearly mapped with a spatial resolution of about 100 nm, revealing the facet-dependent and defect-dependent spatial heterogeneity (Fig. 8b–d) [40]. Compared with localization-based super-resolution ECL imaging, the SRRF-ECL obtains a relatively low spatial resolution of about 100 nm. Balancing the spatial and temporal resolution in ECL super-resolution microscopy remains a challenge.

The emerging ECL super-resolution imaging techniques draw inspiration from fluorescence-based super-resolution microscopy to overcome the optical diffraction limit. Reference to the principles of SMLM, single-molecule localization-based ECL imaging achieves nanoscale spatial resolution by temporally isolating and precisely localizing stochastic single-molecule ECL events. Fluctuation-based ECL imaging applies statistical analysis of ECL intensity fluctuations and temporal correlation, which is conceptually consistent with the SOFI/SRRF principles in fluorescence microscopy. While single-molecule localization provides high spatial precision, fluctuation analysis offers enhanced temporal resolution and insight into ensemble behavior. These ECL super-resolution techniques promise to map nanoscale electrochemical processes, single-particle active sites, and subcellular structures. The appropriate method can be selected according to the actual application. In addition to this physical feature-based super-resolution imaging, a future direction may be to combine ECL with adaptive optical imaging or computational algorithms to

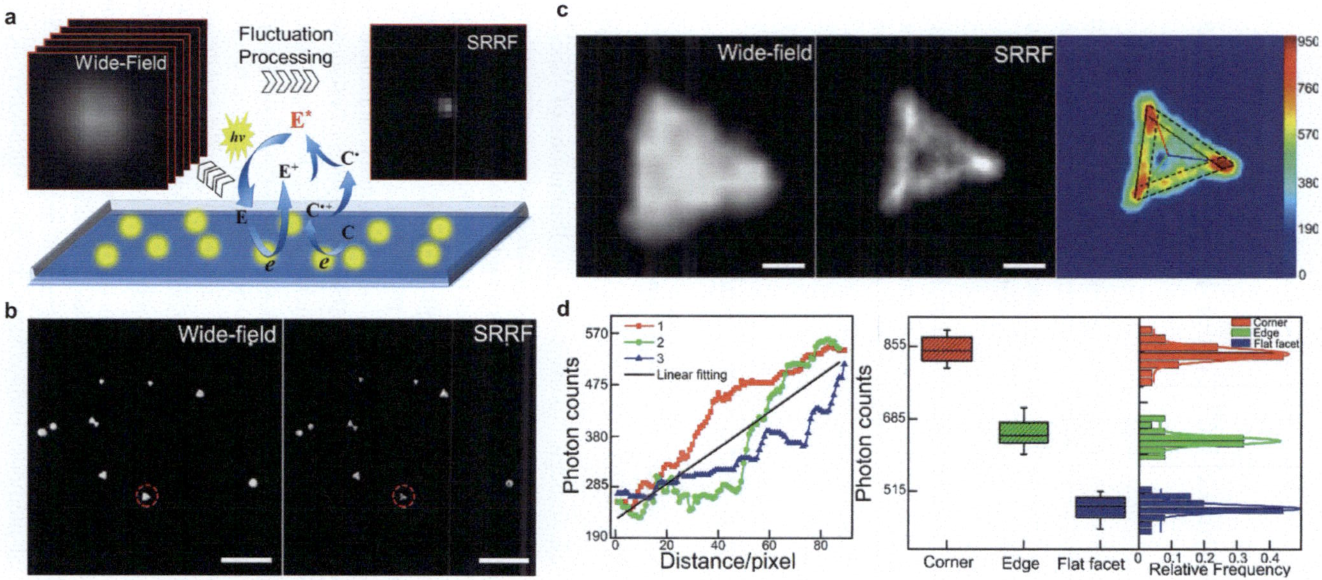

Fig. 8 SRRF super-resolution ECL imaging of Au nano-catalysts [40]. **a** Schematic of SRRF-ECL imaging. **b** Wide-field and SRRF ECL images of Au nanoplates. Scale bars: 5 μm. **c** Enlarged Au nanoplates labeled with red circles in "**b**" and its ECL intensity mapping. Scale bars: 200 nm. **d** Intensity profile and photon counts distribution of lines labeled in "**c**". (**a**–**d** are all adapted with permission from Ref. [40]). Copyright © 2021, American Chemical Society

improve spatiotemporal resolution, referencing to the development of fluorescence microscopy.

4 Deep Learning-Based ECL Super-Resolution Microscopy

Despite the recent advancements in physical feature-based ECL super-resolution imaging, these approaches remain constrained by limited temporal resolution, high computational complexity, and sensitivity to photon counts. As a data-driven computational approach, deep learning has already opened new avenues for accelerating image reconstruction and achieving higher spatial resolution in super-resolution fluorescence microscopy. By learning from massive training data, deep learning models can potentially extract and reconstruct high-resolution information from low-photon ECL images. The following subsections present a structured overview of deep learning for super-resolution microscopy and provide application examples of deep learning-enhanced ECL microscopy (DEECL).

4.1 Deep Learning-Based Super-Resolution Microscopy

4.1.1 Advances in Deep Learning-Based Super-Resolution Fluorescence Microscopy

Deep learning offers new strategies to overcome the limitations of super-resolution microscopy. Instead of relying on traditional mathematical reconstruction and large frames, recent approaches use data-driven learning to directly infer high-resolution outputs from low-quality inputs. These approaches can be divided into three main categories, including accelerating image reconstruction, reducing the number of input frames, and improving molecular localization precision.

Accelerating image reconstruction. One of the primary challenges in super-resolution microscopy is the time-consuming PSF calculation of image reconstruction, particularly in SMLM. Recent advancements in deep learning have considerably accelerated this process. In 2018, Nehme et al. introduced the Deep-STORM method [39], which employs a simple encoder–decoder architecture based on CNNs to rapidly reconstruct images (Fig. 9a). This approach greatly accelerates image reconstruction and enables real-time imaging while maintaining a high SNR. On the basis of Deep-STORM, Li et al. introduced the Deep Recurrent-Supervised Network-STORM (DRSN-STORM) in 2021 [50], which incorporates recurrent neural networks (RNNs) to capture temporal correlations across image frames (Fig. 9b). This approach reduces image reconstruction times by at least 40% compared to traditional methods. Unlike static models, DRSN-STORM exploits sequential data to optimize the reconstruction process by identifying temporal features that improve both speed and accuracy.

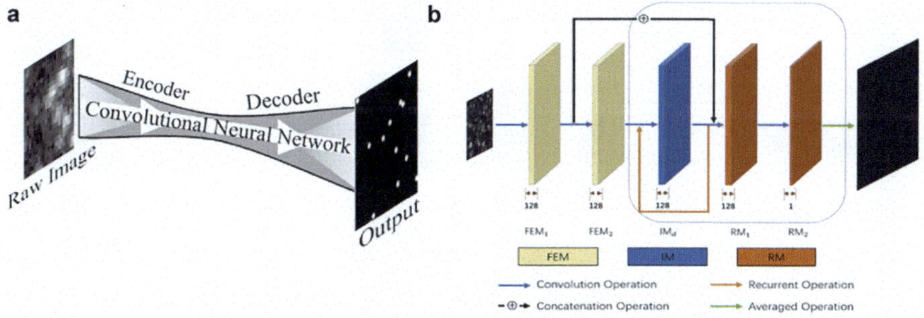

Fig. 9 Representative deep learning model accelerated image reconstruction in super-resolution fluorescence microscopy. **a** Deep-STORM network architecture [39]. (Nehme et al., 2022, licensed under CC BY). **b** DRSN-STORM network architecture [50]. (Li et al., 2021, licensed under CC BY)

Reducing the number of required input frames. Another challenge in super-resolution microscopy is the need for a large number of frames to achieve high-quality reconstructions. This not only increases the overall imaging time but also amplifies the risks of phototoxicity and photobleaching, especially in live-cell imaging. To address this problem, Ouyang et al. (2018) introduced ANNA-PALM (artificial neural network accelerated PALM), which is designed for reducing the required number of frames for high-quality reconstruction (Fig. 10a) [24]. Its network employs a multi-scale loss function to ensure that the generated super-resolution images closely match those reconstructed from the dense dataset. This approach allows for the generation of high-quality images from sparse SMLM data with only 300 frames. Compared to the thousands typically required, it offers substantial improvements in both imaging speed and phototoxicity reduction. In 2023, Saguy et al. introduced the Data Base Link (DBlink) network [51], combining CNN and long short-term memory (LSTM) networks, which further reduced the number of frames required by capturing the temporal correlations between consecutive frames (Fig. 10b). This method efficiently captures long-term dependencies in time-series data, which makes it particularly effective for real-time, high-speed imaging of dynamic biological processes.

Improving molecular localization precision. The localization precision of individual molecules is essential for obtaining high-resolution images, particularly in dense molecular environments where overlapping signals can lead to inaccuracies in localization. To meet this challenge, Cascarano et al. proposed the DeepCEL0 network, which uses a CNN backbone and adopts CEL0 regularized deconvolution as the loss function [52]. This design combines the high-precision molecular localization of CEL0 with the fast reconstruction speed of DeepSTORM. DeepCEL0 achieves high-accuracy localization images without increasing computational cost, and the reconstruction results are shown in Fig. 11a. In addition to this, deep learning can also extend the dimension of super-resolution reconstruction. Nehme et al. introduced DeepSTORM3D that integrates CNN with PSF engineering (Fig. 11b) [53]. This method can also address the challenge of high-density molecule localization

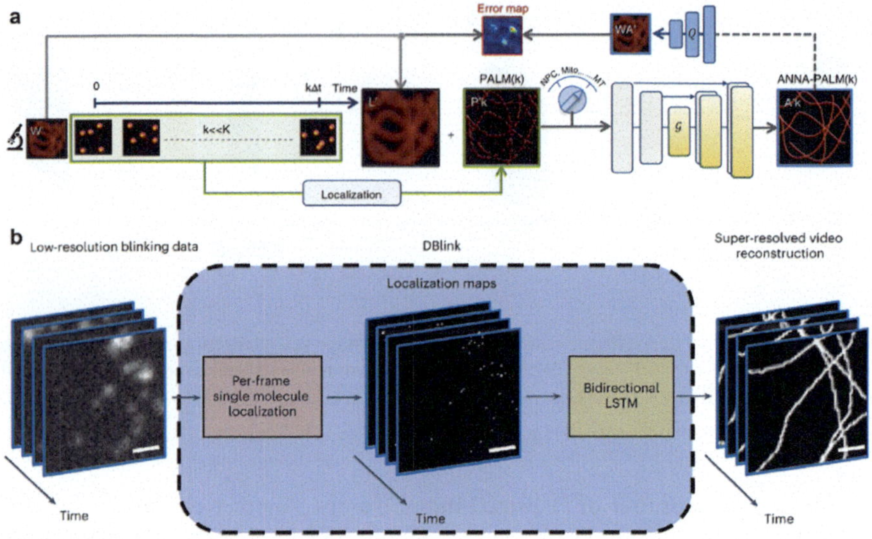

Fig. 10 Representative deep learning model reducing frame requirements in super-resolution microscopy. **a** ANNA-PALM network architecture [24]. Copyright © 2018, Springer Nature. **b** DBlink network architecture [51]. Copyright © 2023, Springer Nature

by analyzing overlapping PSFs across large axial ranges. Furthermore, the CNN can decode low-resolution 2D images to recover precise 3D positions.

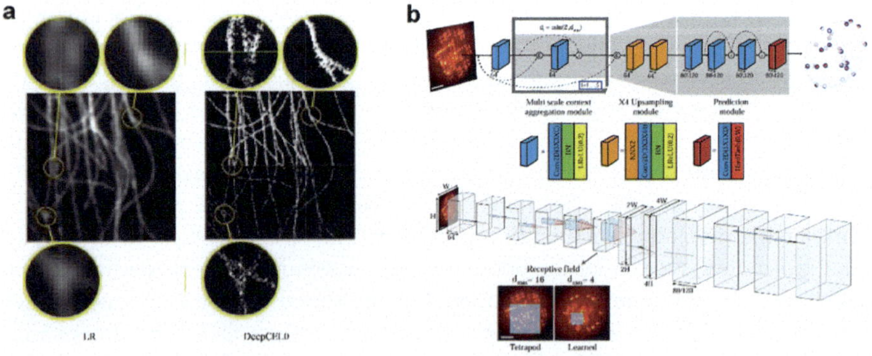

Fig. 11 Representative deep learning model enabling high-precision molecular localization in super-resolution microscopy. **a** Image reconstruction results of microtubules using DeepCEL0 [52]. Copyright © 2021, Oxford University Press. **b** DeepSTORM3D network architecture [53]. Copyright © 2020, Springer Nature

4.1.2 Transportability of Deep Learning-Based Super-Resolution Imaging: From Fluorescence to ECL

Deep learning has greatly improved the spatiotemporal resolution of super-resolution fluorescence imaging. Based on the same principle of processing the luminescence images, the application of deep learning to super-resolution fluorescence imaging has the potential to be transferred to ECL imaging. For instance, the application of CNNs to enhance localization accuracy can similarly be employed for ECL images. GANs offer a promising approach for denoising ECL signals and restoring image contrast under low photon emission. Additionally, temporal models such as recurrent neural networks (RNNs) and long short-term memory (LSTM) networks can also be applicable for multi-frame enhancement in dynamic ECL imaging.

However, the application of deep learning in ECL super-resolution imaging has not been explored until now due to the lack of large datasets for model training. With the development of the aforementioned physical-feature-based super-resolution ECL imaging techniques, it is possible to obtain a large number of training datasets. In addition, leveraging the synthetic data generation techniques can also be used to expand training datasets, such as simulating ECL signals based on theoretical models.

Although no studies have yet applied deep learning for super-resolution ECL imaging, deep learning has been applied to traditional ECL microscopy for improving the temporal resolution as a proof of concept, which will be discussed in the following subsection [41].

4.2 Deep Learning-Enhanced ECL Microscopy

In 2023, Zhao et al. proposed DEECL, a deep learning-driven approach for enhancing the acquisition speed of conventional ECL microscopy [41]. The proposed deep learning model helps to obtain the same image quality of ECL cell images at short exposure times as that obtained at long exposure times (Fig. 12a). With improved temporal resolution, this method enables high-throughput ECL imaging and dynamic cell imaging. In the following, the principles of DEECL and its performance evaluation are discussed.

4.2.1 Fundamental Principles of DEECL

In conventional ECL imaging, it typically requires tens of seconds to accumulate photons for high SNR. Such long exposure time limits the application of ECL for dynamic or high-throughput imaging. DEECL microscopy was developed to overcome this problem by introducing a computational strategy. The core of DEECL lies in a CNN architecture that learns a nonlinear mapping from noisy, low-photon ECL inputs to high-quality target images. This learning is supervised and based

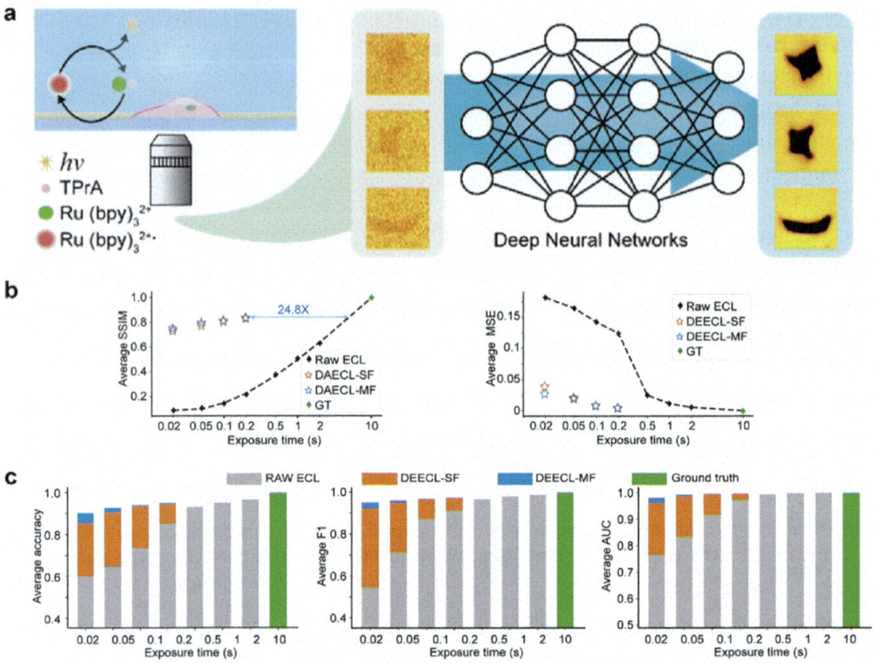

Fig. 12 The principle and performance of DEECL in cell imaging [41]. **a** Illustration of DEECL for cell imaging. **b** Average SSIM and MSE evaluation under different exposure times. **c** Average accuracy, average F1-score and average AUC evaluation under different exposure times. (**a–c** are all adapted with permission from Ref. [41]). Copyright © 2023, American Chemical Society

on a paired dataset of short- and long-exposure ECL images. Through training, the network captures both spatial patterns and semantic features from low-photon inputs to infer high-quality ECL images.

DEECL is implemented in two modalities: DEECL-SF (single-frame) and DEECL-MF (multi-frame). The DEECL-SF model takes a single short-exposure frame as input. In contrast, DEECL-MF takes the current frame and multiple previous frames as input that increases the available information for inference. This multi-frame approach takes advantage of the similar information between adjacent frames to better restore fine structural details. From a computational perspective, DEECL integrates key components typical of deep image restoration pipelines, including successive convolutional and deconvolutional layers, skip connections, and ReLU activations. The architecture encodes features at multiple spatial scales to enable local and global aggregation. The training process minimizes the pixel-wise mean squared error (MSE) loss between the reconstructed image and the ground truth, thus ensuring that the output preserves both contrast and structure fidelity. Besides, synthetic data augmentation strategies are used to simulate motion and deformation, which improves the robustness of the network for dynamic imaging tasks.

4.2.2 Evaluation of DEECL

The performance of DEECL microscopy was evaluated through multiple metrics to assess its effectiveness in improving temporal resolution. The primary evaluation criteria include image similarity and pixel-wise reconstruction accuracy.

Image quality evaluation. To quantitatively evaluate the image quality of the DEECL reconstructions, Structural Similarity Index (SSIM) and Mean Squared Error (MSE) were employed. These metrics provide a measure of how closely the reconstructed images match the ground truth images (Fig. 12b). The SSIM evaluates the perceived structural similarity between the reconstructed image and the ground truth. A higher SSIM indicates greater similarity between the reconstructed image and the reference image. The MSE, on the other hand, calculates the pixel-wise difference between the reconstructed image and the ground truth, where lower MSE values indicate better performance. The DEECL method consistently achieved higher SSIM values compared to the raw short-exposure ECL images, effectively enhancing the image quality (Fig. 12b). For example, when the exposure time was 0.2 s, DEECL-MF showed SSIM values comparable to those obtained from ground truth images acquired under exposure times of 4.96 s, which indicates a ~ 24.8-fold speed-up.

Classification performance. To evaluate the reconstruction quality of DEECL, classification metrics including accuracy, F1-score, and AUC were applied by binarizing ECL images based on intensity thresholds (Fig. 12c). With decreased exposure times, the reconstruction advantage of DEECL becomes more evident. For instance, DEECL-MF improved classification accuracy from 0.420 to 0.878 at an 0.02 s exposure. This performance improvement was consistent across all tested exposure times, demonstrating that DEECL is effective in recovering structural features even under low photon emission conditions.

DDECL is the first example of the integration of conventional ECL images with deep learning, demonstrating excellent temporal resolution enhancement. Looking forward, combining deep learning with super-resolution ECL imaging is expected to achieve higher spatiotemporal resolution, which will help to super-resolution dynamic observation.

5 Application of Super-Resolution ECL Imaging

Reference to the principles of super-resolution fluorescence microscopy, both physically and computationally based ECL microscopy has led to significant progress in improving spatiotemporal resolution. This advancement has promoted the development of applications in the fields of cell biology, material characterization, and chemical reaction kinetics observation. Detailed examples are introduced below.

5.1 Single-Cell Structure Observation

As the basic constituent of living organisms, single cells play an indispensable role in life activities. Studying the heterogeneity of cellular structures at the single-cell level is of great significance in revealing the basic principles of life activities [54]. ECL imaging demonstrates unique advantages in cellular studies, particularly in high sensitivity and spatiotemporal resolution [9, 26]. In detail, ECL eliminates the requirements for external laser sources and thereby effectively avoids autofluorescence, photobleaching, and phototoxic damage to live cells. Compared with fluorescence, ECL has nearly zero background interference and an improved SNR, enabling single-molecule-level detection of biomarkers on the single cell [55]. In addition, high spatial resolution helps to precisely observe subcellular structures and dynamics, which may help us understand the details of organelle interactions. These advantages make ECL super-resolution imaging a powerful tool for revealing cellular heterogeneity and subcellular molecular mechanisms. Dong et al. investigated the distribution of cell adhesion on the electrode with localization-based super-resolution ECL imaging with a sub-diffraction spatial resolution (Fig. 13a) [35]. Thanks to the low phototoxic damage of ECL, they further explored the dynamic process of cell adhesion and analyzed the motion of adhesive sites with a temporal resolution of 12 s, confirming the reliability of super-resolution ECL for cellular dynamic imaging (Fig. 13b, c) [35]. In addition to single-cell imaging, Zhou et al. investigated bacterial morphology with super-resolution ECL. The intensity differences between bacterial division gaps and bacterial junctions can be clearly observed (Fig. 13d) [56]. This research expands the biological application of localization-based super-resolution ECL imaging from single cell to single bacterial. It is expected to enable finer observation of smaller organisms in the future. The spatiotemporal controllability of ECL signals can further extend its capability in cell imaging. For instance, alternating applied voltages can selectively trigger probes with different emission wavelengths, thus facilitating multichannel super-resolution imaging without additional optical filters [8].

Except for localization-based super-resolution ECL imaging methods, deep learning-enhanced computational imaging methods have also contributed significantly to the improvement of cellular imaging resolution. The deep learning-enhanced ECL microscopy has improved the temporal resolution by 100-fold compared to conventional ECL microscopy. Such a high temporal resolution enables the observation of fast cellular dynamic processes. In the future, we believe that as super-resolution ECL imaging technology matures, it will be easy to obtain plenty of super-resolution training datasets for deep learning-enabled super-resolution ECL imaging.

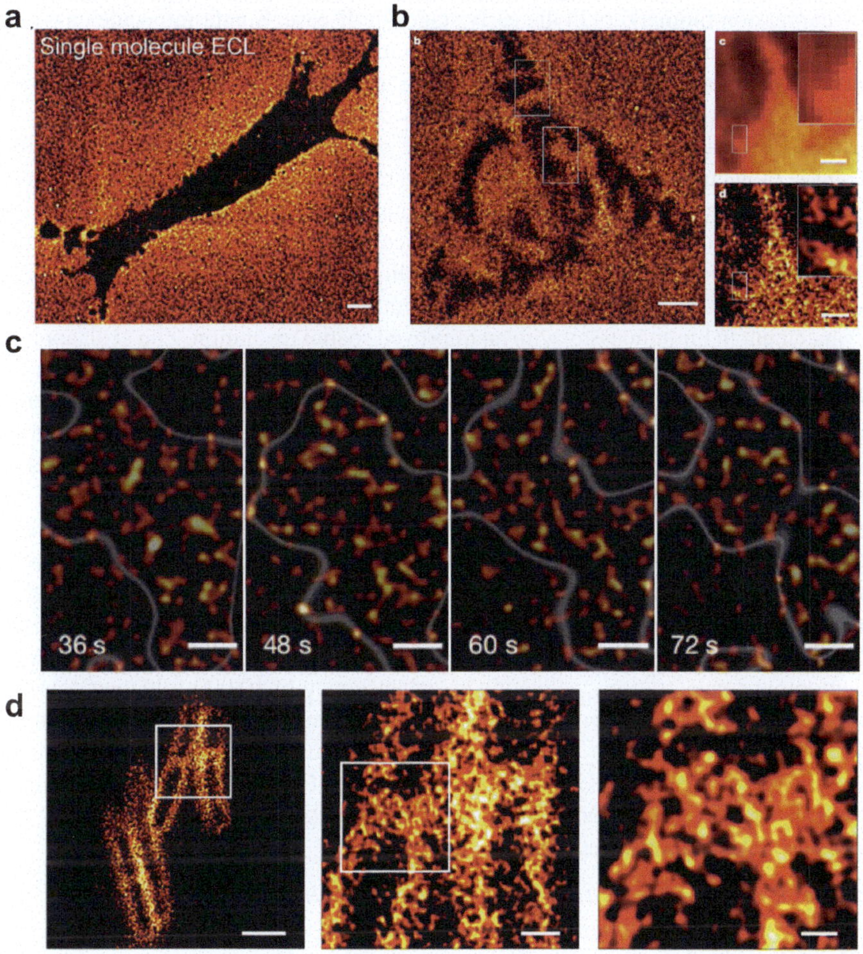

Fig. 13 Super-resolution ECL imaging of biology samples. **a** Super-resolved ECL image of cell adhesion in a fixed cell [35]. Scale bar: 5 μm. **b** Super-resolved ECL images of cell adhesion in a live cell [35]. Scale bar: 5 μm. **c** Dynamic visualization of the cell adhesions at different moments (36 s, 48 s, 60 s, and 72 s) for the selected regions (bottom white box) in "**b**" [35]. (**a**–**c** are all adapted with permission from Ref. [35]) Copyright © 2021, Springer Nature. **d** Super-resolved ECL images of bacteria membrane [56]. Scale bars: 2 μm, 500 nm and 200 nm respectively. Copyright © 2023, American Chemical Society

5.2 Single-Molecule Bioassay

The detection of low-abundance biomarkers is critical for early clinical diagnosis and monitoring disease progression. However, conventional immunoassays typically detect biomarkers at the ensemble level, which presents a challenge in detecting low-expression biomarkers in the early stages of disease. Single-molecule enzyme-linked immunoassay based on fluorescence has significantly improved detection sensitivity and provides an effective method for measuring protein biomarkers with extremely low concentrations in serum. However, fluorescence-based detection is restricted by laser effects that limit absolute sensitivity and quantitative counting of single molecules remains a challenge. The unique electrochemical-driven nature of the ECL provides distinct advantages in analytical applications, particularly in biosensing and surface imaging where electrical control can improve sensitivity and spatial resolution. This unique mechanism allows ECL to have a wide application in in-vitro bioassays. Particularly, super-resolution ECL imaging will lead to a great advancement in immunoassay with an improved spatial resolution. Zhu et al. applied localization-based super-resolution ECL imaging to a conventional immunoassay model (Fig. 14a, b) [57]. Single-molecule localization not only provides spatial information of detected biomarkers with high resolution, but also can be used to extract quantitative insight for molecular quantification. In general, the number of localized molecules represents the direct number of molecules. However, the repeated ECL reactions may result in overcounting of labeled molecules, which can be corrected with a reasonable spatiotemporal clustering correction. The combination of single-molecule localization and spatiotemporal clustering has enabled the precise quantification of a widely used biomarker–CEA, with single-molecule sensitivity. This method greatly improves the limit of detection (LOD) of CEA to 67 attomole concentrations, achieving an enhancement of 10,000-fold over conventional ECL immunoassay while also providing high spatial resolution. Such a high spatial resolution is absent in other immunoassay methods. The improvement in detection sensitivity and LOD offers great potential for early diagnosis and monitoring of various cancers. Moreover, the spatial resolution is potentially useful for multiplexed biomarker detection.

5.3 Chemical Active Site Characterization

ECL directly links light emission to chemical reactivity according to its luminescence mechanism, thus providing a unique chemical heterogeneity dimension rather than just spatial information. Therefore, ECL has become an essential analytical tool for mapping electrochemical activity. Integration of ECL with super-resolution imaging overcomes the optical diffraction limit and addresses the ensemble-averaging limitations of conventional ECL imaging. Dong et al. mapped the distribution of reactive sites on gold nanosheets using a localization-based super-resolution ECL imaging method in 2022 [28]. The ECL intensity and emission rate are heterogeneous for different active sites. Through spatiotemporal isolation of individual ECL events, super-resolved activity maps can be reconstructed with a resolution of 37 nm and correlated with structural defects (Fig. 14c). It distinguishes the advantages of ECL's chemical dimension and label-free detection capability for real-time visualization of dynamic processes in chemical reactions. This super-resolution ECL method revealed

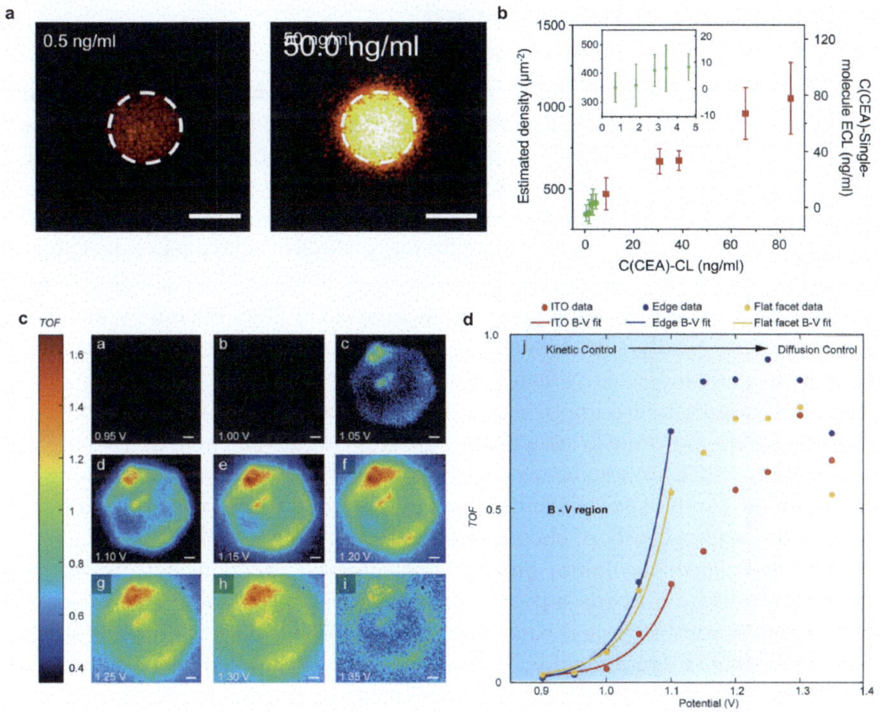

Fig. 14 Super-resolution ECL image for immunoassay and reaction kinetics analysis on Au plates. **a** Super-resolved ECL image of magnetic beads with a CEA concentration of 0.5 ng/ml and 50 ng/ml. Scale bar: 3 μm [57]. **b** CEA concentrations in clinical samples are measured with single-molecule ECL bioassay and standard commercial chemiluminescence for correlation [57]. (**a**, **b** are adapted with permission from Ref. [57]). Copyright © 2022 Wiley-VCH GmbH. **c** TOF mapping on the Au plate under different voltages. Scale bar: 500 nm [28]. **d** B-V equation fitting of TOF between 0.9 and 1.1 V [28]. (**c**, **d** are adapted with permission from Ref. [28]). Copyright © 2022 Wiley-VCH GmbH

heterogeneous Butler-Volmer (B-V) kinetics at single-molecule level, where differences in charge transfer coefficients can be obtained in different nanosheet regions (Fig. 14d). In addition, the ECL-SRRF method further enhances both the spatial and temporal resolution of the reactive site observation, successfully resolves the electrochemical activity gradients along 1D nanowires and 2D material edges [40]. Both these two super-resolution ECL techniques provide a convenient method for in-situ characterization of catalysts, which helps to reveal the defect-dependent surface activity and rationalize catalyst design.

6 Conclusion and Ways Forward

Technological breakthroughs in super-resolution fluorescence microscopy inspired super-resolution ECL imaging. Referring to the SMLM and fluctuation-based super-resolution fluorescence microscopy, single-molecule localization-ECL, and SRRF-ECL have been demonstrated to break the optical diffraction limit and achieve spatial resolutions of at least 22 nm, which is comparable to fluorescence techniques. The integration of deep learning may potentially further improve the temporal resolution of super-resolution ECL microscopy. Improvements in the spatial resolution of ECL imaging have promoted its widespread applications in biology, chemistry, and material science. Specifically, super-resolution ECL helps to reveal finer cellular/bacterial structures at the nano-scale, which are absent in conventional ECL cell imaging. Super-resolution ECL also improves the detection sensitivity of in vitro immunoassays to the single-molecule level, with a 10,000-fold improvement in LOD compared to conventional ECL immunoassays. Moreover, super-resolution ECL imaging can probe individual active sites of chemical reactions in solution and catalytic interfaces, revealing the heterogeneity of chemical reactions at the nanoscale.

While ECL super-resolution imaging has achieved excellent performance for nanoscale observation, there still exist some challenges including purchasing the ultimate spatial resolution and balancing the spatial and temporal resolution. The inherent low photon emission of ECL limits the further improvement of spatial resolution. Achieving a strong and stable ECL signal is essential for enhancing the resolution. Developing novel ECL probes with high emission intensity is potential to realize molecule-scale resolution in the future. While ECL achieves acceptable spatial resolution, its temporal resolution remains constrained by the stochastic electrochemical reactions. It usually requires prolonged imaging durations for improved spatial resolution. SRRF-ECL has solved this problem to some extent, future developments of such fluctuation-based methods or deep learning-based methods could balance the spatial and temporal resolution in ECL imaging.

Another challenge in super-resolution ECL imaging is specific imaging under complex conditions. Current super-resolution reconstructions are all performed on unlabeled systems, which greatly rely on the adhesion of the imaging targets to the electrode surface, thus lacking specificity. Specific super-resolution ECL imaging remains elusive due to the limited labeling efficiency and lower emission intensity of the labeled probe. Moreover, some of the labeled probes may be far away from the electrode where the excitation efficiency is greatly reduced. To improve the ECL efficiency on specific labeling systems. A series of nanoparticle ECL probes has been developed, which has enabled specific imaging of single molecules such as proteins on cell membrane. However, it also requires prolonged exposure times to accumulate SNR, which is adverse for super-resolution imaging. Based on these limitations, super-resolution imaging of specific target molecules under complex conditions has not been well explored. We believe that with the future improvement in high-efficiency labeling and the ECL imaging system, the specificity, spatial resolution, and temporal resolution can be further improved, thus expanding ECL's application for specific molecule observation in complex conditions.

Based on achievements and challenges of super-resolution ECL imaging, future advances can focus on several key areas, including chemically or computationally enhancing the ECL efficiency, developing multi-target imaging, and multi-mode imaging. Chemically improving the ECL efficiency includes designing new probes, optimizing existing probes for improved quantum yields and stability, and exploring new co-reactants that can react more effectively with ECL probes. Computationally improving the ECL efficiency can be realized by deep learning and AI. For example, the SNR of a single frame can be improved with deep learning thus helping super-resolution reconstruction. Due to the absence of multi-channel probes with distinct emission profiles, ECL imaging is currently limited to single-target detection. A wide range of probes with distinguishable emission wavelengths needs to be developed for multiplexed imaging in the future. Finally, combining super-resolution ECL with other analytical techniques such as fluorescence can improve detection capabilities and provide complementary information about the analytes since ECL has ultrahigh interfacial sensitivity, which is absent in other imaging techniques. Developing multi-mode imaging that consists of ECL and other complementary imaging modalities can extend our understanding of complex molecular reaction mechanisms.

While super-resolution ECL imaging currently enables nanoscale observation of imaging targets, achieving wide applications in both biology and chemistry. Thereis a big gap between the super-resolution ECL imaging and super-resolution fluorescence imaging especially in nanoscale cell imaging. Super-resolution fluorescence imaging has realized the real-time, multi-color, 3D dynamic imaging of specific molecules and allowed observation of nanoscale organelle interactions. However, super-resolution ECL imaging remains on the whole-cell level, only realizing 2D reconstruction of the cell adhesion, and requires long imaging times. The high spatiotemporal resolution, specific, 3D and multiplexed imaging of super-resolution ECL microscopy remains challenging, highlighting the need for the development of more efficient ECL systems and super-resolution reconstruction methods.

Future developments of ECL super-resolution imaging may enable ECL distinguish a comparable performance to fluorescence, even provide a complementary view to fluorescence due to its surface sensitivity. Moreover, ECL can also provide an additional chemical view, which is absent in other super-resolution techniques that can help to characterize the reaction activity of the imaging targets. We believe that as the performance of the super-resolution ECL microscope is effectively improved, it will open new avenues for a deeper understanding of single-molecule behavior in the future.

References

1. Yang, F., et al.: The molecular structure of green fluorescent protein. Nat. Biotechnol. **14**, 1246–1251 (1996)

2. Sharma, A., et al.: Theranostic fluorescent probes. Chem. Rev. **124**, 2699–2804 (2024)
3. Balzarotti, F., et al.: Nanometer resolution imaging and tracking of fluorescent molecules with minimal photon fluxes. Science **355**, 606–612 (2017)
4. Weber, M., et al.: MINSTED nanoscopy enters the Angstrom localization range. Nat. Biotechnol. **41**, 569–576 (2023)
5. Reinhardt, S.C.M., et al.: Ångström-resolution fluorescence microscopy. Nature **617**, 711–716 (2023)
6. Richter, M.M.: Electrochemiluminescence (ECL). Chem. Rev. **104**, 3003–3036 (2004)
7. Knezevic, S., et al.: Electrochemiluminescence microscopy. Angew. Chem. Int. Ed. **63**, e202407588 (2024)
8. Guo, W., et al.: Potential-resolved multicolor electrochemiluminescence for multiplex immunoassay in a single sample. J. Am. Chem. Soc. **140**, 15904–15915 (2018)
9. Voci, S., et al.: Surface-confined electrochemiluminescence microscopy of cell membranes. J. Am. Chem. Soc. **140**, 14753–14760 (2018)
10. Zhu, M.J., et al.: Electrogenerated chemiluminescence imaging of electrocatalysis at a single Au-Pt janus nanoparticle. Angew. Chem. Int. Ed. **57**, 4010–4014 (2018)
11. Gwosch, K.C., et al.: MINFLUX nanoscopy delivers 3D multicolor nanometer resolution in cells. Nat. Methods **17**, 217–224 (2020)
12. Sahl, S.J., et al.: Direct optical measurement of intramolecular distances with angstrom precision. Science **386**, 180–187 (2024)
13. Miao, W., et al.: Electrogenerated chemiluminescence 69: the tris(2,2'-bipyridine)ruthenium(II), ($Ru(bpy)_3{}^{2+}$)/trinpropylamine (TPrA) system revisited: a new route involving $TPrA^{\cdot+}$ cation radicals. J. Am. Chem. Soc. **124**, 14478–14485 (2002)
14. Gustafsson, M.G.L.: Nonlinear structured-illumination microscopy: wide-field fluorescence imaging with theoretically unlimited resolution. Proc. Natl. Acad. Sci. **102**, 13081–13086 (2005)
15. Willig, K.I., et al.: STED microscopy reveals that synaptotagmin remains clustered after synaptic vesicle exocytosis. Nature **440**, 935–939 (2006)
16. Betzig, E., et al.: Imaging intracellular fluorescent proteins at nanometer resolution. Science **313**, 1642–1645 (2006)
17. Rust, M.J., et al.: Sub-diffraction-limit imaging by stochastic optical reconstruction microscopy (STORM). Nat. Methods **3**, 793–795 (2006)
18. Jungmann, R., et al.: Multiplexed 3D cellular super-resolution imaging with DNA-PAINT and exchange-PAINT. Nat. Methods **11**, 313–318 (2014)
19. Dertinger, T., et al.: Fast, background-free, 3D super-resolution optical fluctuation imaging (SOFI). Proc. Natl. Acad. Sci. **106**, 22287–22292 (2009)
20. Laine, R.F., et al.: High-fidelity 3D live-cell nanoscopy through data-driven enhanced super-resolution radial fluctuation. Nat. Methods **20**, 1949–1956 (2023)
21. Stokes, J.M., et al.: A deep learning approach to antibiotic discovery. Cell **180**, 688-702.e613 (2020)
22. Scarselli, F., et al.: The graph neural network model. IEEE Trans. Neural Networks **20**, 61–80 (2009)
23. Feng, J.: Electrochemistry probed one molecule at a time. Curr. Opin. Electrochem. **34**, 101000 (2022)
24. Ouyang, W., et al.: Deep learning massively accelerates super-resolution localization microscopy. Nat. Biotechnol. **36**, 460–468 (2018)
25. Dong, J., Feng, J.: Electrochemiluminescence from single molecule to imaging. Anal. Chem. **95**, 374–387 (2023)
26. Valenti, G., et al.: Single cell electrochemiluminescence imaging: from the proof-of-concept to disposable device-based analysis. J. Am. Chem. Soc. **139**, 16830–16837 (2017)
27. Deiss, F., et al.: Multiplexed sandwich immunoassays using electrochemiluminescence imaging resolved at the single bead level. J. Am. Chem. Soc. **131**, 6088–6089 (2009)
28. Dong, J., et al.: Operando imaging of chemical activity on gold plates with single-molecule electrochemiluminescence microscopy. Angew. Chem. Int. Ed. **61**, e202200187 (2022)

29. Weber, M., et al.: MINSTED fluorescence localization and nanoscopy. Nat. Photonics **15**, 361–366 (2021)
30. Li, X., et al.: Three-dimensional structured illumination microscopy with enhanced axial resolution. Nat. Biotechnol. **41**, 1307–1319 (2023)
31. Schermelleh, L., et al.: Super-resolution microscopy demystified. Nat. Cell Biol. **21**, 72–84 (2019)
32. Scheiderer, L., et al.: MINFLUX achieves molecular resolution with minimal photons. Nat. Photonics **19**, 238–247 (2025)
33. Basak, S., et al.: Super-resolution optical fluctuation imaging. Nat. Photonics **19**, 229–237 (2025)
34. White, E.H., et al.: Chemilunimescence of luminol: the chemcial reaction. J. Am. Chem. Soc. **86**, 940–941 (2002)
35. Dong, J., et al.: Direct imaging of single-molecule electrochemical reactions in solution. Nature **596**, 244–249 (2021)
36. Liu, G., et al.: Potential-resolved electrochemiluminescence nanoprobes for visual apoptosis evaluation at single-cell level. Anal. Chem. **91**, 6363–6370 (2019)
37. Ding, Z., et al.: Electrochemistry and electrogenerated chemiluminescence from silicon nanocrystal quantum dots. Science **296**, 1293–1297 (2002)
38. Li, B., et al.: High electrochemiluminescence from $Ru(bpy)_3{}^{2+}$ embedded metal-organic frameworks to visualize single molecule movement at the cellular membrane. Adv. Sci. **9**, e2204715 (2022)
39. Nehme, E., et al.: Deep-STORM: super-resolution single-molecule microscopy by deep learning. Optica **5**, 458–464 (2018)
40. Chen, M.-M., et al.: Super-resolution electrogenerated chemiluminescence microscopy for single-nanocatalyst imaging. J. Am. Chem. Soc. **143**, 18511–18518 (2021)
41. Zhao, P., et al.: Deep learning enhanced electrochemiluminescence microscopy. Anal. Chem. **95**, 4803–4809 (2023)
42. Henriques, R., et al.: PALM and STORM: unlocking live-cell super-resolution. Biopolymers **95**, 322–331 (2011)
43. Shroff, H., et al.: Live-cell photoactivated localization microscopy of nanoscale adhesion dynamics. Nat. Methods **5**, 417–423 (2008)
44. Bates, M., et al.: Multicolor super-resolution fluorescence imaging via multi-parameter fluorophore detection. Chem. Phys. Chem. **13**, 99–107 (2012)
45. Huang, B., et al.: Three-dimensional super-resolution imaging by stochastic optical reconstruction mcroscopy. Science **319**, 810–813 (2008)
46. Huang, B., et al.: Whole-cell 3D STORM reveals interactions between cellular structures with nanometer-scale resolution. Nat. Methods **5**, 1047–1052 (2008)
47. Eilers, Y., et al.: MINFLUX monitors rapid molecular jumps with superior spatiotemporal resolution. Proc. Natl. Acad. Sci. **115**, 6117–6122 (2018)
48. Geissbuehler, S., et al.: Mapping molecular statistics with balanced super-resolution optical fluctuation imaging (bSOFI). Opt. Nanosc. **1**, 4 (2012)
49. Gustafsson, N., et al.: Fast live-cell conventional fluorophore nanoscopy with ImageJ through super-resolution radial fluctuations. Nat. Commun. **7**, 12471 (2016)
50. Li, J., et al.: Spatial and temporal super-resolution for fluorescence microscopy by a recurrent neural network. Opt. Express **29**, 15747–15763 (2021)
51. Saguy, A., et al.: DBlink: dynamic localization microscopy in super spatiotemporal resolution via deep learning. Nat. Methods **20**, 1939–1948 (2023)
52. Cascarano, P., et al.: DeepCEL0 for 2D single-molecule localization in fluorescence microscopy. Bioinformatics **38**, 1411–1419 (2021)
53. Nehme, E., et al.: DeepSTORM3D: dense 3D localization microscopy and PSF design by deep learning. Nat. Methods **17**, 734–740 (2020)
54. Sigal, Y.M., et al.: Visualizing and discovering cellular structures with super-resolution microscopy. Science **361**, 880–887 (2018)

55. Liu, Y., et al.: Single biomolecule imaging by electrochemiluminescence. J. Am. Chem. Soc. **143**, 17910–17914 (2021)
56. Zhou, Y., et al.: Imaging of single bacteria with electrochemiluminescence microscopy. J. Am. Chem. Soc. **145**, 8947–8953 (2023)
57. Zhu, W., et al.: Quantitative single-molecule electrochemiluminescence bioassay. Angew. Chem. Int. Ed. **62**, e202214419 (2023)

Biosensors-Immunoassay by ECL Imaging and Microscopy

Yu Zheng, Yuriy Tymofiiovych Zholudov, Dmytro Viktorovych Snizhko, Kateryna Muzyka, Yiran Guan, and Guobao Xu

Abstract This chapter focuses on the application of ECL imaging for quantitative detection of biomarkers and multiplexed assays. We will firstly introduce the basic concepts of ECL imaging biosensors and their structure. Secondly, we will discuss the application of ECL imaging immunosensors for biomarker detection, categorize and list biomarkers with respect to different diseases and their relevance, and explore the use of ECL immunosensors in biomarker detection, highlighting their broad application prospects in medical and non-medical fields. Subsequently, we will focus on the recent advances in multiplex ECL detection, including its combination with microfluidic chips, smart sensors, and multifunctional nanomaterials. Finally, this chapter will also analyze the challenges of ECL imaging in biomarker quantification and multiplexed assays in the context of specific application cases and consider the prospects for its future development, particularly in portable devices, high-throughput analyses, and for various fields such as precision medicine, environmental monitoring, and food safety.

Keywords Electrochemiluminescence (ECL) imaging · Biosensors · Immunoassay · Biomarker detection · Multiplex detection · Nanomaterials · Microfluidic chips · Clinical diagnosis

---

Y. Zheng · Y. Guan · G. Xu (✉)
State Key Laboratory of Electroanalytical Chemistry, Changchun Institute of Applied Chemistry, Chinese Academy of Sciences, Changchun, Jilin, China
e-mail: guobaoxu@ciac.ac.cn

Y. Zheng · G. Xu
School of Applied Chemistry and Engineering, University of Science and Technology of China, Hefei, Anhui, China

Y. T. Zholudov · D. V. Snizhko · K. Muzyka
Laboratory of Analytical Optochemotronics, Kharkiv National University of Radio Electronics, Kharkiv, Ukraine

K. Muzyka
DyNaMo, INSERM Marseille, Marseille, France

F. Paolucci et al. (eds.), *Electrogenerated Chemiluminescence Microscopy and Imaging*,
https://doi.org/10.1007/978-3-032-14602-1_7

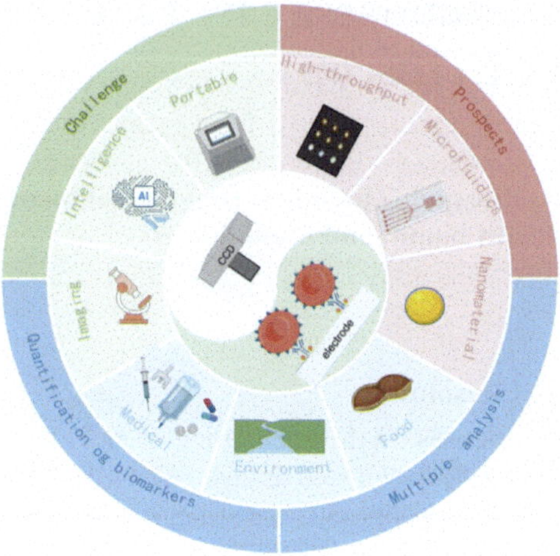

1 Introduction

With the continuous development of biomedical research, the detection of biomarkers has become increasingly important in clinical diagnosis, disease monitoring, and pharmacology, as they act as signaling molecules for diseases and can reflect an individual's physiological or pathological state. Cancer, cardiovascular disease, renal disease, and many other complex disorders have corresponding biomarkers, which provide a basis for early detection and monitoring of disease progression [54]. Therefore, their detection is essential for early diagnosis, assessing treatment efficacy, and prognosis prediction of diseases. However, traditional biomarker detection methods have many shortcomings in sensitivity, specificity, and operational simplicity, which are challenging to meet the needs of modern medicine for rapid, accurate, and low-cost detection [139].

Although traditional assays (e.g., enzyme-linked immunosorbent assay, radioimmunoassay, etc.) have been widely used in the detection of biomarkers, they are usually not sensitive enough when confronted with low concentrations of biomarkers, which can easily lead to false-negative results. This is particularly challenging in the early diagnosis of disease, where biomarker concentrations are often very low. In order to improve sensitivity, complex sample pre-treatment steps are often required, which are time-consuming and prone to increased operational errors. Long assay duration, e.g., enzyme-linked immunosorbent assays (ELISA), often takes several hours to obtain results and is a major limitation for clinical applications where rapid decision-making is required. In some acute or infectious disease scenarios, it is critical to get results quickly; however, traditional methods often do not satisfy these requirements.

As the need for precision and personalized medicine continues to grow, these limitations emphasize the urgent need for developing novel assays with higher sensitivity, shorter response time, and the ability to detect multiple markers simultaneously.

In recent years, the rapid development of biosensors has opened up the possibility of solving these problems. Biosensors are capable of converting biometric events into detectable signals with the advantages of high sensitivity, high selectivity, and integrability [42, 71, 93]. In particular, the application of electrochemiluminescence (ECL) technology has led to a completely new stage of biomarker detection [35]. ECL has been employed in both in vitro diagnosis (IVD) and cell imaging, enhancing its application in immunoassays [72]. The advantages of ECL technology include high sensitivity, low background interference, and good signal controllability, making it suitable for the analysis of complex biological samples. It combines the dual advantages of electrochemistry and chemiluminescence and achieves the quantitative detection of specific molecules through the excitation of luminescent molecules by electrochemical reactions to produce analytical signals.

ECL immunoassay sensors are thus increasing attention [29, 124, 136]. The technical means for the commercial applications of electrochemiluminescence based on biomarkers are mainly provided by two companies, Roche Diagnostics and Meso Scale Discovery. It is widely applied in clinical diagnosis, and the number of tests globally reaches approximately 2 billion times per year [153, 22]. A prime example of commercial ECL application is the cobas® ECL immunoassay platform by Roche Diagnostics, which is used globally for high-throughput, automated biomarker quantification in clinical settings. It's widely used in oncology, cardiology, endocrinology, infectious disease screening, and critical care [28, 45].

However, conventional ECL immunoassays typically rely on single-signal outputs, which can be affected by numerous intrinsic or extrinsic factors such as electrode modifications, instrument efficiency, and environmental conditions. In addition, single-signal assays are not sufficient to diagnose or predict certain diseases, as some diseases are associated with more than one biomarker.

To overcome the above problems, various ECL strategies with multiple signal outputs have been developed, such as ratiometric ECL, ECL multiplex immunoassay (MIA), and ECL imaging. ECL ratiometric sensors can provide more accurate results in complex environments because interference can be eliminated by self-calibration with two emissions (Fig. 1b). ECL MIA is capable of detecting multiple targets simultaneously to meet stringent diagnostic criteria (Fig. 1c). ECL imaging analysis is a rapidly evolving bioanalytical technology in the field with high spatial and temporal resolution, high throughput, and visualization (Fig. 1d) [20].

ECL imaging is particularly suitable for the detection of biomarkers at low concentrations, as the ECL signal is generated only in the electrochemical reactions occurring near the electrodes and does not depend on an external light source, minimizing background noise [61], which improves the signal-to-noise ratio. This is particularly important for the detection of markers in complex biological samples, e.g., by precisely controlling the electrode potential, the ECL technique can provide reliable quantitative results for the simultaneous detection of multiple target analytes by using different metal-coordinated luminophores when multiple interfering substances may

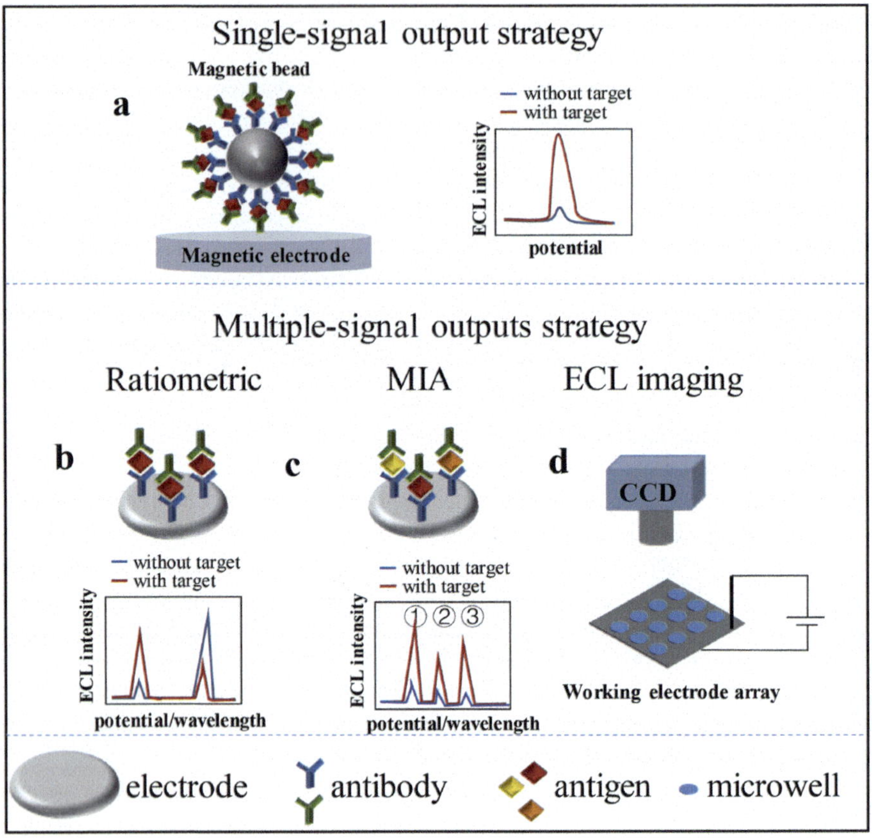

Fig. 1 Summary of signal output strategies for ECL immunoassay: **a** single-signal output strategy and multiple-signal output strategy, including **b** ECL ratiometric, **c** ECL MIA, and **d** ECL imaging [20]

be present in matrices such as blood and urine [150]. The ECL technology can provide reliable quantitative analysis results. In addition, the combination of ECL and microscopic imaging techniques enables high sensitivity and spatial resolution detection of biomolecules, especially excelling in tracking molecular interactions in cell and tissue samples [1, 75]. ECL imaging technology is capable of not only endpoint detection but also real-time monitoring, especially for applications in dynamic biological processes. For example, real-time monitoring of cancer cells' proliferation or drug response can help physicians better grasp the progress of a patient's disease [105].

The development of ECL imaging technology has benefited from advances in nanomaterials, microfluidics, and signal amplification strategies. By introducing functionalized nanomaterials, the intensity and the stability of ECL signals can be significantly improved, enabling the detection of ultra-low concentration analytes [5, 74]. In addition, the combination of ECL imaging and microfluidic chips enables an integrated, high-productivity detection system, which has important application value in clinical diagnosis and prompt point-of-care testing (POCT), especially in multiplexed testing, and provides a new way of thinking about precise diagnosis and monitoring of complex diseases [26, 55, 137].

ECL technology shows great potential in the early diagnosis of major diseases such as cancer, cardiovascular diseases, and infectious diseases. In ECL immunoassays, specific biomarkers can be quantitatively detected in single cells or serum samples through ECL imaging [75]. ECL immunosensors have been used to detect cancer-related biomarkers such as carcinoembryonic antigen (CEA) [61, 110, 138] and prostate-specific antigen (PSA) [20, 68], which have significant advantages in improving early disease detection and reducing misdiagnosis. In addition, ECL is also used to detect important markers such as troponin in cardiovascular disease, providing an important reference for patient monitoring and treatment [96, 104, 119]. In addition to traditional disease diagnosis, ECL technology has been gradually expanded to environmental monitoring, food safety, drug screening, and other areas [155]. Through the development of multifunctional nanomaterials, ECL sensors are able to detect heavy metal pollutants in the environment, pesticide residues in food, etc. in an ultra-sensitive manner [62]. Therefore, biosensors based on ECL imaging analysis not only provide strong support for realizing the needs of modern medicine for high sensitivity, high specificity, and rapid detection, but also have great potential for application in non-medical fields (Table 1).

Based on these backgrounds, this chapter will focus on the application of ECL imaging technology in quantitative detection of biomarkers and multiplexed detection. We will first introduce the basic concepts of ECL imaging biosensors and their roles in biomarker detection; we will discuss in detail the fundamentals of ECL imaging technology, especially its applications in diagnosis of major diseases such as cancer, cardiovascular diseases, and neurodegenerative diseases. Subsequently, we will focus on the latest advances of ECL in multi-detection, including the combined application of microfluidic chips, smart sensors, and multifunctional nanomaterials. Finally, this chapter will also analyze the advantages and challenges of ECL imaging in biomarker quantification and multiplexed detection with specific application cases and look forward to its future development, especially its potential applications in precision medicine, environmental monitoring, and food safety.

2 Biosensor Immunoassay

2.1 Definition of ECL Imaging Biosensors

ECL imaging is an imaging technique that combines an electrochemical reaction with the emission of a light signal, and its unique principle gives it great potential for biomarker detection and multiplex assays [105, 138]. The basic principle is that a

Table 1 Milestones of qualitative changes in biosensors-immunoassay by ECL imaging and microscopy

Year	Milestone	Qualitative change/impact	References
1970s	Early development of electrochemiluminescence (ECL)	– Initial discovery of ECL phenomenon in organic and inorganic compounds – First studies on the use of ECL for analytical applications	[13, 38, 39, 101]
1980s	First ECL-based immunoassays	– Introduction of ECL for biomarker detection – Use of ruthenium-based luminophores for early immunosensing applications – Early ECL imaging: First use of ECL to characterize electrode heterogeneity and visualize active regions, laying foundation for ECL optical characterization	[16, 97, 94]
1990s	Commercialization and clinical applications	– Roche Diagnostics launched ECL-based immunoassay systems for clinical diagnostics (e.g., Elecsys platform) – ECL immunoassays began to replace ELISA in many clinical labs due to higher sensitivity and automation – ECL imaging advancement: ① Time-resolved Pt electrode current density imaging (0.3 μm spatial resolution, 30 ms temporal resolution); ② ECL-SECM fusion to build scanning imaging framework; ③ High-frequency ECL for microelectrode morphology characterization	[21, 108, 152–128]
2000s	Miniaturization & multiplexing of ECL biosensors	– Integration of microfluidics with ECL imaging for high-throughput screening – First multiplexed ECL assays enabling detection of multiple biomarkers simultaneously – Advancements in nanomaterials (gold nanoparticles, carbon nanotubes) to enhance ECL signals	[9, 53, 84, 152, 52]

(continued)

Table 1 (continued)

Year	Milestone	Qualitative change/impact	References
2010s	ECL bioimaging and advanced nanotechnology integration	– Introduction of ECL microscopy for single-cell analysis and tissue imaging – Development of ratiometric ECL sensors for more precise quantification – Incorporation of quantum dots, MOFs, and 2D nanomaterials to enhance signal stability and sensitivity	[30, 64, 90, 100, 131, 144, 145]
2020s	AI-driven ECL analysis and point-of-care testing	– Application of machine learning to optimize ECL data interpretation – Integration with portable biosensors for real-time and on-site detection – Use of wearable ECL sensors for continuous health monitoring – Expansion into environmental monitoring, food safety, and forensic applications	[8, 12, 73, 77, 88, 147, 89]

voltage applied between the electrodes induces a redox reaction of a specific chemical, resulting in the production of excited molecules, which return to the ground state releasing energy that is emitted as light. The intensity of the ECL signal is proportional to the concentration of the reactant and can therefore be used for highly sensitive imaging [54]. The signal is acquired using an image sensor device (e.g., based on a charge-coupled device (CCD) or a complementary metal–oxide–semiconductor (CMOS) technology) of a camera and quantitatively analyzed by image processing software.

A biosensor is a device for detecting biomolecules or biological signals based on the principle of converting biorecognition events into detectable signals by using specific interactions between biosensitive elements (e.g., enzymes, antibodies, or DNA probes) and target molecules [49]. Typically, biosensors read these signals physically, chemically, or electrochemically to enable quantitative or qualitative analysis of biomolecules. Immunoassays are analytical methods based on the specific binding between antigens and antibodies. By using antibodies as recognition elements, immunoassays are able to detect and recognize target molecules (e.g., proteins or pathogens) with a high degree of specificity. When a target molecule binds to an antibody, a detectable signal is typically generated, which can be measured by various methods (e.g., optical, chemical, or electrochemical means) [50]. In biosensors, sensors based on antigen–antibody binding are called immunosensors, with the antibody layer referred to as the immunoaffinity layer.

The principles of immunosensors and ECL imaging are combined with each other to form a powerful detection tool, the ECL imaging immunosensor.

2.2 Structure of ECL Imaging Biosensors

ECL imaging biosensors typically consist of several key components:

Electrode materials The core of the ECL sensor is the working electrode modified with luminophores, and the choice of electrode materials is crucial for the efficiency of the ECL reaction and the stability of the signal. Commonly used electrode materials include gold, platinum, glassy carbon, indium tin oxide (ITO), etc. These materials have excellent conductivity and chemical stability, and can provide ECL reaction efficiency and signal stability [51, 57]. By modifying the surface of the electrodes with nanomaterials (e.g., gold nanoparticles, carbon nanotubes, or biorecognition molecules), the conductivity and reactivity of the electrodes can be enhanced, thus increasing the intensity of the ECL signal [36, 62, 103].

Luminophores The most commonly used luminophores for ECL reactions include trisbipyridine ruthenium ($Ru(bpy)_3^{2+}$) and luminol [51]. When a potential is applied to the electrode, the luminophores react electrochemically on the surface of the electrode, further generating excited-state molecules in reactions with co-reactants, and subsequently release energy through luminescence [25]. The choice of the luminophore plays a key role in the sensitivity and detection range of the ECL sensor [57].

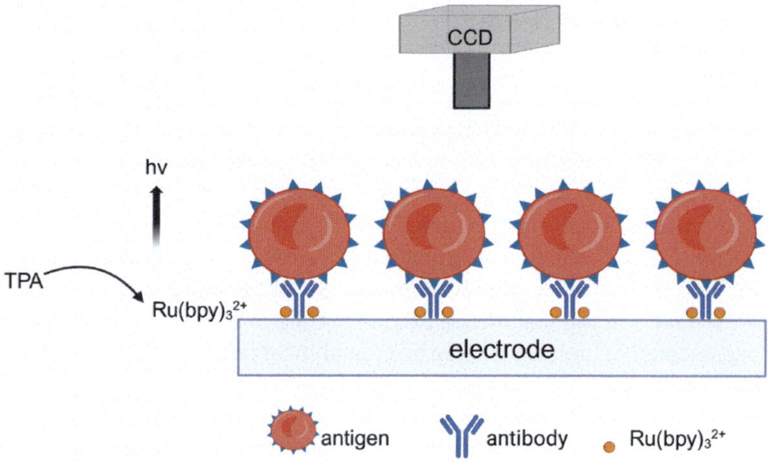

Fig. 2 ECL imaging immunosensor

Antibody–antigen system The ECL immunosensor uses the specific binding of antibody–antigen to achieve recognition and capture of target molecules. When the target antigen binds to the antibody immobilized on the electrode surface, the ECL signal changes, and the sensor achieves quantitative detection of the target based on the ECL intensity [63, 122].

Imaging biosensors and digital cameras A combination of imaging biosensors and digital cameras is used to capture ECL signals to enable imaging and analysis of samples. By using digital cameras, spatially resolved biomolecular detection can be performed in cell and tissue samples, which is especially important for real-time tracking of molecular interactions [40]. And this approach also allows signals from multiple targets to be captured and analyzed simultaneously [67] (Fig. 2).

3 Application of ECL Immunosensors in Biomarker Detection

Biomarkers are a class of substances (genes, proteins, polysaccharides, etc.) produced by normal physiological processes or biochemical pathways in the body that have valuable medical or biological significance [80]. Pathology-associated biomarkers can provide physicians with in-depth information about the health status of the body with minimal disruption and less effort or warn of possible diseases that may be evolving, which is important for patients to receive medical treatment that effectively controls their disease [141].

In addition, there are many biomarkers that can be detected by ECL immunosensor in non-medical fields, such as environment and food testing, which greatly extends the potential of ECL imaging immunosensor applications.

3.1 Cancer Marker Testing

Cancer remains one of the major threats to human health globally, with high prevalence and mortality rates [117]. The prevalence and mortality rates remain high. Although significant advances have been made in cancer treatment in recent years, these advances have focused on individual cases of early-stage, non-metastatic cancers [82]. For the majority of patients, the diagnosis of the disease is not always straightforward. For the majority of patients, the overall prevalence and mortality rates have not been significantly reduced, as most of them are in the middle to late stages or even have malignant metastases at the time of initial diagnosis. Therefore, finding representative biomarkers, sensitive analytical reaction, and practical assays is essential for the early diagnosis of cancer [54, 64].

Various standard techniques for the detection of cancer biomarkers are shown in Fig. 3, and among all these potential techniques, the ECL biosensors have been shown to have a better potential for clinical applications [66]. ECL technology is widely used in the detection of cancer markers, especially for the quantification of cancer-related proteins and circulating tumor cells (CTCs). Common protein cancer markers, such as CEA (a common marker for many types of cancers such as colorectal, breast, and lung cancers), prostate specific antigen (PSA), and glycan antigens (e.g., CA125, CA15-3), can be ultrasensitively detected by ECL sensors [24, 35, 61]. These biomarkers are crucial for cancer diagnosis and monitoring. Studies have shown that using ECL technology, the detection limit for CEA can reach picomolar levels, which is significantly better than the detection limit of conventional methods. In addition, ECL immunoassays enable a comprehensive assessment of a patient's tumor status through the simultaneous detection of multiple tumor markers.

For example, ECL immunosensors based on ruthenium complexes have demonstrated excellent performance in the detection of a wide range of cancer markers. Through signal amplification strategies (e.g., tyramine signal amplification [14], two-potential ECL [138] and the introduction of nanomaterials (e.g., gold nanoparticles, carbon nanodots, etc.), ECL sensors achieve ultra-low detection limits for these biomarkers. For example, a label-free ECL sensor based on gold nanoparticles and aptamers is capable of detecting CTCs with high sensitivity, which has essential applications in early diagnosis and prognostic monitoring of cancer [7, 15].

There is a wide variety of cancer markers, and below we broadly classify them into protein markers, carbohydrate markers, and cell-associated markers (Fig. 4).

3.1.1 Protein-Based Cancer Markers

Protein markers are particularly significant in the early diagnosis of cancer, and ECL technology combined with immunoassays is able to sensitively detect protein-based markers, such as alpha-fetoprotein (AFP), CEA, PSA, and human epidermal growth factor receptor 2 (HER2).

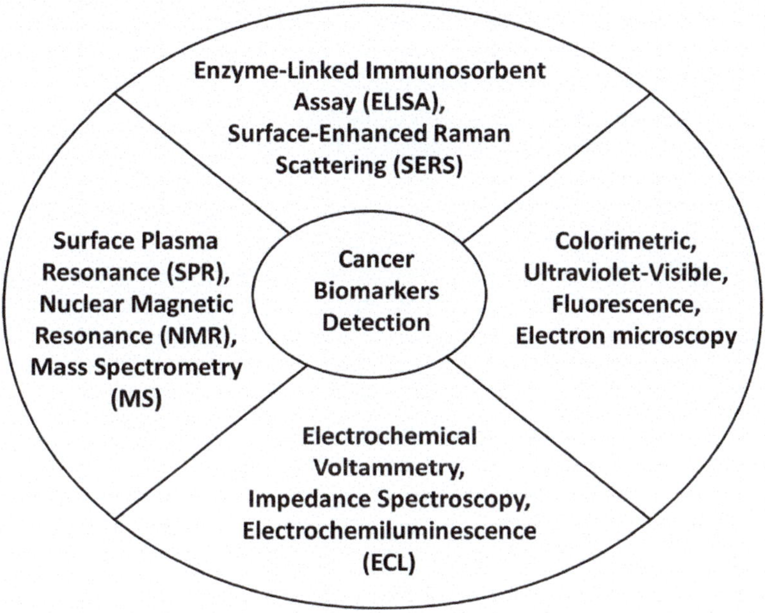

Fig. 3 Schematic representation of standard techniques used to detect cancer biomarkers [78]

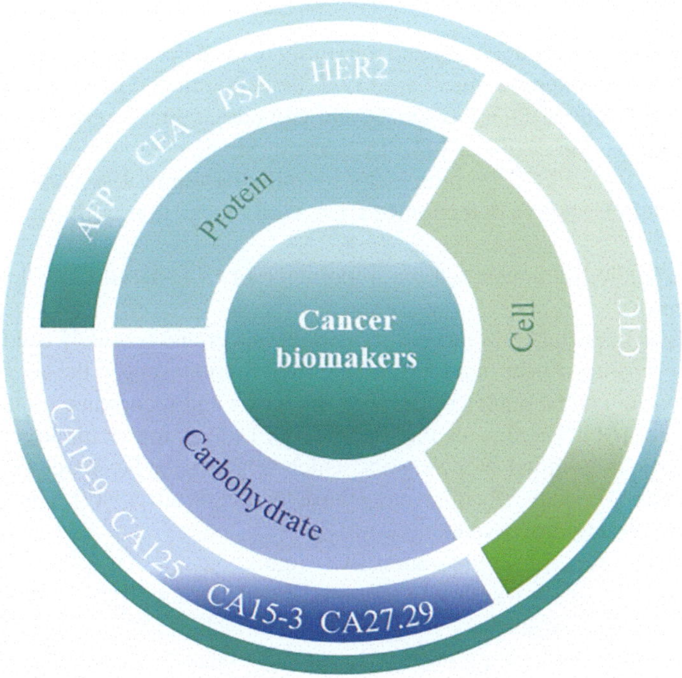

Fig. 4 Classification of cancer biomarkers

AFP It is an important marker for the diagnosis of primary liver cancer. The ECL sensor enables highly sensitive detection of AFP in complex serum samples, especially for low concentrations of AFP [25].

Using gold nanoparticles and carbon quantum dots as signal amplification materials, Wang et al. successfully constructed an ECL sensor for the detection of AFP with a detection limit of picomolar level [74, 156]. In addition, Wu et al. [129] developed an ECL imaging-based microfluidic chip platform for the simultaneous detection of multiple cancer markers, including AFP, which achieved signal amplification through the ECL resonance energy transfer (ECL-RET) mechanism and exhibited high sensitivity and specificity. These methods show good accuracy in clinical human serum samples and provide a powerful tool for early diagnosis of AFP-related cancers. In addition, by using a bipolar electrode (BPE) device, Li et al. achieved a sensitive detection of AFP on the surface of HepG2 cells using $Ru(bpy)_3{}^{2+}$-coated SiO_2 nanoparticles as ECL emitters, with a LOD of 6.71 pg/cell [58].

CEA It is a broad-spectrum protein biomarker for a variety of malignant tumors and is commonly associated with pancreatic, colorectal, and breast cancers [120].

MOF-based ECL sensor for simultaneous detection of lung cancer markers significantly improves detection efficiency [78]. Gao et al. prepared an aggregation-induced emission (AIE) active Zn^{2+}-doped gold nanoclusters (Zn^{2+}-AIE-Au NCs) and used them for real-time multiplexed detection with a sensitivity of 0.3 pg/mL for CEA [157]. In addition, Zhang et al. employed silica inverse opal photonic crystals (SIOPC) to enhance the ECL output with a detection limit of 0.032 pg/mL for CEA [149].

In order to improve the accuracy and reliability of the detection, Huang et al. developed a potential/wavelength ratio-based ECL detection method combining luminol and CdS quantum dots (QDs) as anode and cathode emitters, respectively, which showed high detection accuracy [41]. In contrast, Shang et al. designed a potentiodynamically resolved ECL immunoassay by utilizing luminol as the sole emitter, achieving an ultrasensitive detection of CEA with a detection limit of 87.1 ag/mL [98].

All these studies demonstrated the importance of ECL-based CEA detection in cancer diagnosis.

PSA It is an important biomarker for prostate cancer and is widely used for early screening of prostate cancer in men. With the ECL immunosensor, the concentration of PSA in serum can be detected for early diagnosis and recurrence monitoring of prostate cancer [61]. There is also a label-free ECL detection technology based on gold nanoparticle-modified ECL sensors, which achieves the label-free detection of PSA through aptamers, greatly improving the detection sensitivity [7]. The sensitivity of detection is greatly improved.

Qin et al. developed a double-burst ECL immunoassay that achieved a highly sensitive detection of PSA with a LOD of 2.2 fg/mL by modifying gold nanoclusters (ATT-Au NCs) as emitters [86]. Wang et al. achieved a sensitive detection of PSA with a LOD of 3.0 pg/mL by designing a label-free aptasensor based on luminol-functionalized molecularly imprinted polymers (MIPs) of electropolymerized dopamine films [118].

To further improve the sensitivity, a label-free ECL immunosensor based on Ru@SiO_2-PEI-Au@Pt hybrid nanostructures was developed, achieving a low detection limit of 6.6 fg/mL for PSA [33]. In addition, Zhao et al. employed metal organic gel Cys-[Ru(dcbpy)] as ECL emitters, ZnS quantum dots as a quencher, and aminated mesoporous silica nanocontainers (SiO_2-NH2) as ECL bursters, which significantly improved the sensitivity of PSA detection through the RET mechanism, with a low LOD of 1.01 fg/mL [151].

Human epidermal growth factor receptor 2 (HER2) It is an important marker for breast cancer, and in some breast cancer patients, HER2 protein is overexpressed and is an important reference marker for targeted therapy.

By combining ECL with nanomaterials, researchers have developed an efficient immunosensor for the detection of breast cancer markers such as CEA, cancer antigen 15-3 (CA15-3), and HER2. These biomarkers are key indicators for the diagnosis and monitoring of breast cancer, and the signal amplification effect is remarkable through the rationally designed nanoplatform, which improves the sensitivity and selectivity of detection [24].

In addition, the ECL sensor also enables highly sensitive quantitative detection of HER2 proteins by using a combination of ruthenium complexes and aptamers [25]. This label-free approach reduces errors in conventional antibody detection, and the developed sensor has a very low detection limit, making it suitable for early breast cancer screening.

Neuron-specific enolase (NSE) It is widely used in the detection of neurodegenerative diseases and small cell lung cancer. ECL technology provides an effective tool for comprehensive disease diagnosis by enhancing electrochemical activity and allowing simultaneous detection of NSE and other markers [138].

Dong et al. developed a dual-signal ratiometric ECL platform based on $Ru(bpy)_3^{2+}$-functionalized Zr-MOF, which enabled the efficient detection of NSE with a detection limit of 41 fg/mL [19]. Han et al. used nitrogen-doped carbon quantum dots (N-CQDs) as ECL emitters in combination with molybdenum sulfide/ferric oxide (MoS_2@Fe_2O_3) co-reaction promoter to achieve an ultrasensitive detection of NSE with a detection limit of 6.30 fg/mL [37]. The results showed that the detection limit of NSE was 6.30 fg/mL.

In addition, Yang et al. constructed a conductivity-based NSE assay using AIE-activated biocompatible emitters, which further improved the sensitivity and accuracy of the assay [142].

CYFRA 21-1 It is one of the important markers for non-small cell lung cancer (NSCLC).

Yang et al. constructed an AIE-activated ECL emitter for CYFRA 21-1 detection by encapsulating $Ir(ppy)_3$ into the desferritin cavity, achieving high sensitivity and accuracy [142]. Li et al., on the other hand, developed a BPE-assisted ECL-based immunoassay for the sensitive detection of CYFRA 21-1, which showed good detection performance [56].

3.1.2 Sugar-Based Cancer Markers

Glycoprotein markers, such as the glycan antigens CA19-9 and CA125, are widely used in the diagnosis of pancreatic, ovarian, and other cancers, and the ECL immunosensors are able to bind to these glyco-biomarkers with specific antibodies, enabling their ultra-sensitive detection. CA15-3 and CA27.29: These glycoprotein biomarkers are often elevated in breast cancer patients and can be used for early screening and recurrence monitoring [35]. These biomarkers are often elevated in breast cancer patients and can be used for early screening and recurrence monitoring.

Hao Yang's team developed an ECL imaging immunosensor for the detection of the glycan antigen CA19-9 using a handwritten paper-based electrochemical platform with gold nanoparticles covered with trisbipyridine ruthenium. The sensor was attached to the detection antigen CA19-9 by adsorbing a detection antibody, and the ECL signal showed a logarithmic relationship with the concentration of CA19-9, which achieves the advantages of high sensitivity and low cost [74].

3.1.3 Cell-Related Cancer Markers

CTCs are cells shed from tumors into the bloodstream, and their presence is strongly associated with cancer metastasis and recurrence. ECL technology has shown good potential for CTC detection and is capable of sensitively detecting individual CTCs.

By combining magnetic nanoparticles and quantum dots, the ECL sensor is able to recognize and capture CTC, enabling its precise detection at low concentrations. Combined with microfluidic chip technology, the ECL system enables rapid detection of CTC in complex blood samples [63].

3.2 Application of ECL in the Detection of Cardiovascular Disease Markers

Early diagnosis of cardiovascular disease is critical to preventing cardiovascular events (e.g., myocardial infarction, heart failure, etc.), and ECL technology provides new means of detecting markers of myocardial injury (e.g., troponin I and creatine kinase MB). Studies have shown that ECL immunoassays can provide accurate information about myocardial injury in a short period of time, helping physicians make timely diagnostic and therapeutic decisions [121].

The detection of cardiovascular biomarkers, such as troponin, myoglobin, and CRP biomarkers, can provide valuable insights into the severity and prognosis of cardiovascular disease. Below, we have categorized them into protein-based cardiovascular markers, metabolic cardiovascular markers, and nucleic acid-based cardiovascular markers.

3.2.1 Protein-Based Cardiovascular Markers

Protein-based cardiovascular markers are the main indicators of cardiac impairment, and ECL sensors play an important role in the detection of protein markers in cardiovascular disease.

Troponin I (cTnI) cTnI is a specific marker of myocardial injury.

Ming Wang's team used ECL technology combined with electrode modification with gold nanoparticles and carbon nanotubes to achieve ultrasensitive detection of cTnI with a detection limit up to the pg/mL level [63]. Mohammad Saremi et al. designed a nanoparticle-based ECL aptasensor relying on polymerization-induced ECL enhancement for the detection of cardiac troponin I (CTnI), a biomarker of acute myocardial infarction. The sensor enhances the sensitivity and signal stability of the assay through the use of gold nanoparticles, nitrogen-doped graphene, and cyclometallated iridium(III) polymer nanoparticles. The literature reports that the sensor exhibits a good linear response to cTnI concentrations from 0.1 pM to 10 nM, with a detection limit as low as 0.02 pM, making it suitable for the early diagnosis of acute myocardial infarction [96].

Creatine kinase type MB (CK-MB) CK-MB is used to diagnose myocardial infarction. Through ECL technology combined with immunoassay, the researchers developed a sensor based on cadmium sulfide quantum dots, which is capable of sensitively detecting CK-MB with a very low limit of detection [74].

BNP heart failure is the result of systolic or diastolic dysfunction of the heart and is often considered the end stage of many cardiovascular diseases [140]. BNP is a crucial biomarker of heart failure and is of prognostic importance in heart failure and other cardiovascular diseases [120]. By combining carbon quantum dots with specific antibodies, the ECL sensor can detect BNP at very low concentrations with significantly higher sensitivity than conventional methods.

Multi-parameter analysis-based ECL sensors enable the simultaneous detection of multiple biomarkers (e.g., cardiac gonadotropin BNPT and cardiac troponin I, cTnI) via multivariate linear algebraic equations. This strategy utilizes nucleic acid dendrimers and amplification methods (e.g., hybrid chain reaction HCR and rolled-loop amplification RCA) to achieve significant signal enhancement, resulting in more sensitive and accurate assays [11]. Ji et al. [43] designed a dual-signal ECL immunoassay using $Ru(bpy)_3^{2+}$@HKUST-1 as the anodic ECL emitter and Ce_2SnO_{27} nanocubes as the cathodic emitters, in which the detection accuracy of NT-proBNP was well improved thanks to the self-calibration of the anodic and cathodic ECL responses.

3.2.2 Metabolic Cardiovascular Markers

Metabolic markers (e.g., cholesterol, lipoproteins, glucose, etc.) play a leading role in predicting the risk of cardiovascular disease, and ECL technology is able to efficiently detect metabolic markers by combining electrode surface modification with nanomaterials.

Low-density lipoprotein cholesterol (LDL-C) LDL-C is a risk factor for atherosclerosis.

A label-free ECL immunosensor for the detection of LDL and ox-LDL was developed by Lian et al. [130]. They immobilized gold-cobalt alloy nanoparticles (Au–Co NPs) on APTMS-silanated electrodes with antibodies against LDL and ox-LDL. The limits of detection were calculated to be 0.256 pg/mL and 0.330 pg/mL, respectively, with linear ranges of 0.420–100 pg/mL and 0.500–60 pg/mL, respectively, as measured by ECL.

CRP It is one of the acute phase protein one of the most important indicators. Its abnormal elevation is associated with a variety of inflammatory conditions and diseases such as myocardial infarction and atherosclerosis (AS) [146]. Current studies have shown that CRP is mainly produced in the liver in response to interleukin 6 (IL-6), a soluble mediator that has an important role in inflammation, immune response, and hematopoiesis, with pleiotropic effects on inflammation, immune response, and hematopoiesis [113]. During the inflammatory response, leukocytes phagocytose bacteria and lead to the production of interleukin release, which leads to a rapid increase in CRP. Studies have shown that CRP is also associated with the formation of atherosclerosis, and large deposits of CRP have been found in human AS plaques. During AS plaque formation, CRP, its complementary complexes, and other cells are deposited on the arterial wall. CRP can interact with lipoprotein binding, activating the complement system and generating large amounts of inflammatory mediators, inducing oxygen free radical release, damaging the endothelium, and mutating the plaque. In healthy people, CRP levels are generally below 10 mg/L but increase exponentially when inflammation occurs, reaching 350–40,010 mg/L [116]. Therefore, CRP has been designated as a key component of cardiovascular disease and acute myocardial infarction and acute myocardial infarction [4]. As these diseases usually begin with low CRP levels, the development of highly sensitive and specific CRP assay methods is important for clinical diagnosis [91].

An ECL immunosensor based on a composite of titanium nanotubes (TiNTs) and platinum nanowires (PtNWs) was used for the detection of C-reactive protein (CRP). The sensor employs trisbipyridine ruthenium as a luminophore and detects the CRP concentration by ECL signals in the range of 0.05–6.25 ng, with a detection limit of 0.011 ng. This sensor has the advantages of high sensitivity, good selectivity, and low cost, which makes it suitable for rapid clinical detection and disease prevention [91].

3.3 Kidney Disease Marker Tests

Early diagnosis of kidney disease is essential to prevent renal failure and improve patient prognosis. ECL technology is a valuable tool for detecting markers of kidney disease through its high sensitivity and low background noise [102].

3.3.1 Creatinine Test

Creatinine is an important biomarker in assessing renal function, and traditional assays include enzymatic, chemical colorimetric, and fluorescent assays, which often suffer from insufficient sensitivity.

The ECL-based creatinine sensor significantly improves the selectivity and sensitivity to creatinine by introducing copper(II) complexes to modify the electrodes, with detection limits down to the nanomolar level. The rapid response capability of this sensor makes it promising for important applications in the early screening of kidney disease. In addition, the creatinine detection platform can be combined with other renal function markers to form a multi-analysis system, enabling simultaneous detection of multiple indicators in patients with kidney disease [2]. The platform can also be combined with other renal function markers to form a multi-analysis system to achieve simultaneous detection of multiple indicators in kidney disease patients.

3.4 Neurological Disease Marker Tests

Early diagnosis and monitoring of neurological diseases (e.g., Alzheimer's disease, Parkinson's disease, etc.) is one of the focuses of neuroscience research. Markers of neurological diseases usually focus on specific proteins or nucleic acids, such as Aβ proteins, Tau proteins, etc. By detecting these neuron-specific markers, ECL technology can provide the possibility of early diagnosis and provide a basis for clinical intervention.

3.4.1 Detection of Alzheimer's Disease Markers

AD has become a worldwide brain disease that threatens the health of older people [44]. It is crucial to protect them from AD and to get timely treatment at an early stage. *Amyloid-β* protein (Aβ1-42) is considered to be the most critical and predictive protein biomarker for the diagnosis of AD.

Qin et al. [87] presented a dual-wavelength ratiometric ECL immunoassay for Aβ1-42 *analysis* based on two ECL emitters, including Ru@TiO_2@Au (620 nm) with different emission wavelengths, and AuNPs modified gCN_{34} (460 nm). By virtue of the ECL-RET between these two emitters, sensitive detection of Aβ1-42 was achieved in cerebrospinal fluid with an ultra-low LOD of 2.6 fg/mL. In addition, Tan et al. [92] developed an advanced ECL aptamer sensing system by affinity screening of junction aptamers and size screening of silica nano-channels. The sensing mechanism was operated by a spatial site barrier formed by the specific interaction of Aβ1-42 on the arrayed nano-channels with the aptamer, which impeded the luminol mass transfer and led to the *reduction of* ECL corresponding to the concentration of Aβ1-42. Recently, Alzheimer-associated neuronal filament protein (AD7c-NTP) has been identified in the cerebrospinal fluid, blood, and urine of AD patients, and may be considered as a new and reliable biomarker for AD diagnosis [133]. However, only an ELISA was used to detect AD7c-NTP. In order to obtain higher sensitivity than ELISA, the superior ECL technique is more suitable for ultra-trace analysis of AD7c-NTP. Liang et al. [65] recently constructed a dry chemistry-based BPE device for ECL determination of AD7c-NTP. The device consisted of a screen-printed fiber chip filled with Ru(II)-poly-L-lysine complex for immunoreactivity, which allowed a self-enhanced ECL assay to be completed in less than 6 min, with a LOD of 0.15 pg/mL sample in human urine.

In addition, through ECL technology, the researchers developed an ECL sensor platform based on functionalized nanoparticles capable of simultaneously detecting Aβ and Tau proteins [2]. This multi-detection platform helps early detect pathological changes in Alzheimer's disease and provides data to support treatment.

3.4.2 Detection of Parkinson's Disease Markers

Alpha-synuclein is an important marker for Parkinson's disease.

The ECL sensor enables ultra-sensitive detection of alpha-synuclein in cerebrospinal fluid, helping clinicians to diagnose Parkinson's disease at an early stage [7].

3.4.3 Detection of Neurodegenerative Diseases and Neuroinflammation

NSE is widely used in the detection of neurodegenerative diseases and small-cell lung cancer. ECL technology provides an effective tool for comprehensive disease diagnosis by enhancing electrochemical activity and allowing simultaneous detection of NSE and other markers [138].

In addition, ECL technology has important applications in the detection of markers associated with neuroinflammation (e.g., matrix metalloproteinase MMP-3) [63]. This test not only helps doctors to determine the progress of the disease but can also be used to assess the effectiveness of treatment.

3.5 Infectious Disease and Pathogen Detection

Rapid diagnosis of infectious diseases is of great significance for epidemic control and public health protection. The application of ECL technology in the diagnosis of infectious diseases is not limited to detecting a single pathogen; it can simultaneously detect multiple pathogens via a multi-analysis platform.

3.5.1 Multi-Pathogen Detection

Researchers have developed an ECL-based multi-pathogen analysis platform that enables simultaneous detection of multiple pathogens (e.g., viruses, bacteria, fungi) by using different ECL probes. This system is capable of providing results for multiple pathogens in a relatively short period and is suitable for large-scale infectious disease screening [5].

Viruses Simultaneous detection of influenza virus, *Streptococcus Pneumoniae* and *Staphylococcus Aureus* was achieved using CdSe and CdTe nanoparticles labeled with different ECL probes. This technology provides strong supports for rapid disease diagnosis and epidemic control [5].

Mycotoxin ECL imaging technology in mycotoxin detection, through nanomaterials, enzyme amplification, and DNA chain reaction, significantly improved the sensitivity and efficiency of the test [36].

Anthrax A bimodal fluorescence and electrochemiluminescence (R- FL/ECL)-based dual-mode detection platform using terbium and europium-based metal–organic frameworks (Ln-MOFs) has been developed for the detection of the anthrax biomarker, 2,6-pyridine dicarboxylic acid (DPA), in a recent study. This dual-modal platform significantly improved detection sensitivity and selectivity, enabling rapid and reliable DPA detection in water and serum samples [134]. The bimodal platform significantly improves the sensitivity and selectivity of the assay.

3.5.2 Detection of Infectious Diseases

ECL technology has demonstrated superiority in the detection of infectious diseases, especially in the early detection of infectious diseases such as novel coronaviruses (SARS-CoV-2). ECL immunoassays against specific viral antigens or antibodies enable rapid and accurate identification of infected individuals and support public health surveillance. ECL-based biosensors have been used for the detection of viruses such as SARS-CoV-2, influenza virus, HIV, HPV, and hepatitis B, demonstrating their advantages of high sensitivity and low background noise in infectious disease surveillance [104].

Yuansheng Li et al. developed an ECL imaging immunosensor based on m-MOFs for the detection of SARS-CoV-2 nucleocapsid proteins, which achieved the advantages of high sensitivity and low detection limit (56 fg/mL) by designing without additive co-reactants [59].

3.6 *Other Disease Detection*

3.6.1 Sepsis Detection

Sepsis is a systemic response to infection, usually caused by bacteria, and results in life-threatening inflammation throughout the body. Calcitoninogen (PCT) secreted by thyroid C-cells shows a strong correlation with the severity of sepsis and is recognized as a representative biomarker for the early diagnosis of sepsis.

Xu et al. [135] developed a sandwich quenching ECL immunosensor based on ferrocenecarboxylic acid (Fca) encapsulated into silica NP as a quencher of the ECL reaction. At the same time, n-butyl diethanolamine, a co-reactant, co-doped in the same silica NP bursts the ECL reaction. The ECL immunosensor was used to detect PCT, which alters the balance between bursting and quenching in the ECL reaction. In this assay, good sensitivity for PCT detection was achieved by an electron transfer-mediated bursting mechanism, which helped to achieve a low LOD of 0.85 pg/mL. The ECL-RET has also been substituted for the quenching approach, which is considered a highly sensitive technique for PCT detection. Song et al. [106] utilized CNTs and gC_3N_4-CNT@Au as ECL donors and CuO@ PDA as acceptors, respectively. The RET from gC_3N_4 -CNT@Au to CuO@PDA was highly efficient and could sensitively detect PCT as low as 25.7 fg/mL in human serum. In addition, Wang et al. [126] prepared a dual-emission ECL immunosensor using MnO_2 nanoflowers (NFs) and ZnS QDs as dual ECL emitters and $S_2O_8{}^{2-}$ as co-reactants, which showed good linearity in the range of 0.1 pg/mL to 100 ng/mL. The LOD was 0.033 pg/mL for PCT detection in serum. Recently, Song et al. [107] designed a ternary ECL sensing system using lipoic acid (LA)-capped Ag NC as the cathodic ECL emitter in an electrolyte containing $S_2O_8{}^{2-}$. In this assay, ligand-to-metal charge transfer between LA and Ag NC induced the desired ECL emission, thereby conferring a low LOD of 3.56 fg/mL to the PCT immunosensor.

3.6.2 Diabetes Testing

C-peptide is a pancreatic beta-cell byproduct of the insulin secretion process; its levels in body fluids are strongly associated with diabetes.

Literature [70] an immunosensing strategy based on label-free ECL signaling is reported for the simple and effective detection of C peptides with high sensitivity and specificity. The substrate electrode consists of ITO glass as a conductive substrate decorated with AuNP and 3-aminopropyl trimethoxysilane as a linker. The immunosensor, after immobilization of the C-peptide antibody, took full advantage of the binding capacity of the AuNPs and allowed the quantification of C-peptide using luminol as an ECL probe. By measuring ECL inhibition, a calibration could be established to report 0.05 ng/mL and 100 ng/mL with a detection limit of 0.0142 ng/mL. As a proof-of-concept, the proposed strategy is a promising and multifunctional platform for clinical diagnosis, classification, and research of diabetes.

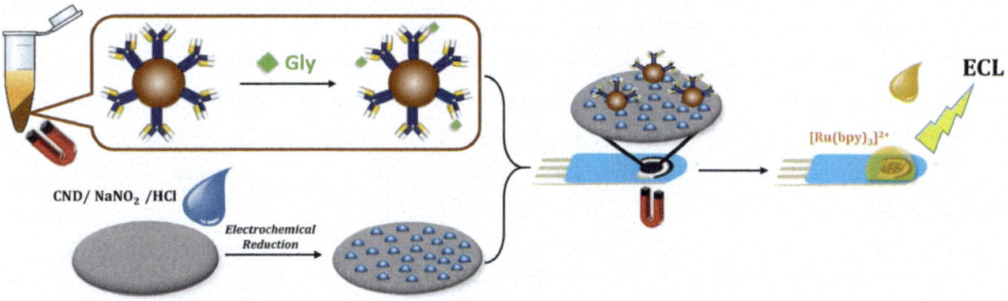

Fig. 5 Magnetic immunosensor for glyphosate [31]

3.7 Applications in Environmental and Food Safety Testing

ECL biosensors have been widely used in environmental monitoring and food safety testing in recent years due to their high sensitivity, specificity, low cost, and fast response.

3.7.1 Detection of Glyphosate

Glyphosate is a widely used non-selective herbicide whose large-scale application in agriculture has caused environmental and health problems because of contamination of water, soil, and food [115, 132].

A label-free ECL immunosensor was developed to achieve sensitive detection of glyphosate. The sensor improves the sensitivity of the ECL signal by means of a screen-printed carbon electrode modified with nitrogen-rich CND and incorporates the selectivity of an anti-glyphosate antibody. The development of this sensor provides a rapid, sensitive, and simple method for the detection of glyphosate in environmental samples and food samples [31, 66] (Fig. 5).

3.7.2 Heavy Metal Ion Detection

Significant progress has also been made in the application of ECL technology in the detection of heavy metal ions.

An ECL sensor based on hollow AuAg nanoboxes has been proposed in the literature for the ultrasensitive detection of lead (Pb^{2+}) and mercury (Hg^{2+}) ions using DNA enzymes and the "T-Hg^{2+}-T" structure. The sensor exploits surface plasmon resonance (SPR) and co-reaction acceleration to provide high sensitivity and specificity of heavy-metal detection in environmental monitoring [62]. The sensor has been tested with real samples.

3.7.3 Genetically Modified Crop Testing

Genetically modified (GM) crops are the first to have their DNA plants genetically modified or inserted with an exogenous gene. GM crops have desirable traits, such as resistance to pests and diseases or increased nutrient content, compared to conventional crops [23].

Zeng Haijuan et al. developed two different models of ECL imaging immunosensors for the detection of CP4-EPSPS proteins in transgenic crops. One is based on nitrogen-doped graphene and g-C_3N_4 composites to construct a signal-attenuated sensor, and the other is constructed by CdSe/ZnS quantum dot-labeled antigen to construct a signal-enhanced sensor, which achieves the advantage of sensitive and quantitative detection of CP4-EPSPS protein. With high sensitivity, stability, and reproducibility, these sensors can be used to effectively regulate transgenic crops [148].

3.7.4 Estrone (E1) Test

E1 is an important environmental endocrine disruptor (EDC) found in the environment, including water and soil, and may have a negative impact on the reproduction and development of organisms. E1 is also used in the assessment of human health, particularly in the early diagnosis of menopause in women and prostate cancer in men [76, 95, 109].

The current national standard detection methods, such as high-performance liquid chromatography–mass spectrometry (LC–MS/MS), are sensitive but expensive and operationally complicated. To this end, a novel MIP-ECL sensor was constructed by combining MIP with an ECL sensor for sensitive, real-time detection of E1 in environmental samples, which not only improves the specificity and sensitivity of the detection, but also achieves effective separation, enrichment, and quantification of trace targets in complex wastewater [10]. The sensor is designed for sensitive real-time detection of E1 in environmental samples.

3.7.5 Detection of Harmful Residues in Foods

Xia Wei's team developed an immunosensor based on ECL imaging for the detection of harmful residues such as antibiotics in food. By introducing gold nanoparticles on the sensor, the team improved the sensitivity of the ECL signal, which enabled the ultra-sensitive detection of antibiotics, achieving the advantages of high selectivity, low detection limit, and low cost [63].

3.8 Summary

In this section, biomarkers are classified and listed according to different diseases and uses, and the applications of ECL immunosensors in biomarker detection are summarized, which cover quantitative and partially multiplexed analysis applications, reflecting the wide range of applications of ECL immunosensors in both medical and non-medical fields. Applications in multiplexed assays are given more systematically in the following section.

4 Application of ECL Immunosensors in Multiplex Detection

Many of the aforementioned ECL immunosensors can only enable the detection of a single antigen in a single run. Indeed, the rapidly increasing need for simultaneous detection of multiple targets in immunodiagnostics is placing greater demands on the signal resolution strategies of multiple ECL transducers [32]. Depending on the signal resolution strategy, we can categorize ECL immunosensors into various forms, such as electrochemical signal resolution (e.g., current, potential, or impedance changes), wavelength resolution, spatial resolution, temporal resolution, and multiple signal resolution.

4.1 ECL Immunosensors with Different Signal Resolution Strategies

4.1.1 Electrochemical Signal Resolution and Wavelength Resolution

ECL immunosensors based on electrochemical signal resolution are analyzed by detecting electrochemical signals (e.g., currents, potentials, or impedance changes) generated by target molecules during electrochemical reactions. In such sensors, electrochemical signals serve as the primary source of signal output, and ECL signals are excited and captured electrochemically. It provides high sensitivity, low detection limit analysis, and fast response time. Jiang et al. [34] proposed a potential solution for the simultaneous detection of CEA and AFP on the cell surface. CEA and AFP were detected by two-potential scanning using luminol and $Ru(bpy)_3^{2+}$ labeled antibodies as ECL probes (Fig. 6a).

On the other hand, ECL immunosensors based on wavelength-resolved signal discrimination can distinguish between different markers or chemicals by discriminating the wavelength of the ECL emission signal. In such sensors, the ECL signals are set to different wavelengths so that the responses of various target molecules can

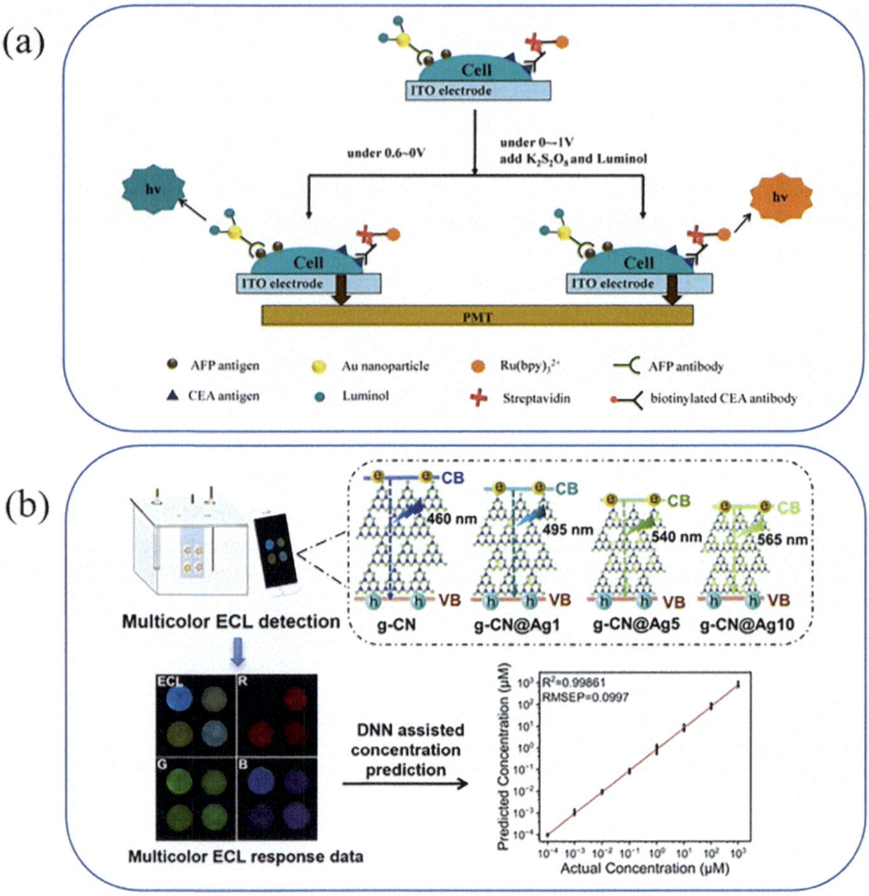

Fig. 6 Multiple ECL signal imaging for quantification: (**a**) schematic of simultaneous detection of CEA and AFP antigens on the cell surface based on potentiometric resolution approach [34], (**b**) schematic of multiple ECL imaging based on wavelength-resolved strategy [60]

be analyzed by wavelength characterization. Wavelength discrimination can effectively distinguish signals from different markers and is suitable for multiple analysis and complex sample detection. Thus, the ECL imaging system quantifies cell surface expression of CEA and AFP. Li et al. [60] reported a facile method to modulate the ECL emission wavelength of g-CN using Ag NPs as modifying materials. The g-CN@Ag nanocomposites with multicolor ECL emission were synthesized. The ECL wavelength of g-CN@Ag can be modulated between 460 and 565 nm by adjusting the content of loaded Ag NPs. The mechanism of multicolor ECL emission is discussed in the work. Then, multicolor ECL detection arrays were fabricated using the synthesized g-CN, g-CN@Ag1, g-CN@Ag5, and g-CN@Ag10 as ECL emitters for the determination of dopamine. A deep neural network (DNN) algorithm was combined to achieve machine learning-assisted multi-parameter concentration prediction of dopamine. This work reports for the first time the multicolor ECL performance of g-CN@Ag and presents the first example of a machine learning-assisted multicolor ECL detection strategy for multiparameter quantification (Fig. 6b).

4.1.2 Spatial and Temporal Resolution

ECL imaging can also combine spatially and temporally resolved techniques to enable multiple detection. By precisely designing the spatial distribution of electrodes, each electrode can be specifically modified for different target substances, thus enabling the detection of multiple markers at the same time. In addition, through a time-resolved approach, the excitation and decay characteristics of different luminophores can be exploited to collect signals at different time points, thus effectively differentiating multiple target analytes. This approach is beneficial for dynamic monitoring of multiple biomarkers in complex diseases and can provide more comprehensive time-dimensional information on disease progression.

Li et al. [17] designed an ECL immunosensor with multiple independent electrode regions, each capable of capturing different targets, enabling spatially resolved multiplexed detection. In addition, the accuracy and sensitivity of multiplexed analysis can be further improved using time-resolved ECL signal acquisition, for example, good results have been achieved in the detection of different markers in single cells [46]. A multiplexed detection platform based on a SPCE and a portable ECL readout device is mentioned in the literature. The platform enables the simultaneous detection of three mTBI-related biomarkers (GFAP, h-FABP, and S100β) through a spatially resolved ECL method. This multi-detection capability facilitates the rapid detection of multiple biomarkers on a single electrode, offering the potential for rapid diagnosis of traumatic brain injury, especially in field diagnostic (POC) applications.

4.1.3 Multiple Signal Resolution

This sensor uses a combination of multiple signal resolution strategies (e.g., combining signals of time, wavelength, intensity, etc. simultaneously) to improve analytical accuracy and multiplexing capabilities. Due to its ability to detect multiple targets simultaneously, it is suitable for multiple immunoassays or the detection of composite samples.

A BPE-ECL sensor based on $CuMoO_x$ electrocatalyst is also available for simultaneous detection and imaging of tetracycline (TET) and lincomycin (LIN). The sensor enables the simultaneous detection and imaging of TET and LIN using a spatial-potential-color resolution approach. The $CuMoO_x$ electrocatalyst catalyzes the electrochemical reduction of H_2O_2, which significantly enhances the ECL signal. This study demonstrates the potential of BPE-ECL sensors for microchip and in situ rapid detection (point-of-care testing, POCT) applications [55] (Fig. 7).

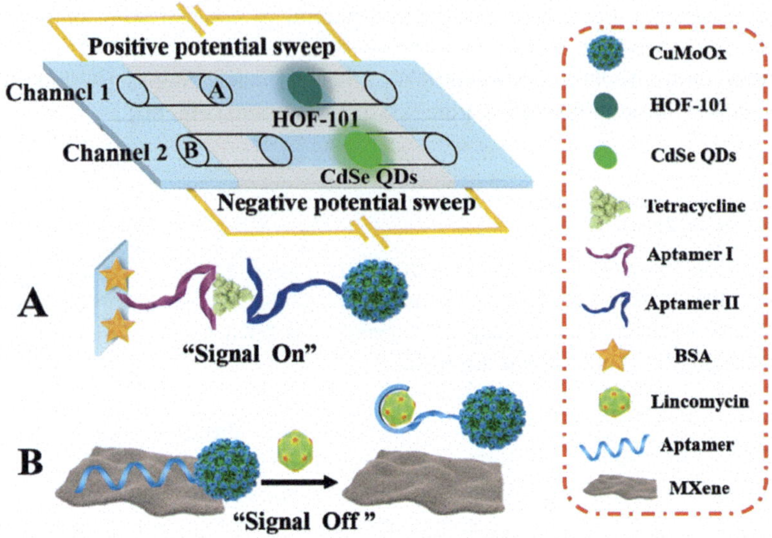

Fig. 7 Schematic diagram of the Spatial-Potential-Color-Resolved BPE-ECL biosensor using CuMoOx electrocatalyst for the simultaneous detection and imaging of TET and LIN[1]

4.2 Enhancement of Multi-detection Capabilities

Based on the above multiple strategies, ECL imaging has a powerful multi-detection capability, especially when combined with microfluidic chips, smart sensors, and multifunctional nanomaterials; its detection efficiency and sensitivity have been significantly improved. The combination of these technologies not only enables ECL imaging to detect multiple target molecules simultaneously but also improves the automation of the system and the data analysis capability, which greatly expands the application scope of ECL imaging, especially in biomarker quantification, environmental monitoring, and disease screening.

4.2.1 Combination of Microfluidic Chips and ECL Imaging

The combination of ECL imaging and microfluidic chips allows for greater automation and integration of multiplexed assays. By integrating multiple reaction channels on a single microfluidic platform, analysis of multiple target biomarkers can be performed simultaneously. The microfluidic chip not only allows precise control of fluid flow and reaction conditions, but also reduces the amount of reagents used and the reaction time, thus making the assay more efficient and economical. This approach is particularly suitable for situations that require rapid response and on-site diagnosis, such as mass screening during pandemic outbreaks [150].

[1] The sensing process of TET(A) and LIN(B) [55].

Microfluidic chips combined with ECL technology enable precise fluid control and reaction integration, thus reducing reagent and sample consumption while maintaining high sensitivity and throughput detection capabilities [138].

A microfluidic-based ECL biosensor has successfully achieved simultaneous detection of multiple cancer markers, such as CEA and glycan antigen (CA125) [105]. The combination of ECL imaging sensors with microfluidic chips is an area with excellent growth potential. By combining ECL imaging with microfluidic chips, sensors can enable multi-target detection in tiny samples while reducing sample consumption and assay costs. For example, Chen et al. developed a rotating paper disk-based ECL multistep assay device that detects CEA and PSA through simple manual operation. These portable and cost-effective devices demonstrate great potential for on-site analysis and immediate diagnosis [112].

4.2.2 Smart Sensors in ECL Imaging

The combination of smart sensors and ECL imaging technology greatly improves the level of automation and real-time analysis of detection. By linking with ECL signals, smart sensors can acquire and analyze detection data of multiple target molecules in real time. Incorporating artificial intelligence and machine-learning algorithms in the smart ECL sensor assists in automatically identifying and processing multiple signals in complex samples and automatically adjusting to real-time data.

Smart sensors can automatically analyze ECL signals through algorithms and optimize the detection process based on real-time feedback, reducing human intervention. For example, using machine learning techniques to analyze ECL signal changes, AI technology can significantly improve the efficiency of analyzing ECL detection data, especially in multi-detection and high-throughput data processing. Through machine learning algorithms, complex ECL data can be automatically analyzed to further improve detection accuracy and speed [79] and thus more rapidly and accurately detect multiple biomarkers [60, 79]. The ECL is a powerful tool for the detection of multiple biomarkers in a fast and accurate way.

Li et al. achieved highly sensitive detection of dopamine by g-CN@Ag nanocomposites combined with DNN algorithms for multi-parameter concentration prediction [60]. By incorporating AI technology, the researchers are able to optimize the design of the ECL sensor and predict the best response conditions. For example, in cancer detection, AI can help identify complex data patterns of different markers to improve diagnostic accuracy [122]. In addition, machine learning can help optimize the design of ECL sensors by predicting the best electrochemical reaction conditions and luminescent material combinations through data analysis and modeling [15].

4.2.3 Application of Multifunctional Nanomaterials in Multiplex ECL Detection

The introduction of nanomaterials has dramatically enhanced the multi-detection capability of ECL imaging technology. Multifunctional nanomaterials (e.g., gold nanoparticles, carbon nanotubes, graphene, quantum dots, etc.) can enhance the amplification of ECL signals, improve signal strength and detection sensitivity, as well as enable the sensor to selectively recognize multiple target molecules.

Nanomaterials have significantly improved the performance of ECL sensors by increasing the electrode surface area, enhancing the electron transfer efficiency, and synergizing with ECL luminophores. The application of novel nanomaterials, such as MOFs, has also led to the broader application of ECL technology in multiple detection [62, 78].

Studies have shown that ECL sensors combining quantum dots and nanoparticles can simultaneously detect multiple markers, such as CEA, PSA, etc., with very low detection limits [7, 25].

4.2.4 Implementation of High-Throughput and Multiple Detection

The multi-detection capability of ECL technology is not only applicable to the detection of clinical biomarkers but can also be used for high-throughput screening. In drug screening and clinical diagnostics, ECL imaging sensors combined with microfluidics and smart sensor technology can achieve rapid high-throughput detection of hundreds or thousands of samples, thus significantly improving screening efficiency [10].

By designing multicolor luminescent materials and multi-electrode systems, ECL imaging sensors can achieve simultaneous detection of multiple target molecules on the same platform. Multi-color luminescent materials are designed so that ECL imaging can distinguish multiple target analytes in the same region by using different luminophores emitting different colors. This color difference increases the level of multiple detection and enhances the ability to quantify biomarkers simultaneously. And ECL imaging is combined with a multi-electrode array platform to generate spatially resolved signals for different biomarkers. Each electrode can be modified with a different capture probe or antibody to enable simultaneous detection of multiple analytes. Multi-electrode arrays not only significantly increase the throughput of the assay, but also make it more suitable for large-scale screening efforts by reducing the cost of a single assay. This is particularly important in high-throughput screening and large-scale epidemiological investigations [104]. The technology has greatly improved the efficiency of the assay and has enabled it to be used in large-scale screening, also reducing experimental cost and time.

The dual-potential ECL platform serves as a multi-detection platform for the simultaneous detection of multiple cancer markers such as CEA, NSE, and CYFRA21-1 [138]. ECL imaging enables multiple detection of different viruses, such as SARS-CoV-2 and Hepatitis B virus, by combining metal–organic frameworks and luminescent nanomaterials [104]. In addition, selective excitation of different luminophores can be achieved by adjusting the electrode potential, thus enabling simultaneous detection of multiple analytes without increasing complexity.

4.3 Summary

In conclusion, ECL imaging shows great potential for applications in multiplexed detection. By combining multi-electrode arrays, color-tunable ECL, microfluidic chips, smart sensors, and multifunctional nanomaterials with ECL imaging, the applications of ECL technology in multiplexed detection and high-throughput analysis have been greatly extended. The combination of these technologies not only improves the sensitivity and detection efficiency of ECL sensors but also dramatically enhances their automation and multifunctional detection capabilities. Due to this, the ECL imaging technology has broad application prospects in the fields of medical diagnosis, environmental monitoring, and food safety [25, 62, 138]. In future, through the continuous optimization of multi-detection technology, ECL imaging technology will have a broad application prospect in medical diagnosis, environmental monitoring, and food safety. The integration of more advanced nanomaterials and microfluidic platforms, ECL imaging is expected to further improve the detection throughput and accuracy, and provide a more comprehensive solution for biomedical testing.

5 Future Prospects and Challenges

5.1 Future Prospects

The development of ECL immunosensors has made remarkable achievements, and future research will focus on improving their sensitivity, throughput, and portability. By combining nanomaterials, microfluidic chips, and machine learning, ECL sensors will play an increasingly important role in the fields of early disease diagnosis, personalized medicine, environmental monitoring, and food safety [10, 63, 85]. Especially in developing portable and high-throughput ECL sensors, ECL imaging technology is expected to provide important technical support for rapid diagnosis in resource-limited areas [3]. In addition, the combination of ECL imaging biosensors and machine learning algorithms can further improve the detection accuracy and data analysis capability, and can promote the development of precision medicine [60, 79].

Through continuous innovation and improvement, ECL imaging biosensors will play an increasingly important role in various fields, such as medical diagnosis, environmental protection, and food safety [78, 148]. In addition, as ECL technology continues to mature, its applications in other emerging fields, such as drug screening, neuroscience, and gene editing, can be explored in future. Through multidisciplinary cross-research, ECL technology is expected to promote the development of biomedical and environmental sciences.

5.1.1 Improved Imaging Techniques and Their Applications

ECL imaging technology, as a high-sensitivity and low-background bio-detection tool, shows a broad application prospect in the field of immunosensors. However, after reviewing the information, it is found that not many applications combine ECL imaging with immunosensors nowadays, and the use of CCD, EMCCD (Electron Multiplying CCD), ICCD (Intensified CCD), sCMOS (scientific-grade CMOS), or even smartphones' buit-in CMOS cameras instead of PMTs to capture ECL signals has significant advantageous features in terms of high temporal and spatial resolution, high throughput, and visualization [67].

Different imaging devices have different strengths, as do different luminescence systems. $Ru(bpy)_3{}^{2+}$/TPrA systems stand out due to their ability to regenerate luminescence during the ECL process. This property allows it to be used as an efficient label for most ECL imaging applications [20]. ECL imaging methods, on the other hand, include pseudo-color processes that are not easily detected visually in real-time. Nowadays, true-color ECL imaging systems are receiving more and more attention due to the simultaneous detection of protein biomarkers with different spectral luminophores and the study of electron transfer pathways through the manipulation of applied potentials [6, 48]. These systems have shown significant advantages for spatial discrimination and visual detection in field tests [18]. However, current true-color ECL imaging is usually achieved with classical luminophores such as ruthenium and iridium complexes, and the insolubility of certain complexes in aqueous solutions limit their use in bioanalysis [123]. Therefore, it is essential to find excellent ECL luminophores with good biocompatibility and high ECL efficiency for true-color ECL imaging [125]. New nanotechnological approaches can solve this limitation even for known insoluble luminophores [154].

In future, we can try to combine different imaging devices as well as different luminescence systems with immunosensors to reflect the diversity of ECL imaging immunosensors. In addition, the combination of diverse imaging modes of ECL (e.g., ECL^+ and ECL^-) can further enhance the diversity and application dimensions of bioimaging, such as three-dimensional characterization of biological entities, similar to the function of light sheet fluorescence microscopy [51] (Fig. 8).

In addition, future research could also focus on optimizing ECL imaging systems to improve sensitivity and resolution for high-quality detection in more complex biological samples and establish uniform operating procedures and standardized methodologies to ensure consistency and reproducibility of results between different laboratories. Ultimately, as the technology evolves, ECL imaging has the potential to become a routine clinical testing tool, advancing early diagnosis and personalized treatment.

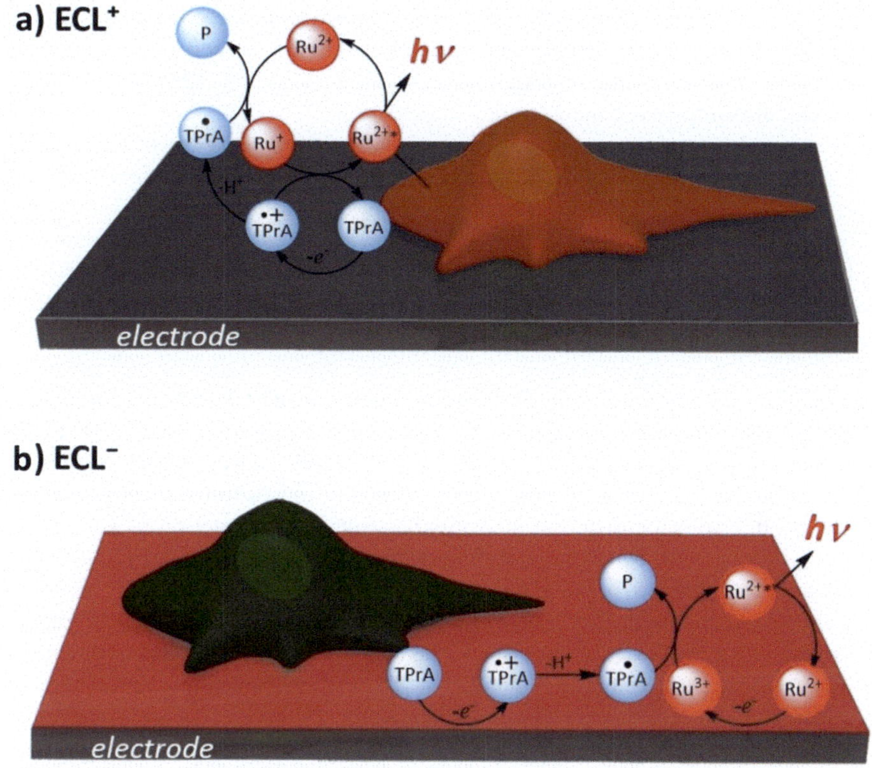

Fig. 8 Schematic diagrams showing the principles of single-cell **a** ECL^+ and **b** ECL^- imaging modes **a** labeled with $Ru(bpy)_3{}^{2+}$ luminescence, and **b** luminescence and co-reactants freely diffusing in solution, respectively [51]

5.1.2 Enhancement of ECL Luminescence

To improve the sensitivity of ECL imaging biosensors, researchers have developed various strategies. For example, nanoparticles such as gold nanoparticles, quantum dots, and MOFs [59] have been widely used to improve the sensitivity and selectivity of ECL signals due to their high surface area and unique electrochemical properties. These nanomaterials not only provide more reaction sites but also significantly improve the efficiency of electron transfer, thus enhancing the luminescence intensity [150]. By introducing nanomaterials onto the electrode surface, the electron transfer efficiency can be increased, significantly enhancing the ECL signal [62, 63].

The application of nanomaterials, such as MOFs and quantum dots, further enhances the sensitivity of the sensors, enabling the ECL sensors to detect the target biomarkers at very low concentrations [78]. It has been shown that the signal intensity and selectivity of ECL sensors can be significantly improved by introducing MOFs, which, with their large specific surface area and porous structure, can effectively enrich target molecules and reduce non-specific adsorption, thereby improving the accuracy of detection [83]. The MOFs have large specific surface area and porous structure, which can effectively enrich the target molecules and reduce non-specific adsorption, thus improving the accuracy of detection.

Signal amplification strategies are also an essential part of ECL imaging techniques. The sensitivity of ECL detection can be greatly improved by using a variety of signal amplification mechanisms, such as enzymatic amplification, nano-enzymatic catalysis, and cyclic regeneration. For example, enzyme-catalyzed reactions can amplify the signal through cycling, enabling detection at very low target analyte concentrations. This multiple amplification strategy has led to a wide range of applications for ECL imaging in the detection of ultra-low concentration biomarkers [36].

In addition, using microfluidic chips in conjunction with ECL imaging can significantly improve the accuracy and speed of detection. For example, it has been shown that efficient mixing and transport of reactants can be achieved by precise fluid manipulation within the microfluidic chip, thus enhancing the overall efficiency of the ECL reaction. Exploring different microfluidic structures' impact on ECL reaction efficiency assists in the development of novel integrated microfluidic-ECL systems for efficient, multiplexed, and real-time biomarker detection [50, 99, 137]. Furthermore, the use of functionalized nanomaterials and nanocomposites will also play a leading role in the creation of systems where energy transfer from electrochemical reaction centers to efficient emission centers will increase the quantum efficiency of analytical ECL reactions and improve ECL imaging ability [114].

Improving mentioned strategies, their combination, and developing novel approaches enhances the ECL imaging technology due to the construction of a luminescence-efficient sensor based on microfluidic chips and machine-learning approaches to achieve smarter and accurate detection.

5.1.3 High-Throughput ECL Imaging Sensor Development

Combining electrode arrays with ECL technology is ingenious for constructing high-productivity ECL imaging immunoassays. Yang et al. [143] designed an aptamer-based gold electrode array combined with a multifunctional signal probe to achieve high-sensitivity, crosstalk-free simultaneous detection of multiple biomarkers. In addition, Kadimisetty et al. [47] fabricated a miniature 3D-printed ECL immunoarray (Fig. 9) for high-throughput detection of multiple proteins by a CCD camera with LOD as low as 85–110 fg/mL, demonstrating the promising application of ECL imaging technology in high-throughput analysis.

5.1.4 Portable ECL Imaging Sensor Development

In future innovations, the mentioned researches make a base for next-generation analytical systems. For example, there have been studies that have successfully

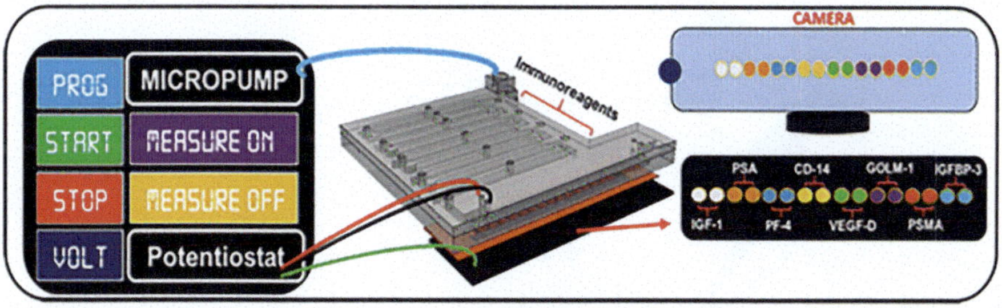

Fig. 9 Automated 3D printed microfluidic arrays that can detect multiple proteins [47]

combined ECL with self-powered systems to achieve efficient light emission without an external power supply [69], which provides ideas for the design of novel portable ECL sensors. By further optimizing the combination of MOFs and co-reactants, more substantial ECL signal amplification can be achieved, improving detection sensitivity. In addition, combining ECL technology with flexible electronics to develop ECL sensors based on flexible substrates can significantly enhance the portability and adaptability of the devices, especially in wearable medical devices and real-time monitoring, showing unique advantages.

In addition, ECL technology can be used in future by integrating it into portable devices and handheld instruments (e.g., smartphones, glucose meters, pH meters, etc.), thereby increasing its utility and portability, especially for rapid on-site testing in resource-limited environments [36]. Paper-based ECL sensors and all-solid-state ECL devices are also characterized by low cost, portability, and high sensitivity, making them ideally suited for on-site testing and POCT (prompt diagnostic) platforms [99].

5.1.5 Expanded Application Areas

Some nanomaterials used in ECL sensors may be potentially biotoxic, limiting their use in clinical diagnostics. Therefore, future research should be devoted to the development of more biocompatible materials to ensure the safety and efficacy of the sensors in practical applications.

Currently, the applications of ECL immunosensors are still mainly in the biological and medical fields, with only a few applications in other fields, such as environmental protection and food safety. For example, an all-solid-state ECL device with the potential to be a portable device can visualize ECL emission in air and can be used as a portable sensor for ambient gas detection or as a luminescent device in dark environments [127]. There are also environmental and food safety testing applications in subsection 3.7 that list some non-medical applications.

With the continuous development of science and technology as well as the continuous improvement of ECL technology, its application in other emerging fields, such as drug screening, gene editing, petrochemicals, automotive machinery, etc., can

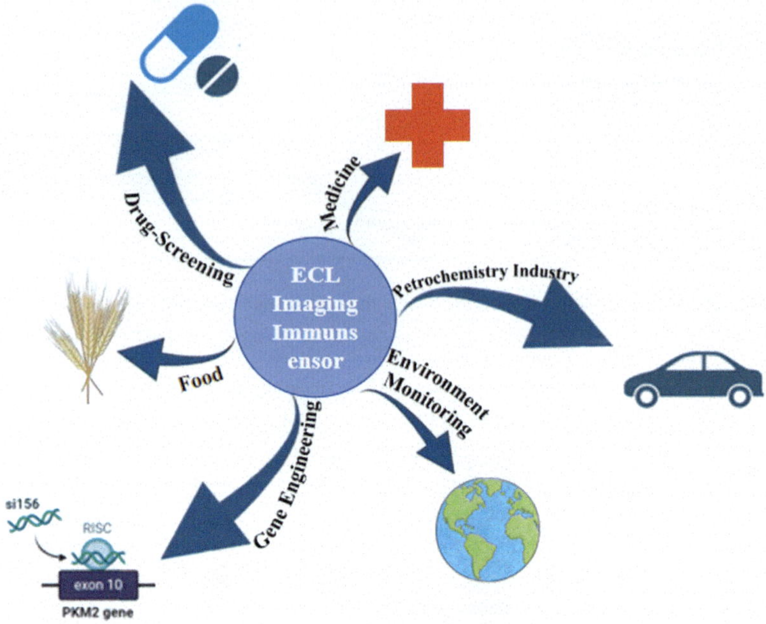

Fig. 10 ECL imaging immunosensor application area map

be explored in future. Under the mutual integration and crossover of various disciplines, ECL imaging immunosensors are expected to be applied to every bit of life and promote the development of science and technology, economy, and academics (Fig. 10).

5.2 Challenges

Although ECL imaging technology shows great potential, its application in the field of immunosensors still faces many challenges.

5.2.1 Sensitivity and Signal Amplification

The number of photons emitted by ECL is low compared to conventional fluorescence methods, so more efficient signal amplification strategies, such as gold nanomaterials, quantum dots, or enzyme-catalyzed amplification, are needed to enhance sensitivity. However, these strategies usually involve complex materials and manipulations that increase the difficulty and cost of fabricating the devices. The structural complexity of quantum dots, for example, still presents some pressing problems—limited charge transfer and non-specific adsorption phenomena. In future, quantum dots are expected to play an important role in single-molecule ECL detection, especially with the demand for higher resolution and sensitivity detection. Therefore, in-depth studies on the surface passivation and structure optimization of quantum dots and their application in ECL systems should be continued to overcome the bottlenecks of existing technologies and to promote the application of ECL imaging in high-precision detection [111].

In future, the development of new nanomaterials or multifunctional nanocomposites and the introduction of more efficient catalytic reaction pathways will be an important direction to improve the efficiency of signal amplification. Combining better biocompatible materials is also expected to enhance the sensitivity while reducing interference with biological systems. In view of the above, carbon quantum dots are of considerable interest due to their biocompatibility and luminescent properties, as well as open platforms for further functionalization.

5.2.2 Antibody Fixation and Immunoaffinity Layer Stability

In ECL immunosensors, a major challenge is ensuring stable antibody immobilization on the electrode surface, especially in complex biological environments. Non-specific adsorption and inactivation of antibodies can significantly affect the performance of the sensor, and more stable surface modification techniques need to be developed to address this issue. More biocompatible materials, such as PEG and nanogold, will be used to modify electrode surfaces. In addition, using genetically engineered modified antibody fragments or binding peptides could also improve the stability and specificity of antibodies, thereby increasing the reliability and sensitivity of the assay.

5.2.3 Sample Complexity and Environmental Impacts

Interference of ECL signals in complex sample matrices is a notable challenge for POCT assays in resource-limited environments. In future, there is a need to further improve the selectivity and interference resistance of sensors to ensure stability and accuracy under a wide range of conditions. Interference from matrix effects can be reduced by developing novel selective recognition elements (e.g., aptamers) and multifunctional nanomaterials. Meanwhile, the integration of microfluidic preprocessing modules can effectively separate interferences in the samples, thus improving the reliability of the detection results.

5.2.4 Miniaturization and Portability of Equipment

Device miniaturization and portability are key to the widespread use of ECL immunosensors in POCT. However, miniaturization usually affects the sensitivity and signal-to-noise ratio of the device, so a balance between performance and portability needs to be found in the device design. At the same time, low-power systems need to be developed to support prolonged use in resource-constrained environments. The power consumption and weight of devices can be reduced by developing new lightweight materials and efficient energy management systems in future. In addition, combining wireless data transmission and cloud-based analysis technology is expected to enable remote diagnosis and monitoring, improves the convenience and utility of the devices.

5.2.5 Crosstalk in Multiple Detection

In multiplex analyses, simultaneous detection of multiple biomarkers may lead to crosstalk of signals and reduce the specificity of the assay. Therefore, electrode design and detection systems need to be optimized to minimize the interactions between different targets and improve the accuracy of multiplex analysis. The effect of crosstalk can be reduced in future by developing spatially resolved multi-electrode arrays or using more selective signal probes. In addition, data analysis using machine-learning and signal-processing algorithms can also be effective in distinguishing and correcting cross-signals in multiplexed assays, improving the specificity and sensitivity of the assays.

6 Summary

This chapter analyzed using ECL imaging for quantitative biomarker detection and multiplexing.

Firstly, the basic concepts of ECL imaging biosensors and their role in biomarker detection are introduced, and the basic principles of ECL imaging technology are discussed in detail. Electrochemical, and particularly ECL biosensors, have developed rapidly over the past decades and remain an active research area in analytical chemistry. The advantages possessed by these biosensors, such as low sample volume, cost-effectiveness, simplicity of operation, high sensitivity, reproducibility, and selectivity, ensure a wide range of potential applications in clinical, environmental, and food safety fields. For electrochemical immunosensors, antibodies recognize analytes, bind them to the electrode, and generate or amplify the molecular label-bound signal production from target analyte. Nanomaterials are widely used in electrode fabrication and synthesis of labeled molecules to increase the conductivity, electron transfer rate, signal generation, and amplification effects. Therefore, the synthesis of efficient nanomaterials and their optimal application in sensors can significantly enhance the overall performance of electrochemical and ECL biosensors. Future studies should continue to explore strategies for the introduction of new efficient nanomaterials, especially the development of composite-conjugated antibodies with a high density of antigen-binding sites to achieve higher sensitivity and better specificity.

With its high sensitivity, wide dynamic range, and good reproducibility, ECL imaging technology shows remarkable potential for biomarker detection in patients with cancer, cardiovascular, and neurodegenerative diseases. In addition, through the cleverly designed multi-detection platform, ECL imaging enables simultaneous detection of multiple markers, which contributes to the in-depth analysis of complex biological systems. These features have made ECL imaging one of the indispensable tools in immunoassays, advancing the research and diagnosis of a wide range of complex diseases in the biomedical field. With the continuous improvement of image sensors and high-efficiency ECL systems, ECL imaging is emerging as one of the most promising techniques for multiplexed immunoassays. ECL imaging has significantly revolutionized the development of POCT by combining multiple advanced technologies, such as BPEs and 3D printing, into miniaturized ECL electrode arrays and employing a smartphone as the detection device. Such innovations enable portable and efficient ECL imaging-based detection in remote areas and resource-limited environments, which is important for early diagnosis of diseases and health monitoring.

Secondly, the application of ECL imaging immunosensors for biomarker detection is explored, especially its application for diagnosis of major pathologies, such as cancer, cardiovascular diseases, and neurodegenerative diseases. The biomarkers are classified and listed by different diseases and their relevance. The applications of ECL immunosensors for biomarker detection are given, which cover quantitative and some multiplexed analysis applications, reflecting the broad application prospects of ECL immunosensors in both medical and non-medical fields. From these numerous use cases, it can be seen that the current ECL technology combined with biosensor platforms has a wide range of promising applications, especially in portable devices and clinical POCT platforms. However, the application of ECL imaging immunosensors is still rather scarce in the current reports.

ECL imaging immunosensor combines the features of high spatial and temporal resolution, high throughput, and provides visualization of ECL emission field. It also combines the advantages of immunosensor, such as low sample volume, cost-effectiveness, simple operation, high sensitivity, reproducibility, and selectivity. All of these factors give the ECL imaging immunosensor significant advantages over other methods for quantitative analysis of biomarkers in various fields. Therefore, the wider use of this method will be an inevitable trend for future progress of analytical science.

Subsequently, we focused on the latest advances of ECL in multiplexed assays, including the combined application of microfluidic chips, smart sensors, and multifunctional nanomaterials as well as the realization of high-throughput multiplexed assays. With the advancement of novel nanomaterials and microfluidics, ECL imaging immunosensors are expected to achieve higher sensitivity, stronger anti-interference capabilities, and a more portable design to meet the needs of biomarker detection in different application scenarios. These developments will not only promote basic scientific research, but also play an important role in public health, personalized medicine, environmental protection, and food safety.

Finally, this chapter also analyzes the challenges of ECL imaging immunosensors in biomarker quantification and multiplexed detection with specific application examples. It also explores future directions, especially their potential applications in precision medicine, environmental monitoring, and food safety.

In future, ECL imaging immunosensors will evolve toward greater efficiency, portability and versatility, which will not only satisfy the current needs for biomarker detection, but also provide new possibilities for personalized medicine and early warning of diseases. To further expand the application of ECL imaging in biosensors, future research should focus on the following aspects. Firstly, upgrading ECL imaging technology and developing its application remains a top priority, using different imaging devices as well as different luminescence systems in combination with immunosensors to reflect the diversity of ECL imaging immunosensors. Secondly, the introduction and optimization of nanomaterials for ECL imaging is also a central topic, especially the synthesis of efficient nanomaterials compatible with ECL systems, which can significantly improve the conductivity and redox activity of electrodes and facilitate ECL signal amplification. The application of nanomaterials, such as MOFs and quantum dots, further enhances the sensitivity of the sensors, enabling the ECL sensors to detect target biomarkers at very low concentrations [78]. In addition, the combination of microfluidics and ECL imaging will significantly advance the miniaturization and portability of ECL sensors, enabling their application in immediate detection devices for personalized medicine and disease monitoring. The development of new technologies for smartphone-sensor integration will greatly enhance the convenience and applicability of ECL imaging in POCT. Finally, the realization of high-throughput detection and the expansion of its application areas will enable more efficient detection and multidisciplinary development. Through the combination of these technologies, ECL imaging has the potential to become a truly universal analytical technique in future, contributing to the improvement of global health and quality of life.

In conclusion, ECL imaging, as an emerging electrochemical assay technique, demonstrates great promise in biomarker quantification, disease diagnosis, and other application areas by virtue of its excellent sensitivity, specificity, and adaptability. Through continuous technological innovation, ECL imaging is expected to play an irreplaceable role in a number of key areas and become an important driving force for the future development of immunosensor technology. In future, with the continuous integration of ECL immunosensors with other advanced technologies, their applications in multiple fields will continue to expand and deepen, and ultimately achieve a comprehensive breakthrough from basic research to practical applications, bringing more well-being and health protection to human society.

Acknowledgements This work was kindly supported by the Ministry of Science and Technology of the People's Republic of China (2023YFE0201800), National Natural Science Foundation of China (No. 22174136), Natural Science Foundation of Jilin Province (No. SKL202302031), and Chinese Academy of Sciences (CAS) President's International Fellowship Initiative (PIFI).

References

1. Abdussalam, A., Xu, G.: Recent advances in electrochemiluminescence luminophores. Anal. Bioanal. Chem. **414**(1), 131–146 (2021). https://doi.org/10.1007/s00216-021-03329-0
2. Afshary, H., Amiri, M.: A fast and highly selective ECL creatinine sensor for diagnosis of chronic kidney disease. Sens. Diagn. **3**, 1562–1570 (2024). https://doi.org/10.1039/d4sd00165f
3. Al Hinaai, M., Khudaish, E.A.: Novel grid surface for creating solid-state electrogenerated chemiluminescence sensor based on Ru(II) complex doped onto AuNPs decorated Ru(II)/poly(tyramine) film. J. Solid State Electrochem. **27**, 2263–2274 (2023). https://doi.org/10.1007/s10008-023-05460-7
4. Ansar, W., Ghosh, S.: C-reactive protein and the biology of disease. Immunol. Res. **56**(1), 131–142 (2013). https://doi.org/10.1007/s12026-013-8384-0
5. Babarniri, B., Bahari, D., Salimi, A.: Highly sensitive bioaffinity electrochemiluminescence sensors: recent advances and future directions. Biosens. Bioelectron. **142**, 11530 (2019). https://doi.org/10.1016/j.bios.2019.111530
6. Barbante, G.J., Kebede, N., Hindson, C.M., Doeven, E.H., Zammit, E.M., Hanson, G.R., Hogan, C.F., Francis, P.S.: Control of excitation and quenching in multi-colour electrogenerated chemiluminescence systems through choice of co-reactant. Chem. Eur. J. **20**(43), 14026–14031 (2014). https://doi.org/10.1002/chem.201403767
7. Barhoum, A., Altintas, Z., Devi, K.S.S., Forster, R.J.: Electrochemiluminescence biosensors for detection of cancer biomarkers in biofluids: princsiples, opportunities, and challenges. Nano Today **50**, 101874 (2023). https://doi.org/10.1016/j.nantod.2023.101874
8. Bhaiyya, M., Rewatkar, P., Pimpalkar, A., Jain, D., Srivastava, S.K., Zalke, J., Kalambe, J., Balpande, S., Kale, P., Kalantri, Y., Kulkarni, M.B.: Deep learning-assisted smartphone-based electrochemiluminescence visual monitoring biosensor: a fully integrated portable platform. Micromachines **15**(8), 1059 (2024). https://doi.org/10.3390/mi15081059
9. Bunyakul, N., Baeumner, A.J.: Combining electrochemical sensors with miniaturized sample preparation for rapid detection in clinical samples. Sensors **15**(1), 547–564 (2014). https://doi.org/10.3390/s150100547
10. Cao, J., Chen, X.-Y.: Stable and reproducible MIP-ECL sensors for ultra-sensitive and accurate quantitative detection of Estrone. Front. Bioeng. Biotechnol. **12**, 1329129 (2024). https://doi.org/10.3389/fbioe.2024.1329129
11. Cao, Y., Zhou, J.L., Ma, Y.W., Zhou, Y., Zhu, J.J.: Recent progress of metal nanoclusters in electrochemiluminescence. Dalton Trans. **51**(23), 8927–8937 (2022). https://doi.org/10.1039/d2dt00810f
12. Ccopa Rivera, E., Swerdlow. J.J., Summerscales, R.L., Uppala, P.P.T., Maciel Filho, R., Neto, M.R.C., Kwon, H.J.: Data-Driven modeling of smartphone-based electrochemiluminescence sensor data using artificial intelligence. Sensors **20**(3), 625 (2020). https://doi.org/10.3390/s20030625
13. Chang, J., Hercules, D.M., Roe, D.K.: Role of triplet state in electrochemiluminescence of rubrene. Electrochim. Acta **13**(5), 1197–2000 (1968). https://doi.org/10.1016/0013-4686(68)80047-7
14. Chen, F.J., Wang, D., Chen, J., Ling, J.J., Yue, H., Gou, L.X., Tang, H.: PtNi nanocubes-catalyzed tyramine signal amplification electrochemiluminescence sensor for nonenzymatic and ultrasensitive detection of hepatocellular carcinoma cells. Sens. Actuators B Chem. **305**, 127472 (2020). https://doi.org/10.1016/j.snb.2019.127472
15. Chen, F.F., Xing, Y.K., Lei, M.Y., Li, L., Luo, L., Liu, J.X., Yang, X.Y., He, X.Q., Xue, J.S., Bu, H.Y., He, Y., Ouyang, X.Y.: Highly sensitive electrochemiluminescence sensing platform based on copper nanoclusters synthesized via a DNA nanoribbon template for the detection of cancer cells. Sens. Actuators B Chem. **420**, 136506 (2024). https://doi.org/10.1016/j.snb.2024.136506

16. Collins, G.E.: Gas-phase chemical sensing using electrochemiluminescence [Article; Proceedings Paper]. Sens. Actuators B Chem. **35**(1–3), 202–206 (1996). https://doi.org/10.1016/s0925-4005(97)80055-x
17. Cuifang, L.Z., Zongyun, F., Zhenxing, L., Wei, W.: Cross-talk-free multiplexed immunoassay using a disposable electrochemiluminescent immunosensor array coupled with a non-array detector. Bios. Bioelect. **27**(1), 141–147 (2011). https://doi.org/10.1016/j.bios.2011.06.031
18. Doeven, E.H., Barbante, G.J., Kerr, E., Hogan, C.F., Endler, J.A., Francis, P.S.: Red–Green–Blue electrogenerated chemiluminescence utilizing a digital camera as detector. Anal. Chem. **86**(5), 27–32 (2014). https://doi.org/10.1021/ac404135f
19. Dong, X., Du, Y., Zhao, G.H., Cao, W., Fan, D.W., Kuang, X., Wei, Q., Ju, H.X.: Dual-signal electrochemiluminescence immunosensor for Neuron-specific enolase detection based on "dual-potential" emitter $Ru(bpy)_3{}^{2+}$functionalized zinc-based metal-organic frameworks. Biosens. Bioelectron. **192**, 113505 (2021). https://doi.org/10.1016/j.bios.2021.113505
20. Du, F.X., Chen, Y.Q., Meng, C.D., Lou, B.H., Zhang, W., Xu, G.B.: Recent advances in electrochemiluminescence immunoassay based on multiple-signal strategy. Curr. Opin. Electrochem. **28**,100725 (2021). https://doi.org/10.1016/j.coelec.2021.100725
21. Erler, K.: Elecsys immunoassay systems using electrochemiluminescence detection. Wien. Klin. Wochenschr. **110**(3), 5–10 (1998)
22. Faatz, E., Finke, A., Josel, H.P., Prencipe, G., Quint, S., Windfuhr, M.: Analytical electrogenerated chemiluminescence, pp. 443–469 (2019). https://doi.org/10.1039/9781788015776-fp001
23. Farías, M.E., Correa, N.M., Sosa, L., Niebylski, A.M., Molina, P.G.: A simple electrochemical immunosensor for sensitive detection of transgenic soybean protein CP4-EPSPS in seeds. Talanta **237**, 122910 (2021). https://doi.org/10.1016/j.talanta.2021.122910
24. Feng, Y.G., Zhu, J.H., Wang, X.Y., Wang, A.J., Mei, L.P., Yuan, P.X., Feng, J.J.: New advances in accurate monitoring of breast cancer biomarkers by electrochemistry, electrochemiluminescence, and photoelectrochemistry. J. Electroanal. Chem. **882**, 115010 (2021). https://doi.org/10.1016/j.jelechem.2021.115010
25. Fu, Y.T., Ma, Q.: Recent developments in electrochemiluminescence nanosensors for cancer diagnosis applications. Nanoscale **12**(26), 13879–13898 (2020). https://doi.org/10.1039/d0nr02844d
26. Fu, W.X., Wang, X.X., Ying, X.D., Sun, T., Wang, Y.F., Wang, J., Su, B.: Electrochemiluminescence lateral flow immunoassay using Ruthenium(II) complex-loaded dendritic mesoporous silica nanospheres for highly sensitive and quantitative detection of SARS-CoV-2 nucleocapsid protein. Adv. Funct. Mater. **34**, 2409632 (2024). https://doi.org/10.1002/adfm.202409632
27. Fu-Ren F., David, F., Allen, C., Bard, J.: Scanning electrochemical microscopy. 37. Light emission by electrogenerated chemiluminescence at SECM tips and their application to scanning optical microscopy. Anal. Chem. **70**(14), 2941–2948 (1998). https://doi.org/10.1021/ac980107t
28. Fung, A.W.S., Knauer, M.J., Blasutig, I.M., Colantonio, D.A., Kulasingam, V.: Evaluation of electrochemiluminescence immunoassays for immunosuppressive drugs on the Roche cobas e411 analyzer. F1000Research **6**, 1832 (2017). https://doi.org/10.12688/f1000research.12775.1
29. Gan, X., Han, D., Wang, J., Liu, P., Li, X., Zheng, Q., Yan, Y.: A highly sensitive electrochemiluminescence immunosensor for h-FABP determination based on self-enhanced luminophore coupled with ultrathin 2D nickel metal-organic framework nanosheets. Biosens. Bioelectron. **171**, 112735 (2021). https://doi.org/10.1016/j.bios.2020.112735
30. Gao, W., Liu, Y., Zhang, H., Wang, Z.: Electrochemiluminescence biosensor for nucleolin imaging in a single tumor cell combined with synergetic therapy of tumor. ACS Sensors **5**(4), 1216–1222 (2020). https://doi.org/10.1021/acssensors.0c00292
31. Guerrero-Esteban, T., Gutiérrez-Sánchez, C., Martínez-Periñán, E., Revenga-Parra, M., Pariente, F., Lorenzo, E.: Sensitive glyphosate electrochemiluminescence immunosensor based on electrografted carbon nanodots. Sensors Actuators B: Chem **330**, 129389 (2021). https://doi.org/10.1016/j.snb.2020.129389

32. Guo, W., Ding, H., Gu, C., Liu, Y., Jiang, X., Su, B., Shao, Y.: Potential-Resolved Multicolor Electrochemiluminescence for Multiplex Immunoassay in a Single Sample. J. Am. Chem. Soc. **140**(46), 15904–15915 (2018). https://doi.org/10.1021/jacs.8b09422
33. Hailaxi, A., Cheng, G., Chai, R., Song, Y., Shen, Z., Li, G.: Novel label-free electrochemiluminescence Immunosensor based on Ru@SiO_2-PEI-Au@Pt hybrid nanostructures for highly sensitive PSA detection. Electroanalysis **38**(2), e70101 (2026). https://doi.org/10.1002/elan.70101
34. Han, F., Jiang, H., Fang, D., Jiang, D.: Potential-Resolved electrochemiluminescence for determination of two antigens at the cell surface. Anal. Chem. **86**(14), 6896–6902 (2014). https://doi.org/10.1021/ac501571a
35. Han, T.T., Ma, C., Wang, L.Y., Cao, Y., Chen, H.Y., Zhu, J.J.: A novel electrochemiluminescence janus emitter for dual-mode biosensing. Adv. Funct. Mater. **32**(24), 2200863 (2022). https://doi.org/10.1002/adfm.202200863
36. Han, T.T., Cui, C., Xing, Y.J., Shi, J.R., Xu, J.H., Zhu, J.J.: Advances of signal amplification strategies and sensing formats in electrochemiluminescence sensors for mycotoxins. TrAC Trends. Anal. Chem. **180**, 117961 (2024). https://doi.org/10.1016/j.trac.2024.117961
37. Han, Y.J., Jia, Y., Du, Y., Li, Y.Y., Ren, X., Ma, H.M., Wu, D., Kuang, X., Fan, D.W., Wei, Q.: Controlled growth of MoS_2 on dendritic ferric oxide to enhance electrochemiluminescence of nitrogen-doped carbon quantum dots for sensitive immunoassay. Anal. Chem. **95**(16), 6655–6663 (2023). https://doi.org/10.1021/acs.analchem.3c00058
38. Hercules, D.M.: Chemiluminescence resulting from electrochemically generated species. Science **145**(363), 808 (1964). https://doi.org/10.1126/science.145.3634.808
39. Hercules, D.M.: Chemiluminescence from electron-transfer reactions. Acc. Chem. Res. **2**(10), 301 (1969). https://doi.org/10.1021/ar50022a003
40. Hu, L., Xu, G.: Applications and trends in electrochemiluminescence. Chem Soc Rev **39**(8), 3275–304 (2010). https://doi.org/10.1039/b923679c
41. Huang, Y., Lei, J.P., Cheng, Y., Ju, H.X.: Ratiometric electrochemiluminescent strategy regulated by electrocatalysis of palladium nanocluster for immunosensing. Biosens. Bioelectron. **77**, 733–739 (2016). https://doi.org/10.1016/j.bios.2015.10.038
42. Jayanthi, V., Das, A.B., Saxena, U.: Recent advances in biosensor development for the detection of cancer biomarkers. Biosens. Bioelectron. **91**, 15–23 (2017). https://doi.org/10.1016/j.bios.2016.12.014
43. Ji, Y., He, S., Chen, Y., Zhang, P., Sun, J., Li, Y., Kuang, K.D., Jia, N.Q.: A sensitive dual-signal electrochemiluminescence immunosensor based on $Ru(bpy)_3{}^{2+}$@HKUST-1 and $Ce_2Sn_2O_7$ for detecting the heart failure biomarker NT-proBNP. J. Mater. Chem. B **11**(12), 2754–2761 (2023). https://doi.org/10.1039/d2tb02555h
44. Jia, Y., Yang, L., Feng, R., Ma, H., Fan, D., Yan, T., Feng, R., Du, B., Wei, Q.: $MnCO_3$ as a new electrochemiluminescence emitter for ultrasensitive bioanalysis of β-Amyloid1–42 oligomers based on site-directed immobilization of antibody. ACS Appl. Mater. Interfaces **11**(7), 7157–7163 (2019). https://doi.org/10.1021/acsami.8b21928
45. Jing, Y., Dalibor, S., Jasmina, V., Neso, S.: Recent advances in electrochemiluminescence immunosensing. Sens. Diagn. **3**, 1887–1898 (2024). https://doi.org/10.1039/d4sd00272e
46. Jović, M., Prim, D., Saini, E., Pfeifer, M.E.: Towards a point-of-care (POC) diagnostic platform for the multiplex electrochemiluminescent (ECL) sensing of mild traumatic brain injury (mTBI) biomarkers. Biosensors **12**(3), 172 (2022). https://doi.org/10.3390/bios12030172
47. Kadimisetty, K., Malla, S., Bhalerao, K.S., Mosa, I.M., Bhakta, S., Lee, N.H., Rusling, J.F.: Automated 3D-Printed microfluidic array for rapid nanomaterial-enhanced detection of multiple proteins. Anal. Chem. **90**(12), 7569–7577 (2018). https://doi.org/10.1021/acs.analchem.8b01198
48. Kerr, E., Doeven, E.H., Barbante, G.J., Hogan, C.F., Bower, D.J., Donnelly, P.S., Connell, T.U., Francis, P.S.: Annihilation electrogenerated chemiluminescence of mixed metal chelates in solution: modulating emission colour by manipulating the energetics. Chem. Sci. **6**(1), 472–479 (2014). https://doi.org/10.1039/c4sc02697g

49. Kim, J.K., Song, H.M., Jun, J.W., Park, H.J., Lim, E.S., Lee, K., Lee, S., Kim, S.: Clinical studies of Ci-5, sol-gel encapsulated multiplex antibody microarray for quantitative fluorometric detection of simultaneous five different tumor antigens. Biochip. J. **13**(4), 378–386 (2019). https://doi.org/10.1007/s13206-019-3409-y
50. Kim, J., Park, M.: Recent progress in electrochemical immunosensors. Biosensors (Basel) **11**(10), 360 (2021). https://doi.org/10.3390/bios11100360
51. Knežević, S., Han, D., Liu, B., Jiang, D., Sojic, N.: Electrochemiluminescence microscopy. Angew. Chem. Int. Ed. Engl. **63**(29), e202407588 (2024). https://doi.org/10.1002/anie.202407588
52. Kukoba, A.V., Bykh, A.I., Svir, I.B.: Analytical applications of electrochemiluminescence: an overview. Fresen. J. Anal. Chem. **368**(5), 439–442 (2000). https://doi.org/10.1007/s002160000548
53. Lakard, B., Herlem, G., de Michel, L., Daniau, W., Martin, G., Jeannot, J.C., Robert, L., Fahys, B.: Miniaturized pH biosensors based on electrochemically modified electrodes with biocompatible polymers. Biosens. Bioelectron. **19**(6), 595–606 (2004).
54. Lei, H.X., Niu, C.C., Li, T., Wan, Y.F., Liang, W.B., Yuan, R., Liao, P.: A Novel electrochemiluminescent immunoassay based on target transformation assisted with catalyzed hairpin assembly amplification for the ultrasensitive bioassay. ACS Appl. Mater. Interfaces **11**(34), 31427–31433 (2019). https://doi.org/10.1021/acsami.9b12428
55. Li, H.K., Cai, Q.Q., Wang, Y.H., Jie, G.F., Zhou, H.: Spatial-Potential-Color-Resolved bipolar electrode electrochemiluminescence biosensor using a CuMoOx electrocatalyst for the simultaneous detection and imaging of tetracycline and lincomycin. Anal. Chem. **96**(18), 7073–7081 (2024). https://doi.org/10.1021/acs.analchem.4c00388
56. Li, X.J., Du, Y., Wang, H., Ma, H.M., Wu, D., Ren, X., Wei, Q., Xu, J.J.: Self-Supply of H_2O_2 and $O_2$by hydrolyzing CaO_2 to enhancethe electrochemiluminescence of luminol based on a closed bipolar electrode. Anal. Chem. **92**(18), 12693–12699 (2020). https://doi.org/10.1021/acs.analchem.0c03170
57. Li, S.J., Liu, Y., Ma, Q.: Nanoparticle-based electrochemiluminescence cytosensors for single cell level detection. TrAC Trends. Anal. Chem. **110**, 277–292 (2019). https://doi.org/10.1016/j.trac.2018.11.019
58. Li, X.X., Qin, X., Tian, Z.Y., Wang, K., Xia, X.H., Wu, Y.F., Liu, S.Q.: Gold nanowires array-based closed bipolar nanoelectrode system for electrochemiluminescence detection of α-fetoprotein on cell surface. Anal. Chem. **94**(20), 7350–7357 (2022). https://doi.org/10.1021/acs.analchem.2c00785
59. Li, Y., Wan, Y., Fu, X., Chen, J., Wu, W., Feng, X., Man, T., Huang, Y., Piao, Y., Zhu, L., Lei, J., Deng, S.: Sub-second electrochemiluminescence imaging assay of SARS-CoV-2 nucleocapsid protein based on reticulation of endo-coreactants. Anal. Chem. **96**(7), 3116–3123 (2024). https://doi.org/10.1021/acs.analchem.3c05388
60. Li, F., Peng, H., Shen, N., Yang, C., Zhang, L., Li, B., He, J.: Electrochemiluminescence in graphitic carbon nitride decorated with silver nanoparticles for dopamine determination using machine learning. ACS Appl. Mater. Interfaces **16**(21), 27767–27777 (2024). https://doi.org/10.1021/acsami.4c03996
61. Li, J.J., Zhao, C.L., Lai, W.J., Li, P.L., Jiang, M.Z., Wang, M., Ma, C.Y., Hong, C.L.: A self-reinforced luminescence signal amplification strategy to construct electrochemiluminescence immunosensor for carcinoembryonic antigen sensing. Microchem. J. **206**, 111263 (2024). https://doi.org/10.1016/j.microc.2024.111263
62. Li, J.X., Yang, H.F., Cai, R., Tan, W.H.: Novel nucleic acid-assisted ion-responsive ECL biosensor based on hollow AuAg nanoboxes with excellent SPR and effective coreaction acceleration. Anal. Chem. **96**(27), 11076–11082 (2024). https://doi.org/10.1021/acs.analchem.4c02231
63. Li, M.X., Shi, J.W., Zhang, Y.H., Cui, S.H., Zhang, L., Shen, Q.M.: ECL cytosensor for sensitive and label-free detection of circulating tumor cells based on hierarchical flower-like gold microstructures. Anal. Chim. Acta **1303**, 342505 (2024). https://doi.org/10.1016/j.aca.2024.342505

64. Li, W.H., Li, C.Y., Zhou, T., Liu, X.H., Liu, X.N., Li, X.H., Chen, D.X.: Role of exosomal proteins in cancer diagnosis. Mol Cancer **16**(1), 145 (2017). https://doi.org/10.1186/s12943-017-0706-8
65. Liang, Y., Xue, K., Shi, Y., Zhan, T., Lai, W., Zhang, C.: Dry chemistry-based bipolar electrochemiluminescence immunoassay device for point-of-care testing of alzheimer-associated neuronal thread protein. Anal. Chem. **95**(6), 3434–3441 (2023). https://doi.org/10.1021/acs.analchem.2c05164
66. Liu, M.L., Chen, B.B., Li, C.M., Huang, C.Z.: Carbon dots prepared for fluorescence and chemiluminescence sensing. Sci. China Chem. **62**(8), 968–981 (2019). https://doi.org/10.1007/s11426-019-9449-y
67. Liu, Y.H., Guo, W.L., Su, B.: Recent advances in electrochemiluminescence imaging analysis based on nanomaterials and micro-/nanostructures. Chin. Chem. Lett. **30**(9), 1593–1599 (2019). https://doi.org/10.1016/j.cclet.2019.05.038
68. Liu, Y., Zhang, N., Pan, J.B., Song, J., Zhao, W., Chen, H.Y., Xu, J.J.: Bipolar electrode array for multiplexed detection of prostate cancer biomarkers. Anal. Chem. **94**(6), 3005–3012 (2022). https://doi.org/10.1021/acs.analchem.1c05383
69. Liu, H., Hussain, A., Zholudov, Y.T., Snizhko, D.V., Sojic, N., Xu, G.: Self-Powered electrochemiluminescence for imaging the corrosion of protective coating of metal and quantitative analysis. Angew. Chem. Int. Ed **304**, 127364 (2024). https://doi.org/10.1002/anie.202411764
70. Liu, X., Fang, C., Yan, J., Li, H., Tu, Y.: A sensitive electrochemiluminescent biosensor based on AuNP-functionalized ITO for a label-free immunoassay of C-peptide. Bioelectrochemistry **123**, 211–218 (2018). https://doi.org/10.1016/j.bioelechem.2018.05.010
71. Lv, W., Ye, H., Yuan, Z., Liu, X., Chen, X., Yang, W.: Recent advances in electrochemiluminescence-based simultaneous detection of multiple targets. Trends. Analyt. Chem. **123**, 115767 (2019). https://doi.org/10.1016/j.trac.2019.115767
72. Ma, X., Gao, W., Du, F., Yuan, F., Yu, J., Guan, Y., Sojic, N., Xu, G.: Rational design of electrochemiluminescent devices. Acc. Chem. Res. **54**(14), 2936–2945 (2021). https://doi.org/10.1021/acs.accounts.1c00230
73. Ma, T., Ren, S., Wang, Y., Yu, H., Li, L., Li, X., Zhang, L., Yu, J., Zhang, Y.: Paper-based bipolar electrode electrochemiluminescence sensors for point-of-care testing. Biosens. Bioelectron. **235**, 115384 (2023). https://doi.org/10.1016/j.bios.2023.115384
74. Martínez-Periñán, E., Gutiérrez-Sánchez, C., García-Mendiola, T., Lorenzo, E.: Electrochemiluminescence biosensors using screen-printed electrodes. Biosensors (Basel) **10**(9), 118 (2020). https://doi.org/10.3390/bios10090118
75. Meng, C., Knežević, S., Du, F., Guan, Y., Kanoufi, F., Sojic, N., Xu, G.: Recent advances in electrochemiluminescence imaging analysis. eScience **2**(6), 591–605 (2022). https://doi.org/10.1016/j.esci.2022.10.004
76. Mezzullo, M., Pelusi, C., Fazzini, A., Repaci, A., Di Dalmazi, G., Gambineri, A., Pagotto, U., Fanelli, F.: Female and male serum reference intervals for challenging sex and precursor steroids by liquid chromatography—tandem mass spectrometry. J. Steroid Biochem. Mol. Biol. **197**, 105538 (2019). https://doi.org/10.1016/j.jsbmb.2019.105538
77. Mohammadniaei, M., Zhang, M., Qin, X., Wang, W., Pia, L., Gürbüz, H., Helalat, S.H., Naseri, M., Sun, Y.: A hand-held electrochemiluminescence biosensor for detection of carcinoembryonic antigen. Talanta **266**(2), 125087 (2023). https://doi.org/10.1016/j.talanta.2023.125087
78. Mohan, B., Kumar, S., Kumar, V., Jiao, T.H., Sharma, H.K., Chen, Q.S.: Electrochemiluminescence metal-organic frameworks biosensing materials for detecting cancer biomarkers. TrAC Trends. Anal. Chem. **157**, 116735 (2022). https://doi.org/10.1016/j.trac.2022.116735
79. Pai, S., Bontempi, D., Hadzic, I., Prudente, V., Sokač, M., Chaunzwa, T.L., Bernatz, S., Hosny, A., Mak, R.H., Birkbak, N.J., Aerts, H.J.W.L.: Foundation model for cancer imaging biomarkers. Nat. Mach. Intell. **6**(3), 354–367 (2024). https://doi.org/10.1038/s42256-024-008 07-9
80. Perera, G.S., Ahmed, T., Heiss, L., Walia, S., Bhaskaran, M., Sriram, S.: Rapid and selective biomarker detection with conductometric sensors. Small **17**(7), e2005582 (2021). https://doi.org/10.1002/smll.202005582

81. Pharr, C.M., Engstrom, R.C., Tople, R.A., Bee, T.K., Unzelman, P.L.: Time-resolved imaging of current density at inlaid disk electrodes. J. Electroanal. Chem. Interfacial Electrochem. **278**(1–2), 119–128 (1990). https://doi.org/10.1016/0022-0728(90)85127-Q
82. Poleszczuk, J., Luddy, K,. Chen, L., Lee, J.K., Harrison, L.B., Czerniecki, B.J., Soliman, H., Enderling, H.: Neoadjuvant radiotherapy of early-stage breast cancer and long-term disease-free survival. Breast Cancer Res. **19**, 75 (2017). https://doi.org/10.1186/s13058-017-0870-1
83. Punj, S., Sidhu, D., Bhattacharya, D., Wang, M., Wong, P.K.: An Electrochemical biosensor platform for rapid immunoanalysis of physiological fluids. IEEE Open J. Nanotechnol. **1**, 31–37 (2020). https://doi.org/10.1109/ojnano.2020.2997296
84. Qi, H., Peng, Y., Gao, Q., Zhang, C.: Applications of nanomaterials in electrogenerated chemiluminescence biosensors. Sensors **9**(1), 674–695 (2009). https://doi.org/10.3390/s90100674
85. Qian, M.P., Zeng, Y., Li, M., Gao, Q., Zhang, C.X., Qi, H.L.: Electrogenerated chemiluminescence biosensor for quantization of matrix metalloproteinase-3 in serum via target-induced cleavage of oligopeptide. Biosensors-Basel **14**(4), 181 (2024). https://doi.org/10.3390/bios14040181
86. Qin, D.M., Meng, S., Wu, Y.S., Mo, G.C., Deng, B.Y.: Aggregation-induced electrochemiluminescence resonance energy transfer with dual quenchers for the sensitive detection of prostate-specific antigen. Sens. Actuators B Chem. **367**, 132176 (2022). https://doi.org/10.1016/j.snb.2022.132176
87. Qin, D., Meng, S., Wu, Y., Mo, G., Jiang, X., Deng, B.: Design of a dual-wavelength ratiometric electrochemiluminescence immunosensor for sensitive detection of amyloid-β protein in human serum. ACS Sustain. Chem. Eng. **9**(22), 7541–7549 (2021). https://doi.org/10.1021/acssuschemeng.1c01237
88. Reza, A., Xinyue, H., Alain, Z., Isabelle, D., Sebastian, W.-H.: Optical image sensors for smart analytical chemiluminescence biosensors. Bioengineering **11**(9), 912 (2024). https://doi.org/10.3390/bioengineering11090912
89. Reza, A.J., Sorina, L., Sebastian, S., Wachsmann-Hogiu, S.: SE-ECL on CMOS: a miniaturized electrochemiluminescence biosensor. Lab. on a Chip. **22**(5), 994–1005 (2022). https://doi.org/10.1039/D1LC00905B
90. Rizwan, M., Mohd-Naim, N.F., Ahmed, M.U.: Trends and advances in electrochemiluminescence nanobiosensors. Sensors **18**(1), 166 (2018). https://doi.org/10.3390/s18010166
91. Rong, Z., Chen, F., Jilin, Y., Yifeng, T.: A C-reactive protein immunosensor based on platinum nanowire/titania nanotube composite sensitized electrochemiluminescence. Talanta **205**, 120135 (2019). https://doi.org/10.1016/j.talanta.2019.120135
92. Rong, T., Yueju, W., Xiaona, M., Huiling, L., Yifeng, T.: A dual-screening electrochemiluminescent aptasensor based on a mesoporous silica nano-sieve for specific detection of amyloid-β monomer. Sens. Actuators B: Chem, **352**(2), 131065 (2021). https://doi.org/10.1016/j.snb.2021.131065
93. Roointan, A., Mir, T.A., Wani, S.I., Matiur, R., Hussain, K.K., Ahmed, B., Abrahim, S., Savardashtaki, A., Gandomani, G., Gandomani, M., Chinnappan, R., Akhtar, M.H.: Early detection of lung cancer biomarkers through biosensor technology: a review. J. Pharm. Biomed. Anal. **164**, 93–103 (2019). https://doi.org/10.1016/j.jpba.2018.10.017
94. Royce, C., Engstrom Kirk W., Scott, J., Jarlais, D.: Characterization of electrode heterogeneity with electrogenerated chemiluminescence. Anal. Chem. **59**(4), 670–673. https://doi.org/10.1021/ac00131a028
95. Salkho, N.M., Paul, V., Kawak, P., Vitor, R.F., Martins, A.M., Al Sayah, M., Husseini, G.A.: Ultrasonically controlled estrone-modified liposomes for estrogen-positive breast cancer therapy. Artif. Cells Nanomed. Biotechnol. **46**(sup2), 462–472 (2018). https://doi.org/10.1080/21691401.2018.1459634
96. Saremi, M., Amini, A., Heydari, H.: An aptasensor for troponin I based on the aggregation-induced electrochemiluminescence of nanoparticles prepared from a cyclometallated iridium(III) complex and poly(4-vinylpyridine-co-styrene) deposited on nitrogen-doped graphene. Microchim. Acta **186**(4), 254 (2019). https://doi.org/10.1007/s00604-019-3352-6

97. Schiffrin, D.J., Wilson, R., Kremeskotter, J.: Electrochemiluminescence based biosensors. Symp. Chem. Biol. Sens. Anal. Electrochem. Methods, Paris, Fr **97**(19), 6132–6620 (1997)
98. Shang, L., Shi, B.J., Zhang, W., Jia, L.P., Ma, R.N., Xue, Q.W., Wang, H.S.: Ratiometric electrochemiluminescence sensing of carcinoembryonic antigen based on luminol. Anal. Chem. **94**(37), 12845–12851 (2022). https://doi.org/10.1021/acs.analchem.2c02803
99. Shen, J., Zhou, T., Huang, R.: Recent advances in electrochemiluminescence sensors for pathogenic bacteria detection. Micromachines **10**(8), 532 (2019). https://doi.org/10.3390/mi10080532
100. Shi, Z., Li, G., Hu, Y.: Progress on the application of electrochemiluminescence biosensor based on nanomaterials. Chin. Chem. Lett. **30**(9), 1600–1606 (2019). https://doi.org/10.1016/j.cclet.2019.04.066
101. Short, G.D., Hercules, D.M.: Electroluminescence of organic compounds. Role of gaseous discharge in excitation process. J. Am. Chem. Soc. **87**(7), 1439 (1965). https://doi.org/10.1021/ja01085a005
102. Snizhko, D., Zholudov, Y., Bilash, O.: Sensor based on diamond-like film modified electrodes for bilirubin detection. In: 2019 IEEE 39th International Conference on Electronics and Nanotechnology (ELNANO), pp. 471–474 (2019)
103. Snizhko, D.V., Zholudov, Y.T., Bilash, O.M., Kukoba, A.V., Rozhitskii, M.M.: Electrochemiluminescence at nitrogen doped diamond-like carbon film electrodes. Russ. J. Electrochem. **50**(3), 260–266 (2014). https://doi.org/10.1134/s1023193514020037
104. Sobhanie, E., Salehnia, F., Xu, G.B., Hamidipanah, Y., Arshian, S., Firoozbakhtian, A., Hosseini, M., Ganjali, M.R., Hanif, S.: Recent trends and advancements in electrochemiluminescence biosensors for human virus detection. TrAC. Trends Anal. Chem. **157**, 116727 (2022). https://doi.org/10.1016/j.trac.2022.116727
105. Solhi, E., Hasanzadeh, M.: Critical role of biosensing on the efficient monitoring of cancer proteins/biomarkers using label-free aptamer based bioassay. Biomed. Pharmacother. **132**, 110849 (2020). https://doi.org/10.1016/j.biopha.2020.110849
106. Song, C., Li, X., Hu, L., Shi, T., Wu, D., Ma, H., Zhang, Y., Fan, D., Wei, Q., Ju, H.: Quench-Type electrochemiluminescence immunosensor based on resonance energy transfer from carbon nanotubes and Au-nanoparticles-enhanced g-C_3N_4 to CuO@Polydopamine for procalcitonin detection. ACS Appl. Mater. Interfaces **12**(7), 8006–8015 (2020). https://doi.org/10.1021/acsami.9b22782
107. Song, X., Zhao, L., Zhang, N,. Liu, L., Ren, X., Ma, H., Kuang, X., Li, Y., Luo, C., Wei, Q.: Ultrasensitive electrochemiluminescence biosensor with silver nanoclusters as a novel signal probe and α-Fe_2O_3–Pt as an efficient co-reaction accelerator for procalcitonin immunoassay. Anal. Chem. **95**(2), 1582–1588 (2023). https://doi.org/10.1021/acs.analchem.2c04673
108. Stockmann, W., Bablok, W., Luppa, P.: Analytical performance of Elecsys 2010–a multicentre evaluation. Wien. Klin. Wochenschr. **110**(3), 40–51 (1998)
109. Su, C., Cui, Y., Liu, D., Zhang, H., Baninla, Y.: Endocrine disrupting compounds, pharmaceuticals and personal care products in the aquatic environment of China: Which chemicals are the prioritized ones? Sci. Total Environ. **720**, 137652 (2020). https://doi.org/10.1016/j.scitotenv.2020.137652
110. Su, Y., Lai, W., Liang, Y., Zhang, C.S.: Novel cloth-based closed bipolar solid-state electrochemiluminescence (CBP-SS-ECL) aptasensor for detecting carcinoembryonic antigen. Anal. Chim. Acta **1206**, 339789 (2022). https://doi.org/10.1016/j.aca.2022.339789
111. Sun, H., Zhou, P., Su, B.: Electrochemiluminescence of semiconductor quantum dots and its biosensing applications: a comprehensive review. Biosensors **13**(7), 708 (2023). https://doi.org/10.3390/bios13070708
112. Sun, X., Li, B., Tian, C., Yu, F., Zhou, N., Zhan, Y., Chen, L.: Rotational paper-based electrochemiluminescence immunodevices for sensitive and multiplexed detection of cancer biomarkers. Anal. Chim. Acta **1007**, 33–39 (2018). https://doi.org/10.1016/j.aca.2017.12.005
113. Tanaka, T., Narazaki, M., Kishimoto, T.: IL-6 in Inflammation, immunity, and disease. Cold Spring Harb. Perspect. Biol. **6**(10), a016295 (2014). https://doi.org/10.1101/cshperspect.a016295

114. Tian, K., Li, D., Tang, T., Nie, F., Zhou, Y., Du, J., Zheng, J.: A novel electrochemiluminescence resonance energy transfer system of luminol-graphene quantum dot composite and its application in H_2O_2 detection. Talanta **185**, 446–452 (2018). https://doi.org/10.1016/j.talanta.2018.03.064
115. Valle, A.L., Mello, F.C.C., Alves-Balvedi, R.P., Rodrigues, L.P., Goulart, L.R.: Glyphosate detection: methods, needs and challenges. Environ. Chem. Lett. **17**, 291–317 (2018). https://doi.org/10.1007/s10311-018-0789-5
116. Vashist, S.K., Venkatesh, A.G., Schneider, E.M., Beaudoin, C., Luppa, P.B., Luong, J.H.T.: Bioanalytical advances in assays for C-reactive protein. Biotechnol. Adv. **34**(3), 272–290 (2016). https://doi.org/10.1016/j.biotechadv.2015.12.010
117. Vrdoljak, E., Bodoky, G., Jassem, J., Popescu, R.A., Mardiak, J., Pirker, R., Cufer, T., Beslija, S., Eniu, A., Todorovic, V., Kubácková, K., Kurteva, G., Tomasevic, Z., Sallaku, A., Smichkoska, S., Bajic, Z., Sikic, B.I.: Cancer control in central and eastern Europe: current situation and recommendations for improvement. Oncologist **21**(10), 1183–1190 (2016). https://doi.org/10.1634/theoncologist.2016-0137
118. Wang, Y.Y., Kan, X.W.: Sensitive and selective "signal-off" electrochemiluminescence sensing of prostate-specific antigen based on an aptamer and molecularly imprinted polymer. Analyst **146**(24), 7693–7701 (2021). https://doi.org/10.1039/d1an01645h
119. Wang, C.X., Li, M.S., Wang, P.J., Liu, D.F.: An electrochemiluminescence biosensor based on boron nitride quantum dots as novel coreactant for quantitative determination of concanavalin A. Microchim. Acta **187**(7), 409 (2020). https://doi.org/10.1007/s00604-020-04385-6
120. Wang, C., Liu, S., Ju, H.: Electrochemiluminescence nanoemitters for immunoassay of protein biomarkers. Bioelectrochemistry **149**, 108281 (2023). https://doi.org/10.1016/j.bioelechem.2022.108281
121. Wang, Y.L., Liu, X.M., Ren, S.W., Cao, J.T., Liu, Y.M.: Etching of Ag nanoparticles triggered bidirectional regulation for electrochemiluminescence ratiometric immunoassay. Anal. Bioanal. Chem. **416**(21), 4759–4767 (2024). https://doi.org/10.1007/s00216-024-05277-x
122. Wang, Y., Luo, J.P., Liu, J.T., Sun, S., Xiong, Y., Ma, Y.Y., Yan, S., Yang, Y., Yin, H.B., Cai, X.X.: Label-free microfluidic paper-based electrochemical aptasensor for ultrasensitive and simultaneous multiplexed detection of cancer biomarkers. Biosens. Bioelectron. **136**, 84–90 (2019). https://doi.org/10.1016/j.bios.2019.04.032
123. Wang, Y.-Z., Xu, C.-H., Zhao, W., Guan, Q.-Y., Chen, H.-Y., Xu, J.-J.: Bipolar electrode based multicolor electrochemiluminescence biosensor. Anal. Chem. **89**(15), 8050–8056 (2017). https://doi.org/10.1021/acs.analchem.7b01494
124. Wang, Y., Zhao, G., Chi, H., Yang, S., Niu, Q., Wu, D., Cao, W., Li, T., Ma, H., Wei, Q.: Self-Luminescent lanthanide metal-organic frameworks as signal probes in electrochemiluminescence immunoassay. J. Am. Chem. Soc. **143**(1), 504–512 (2021). https://doi.org/10.1021/jacs.0c12449
125. Wang, N., Feng, Y., Wang, Y., Ju, H., Yan, F.: Electrochemiluminescent imaging for multi-immunoassay sensitized by dual DNA amplification of polymer dot signal. Anal. Chem. **90**(12), 7708–7714 (2018). https://doi.org/10.1021/acs.analchem.8b01610
126. Wang, N., Yang, J., Luo, Z., Qin, D., Wu, Y., Deng, B.: A dual-emitting immunosensor based on manganese dioxide nanoflowers and zinc sulfide quantum dots with enhanced electrochemiluminescence performance for the ultrasensitive detection of procalcitonin. Analyst **148**(9), 2122–2132 (2023). https://doi.org/10.1039/d3an00153a
127. Wang, W., Fang, H., Deng, Y., Jiang, D., Fang, D.: Visual electrochemiluminescence from an all-solid-state electrochemical cell. Analyst **149**(14), 3721–3724 (2024). https://doi.org/10.1039/d4an00355a
128. Wightman, R.M., Curtis, C.L., Flowers, P.A., Maus, R.G., McDonald, E.M.: Imaging microelectrodes with high-frequency electrogenerated chemiluminescence. J. Phys. Chem. B **102**(49), 9991–9996 (1998). https://doi.org/10.1021/jp982913+
129. Wu, M.S., Shi, H.W., He, L.J., Xu, J.J., Chen, H.Y.: Microchip device with 64-site electrode array for multiplexed immunoassay of cell surface antigens based on electrochemiluminescence resonance energy transfer. Anal. Chem. **84**(9), 4207–4213 (2012). https://doi.org/10.1021/ac300551e

130. Xiaoting, L.Z., Rong, F., Xiaona, T., Yifeng, M., Tu, Y.: Direct electrochemiluminescent immunosensing for an early indication of coronary heart disease using dual biomarkers. Anal. Chim. Acta **1110**, 82–89 (2020). https://doi.org/10.1016/j.aca.2020.03.022
131. Xu, Y., Wang, Z., Ding, C., Luo, X.: Ratiometric antifouling electrochemiluminescence biosensor based on bi-functional peptides and low toxic quantum dots. Sensors Actuators B: Chem. **322**, 128613 (2020). https://doi.org/10.1016/j.snb.2020.128613
132. Xu, J., Smith, S., Smith, G., Wang, W., Li, Y.: Glyphosate contamination in grains and foods: an overview. Food Control **106**, 106710 (2019). https://doi.org/10.1016/j.foodcont.2019.106710
133. Xu, M.-R., Dai, R.-F., Wei, Q.-Q., Wang, J., Feng, Y.-Y., Hu, Y.: Urinary AD7c-NTP evaluates cognition impairment and differentially diagnoses AD and MCI.: Am. J. Alzheimer's Dis. Other Dement **37**, 15333175221115247 (2022). https://doi.org/10.1177/15333175221115247
134. Xu, N., Li, X., Luan, F., Tian, C., Zhang, Z., Chen, L., Zhuang, X.: Ratiometric fluorescent and electrochemiluminescent dual modal assay for detection of 2,6-pyridinedicarboxylic acid as an anthrax biomarker. Anal. Chim. Acta **1288**, 342181 (2023). https://doi.org/10.1016/j.aca.2023.342181
135. Xu, P., Zhang, Y., Li, X., Ren, X., Fan, D., Wang, H., Wei, Q., Ju, H.: Electrochemiluminescence immunosensor based on ferrocene functionalized ZIF-8 quenching the electrochemiluminescence of $Ru(bpy)_3{}^{2+}$-doped silica nanoparticles embodied N-butyl diethanolamine. Sensors Actuators B: Chem. **329**, 129101 (2020). https://doi.org/10.1016/j.snb.2020.129101
136. Xue, J.W., Yang, L., Du, Y., Ren, Y., Ren, X., Ma, H.M., Wu, D., Ju, H.X., Li, Y.Y., Wei, Q.: Electrochemiluminescence sensing platform based on functionalized poly-(styrene-co-maleicanhydride) nanocrystals and iron doped hydroxyapatite for CYFRA 21–1 immunoassay. Sens. Actuators B Chem. **321**, 128454 (2020). https://doi.org/10.1016/j.snb.2020.128454
137. Xue, K.F., Lin, B.Y., Huang, S.J., Cai, B.L., Lai, W., Zhan, T.T., Liang, Y., Zhang, C.S.: A dry chemistry, ultrasensitive microfluidic fiber material-based immunosensor for electrochemiluminescence point-of-care testing of luteinizing hormone. Sens. Actuators B Chem. **393**, 134259 (2023). https://doi.org/10.1016/j.snb.2023.134259
138. Xue, Y.W., Tang, X.H., Shen, Q.R., Yu, S.M., Yu, X.F., Shen, D.Z.: Dual potential- and waveband-resolved electrochemiluminescence platform for simultaneous determination of three lung cancer biomarkers. Sens. Actuators B Chem. **389**, 133916 (2023). https://doi.org/10.1016/j.snb.2023.133916
139. Yaman, D., Jimenez, M., Gonzalez, S.F., Corrigan, D.: Current trends in electrochemical approaches for liver biomarker detection: a mini-review. Analyst **149**(21), 5156–5164 (2024). https://doi.org/10.1039/d4an01109k
140. Yang, L., Jia, Y., Wu, D., Zhang, Y., Ju, H., Du, Y., Ma, H., Wei, Q.: Synthesis and application of CeO(2)/SnS(2) heterostructures as a highly efficient coreaction accelerator in the luminol-dissolved O(2) system for ultrasensitive biomarkers immunoassay. Anal. Chem. **91**(21), 14066–14073 (2019). https://doi.org/10.1021/acs.analchem.9b03796
141. Yang, L., Li, J.H.: Recent advances in electrochemiluminescence emitters for biosensing and imaging of protein biomarkers. Chemosensors **11**(8), 432 (2023). https://doi.org/10.3390/chemosensors11080432
142. Yang, L., Sun, X., Wei, D., Ju, H.X., Du, Y., Ma, H.M., Wei, Q.: Aggregation-Induced electrochemiluminescence bioconjugates of apoferritin-encapsulated iridium(III) complexes for biosensing application. Anal. Chem. **93**(3), 1553–1560 (2021). https://doi.org/10.1021/acs.analchem.0c03877
143. Yang, X., Zhao, Y., Sun, L., Qi, H., Gao, Q., Zhang, C.: Electrogenerated chemiluminescence biosensor array for the detection of multiple AMI biomarkers. Sensors Actuators B: Chem. **257**, 60–67 (2017). https://doi.org/10.1016/j.snb.2017.10.108
144. Yao, J., Li, L., Li, P., Yang, M.: Quantum dots: from fluorescence to chemiluminescence, bioluminescence, electrochemiluminescence, and electrochemistry. Nanoscale **9**, 13364–13383 (2017). https://doi.org/10.1039/c7nr05233b
145. Yao, Y., Li, H., Wang, D., Liu, C., Zhang, C.: An electrochemiluminescence cloth-based biosensor with smartphone-based imaging for detection of lactate in saliva. Analyst **142**, 3715–3724 (2017). https://doi.org/10.1039/c7an01008g

146. Yeboah, J., McClelland, R.L., Polonsky, T.S., Burke, G.L., Sibley, C.T., O'Leary, D., Carr, J.J., Goff, D.C., Greenland, P., Herrington, D.M.: Comparison of novel risk markers for improvement in cardiovascular risk assessment in intermediate-risk individuals. JAMA **308**(8), 788–795 (2012). https://doi.org/10.1001/jama.2012.9624
147. Ying, X., Zhou, L., Fu, W., Wang, Y., Su, B.: Electrochemiluminescence devices for point-of-care testing. Sens. Diagn. **2**, 480–491 (2023). https://doi.org/10.1039/d2sd00232a
148. Zeng, H., Zhang, M., Liu, H., Liu, J., Zhu, L., Feng, D., Wang, J.: Two electrochemiluminescence immunosensors for the sensitive and quantitative detection of the CP4-EPSPS protein in genetically modified crops. Food Chem. **428**, 136818 (2023). https://doi.org/10.1016/j.foodchem.2023.136818
149. Zhang, X., Wang, Z.J., Wang, X., Zhang, Y.H., Qu, J., Ding, S.N.: Band-Edge effect-induced electrochemiluminescence signal amplification based on inverse opal photonic crystals for ultrasensitive detection of carcinoembryonic antigen. Anal. Chem. **94**(27), 9919–9926 (2022). https://doi.org/10.1021/acs.analchem.2c01986
150. Zhang, Y., Guo, Y.H., Yang, H., Miao, X.M., Feng, Q.M.: DNA tetrahedral scaffold-corbelled self-feedback circuit for dual-mode ratiometric biosensing with Ru@COF-LZU1 accelerator. Biosens. Bioelectron. **261**, 116520 (2024). https://doi.org/10.1016/j.bios.2024.116520
151. Zhao, L., Song, X., Fan, D., Liu, X., Wang, H., Wei, Q., Wu, D.: Highly efficient signal on/off electrochemiluminescence gel aptasensor based on a controlled release strategy for the sensitive detection of prostate specific antigen. Anal. Chem. **95**(13), 5695–5701 (2023). https://doi.org/10.1021/acs.analchem.2c05655
152. Zhao, C., Thuo, M.M., Liu, X.: A microfluidic paper-based electrochemical biosensor array for multiplexed detection of metabolic biomarkers. Sci. Technol. Adv. Mater. **16**(4), 049501 (2013). https://doi.org/10.1088/1468-6996/14/5/054402
153. Zhiyuan, C.Y., Haiyan, S., Bin, Q., Xiaogang, S., Peng, X.: Quantum dots with highly efficient stable and multicolor electrochemiluminescence. ACS Central Sci. **6**(7), 1129–1137. https://doi.org/10.1021/acscentsci.0c00484
154. Zholudov, Y., Snizhko, D., Kukoba, A., Bilash, H., Rozhitskii, M.: Aqueous electrochemiluminescence of polycyclic aromatic hydrocarbons immobilized into Langmuir–Blodgett film at the electrode. Electrochimica Acta **54**(2), 360–363 (2008). https://doi.org/10.1016/j.electacta.2008.07.069
155. Zhou, Z.Y., Xu, L.R., Wu, S.Z., Su, B.: A novel biosensor array with a wheel-like pattern for glucose, lactate and choline based on electrochemiluminescence imaging. Analyst **139**(19), 4934–4939 (2014). https://doi.org/10.1039/c4an00687a
156. Zhu, X.Q., Zhai, Q.F., Gu, W.L., Li, J., Wang, E.K.: High-Sensitivity electrochemiluminescence probe with molybdenum carbides as nanocarriers for α-fetoprotein sensing. Anal. Chem. **89**(22), 12108–12114 (2017). https://doi.org/10.1021/acs.analchem.7b02701
157. Zou, G.Z., Gao, X.W., Liu, X.C., Zeng, Y., Zhang, Q.Q., Zhang, B.: Spectrum-Resolved electrochemiluminescence to multiplex the immunoassay and DNA probe assay. Anal. Chem. **94**(45), 15801–15808 (2022). https://doi.org/10.1021/acs.analchem.2c03579

Electrochemiluminescence Bio-Imaging

Sara Knežević, Dongni Han, Xuedong Huang, Baohong Liu, Dechen Jiang, and Neso Sojic

Abstract Electrochemiluminescence (ECL) is a well-established analytical method, which is evolving into a powerful optical microscopy technique. The orthogonality of the electrochemical trigger and optical readout distinguishes it from classic microscopy since it does not require any photo-excitation. ECL was used to dynamically image local changes in (electro)chemical activity and a wide range of objects, including those with electrocatalytic, insulating, or biological properties. By playing with the electrogenerated light and the surrounding darkness, various approaches have been developed to image biological entities such as cells, bacteria, mitochondria, spheroids, tissue sections, etc. and cell-like microreactors (e.g., liposomes, droplets, etc.).

Keywords Electrogenerated chemiluminescence · Electrochemiluminescence · Microscopy · Single biological entities · Single-molecule imaging

1 Introduction

Since the initial reports [1, 2], electrochemiluminescence (ECL) has captivated scientific interest with its elegant integration of electrochemistry, offering precise and controlled reactivity and signal activation, with the sensitive and spatiotemporally

S. Knežević · N. Sojic (✉)
University of Bordeaux, Bordeaux INP, CNRS, UMR 5255, ENSMAC, Pessac, France
e-mail: neso.sojic@u-bordeaux.fr

S. Knežević
e-mail: neso.sojic@enscbp.fr

D. Han · D. Jiang
State Key Laboratory of Analytical Chemistry for Life Science and School of Chemistry and Chemical Engineering, Nanjing University, Nanjing, China

X. Huang · B. Liu
Department of Chemistry, State Key Laboratory of Molecular Engineering of Polymers, Fudan University, Shanghai, China

F. Paolucci et al. (eds.), *Electrogenerated Chemiluminescence Microscopy and Imaging*,
https://doi.org/10.1007/978-3-032-14602-1_8

resolved optical (luminescent) signal. Such a dual nature of ECL enables one to simultaneously and dynamically follow the faradaic information, representing the global electron flow through the working electrode, and to localize events near the electrode surface by mapping the light emission resulting from the ECL reactions [3].

In 1964, Hercules captured the ECL emission arising from the electrolysis of different aromatic hydrocarbons, obtaining spatially resolved data [4]. This pioneering work enabled postulating and generalizing the mechanism and characteristics of annihilation ECL [5–9], demonstrating the potential of ECL for investigating the (electro)chemical reactions. More broadly, it highlighted the value of ECL imaging in visualizing and studying processes near the electrode surface, laying the foundation for modern ECL research and applications.

1.1 ECL and ECL Microscopy—Capturing the Local (Electro)Chemical Reactivity

ECL is triggered electrochemically, producing light when a luminophore undergoes an exergonic chemical reaction involving intermediates electrogenerated at the electrode surface. This process reflects, in real time, the system's local (electro)chemical reactivity. Thus, when the emitted light is captured through optical microscopy, ECL becomes a powerful imaging tool that can visualize objects with electroactivity distinct from that of the electrode or those that modify the system's reactivity by catalyzing or impeding (electro)chemical processes including the mass transport of reagents.

Early ECL imaging studies utilized this electrochemically activated emission, combined with electrochemical measurements, to investigate the reactivity at different electrode materials [10], the kinetic and structural heterogeneity of electrode surfaces [11–13], and adsorption at electrodes [14]. Building on these seminal works, contemporary research uses ECL to visualize or monitor locally enhanced electric fields [15], catalytic sites [16–20], and stochastic events [21–24]. By capturing 2D ECL information through optical imaging, these studies reveal objects and surface features that influence local reactivity by participating in, catalyzing, impeding, or otherwise altering electrochemical processes.

Moreover, the ability to extract valuable information from its orthogonal experimental descriptors—current and light intensity—distinguishes ECL as a powerful method for probing fundamental physicochemical properties and reactivity across diverse systems [25–27]. Applications include the study of reaction mechanisms and kinetics [28–41], mass transport phenomena (e.g., solution hydrodynamics, ionic migration, diffusion) [34–36, 42–45], and reactivity in confined spaces [22, 46–51] and at interfaces [36, 52–61]. For example, Amatore and coworkers imaged the thickness of the ECL-emitting layer that was modulated by pulsing the electrochemical potential at different pulse lengths [45]. This approach allows for diffusional decoupling between microelectrodes in an array [62–65], without physically isolating the individual electrodes. Similarly, mapping the ECL generated at double-band microelectrodes, combined with simulation approaches, enabled the study of ionic transport (governed by diffusion and migration), as well as optical effects influencing the output in confocal microscopy [42, 43].

In a separate line of investigation, the influence of hydrodynamics and polarization effects on the ECL process and its dominant reaction mechanisms were studied by capturing the emission arising from rotating disk electrodes in both conventional [35] and bipolar [34] electrochemical setups. Conducted more than 30 years apart, these studies highlight the versatility and usefulness of ECL imaging in providing valuable insights into complex systems and interconnected phenomena.

Finally, resolving ECL emission, not only at the electrode surface, but also in its spatial extension into the solution, reveals critical details about the mechanisms and kinetics within analyzed systems. For instance, resolving ECL at micrometric beads decorated with $[Ru(bpy)_3]^{2+}$ enables the investigation of how solution composition and experimental conditions—such as pH, ionic strength, coreactant properties, bead size, redox mediators, electrode potential—affect reactivity, stability, and type of dominant intermediate radicals, and the ECL generation mechanism [28–33, 66–69].

ECL can also serve to probe phenomena associated with confinement and interfacial effects, which drive locally enhanced reactivity at liquid/liquid [37, 56, 57] and gas/liquid interfaces [36, 52–54, 61]. This enables the exploration of processes that are significant for fundamental research, industry, and biology. For example, ECL imaging demonstrated the interface of gas bubbles in aqueous electrolytes stabilizes hydroxyl radicals, enhancing the production of reactive oxygen species and amplifying the ECL in luminol-based systems [36, 53]. When those electrogenerated bubbles contain chlorine, their dissolution produces highly oxidative hypochlorite species, which mediate ECL reactivity, sustaining it for tens of seconds. Even after the bubbles detach from the electrode, this reactivity extends the ECL emission several millimeters into the solution [36]. These findings not only advance the traditionally transient and surface-confined ECL method but also hold potential for studying and developing various industrially significant systems involving bubble generation at electrode surfaces.

All these examples illustrate that ECL, as a complex process involving multiple electrochemical and chemical steps (detailed in Chap. 5), sensitively reflects the local variations in experimental conditions, solution composition, electrode activity (detailed in Chap. 2), and other factors influencing species stability, transport, and the energetics and kinetics of reactions. This makes ECL an ideal tool for *operando* monitoring of reactions and phenomena that are difficult to observe or resolve spatially and temporally with other electrochemical or optical methods. However, a more comprehensive exploration of these processes and a wider range of ECL applications are presented in other chapters. In contrast, this chapter focuses on the use of ECL in bioimaging.

1.2 ECL Bioimaging—Different Modes of ECL Imaging

Apart from using ECL emission to locally map reactivity, one can exploit the inhomogeneity of the ECL signal—arising from locally hindered or enhanced reactions, as well as uneven distributions of luminophores or coreactants—to achieve highly sensitive, quick, and dynamic imaging of micro/nanoscopic entities [70]. Therefore, successful ECL imaging methods rely on a deep understanding of the underlying electrochemical, photochemical, and chemical reactivities, enabling the resolution of single (biological) objects and, in specific cases, even individual molecules [71, 72] (detailed in Chap. 7).

ECL imaging methods are generally classified into two modes: positive ECL (ECL^+) and negative or "shadow" ECL (ECL^-), based on the spatial arrangement between ECL light and surrounding darkness. In ECL^+, objects of interest—such as cells, organelles, or bacteria—appear as bright entities against a dark background (Fig. 1a), whereas ECL^- (vide infra also called SECL) depicts them as dark objects against a bright background (Fig. 1b).

ECL^+ is typically achieved by decorating the objects with an ECL emitter (e.g., molecular luminophore or nanomaterial) [73, 75, 76] or, less commonly, with an ECL coreactant [77–79]. In some cases, biological molecules can participate in the ECL process as coreactants, enabling the visualization of cells or organelles in ECL^+ mode [80, 81]. However, although ECL^+ can also arise from an object's higher ECL activity compared to the electrode (due to its electrocatalytic properties) [19, 82], this is not the case in bioimaging contexts and will not be discussed further in this chapter. Therefore, as with photoluminescence (PL) imaging techniques (i.e., fluorescence, phosphorescence), the ECL^+ mode typically requires labeling biological entities, with the resulting image reflecting the spatial distribution of these labels. Still, while PL relies on external light to excite the luminophores, ECL is driven by an initial electrochemical step. The fundamental difference in the luminophore excitation mechanism makes these two methods complementary, having different

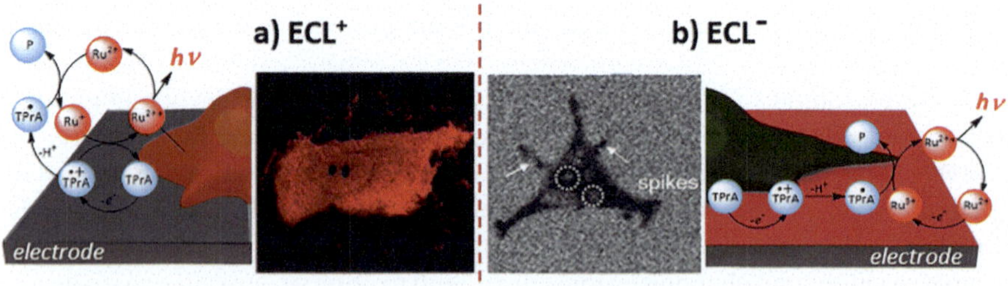

Fig. 1 Schematics and ECL micrographs showing **a** ECL^+ and **b** ECL^- imaging modes of single cells (**a**) labeled with $[Ru(bpy)_3]^{2+}$ luminophores and (**b**) label-free, when both the luminophore and coreactant are freely diffusing in solution. Simplified mechanisms underlying ECL emission are depicted for each mode. White arrows in micrograph (**b**) indicate spikes extending from the cell margin, which appear as dark regions in the ECL images. Reproduced with permission from Ref. [73, 74].

characteristics and offering distinct insights into the imaged system. Unlike PL, ECL emission reflects the mass transport and redox reactivity of the ECL reagents, enabling it to capture dynamic processes within the analyzed system. Moreover, since ECL does not rely on incident light, it achieves near-zero background and high signal-to-noise ratio, delivering excellent contrast and clear visualization of thin peripheral cellular regions, while avoiding the issues associated with PL, such as photobleaching, phototoxicity, background scattered light, or autofluorescence of the biological samples. However, its limitations include lower signal intensities compared to PL and the inability to achieve axial resolution, as ECL reactions are confined to the micrometric region near the electrode surface.

ECL^-, on the other hand, reflects local variations in electrochemical reactivity, particularly the ones arising due to diffusional hindrance of ECL reagents by the biological entities being imaged. This imaging mode was initially introduced by Su and coworkers for latent fingerprint imaging [83–85] and was later adapted to study cells [74, 86], mitochondria [87], and bacteria [88] on electrode surfaces. The analyzed biological entities impede the heterogeneous electron transfer that activates ECL or hinder the diffusion of the ECL reagents to the electrode (and their subsequent reactions), resulting in the locally decreased or completely absent ECL emission. Thus, ECL^- is characterized by an inverted distribution of light and dark areas—dark at the location of the insulating objects and bright on the bare electrode surface initiating the ECL reactions (Fig. 1b). In this regard, it is similar to fluorescence-based label-free microscopy techniques like super-resolution shadow imaging (SUSHI) [89, 90], or stimulated emission depletion (STED) microscopy [91, 92], where the diffusible luminophores do not penetrate cellular membranes. Moreover, considering that the ECL^- image reflects changes in local electroactivity, its detection principle resembles the negative feedback mode used in scanning electrochemical microscopy (SECM). But, unlike SECM [93], which uses an electrochemical probe to scan the surface, ECL^- provides global electrochemical information in a single image, increasing the imaging speed and allowing the analysis of dynamic processes in living cells, such as cellular division, migration, and morphological changes [74, 94].

Much like ECL^+, ECL^- is highly sensitive to small morphological details at the cellular periphery, achieving excellent imaging resolution, constrained by the diffraction limit of optical microscopes. The technique also offers the flexibility to adjust the thickness of the reaction layer by varying luminophore and coreactant concentrations, allowing for axial resolution of the imaged entities [86, 95]. Moreover, by applying signal-processing techniques and statistical analysis, it becomes possible to achieve super-resolution ECL^- microscopy [71, 82, 96]. However, as a label-free technique, ECL^- lacks selectivity for specific proteins or other biological structures that can be selectively labeled and visualized using PL or ECL^+ methods.

Both ECL imaging modes have unique advantages and drawbacks, offering complementary insights into the systems under investigation. Together, they enable selective imaging of specific entities, reactive sites, and molecules, the investigation of dynamic biological processes, transport across cellular membranes, and the analysis of the extracellular space and discrete events in solution. This chapter aims to present the state-of-the-art developments and prospects in ECL bioimaging and highlights the value of approaches that integrate both ECL imaging modes with other techniques. Additionally, it explores pioneering bioanalytical applications of ECL, illustrating their potential to advance current methodologies.

2 ECL Imaging of Biological Entities

2.1 ECL Imaging of Cells and Organelles

Single-cell analysis has become an essential tool in biomedical research, enabling precise characterization of cellular genomes, transcriptomes, proteomes, and metabolomes. By overcoming the limitations of traditional bulk analysis, which often masks cellular heterogeneity, it provides deeper insights into intercellular variations. Imaging techniques are particularly instrumental in single-cell studies, offering a powerful means to investigate cellular structure, function, and interactions within the microenvironment. Over the past decade, ECL microscopy has emerged as an innovative approach for single-cell imaging, garnering considerable attention.

Initially inspired by fluorescence imaging paradigms, where cells are commonly labeled with fluorescent probes, researchers explored the application of ECL luminophores for cell surface labeling, leading to the development of a positive ECL imaging strategy. In 2017, Valenti et al. introduced a single-cell ECL imaging method in which the luminophore $[Ru(bpy)_3]^{2+}$ was immobilized on cell membrane proteins via a biotin-streptavidin linkage (Fig. 2a) [75]. Upon reactions with the electrochemically generated tri-*n*-propylamine (TPrA) radicals, ECL emission was induced. The contrast between the labeled cell regions and the blank electrode surface precisely mapped the adhesion position of single cells at the interface. After optimizing TPrA permeability at the cell–electrode interface, this strategy demonstrated superior single-cell contouring compared to fluorescence imaging, establishing itself as a promising surface-confined ECL microscopy technique [73].

This imaging strategy operates within a heterogeneous ECL reaction system, where the coreactant TPrA undergoes direct oxidation at the electrode surface, generating $TPrA^{+\bullet}$. This radical subsequently deprotonates to form the neutral radical $TPrA^{\bullet}$. The excitation of luminophores on the cell surface occurs through two subsequent electron transfer reactions: initial reduction by $TPrA^{\bullet}$ followed by a highly exergonic oxidation by the cation radical. As a result, both TPrA oxidation and deprotonation are key processes that critically impact ECL emission. Ensuring stable and efficient ECL emission is therefore essential for the effectiveness of this imaging approach.

However, as the reaction progresses, the oxidation rate of TPrA declines due to progressive passivation of the electrode surface and mass transport limitations, leading to a decrease in cellular ECL intensity. To counteract this effect, a cathodic

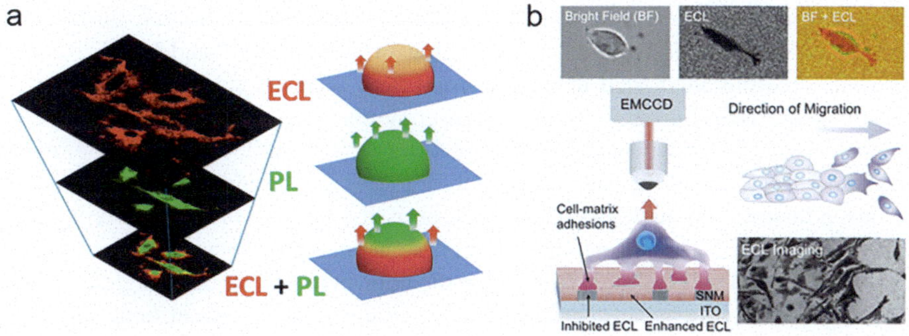

Fig. 2 **a** Imaging of cell membrane proteins using the positive ECL (ECL^+) strategy. **b** Visualization of cell–matrix adhesions and live-cell migration using the negative ECL (ECL^-, also called SECL) strategy. Reproduced with permission from Valenti et al. 2017 and Ding et al. 2020 [74, 75]

treatment was reported by applying short cathodic pulses between imaging intervals to remove adsorbed electrogenerated species from the glassy carbon electrode, thereby restoring ECL emission [97]. Meanwhile, Zhu and coworkers enhanced ECL intensity by leveraging a laser-induced photothermal effect. Localized heating accelerated the ECL reaction and induced thermal convection, where the resulting temperature gradient facilitated both TPrA oxidation and mass transport, achieving a 20-fold increase in imaging contrast [98, 99]. Additionally, an original approach based on the photoelectrochemistry on semiconductors, termed photoinduced ECL (PECL) [100, 101], utilized photovoltage generated at a photoanode to lower the onset potential for both TPrA oxidation and ECL emission, further optimizing single-cell imaging [102].

Beyond improving coreactant oxidation, researchers also explored luminophore excitation. A direct correlation between ECL signal decay and photoluminescence loss due to photobleaching of membrane-bound luminophores was established, highlighting the impact of irreversible luminophore degradation on ECL emission [103–108]. This insight not only deepened the understanding of ECL signal stability but also paved the way for employing ECL imaging in the study of molecular dynamics and mobility.

Compared to the positive ECL strategy, negative ECL offers inherent label-free advantages, making it particularly well-suited for monitoring dynamic changes in live cell morphology. In negative-mode ECL imaging, luminophores (often accompanied by coreactants) are freely dispersed in solution, allowing them to diffuse toward the electrode surface, undergo redox reactions, generate excited-state species, and emit photons upon relaxation. However, when an insulating analyte obstructs direct contact between free luminophores and the electrode, ECL emission in that region is significantly suppressed. This results in an image where the target appears as a dimmed area against a brighter background, resembling a "shadow" effect, hence the term SECL (equivalent to ECL^-).

As already mentioned, the first application of SECL in cellular imaging was reported by Su and coworkers for fingerprint analysis. They later used this strategy to analyze single live-cell adhesion on an electrode surface [74]. In their study, cell–substrate adhesion sites blocked the oxidation of $[Ru(bpy)_3]^{2+}$ and HEPES at the electrode, creating shadowed areas that contrasted with the unadhered regions of the cell membrane (Fig. 2b). SECL microscopy enabled spatial mapping of cell–substrate adhesion and provided dynamic insights into collective cell migration. They further optimized the method by tuning luminophore and coreactant concentrations, adjusting the thickness of the ECL emission layer to match different adhesion structures [86]. This refinement allowed for sequential and selective imaging of cell–substrate and cell–cell junctions from the bottom to the top of a single live cell, providing a comprehensive visualization of cellular morphological heterogeneity. In addition, Zhang and coworkers employed SECL imaging to directly capture various morphological features of live cells, including cell shape, area, average cell boundary, and intercellular distances, demonstrating the method's strong capability for extracting cellular information [94]. Notably, these studies highlight the practical adjustability of the ECL emission layer height by modifying the system component concentrations, suggesting that SECL microscopy has the potential to achieve three-dimensional cellular imaging similar to light-sheet fluorescence microscopy. This adaptability positions SECL as a powerful, structure-specific imaging tool.

In surface-sensitive SECL imaging, signal distribution is governed by substrate properties and the adhesive capacity of the cell surface [109]. Researchers have therefore explored these interactions to extract cellular information. For instance, Su and coworkers investigated how substrate topography affects cell–substrate adhesion and collective migration, revealing that fibroblasts and epithelial cells exhibit microtopography-dependent adhesion behaviors [110]. They also examined the role of integrin subunits in sensing chemically modified surfaces, assessing how biochemical microenvironments influence cellular behavior [111]. Furthermore, by modifying the electrode with different activation stimuli, the specific recognition between T-cell receptors and peptide-MHC complexes, as well as the regulatory functions of co-receptors were characterized [112]. These studies provide valuable insights into cell–microenvironment interactions and underscore the broad potential of SECL-based label-free microscopy for advanced cellular imaging applications.

Given the high signal-to-noise ratio of positive ECL and the label-free advantage of negative ECL, there has been growing interest in integrating both strategies for enhanced single-cell analysis. Recent studies have explored such combinations, leveraging the strengths of each approach. For instance, $[Ru(bpy)_3]^{2+}$ has been employed as the luminophore for positive ECL, immobilized on cell membrane proteins, while $[Ir(sppy)_3]^{3-}$ serves as the negative ECL luminophore, freely diffusing in bulk solution with the coreactant [113]. By spectrally resolving distinct ECL emission wavelengths, this strategy enables simultaneously the visualization of $[Ru(bpy)_3]^{2+}$ distribution on the cell membrane and mapping the entire cell footprint in ECL^- mode. This dual-mode, dual-color ECL microscopy significantly enhances analytical capabilities. Moving forward, the integration of ECL with immunolabeling strategies is expected to drive the development of more information-rich multimodal ECL microscopy.

In the aforementioned studies, the labeling of ECL luminophores was non-specific, meaning that membrane proteins on the cell surface were indiscriminately conjugated with luminophores. Consequently, the extracted information reflected single-cell morphology rather than molecular-level insights. However, intracellular molecular analysis offers invaluable information for understanding cellular processes. To develop ECL imaging techniques capable of molecular resolution, researchers have explored various approaches. Among them, two prominent imaging systems have emerged: the luminol/reactive oxygen species (ROS) system and the $[Ru(bpy)_3]^{2+}$/coreactant system. The former is widely utilized due to the ubiquity of ROS as byproducts in biochemical processes, while the latter benefits from the excellent conjugation flexibility and regeneration capability of $[Ru(bpy)_3]^{2+}$.

In the presence of ROS such as hydrogen peroxide, superoxide, and hydroxyl radicals, luminol and its derivatives exhibit significantly enhanced ECL signals, prompting researchers to leverage this system for visualizing intracellular ROS (primarily hydrogen peroxide) efflux. By integrating luminol-based ECL with enzyme-catalyzed reactions that produce hydrogen peroxide, a wide range of biomolecules can be detected. For example, cholesterol oxidase reacts with active cholesterol on the cell membrane, generating hydrogen peroxide and amplifying ECL intensity (Fig. 3a) [81]. The signal strength correlates with the amount of active cholesterol, while the high standard deviation reflects single-cell heterogeneity. Similarly, glucose oxidase catalyzes glucose released from cells, leading to enhanced ECL signals that enabled the rapid visualization of intracellular glucose in 64 single cells within one minute [114]. This method effectively revealed the substantial heterogeneity in intracellular glucose levels between normal and starved cells. Additionally, certain cell types release hydrogen peroxide under oxidative stress. By depositing $LiFePO_4$ nanoparticles onto an ITO electrode, hydrogen peroxide efflux from MCF-7 cells under oxidative stress at a low potential of 0.5 V was successfully visualized [115]. Such minimal electrochemical perturbation allowed for real-time, accurate observation of cellular activity.

These label-free ROS detection strategies offer rapid and high-throughput capabilities. However, a major limitation of label-free detection is signal diffusion. Hydrogen

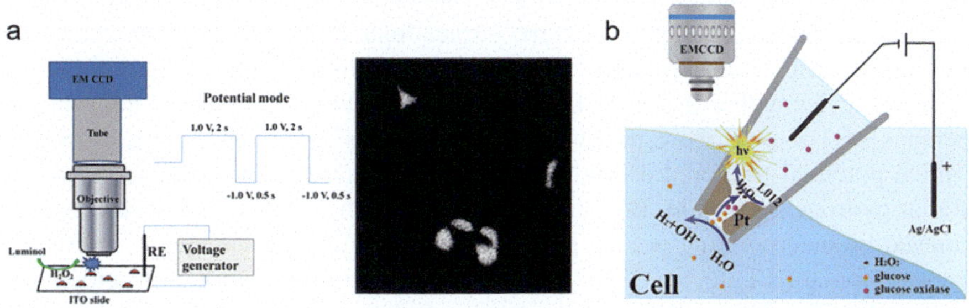

Fig. 3 **a** ECL imaging-based analysis of active membrane cholesterol in a single cell. **b** Schematic illustration of bipolar ECL imaging for wireless detection of intracellular hydrogen peroxide in a single cell. Reproduced with permission from Zhou et al. 2015 and Wang et al. 2020 [46, 81]

peroxide, either secreted by cells or generated through enzymatic reactions, rapidly diffuses into the surrounding solution, causing signal blurring and spatial distortion in ECL imaging. To address this challenge, researchers have developed physical confinement strategies using nanoporous structures to limit lateral diffusion of reactants. A vertically ordered silica nanopore array with a pore size of 2–3 nm was immobilized on an ITO electrode [116]. When cells were cultured on this nanoporous coating, hydrogen peroxide diffused into the confined pores, where it reacted with luminol to generate ECL signals, enabling submicron-resolution mapping of hydrogen peroxide efflux from cells. A similar approach involved modifying FTO electrodes with chitosan and titanium dioxide nanoparticles [117]. The permeable chitosan layer alleviated steric hindrance at the electrode-cell interface, while titanium dioxide nanoparticles enhanced ECL signal intensity. Together, these components facilitated high signal-to-noise-ratio imaging of cell-derived hydrogen peroxide.

Beyond nanopore coatings, electrode structures with intrinsic pore channels have also been fabricated. For example, a gold microtube electrode ensemble was employed as a working electrode to generate spatially confined ECL signals [118]. Hydrogen peroxide efflux from living cells diffused into the microtubes, producing localized ECL signals that revealed subcellular ROS heterogeneity. Xia and coworkers proposed an ECL imaging strategy based on a bipolar electrochemical system and a nanoelectrode array [119]. The spatial confinement within a platinized gold nanoelectrode array led to high oxygen concentrations in the sealed space, inducing significant oxygen reduction currents at the cathode and ECL emission at the anode. This system enabled high-contrast ECL imaging of cell adhesion regions. Similarly, a bipolar ECL setup based on a gold nanorod array successfully visualized hydrogen peroxide release from live cells, where the separation of sensing and reporting electrodes effectively minimized signal interference [120]. These innovative designs have endowed ECL-based single-cell imaging with exceptional resolution and sensitivity, paving the way for highly precise molecular visualization at the single-cell level.

In addition to ROS released by cells, ROS in living cells are essential for cellular signaling and physiological regulation, providing a key avenue for studying the activation mechanisms of signaling pathways. Jiang and coworkers utilized the spatial flexibility of nanoelectrodes to achieve intracellular ROS visualization. A microelectrode system was designed by filling capillaries with chitosan and luminol while coating surfaces with a thin gold layer [121]. These microelectrodes, featuring tip openings of 1–2 μm, were inserted into living cells, where the application of a positive potential triggered a bright ECL signal, indicating the presence of intracellular hydrogen peroxide. Building on this approach, a further enhancement was achieved by incorporating porous Pt deposits at the tip of a nanopipette, creating a wireless bipolar ECL device (Fig. 3b) [46]. Once inserted into a cell, the nanopipette facilitated electrochemical separation of intracellular molecules, which subsequently reacted with luminol derivatives and oxidases within the confined space, generating a bipolar ECL signal. This innovative platform enabled the wireless detection of intracellular hydrogen peroxide, glucose, and sphingomyelinase activity, providing a powerful tool for high-resolution ECL imaging of biomolecules at the single-cell level.

Beyond the luminol/ROS system, the $[Ru(bpy)_3]^{2+}$-based ECL system and its coreactants have significantly advanced molecular imaging. Among these strategies, label-free methods have attracted interest due to their minimal interference with cellular physiological processes. To facilitate membrane protein imaging without labeling, an ECL capacitance microscopy was developed [122]. Antigen–antibody binding reduces local capacitance, leading to a larger potential drop across the double layer at the electrode surface. This drop, in turn, enhances ECL emission in regions where immune complexes are present, effectively distinguishing antigen-binding sites from the surrounding membrane. This technique enables the visualization of antigens at concentrations as low as 1 pg, providing a promising approach for detecting low-abundance biomarkers on the cell surface.

Additionally, certain biomolecules containing amine groups, such as DNA and RNA, can function as coreactants for $[Ru(bpy)_3]^{2+}$. Zhu and coworkers introduced a label-free cellular structural imaging method that utilizes amine-rich biomacromolecules as coreactants [80]. In this approach, intracellular biomolecules act as consumable coreactants, reacting with electrochemically generated $[Ru(bpy)_3]^{3+}$ to drive the ECL process. The resulting intensity-graded ECL images allow for the identification of intracellular structures rich in DNA and RNA, such as the nucleolus and endoplasmic reticulum. This bio-coreactant-enhanced ECL microscopy presents a powerful strategy for studying intracellular biomolecules, cellular architecture, and even yeast metabolism [123].

Building on the paradigm of immunofluorescence imaging, the use of ECL luminophores as molecular labels has emerged as a strategy for cellular imaging. Unlike fluorescence-based techniques, ECL immunoassays eliminate background signal interference due to the absence of external light excitation, enabling exceptionally high sensitivity and signal-to-noise ratios. However, a fundamental limitation of ECL is its inherently weak luminescence, which often necessitates the use of highly sensitive EMCCD cameras and prolonged exposure times (several seconds) to capture sufficient signal, thereby constraining its broader application. To overcome this challenge, various strategies have been developed to enhance ECL emission intensity.

One approach focuses on increasing the local concentration of ECL luminophores. A $[Ru(bpy)_3]^{2+}@SiO_2/Au\text{-}Ab_2$ nanocomposite probe was synthesized, where silica nanoparticles served as a high-capacity scaffold for loading $[Ru(bpy)_3]^{2+}$, while gold nanoparticles further amplified the ECL signal [124]. This nanostructured probe specifically recognized and bound to prostate-specific antigen on the cell membrane via antibody interactions. When integrated with a bipolar ECL system, it enabled wireless, highly sensitive immunodetection of tumor markers on single-cell membranes.

An alternative strategy involves enriching the coreactant rather than the luminophore. A gold nanoflower probe functionalized with guanine-rich single-stranded DNA (G-ssDNA) was designed as a $[Ru(bpy)_3]^{2+}$ coreactant to target carcinoembryonic antigen overexpressed on MCF-7 cells [77]. The high-index facets of the gold nanoflower, combined with the biocompatible G-ssDNA coreactant, synergistically enhanced $[Ru(bpy)_3]^{2+}$ excitation efficiency, enabling high-intensity and stable ECL emission.

Ju and coworkers took a different approach by encapsulating both the luminophore $[Ru(bpy)_3]^{2+}$ and the coreactant persulfate ($S_2O_8^{2-}$) within silica nanoparticles, significantly boosting cathodic ECL intensity [125]. Since live cells exhibit better tolerance to cathodic potentials than anodic ones, this ECL probe successfully enabled in situ imaging of glycoproteins on live cell membranes.

Beyond conventional luminophores, embedding coreactants into polymer dots (Pdots) has proven to be another effective strategy for enhancing ECL intensity. Pdots embedded with triethylamine were conjugated to human epidermal growth factor receptor 2 (HER2) on the cell membrane, producing a strong HER2-related ECL signal upon electrochemical stimulation (Fig. 4a) [79]. Additionally, functionalizing ITO electrodes with a dopamine aptamer-modified, coreactant-loaded Pdot film allowed high-throughput, in situ visualization of dopamine release from single living cells [126]. These innovative methodologies hold significant potential for single-cell analysis applications.

The integration of ECL imaging with therapeutic interventions has also gained attention. Gold nanocages loaded with phorbol myristate acetate (PMA) were employed for intracellular miRNA-21 imaging [127]. Following imaging, the photothermal effect of gold nanocages, combined with ROS generation, efficiently induced apoptosis. Furthermore, photodynamic therapy (PDT), which relies on light irradiation and oxygen interactions, has been incorporated into ECL-based cell imaging. Xu and coworkers reported an ECL-driven PDT system in which $[Ru(bpy)_3]^{2+}$/TPrA-generated ECL served both as an optical indicator for monitoring thermal cell degradation and as an excitation source to activate photosensitizers for ROS-mediated cytotoxicity, achieving a synergistic diagnostic–therapeutic effect (Fig. 4b) [128]. Additionally, by introducing nanoparticles co-loaded with anticancer drugs and PMA into cells, simultaneous nucleolar protein ECL imaging and tumor cell apoptosis were achieved [129]. These examples highlight that ECL

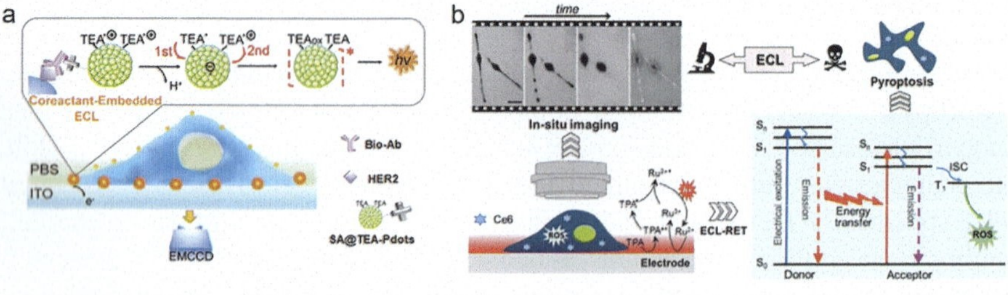

Fig. 4 **a** Schematic illustration of a coreactant-embedded ECL mechanism and microimaging system. **b** Schematic representation of in situ imaging of the ECL-PDT process at the single-cell level using ECL microscopy. Reproduced with permission from Wang et al. 2021 and Chen et al. 2022 [79, 128]

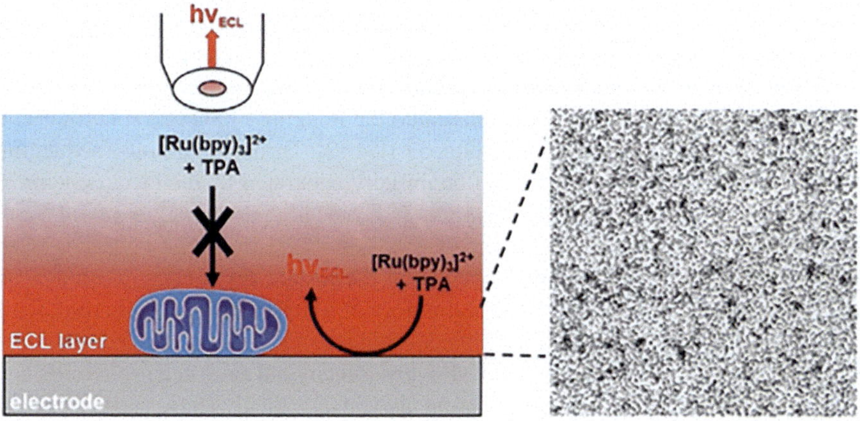

Fig. 5 Label-free shadow ECL microscopy for imaging single mitochondria. The ECL image on the right was captured in a solution containing 10 pM $[Ru(bpy)_3]^{2+}$. Reproduced with permission from Ma et al. 2021 and Descamps et al. 2023 [87, 130]

cell imaging extends beyond simple detection and visualization, offering broader functional capabilities that lay the groundwork for potential clinical applications.

In recent years, ECL microscopy has also increasingly been applied to the visualization of submicron-scale biological structures. Mitochondria undergo dynamic processes such as fission, fusion, and transport, which are highly dependent on cellular state. Understanding these processes is crucial for both fundamental biological research and medical investigations. SECL provides a powerful approach for the direct visualization of individual mitochondria with exceptional contrast. In SECL, the ECL emission layer is confined to the micrometer scale, allowing for precise localization of single mitochondria on the electrode surface (Fig. 5) [87]. This spatial confinement enhances imaging contrast, enabling clear differentiation of mitochondrial structures. Additionally, due to the extreme dilution of luminophores and the cumulative nature of ECL photon emission, this highly sensitive imaging method has been successfully demonstrated at luminophore concentrations as low as the picomolar level, ensuring good biocompatibility [130]. Given its versatility and outstanding sensitivity, SECL holds significant potential for high-resolution subcellular structure imaging.

2.2 *Spheroids and Tissues*

Apart from single-cell/organelles, three-dimensional (3D) cell spheroids can effectively mimic *in vivo* behavior and replicate the cellular response and metabolic activity in physiological contents. And diverse microscopic techniques, such as histological techniques, fluorescent and SECM, have been developed to obtain specific bioinformation and biofunction from this promising cellular model but with various restriction. ECL imaging supplies a new route to monitor in situ cellular behavior of 3D living spheroids with low background by avoiding autofluorescence. Shiku and coworkers for the first time demonstrated the imaging of 3D cell spheroids and their respiratory activity by ECL technique [131, 132]. Through utilizing a sequential potential step, hydrogen peroxide can be initially generated by the electrochemical reduction of dissolved O_2 in the solution, and served as an enhancer of L-012 to produce luminescence (Fig. 6a). Since O_2 is consumed by living cell spheroids, the in situ ECL intensity majorly depends on the instant O_2 concentration around the spheroids and demonstrates the respiratory activity of a single spheroid. This ECL technique succeeds to identify the differences in the respiratory activity of chondrogenic differentiation-induced MSC (mesenchymal stem cell) spheroids and control spheroids. And time-lapse imaging also unveils the dynamic variation in the respiratory activity of the spheroids.

Besides the consumption of O_2 through respiratory activity, the negative distribution or decreased intensity of ECL signal around spheroids also partially originated from the secreted molecules of the spheroids during incubation. By removal of spheroid after the measurement of respiratory activity, Shiku and coworkers also visualized the cellular membrane components and secreted molecules which reflect the local circumstances of spheroids [133, 134]. Combining other techniques, such as microfluidic analysis, RT-qPCR, and immunohistochemical analysis, they subsequently discovered the change of metabolic activity and angiogenesis induction capability during osteogenic differentiation, and gained a comprehensive understanding of MSC spheroids [135]. Dick and coworkers reported the real-time imaging of endogenous hydrogen peroxide produced in mammalian spheroids by ECL [136]. Hence, ECL imaging supplies a real-time evaluation of 3D cell spheroids and becomes a powerful tool for digging biological information from cellular tissues.

In terms of analyzing tissue samples, which is of great importance for clinical diagnosis, ECL imaging provides a quick and simple method for visualizing specific biomolecules compared with time-consuming immunohistochemical

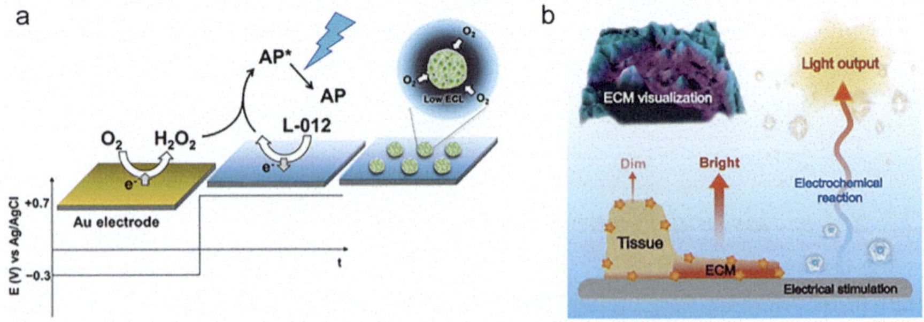

Fig. 6 **a** Schematic illustration of ECL imaging of 3D cell spheroids and their respiratory activity. **b** Schematic illustration of thickness-resolved ECL imaging of ECM at tumor tissues imaging. Reproduced with permission from Kaoru et al. 2021 and Han et al. 2024 [95, 131]

methods. Furthermore, the autofluorescence of tissues can be totally avoided without an external light source by ECL imaging. After the permeabilization process to facilitate tissue labeling by ruthenium-labeled antibody, Jiang and coworkers successfully visualized and recognized the carcinoma and paracancerous tissues of pancreatic ductal carcinoma (PDAC) and lung adenocarcinoma patients by the amount of membrane carcinoembryonic antigen (CEA) on single tissue sections [137]. Near-zero background and extremely high signal-to-noise ratio (S/N) ECL images reveal that CEA proteins mostly exist in the acini and surrounding areas. To further improve the sensitivity and visual contrast, they then report an antigen retrieval-assisted ECL imaging [138]. By utilizing heat-induced epitope retrieval, the binding sites of antigens are re-exposed and allow highly effective interaction between antigen and ruthenium-labeled antibody. Compared with untreated tissue sections, the amount and distribution of CEA antigens at the carcinoma and paracancerous regions are easily detected. This retrieval treatment reduces the occurrence of false-negative phenomena and promotes ECL application in tissues. In addition to antigens, enzymes are vital for understanding tissue physiology and pathology through activity analysis. A label-free enzyme activity imaging method based on the enhancement of luminol ECL by enzyme-catalyzed hydrogen peroxide was proposed [139]. Through the combined action of electrochemical conversion and Fenton reaction-based chemical transformation, long-lived hydrogen peroxide was converted into short-lived ROSs, thereby confining ECL signal diffusion to the nanoscale. This enabled high-resolution, cyclic imaging of multiple enzyme activities within a single tissue section. Combined with colocalization analysis, the method revealed abnormally activated enzymes in PDAC tissues.

The rapid recognition of extracellular matrix (ECM) at tissue sections is also an emerging demand in clinical diagnosis. Traditional staining operation and recognition are time-consuming and require adequate clinical experience. Based on the surface-confined ECL mechanism, Jiang and coworkers developed a thickness-resolved ECL microscopy to image thin-layer ECM at tissue sections for fast histopathological analysis without specific labeling (Fig. 6b) [95]. In detail, the short lifetime and limited diffusion of coreactant radicals determined the thickness of ECL layer. Although the whole tissue section is labeled with biotin-X and streptavidin-modified $[Ru(bpy)_3]^{2+}$ luminophore, only the sections with suitable thickness (e.g., thin-layer ECM) can be lit up by the ECL reaction. This method presents the localization and proportion of the matrix in tissue with high sensitivity and efficient quantification, accompanied with less than 30 min operation. This ECL imaging technique offers an automated protocol to visualize ECM and reflect the structures, tortuosity, and transport properties of tissues, benefiting biomedical research and clinical diagnosis. Liu and coworkers reported the "confocal" ECL 3D imaging of cells and tissue sections allowing the axial position of the ECL focal plane to be adjusted from 0 to 63 μm, and even extended to 400 μm by using a stable coreactant [140]. This technique expands the scope of ECL applications in the field of tissue imaging.

2.3 Bacteria

Identification of microbes like bacteria is necessary to determine appropriate treatment for patients and is crucial for clinical diagnosis. Traditional recognition methods of microorganisms like spectrometry-based approaches require the purification and amplification of biological samples which is time-consuming and lacks spatial information. Feng and coworkers reported shadow ECL imaging for recognition of different species of bacteria [88]. Through shadow ECL imaging, single bacterial morphologies were easily obtained and evaluated. Notably, with the adsorption of $[Ru(bpy)_3]^{2+}$ to the surface of bacteria, ECL imaging mode can transfer from negative to positive contrast which can be further employed for bacteria structures analysis by single molecule ECL detection. In addition, with the help of deep learning classification, direct bacterial counting and identification with an accuracy of 90.5% were achieved.

Since PDT is an efficient treatment for killing cancer cells and pathogenic bacteria, ECL as an electrochemically-trigerred light source from an electric-driven luminous system can be used to excite photosensitizer for producing ROS and antibacterial therapeutics. Wang and coworkers developed the first PDT system driven by electrochemistry, which opens new fields for ECL beyond sensors [141]. The spectral overlap ensures the energy transfer from ECL to photosensitizer and then sensitizes oxygen molecule into ROS. By introducing hydrogel in this system, long-lasting ECL signal can be obtained for more than 10 min upon a 5 s electrochemical pulse, because the redox reactions slow down in hydrogel. Therefore, the reported controllable and convenient ECL-therapeutics system proposed a practical application and opened a new model for PDT.

2.4 Liposomes, Vesicles, and Model Systems

Multiphase environments and reactions occur, and often drive reactivity in biological systems, highlighting the vital importance of understanding the unique physicochemical properties at the phase boundary. ECL is a surface-confined process taking place at the interface of liquid solution and electrode. Hence, this technique is uniquely suitable to map interfaces in multiphase systems, such as liquid/liquid and liquid/solid interfaces. Dick and coworkers constructed and designed a series of interfaces and visualized ECL reactivity at the phase boundaries [23, 56]. By taking advantage of solubilities of the ECL luminophore and the coreactant in different liquid phases, the methods they developed provided an effective way to evaluate the geometry, contact radii, boundary thickness, size, phase content, etc. of various microdroplets [54, 55, 57] (detailed in Chap. 11). These fundamental understandings and experimental tools can promote the application of ECL imaging in biological entities, like artificial cells and liposomes involving phase boundary chemistry.

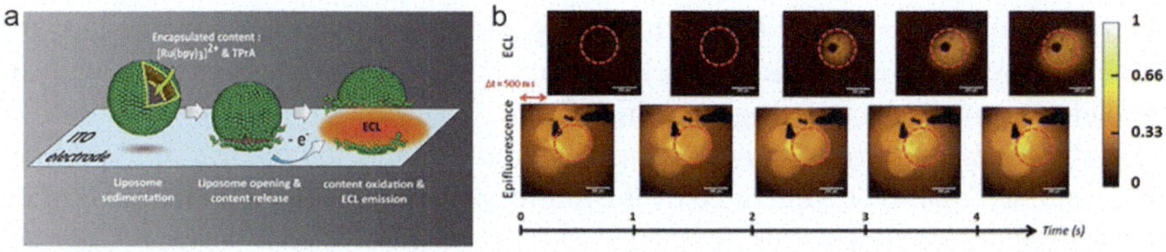

Fig. 7 **a** Schematic illustration of liposome permeabilization by the polarization of ITO electrode and ECL generation by the released reagent. **b** ECL and PL imaging to monitor the liposome permeabilization. Reproduced with permission from Trad et al. 2022 [143]

Liposomes are vesicles composed of a phospholipid bilayer. The lipid membrane permeability or opening are investigated as a model of active and passive lipid barrier crossing. Electrochemical detection of single liposome collisions gathers quantitative information on the number of species released upon liposome opening but lacks spatial resolution [142]. In contrast, ECL imaging combines the advantages of sensitive electrochemical detection and spatially resolved optical imaging. Buriez and coworkers firstly reported a novel approach to observe the delivery of single giant liposome content by ECL imaging (Fig. 7) [143]. With $[Ru(bpy)_3]^{2+}$ and coreactant encapsulated in liposome, ECL generation was recorded as ECL-active species were released from the ruptured liposomes and then oxidized at the polarized ITO surface. While PL microscopy provided information about the localization and status of the liposome content before and during permeation, ECL imaging monitored the efflux of matter from the liposome with high spatial and temporal resolution. Combined with quantitative amperometric signals, complementary PL and ECL information allowed for the real-time characterization of single liposome permeation or opening events.

Antimicrobial peptides (AMPs) provide another route for permeabilization of liposome. To avoid any permeabilization by the polarization of electrode surface, the liposomes were initially deposited on an ITO electrode coated with poly-L-lysine [144]. Then, the addition of different concentrations of melittin, a kind of AMP, triggered the permeabilization of lipid membrane. The released liposome content, like ECL reagents, subsequently reacted on the electrode and finally generated luminescence signals. The ECL and PL imaging displayed a whole process of single liposome permeabilization event by AMPs. Therefore, this multidimensional measurement becomes a complementary biophysical technique to investigate cell, membrane, or vesicle permeabilization [145–147].

Based on this system, enzymatic reactions confined in giant liposome were also imaged by the same group [148]. Since NADH with aliphatic tertiary amine group can substitute TPrA as a biocompatible coreactant with $[Ru(bpy)_3]^{2+}$ while NAD^+ is unable to undergo ECL, enzymatic reactions such as glucose dehydrogenase (GDH)-catalyzed oxidation of glucose can be characterized by ECL imaging. By loading ECL and catalytic reagents (GDH, glucose, NAD^+, and $[Ru(bpy)_3]^{2+}$) into single giant liposome simultaneously, they assessed the completion of GDH-promoted oxidation of glucose according to generated NADH by real-time monitoring of ECL, PL, and amperometric signals. Considering that many enzymatic reactions involve NADH/ NAD^+ redox couple, this system develop a broad path for imaging and post-analysis of metabolic processes in living or artificial cells.

Finally, ECL^- microscopy is also employed to monitor the release of giant liposome content [149]. ECL generated by the luminophore and coreactant in the solution was gradually blocked when the liposome approached the electrode surface. Upon the rupture of the liposome membrane and release of internal ECL-inactive solution triggered by polarization of the ITO electrode, a gravity-driven spreading process initially leads to ECL decrease, and a diffusive regime of active luminophores recovers ECL. Therefore, multiple models of ECL imaging effectively assess the permeabilization of liposome and possess attractive applications in molecule release and delivery.

2.5 *Single Molecules*

Single molecule detection can provide far richer information than bulk level and is of great importance to explore the biological information or structure–activity relationship of catalysts. Hence, various optical and electrochemical techniques have been applied in single molecule detection, such as single molecule fluorescent imaging. However, in terms of ECL imaging, it is difficult to image or capture an individual single molecule from a single luminophore due to low luminous efficiency.

Feng and coworkers (more details in Chap. 7) are the pioneers to image single event from electrochemical reaction in aqueous solution (Fig. 8a) [71]. They reported that through the dilution in space and time, single photon from individual electrochemical reaction can be recorded after the optimization of EMCCD parameters. And this method, which relies on the stochastically collision of luminophore/coreactant, allows to reconstruct super resolution ECL images and reaches a spatial resolution down to 22 nm. Hence, they applied this single molecule detection in SECL imaging of single cell and evaluated the velocity for the movement of cellular adhesive regions. It potentially can present the specific cellular structure by labeling with ruthenium-based probes and be an alternative route of single molecule fluorescent imaging without any background. And they also used single molecule ECL and super resolution ECL imaging to resolve both the electrochemical reactivity of gold microplates and the quantification of single biomolecules on magnetic beads [96, 150]. The high spatial resolution of ECL images build better understanding of the structure–activity relationship of electrocatalysts.

Another significant strategy, which is raised up by Liu and coworkers, is based on the enrichment of luminous molecules into a single nanoemitter and nanoconfinement effect. To construct an effective nanoemitter, $[Ru(bpy)_3]^{2+}$-doped silica/ Au nanoparticles (RuDSNs/AuNPs) carrying tens of thousands luminophores were firstly selected as ECL nanoemitter [151]. The ECL emission is confined to the surface of the nanoemitter and remarkably enhanced by Au nanoparticles. Specific membrane proteins were visualized on the surface of a single cell after labeling with

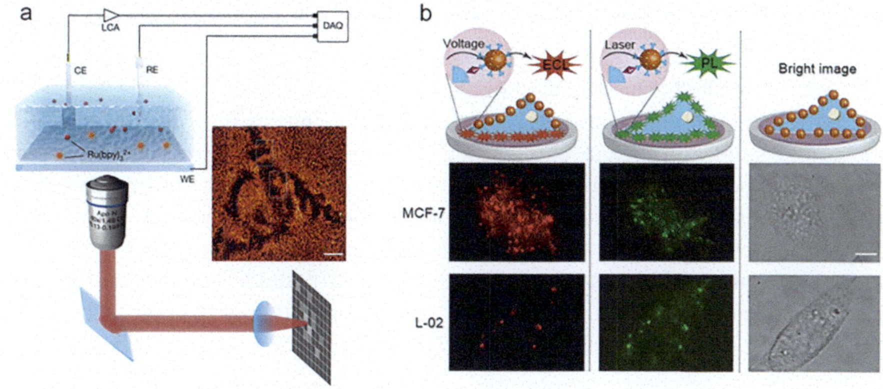

Fig. 8 **a** Single molecule ECL microscopy for the observation of stochastic reactions in solution. **b** Single biomolecule imaging by RuDSNs/AuNPs nanoemitter. Reproduced with permission from Dong et al. and Liu et al. [71, 151]

antibody-functionalized nanoemitters (Fig. 8b). Compared with wide-field fluorescence microscopy, the ECL images have high S/N ratio to present the distribution of membrane proteins at the single-molecule level.

To further improve the S/N ratio of single biomolecule ECL imaging, nanoconfinement effect, where molecular reaction can be significantly accelerated in nanoscale space, was introduced to design the ECL nanoemitter [76, 152, 153]. Since the pore size of mesoporous/microporous nanomaterials is in accordance with the size of $[Ru(bpy)_3]^{2+}$, various $[Ru(bpy)_3]^{2+}$-doped nanoemitters, such as silica, zeolite, and metal–organic framework nanoparticles, were developed by the same group to probe single protein on the surface of single cell. As a result, nanoemitters with pores obtain extremely higher ECL intensity compared with RuDSNs. Moreover, due to the molecular transport capability of pore channel, these nanoemitters hold the characteristics of stable ECL signal (over 1 h), which is beneficial for monitoring long-time single protein movement. Hence, the movement of tyrosine-protein-kinase-like7 proteins at the cellular membrane was observed for long-term and dynamic mapping was achieved which offered fundamental and insight information about cells and small biomolecules [76]. In addition, they also established a dual-mode imaging platform by ECL nanoprobes to realize accurate and reliable detection of SARS-CoV-2 spike proteins through single molecule counting process [154]. Accordingly, this approach based on nanoconfinement effect opens a new route to design ECL nanoprobes and gain in-depth biological information from cells or biomolecules.

3 Novel Approaches and Perspectives of ECL Bioimaging

The well-established, classic ECL imaging approaches offer valuable tools for exploring the morphology, surface properties, and dynamic changes occurring within and around single biological entities and their populations. However, the ever-growing demands of modern bioanalysis continue to drive research toward expanding the capabilities of ECL imaging. The field is undergoing constant innovation, progressing toward wireless detection of intracellular molecules, super-resolution and multimodal imaging, better spatiotemporal control and simplified (and miniaturized) instrumentation. With this in mind, this section highlights some of the most innovative and influential studies driving ECL methods toward these objectives.

3.1 Remote Detection of ECL Signals

While classic ECL analysis provides plenty of information about the systems under study, it requires specific experimental conditions, involving bulky electrochemical setups, biasing the system to high electrode potentials and the use of concentrated, potentially toxic reagents. These conditions cannot only disrupt the system being studied but also make it impossible to analyze living biological entities. To address these challenges, several research groups have developed approaches for wireless generation of ECL[155] with the simultaneous remote readout, aiming to minimize electrochemical and chemical interference with biological systems and potentially enable in vitro, or even in vivo, monitoring of target biomolecules.

This includes the use of a gold-coated optical fiber bundle, which serves a dual purpose: generating ECL through bipolar electrochemistry and guiding the resulting signal to the detector [156, 157]. When polarized in an external electric field, the fiber initiates light emission through bipolar electrochemistry at one end—the analysis site—and transmits it to the opposite end for detection. This setup facilitates wireless bioanalysis, eliminating the need for electrical connections or optical detection at the analysis site. Moreover, since an optical fiber bundle consists of multiple individual fibers, each capable of independently triggering and detecting ECL signals, this method provides a high-throughput approach for screening and quantifying biomarkers. Similarly, molecular wires deposited on the electrode surface can act as both ECL emitters and optical waveguides [158, 159]. ECL generated at one end of the molecular wire remains confined within, propagating along its length, with brighter emissions occurring at the terminals. This enables remote emission in non-conductive regions. Furthermore, when the wire is supported by a micropipette electrode, its ECL transmission can be precisely controlled in air, minimizing optical loss [159]. As the generation of ECL in the molecular wire relies on strongly reducing radicals (i.e., $TPrA^{\bullet}$) which are scavenged by O_2, leading to the quenching of the ECL, the system is well-suited for oxygen detection in air.

Furthermore, in situ ECL imaging of intracellular molecules has been demonstrated using a nanopipette with a porous Pt layer at its tip (Fig. 3b). Serving as a bipolar electrode, the nanopipette enabled the electrochemical sorting of intracellular molecules such as hydrogen peroxide and glucose. Inside the pipette, these molecules interacted with a luminol-based emitter and oxidases, producing wireless bipolar ECL signals [46]. In another study, trying to achieve improved throughput of intracellular analysis, Wang et al. used conductive carbon nanotubes as bipolar electrodes for ECL detection of antigens within cell nuclei [160]. Thanks to their excellent cellular permeability and biocompatibility, antibody-functionalized nanotubes could penetrate cells and bind to intracellular antigens. When exposed to an external electric field, these nanotubes polarized and generated ECL within cell nuclei via bipolar electrochemistry.

Overall, these original approaches provide opportunities for contactless and non-invasive electrochemical analysis, such as intracellular biosensing and remote imaging, allowing the study of (bio)chemical systems without causing (much) electrical or chemical interference.

3.2 *Super-Resolution ECL*

Most instrumental techniques, including electrochemistry, provide insights into the global reactivity of a system, offering averaged measurements of key parameters such as kinetic constants and catalytic properties. While valuable, these measurements overlook critical aspects of local heterogeneity, structure–function relationships, and stochastic events within the system. Addressing this gap has led to the emergence of advanced imaging methods like single-molecule optical detection [161, 162] and scanning probe microscopies [163–165]. Similarly, ECL microscopy can localize reaction hotspots [96, 166], image individual entities on electrodes [73, 75, 87, 88], and, under carefully controlled conditions, detect stochastic events such as particle or droplet collisions with microelectrode surfaces [21–23]. However, it remains limited by the diffraction limit of optical microscopy, making it challenging to resolve individual events at the molecular or photon level.

In 2021, Xu's and Feng's groups independently made significant developments in super-resolution ECL microscopy [71, 82], both achieving submicrometer imaging resolution. Xu's team demonstrated the stochastic nature of the ECL events that follow Poisson statistics, fulfilling the requirements for super-resolution radial fluctuation (SRRF) processing [82]. By applying SRRF to ECL signals, they achieved imaging with spatial resolution of approximately 100 nm, revealing surface activities at a subparticle level, such as facet- and defect-dependent reactivity, and dynamic reactivity fluctuations on single Au nanoplates and nanorods. In contrast, Feng's group achieved the detection of single photons arising from individual ECL reactions (Fig. 8a). They did so by diluting the ECL signals both in space and time [71]. Then, they precisely mapped the reactivity by statistically analyzing the repeated ECL events. This allowed for the reconstruction of the ECL images with a spatial resolution of 22 nm, based on accumulated single-photon detections. The method's utility was further displayed through super-resolution imaging of various systems, including microstructured electrodes, single cells (focusing on cellular adhesions and their movement), catalytic sites on Au microplates, magnetic beads used in immunosensing, and bacteria [71, 88, 96, 150].

Both of these super-resolution ECL imaging approaches can provide important fundamental insights into heterogeneity and the nature of (electro)chemical reactions. Furthermore, they hold promising applications in catalysis, biological imaging, and single-entity analysis.

3.3 The Use of Redox Mediators and Bimodal ECL Imaging

ECL performance is often limited by relatively low emission intensities, which encourage the development of strategies aimed at amplifying its output. While many approaches focus on designing more efficient emitters, a recently introduced, experimentally straightforward method for ECL signal amplification involves using redox mediators to enhance the efficiency of underlying ECL reactions.

In a study by Kerr et al. [167], a tenfold enhancement in the ECL intensity of $[Ru(bpy)_3]^{2+}$/TPrA system was reported in the presence of the redox-active $[Ir^{III}(sppy)_3]^{3-}$ complex, while the overpotentials required for generating $[Ru(bpy)_3]^{2+}$ ECL were simultaneously decreased. Such behavior was attributed to the electrooxidation of Ir(III) complex at low anodic potentials, generating Ir(IV), which mediates the oxidation of $[Ru(bpy)_3]^{+}$ and efficiently populates the $[Ru(bpy)_3]^{2+*}$ excited state at lower applied potentials.

Aiming to understand and generalize this behavior, multiple follow-up studies investigated the properties of such Ir(III)/Ru/TPrA systems through ECL microscopy and spectroscopy techniques, using different redox mediators, bead-based ECL setups for commercial ECL immunoassays, and numerical simulations [29–31, 168, 169]. These investigations identified the role of Ir(III) in mediating and catalyzing oxidative processes in ECL coreactant systems, providing mechanistic and kinetic insights essential for predicting and interpreting ECL dynamics in the presence of different mediators and defining the redox characteristics a mediator should have to enhance the ECL. This understanding could potentially inform the design of tailored ECL systems with improved performance. Such a possibility provides a rational route toward enhanced ECL signals, that could lead to lower detection limits in ECL biosensing—specifically ECL immunoassays, facilitating earlier and less invasive disease diagnostics.

Finally, the $[Ir^{III}(sppy)_3]^{3-}$ mediator was used in combination with $[Ru(bpy)_3]^{2+}$ luminophore and TPrA to develop bimodal ECL imaging of single cells [113]. The approach involved labeling the cellular membranes with $[Ru(bpy)_3]^{2+}$ (λ_{max} = 620 nm), while $[Ir^{III}(sppy)_3]^{3-}$ (λ_{max} = 515 nm) and TPrA were freely diffusing in solution. By spectrally and spatially resolving the emission of these two emitters, it is possible to simultaneously capture the red ECL of $[Ru(bpy)_3]^{2+}$ labels, yielding a bright image of cells against the dark background (ECL^+), and the green ECL of $[Ir^{III}(sppy)_3]^{3-}$ mediator, which reflects the local diffusional hindrance of the ECL reagents by the cells and generates a dark image against a bright background (ECL^-). Due to the complementarity of information these two ECL modes provide, these, simultaneously captured, micrographs of the same cells could provide plenty of useful information and a more detailed understanding of the biological systems under investigation.

3.4 *Photoinduced ECL*

Photoelectrochemistry—a light-driven electrochemical process, formally opposite of ECL—offers a way to reduce, or even eliminate, the externally applied potential required for electrochemical reactions. In this technique, illuminating a semiconductor electrode generates minority carriers (electrons or holes) at the semiconductor/electrolyte interface, producing a photovoltage that efficiently drives electrochemical processes. Bard and coworkers reported photoinduced ECL on semiconductors (p-Si, InP, and GaAs) in annihilation mode in dried organic solvents [100, 101]. In 2019, PECL was demonstrated in water-based electrolytes with the model $[Ru(bpy)_3]^{2+}$/TPrA coreactant system at a remarkably low potential of 0.48 V [104]. PECL provides precise spatiotemporal control over emitted light, with emission confined to illuminated spots on the semiconductor electrode [170, 171]. As the excitation light moves, the ECL emission seamlessly follows, enabling dynamic localization of the signal. Additionally, unlike fluorescence, PECL permits backside illumination of the electrode, avoiding the need for light to go through the electrolyte and illuminate the sample, thus avoiding issues such as photobleaching, sample damage, and autofluorescence [105, 108, 171].

With n-Si/SiO_x/metal (metal-isolator-semiconductor) electrodes, the near-infrared irradiation triggered visible PECL in the $[Ru(bpy)_3]^{2+}$/TPrA coreactant system, producing a negative Stokes shift of −230 nm [170]. This energy upconversion occurs due to the electrochemical step initiating the ECL process. Notably, the PECL emission spot was significantly larger than the excitation laser beam diameter [170]. This phenomenon, attributed to the crystalline silicon's physical properties and the diffusion length of photogenerated holes, offers a potential approach for investigating the lifetime of photogenerated minority carriers.

As a phenomenon that involves photoexcitation, PECL can be used for investigating photocatalytic properties of materials, which was demonstrated by monitoring the PECL driven by plasmon-generated hot carriers on single gold nanoparticles [172]. Moreover, the potential of PECL for bioimaging was investigated at n-Si/SiO_x/Ir electrodes stimulated by an incident near-infrared light, achieving successful and high-contrast imaging of single cells, in both $PECL^+$ and $PECL^-$ modes, with a negative Stokes shift of −430 nm [102].

Ultimately, in 2023, all-optical ECL (AO-ECL) was demonstrated as a wireless approach to PECL, requiring no external power supplies [173]. This was achieved

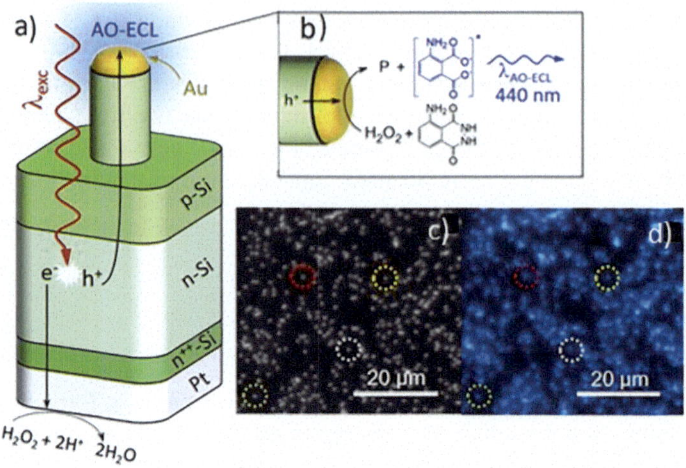

Fig. 9 **a** Schematic illustration of the AO-ECL device with a nanostructured Au-*pnn*$^{++}$Si-Pt junction under illumination. **b** Diagram of hole transfer occurring at the solid–liquid interface at the top of the gold-coated nanopillars, leading to visible photon emission, with "P" representing reaction products. Optical microscopy images of the Au-*pnn*$^{++}$Si-Pt frontside captured under **c** white light and **d** IR illumination triggering ECL. Colored dashed circles identify individual nanopillars in (**c**) and their corresponding bright spots in the AO-ECL image (**d**), highlighting their spatial correlation. Reproduced with permission from Zhao et al. 2023 [173]

using a nanostructured silicon photovoltaic junction modified with catalytic gold coatings (Fig. 9a). The absorbed light drove the oxidation of ECL reagents at the nanostructured photoelectrode-electrolyte interface, producing a blue emission from the luminol/H_2O_2 coreactant system, with an anti-Stokes shift of −610 nm (Fig. 9b). Optical microscopy imaging of the AO-ECL activity revealed that emission spots aligned with the nanoscale architecture of the electrode, with interfacial charge transfer occurring at the tops of gold-capped silicon nanopillars (Fig. 9c and d). With its wireless and all-optical design, AO-ECL significantly simplifies the ECL setup, potentially broadening the applications of ECL and ECL imaging. This approach makes the technology more accessible to non-specialists and paves the way for developing sensitive, portable biosensing devices and imaging technologies.

4 Conclusions and Prospects

This chapter provides a comprehensive overview of the state of the art in ECL imaging, emphasizing its capability to visualize biological entities. Beginning with its historical background, imaging concepts, and commonly employed modes and setups, we explore the advantages and limitations of ECL imaging, particularly its capacity to reveal complex processes and track discrete events. Owing to the orthogonality of its electrochemical trigger and optical readout, ECL dynamically reflects local changes in (electro)chemical activity, allowing for the visualization of a wide range of objects, including those with electrocatalytic or insulating properties, those labeled with luminescent emitters, and even entirely label-free. Here, we look into the application of ECL microscopy for studying biological entities—cells, organelles, bacteria, spheroids, and tissues, as well as cell-like microreactors such as liposomes and droplets.

Innovative approaches in ECL imaging have enabled high spatial resolution of thin cellular regions, such as adhesions and cell–cell junctions, while also allowing for dynamic visualization of their movement or morphological changes under external stimuli [74, 86, 94]. Additionally, catalytic routes to ECL generation have introduced control over the thickness of the emitting layer, expanding this, traditionally surface-confined technique, to tissue imaging [95]. At the same time, the inherent dependence of ECL emission on mass transport has been leveraged to develop different ECL imaging strategies and modes, while the tunable adsorption of ECL reagents can serve to image bacteria without prior labeling [88]. Notably, the clever imaging approach, simultaneously employing Ir(III)- and Ru-based luminophores, has enabled bimodal ECL imaging, allowing single cells to be visualized at the same time in both ECL^+ and ECL^- modes [113]. This possibility holds great potential for providing plenty of information about cellular morphology and heterogeneity, as well as membrane surface properties and transport dynamics.

Despite these advancements, current ECL imaging techniques face limitations due to low emission intensity, which decreases the imaging sensitivity and conditions the use of prolonged exposure times to achieve high-resolution micrographs. To overcome these challenges, significant efforts have been directed toward developing brighter ECL emitters [174], including luminophores and nanomaterials with enhanced quantum yields [151, 175, 176]. In parallel, recent innovations have introduced alternative strategies to improve ECL performance: (i) the use of redox mediators has increased the efficiency of ECL's underlying mechanistic pathways, leading to enhanced emission intensities[168]; (ii) the employment of deep learning has enabled reduced exposure times for capturing ECL images without compromising image quality [177]; (iii) improvements in detector sensitivity and optical setups now allow for single-photon detection, facilitating single-molecule ECL analysis and super-resolution imaging [71].

Another significant direction in ECL research involves the miniaturization and simplification of instrumental setups, aiming to make ECL technology more accessible to non-specialists while reducing electrical and chemical interference with biological systems. The advancements in this area focus on three primary strategies: (i) integrating optical fibers or waveguides for remote ECL generation and signal detection, providing precise and localized imaging while eliminating the interference associated with traditional ECL systems [156, 158, 159]; (ii) employing bipolar electrochemistry to wirelessly generate ECL signals, which removes the need for direct electrical connections and facilitates the miniaturization of ECL setups, supporting ECL reactions on single electrodes [178–181], in nanopipettes or at conductive nanoparticles [46, 160]; and (iii) developing photoinduced (all-optical) ECL, which bypasses the necessity for conventional electrochemical instrumentation, enabling users without electrochemical expertise to perform ECL and achieve precise spatial and temporal control over the resulting emission [173].

Collectively, these advancements significantly enhance the resolution, emission intensity, and biocompatibility of ECL devices and setups. Such innovations establish ECL as a transformative tool for dynamic, high-resolution bioimaging and (bio)molecular detection, enhancing its information domain and expanding its applications across both scientific research and clinical diagnostics. Beyond guiding future research directions, these developments enable a broader range of analytical and imaging applications, including intracellular sensing and ECL microscopy capable of resolving single molecules and reactions in solution. Looking ahead, we anticipate future efforts to focus on refining ECL imaging for even greater sensitivity, enabling detailed investigations of membrane transport properties, cellular heterogeneity, and complex biological processes, ultimately achieving single-biomolecule analysis and opening new frontiers in precision imaging and in vivo molecular diagnostics on complex biological samples.

References

1. Dufford, R.T., Nightingale, D., Gaddum, L.W.: Luminescence of grignard compounds in electric and magnetic fields, and related electrical phenomena. J. Am. Chem. Soc. **49**, 1858 (1927)
2. Harvey, N.: Luminescence during electrolysis. J. Phys. Chem. **33**, 1456 (1929)
3. Knežević, S., Han, D., Liu, B., Jiang, D., Sojic, N.: Electrochemiluminescence microscopy. Angew. Chem. Int. Ed. **63**, e202407588 (2024)
4. Hercules, D.M.: Chemiluminescence resulting from electrochemically generated species. Science **145**, 808 (1964)
5. Hercules, D.M., Lytle, F.E.: Chemiluminescence from reduction reactions. J. Am. Chem. Soc. **88**, 4745 (1966)
6. Visco, R.E., Chandross, E.A.: Electroluminescence in solutions of aromatic hydrocarbons. J. Am. Chem. Soc. **86**, 5350 (1964)
7. Chandross, E.A., Longworth, J.W., Visco, R.E.: Excimer formation and emission via the annihilation of electrogenerated aromatic hydrocarbon radical cations and anions. J. Am. Chem. Soc. **87**, 3259 (1965)
8. Santhanam, K.S.V., Bard, A.J.: Chemiluminescence of electrogenerated 9,10-diphenylanthracene anion radical. J. Am. Chem. Soc. **87**, 139 (1965)
9. Faulkner, L.R., Bard, A.J.: Electrogenerated chemiluminescence. I. Mechanism of anthracene chemiluminescence in N,N-dimethylformamide solution. J. Am. Chem. Soc. **90**, 6284 (1968)
10. Vitt, J.E., Johnson, D.C., Engstrom, R.C.: The effect of electrode material on the electrogenerated chemiluminescence of luminol. J. Electrochem. Soc. **138**, 1637 (1991)
11. Engstrom, R.C., Johnson, K.W., Scott. DesJarlais.: Characterization of electrode heterogeneity with electrogenerated chemiluminescence. Anal. Chem. **59**, 670 (1987)
12. Bowling, R.J., McCreery, R.L., Pharr, C.M., Engstrom, R.C.: Observation of kinetic heterogeneity on highly ordered pyrolytic graphite using electrogenerated chemiluminescence. Anal. Chem. **61**, 2763 (1989)

13. Pharr, C.M., Engstrom, R.C., Klancke, J., Unzelman, P.L.: Determination of microscopic electrode kinetics with electrogenerated chemiluminescence imaging. Electroanalysis **2**, 217 (1990)
14. Engstrom, R.C., Nohr, P.L., Vitt, J.E.: Observation of adsorbate coverage with electrogenerated chemiluminescence imaging. Colloids Surf. A Physicochem. Eng. Asp. **93**, 221 (1994)
15. Cui, C., Jin, R., Jiang, D., Zhang, J., Zhu, J.: Visualization of an accelerated electrochemical reaction under an enhanced electric field. Research **2021**, 1742919 (2021)
16. Chen, Y., Zhao, D., Fu, J., Gou, X., Jiang, D., Dong, H., Zhu, J.-J.: In situ imaging facet-induced spatial heterogeneity of electrocatalytic reaction activity at the subparticle level via electrochemiluminescence microscopy. Anal. Chem. **91**, 6829 (2019)
17. Chen, M., Zhao, W., Zhu, M.-J., Li, X.-L., Xu, C.-H., Chen, H., Xu, J.-J.: Spatiotemporal imaging of electrocatalytic activity on single 2D gold nanoplates via electrogenerated chemiluminescence microscopy. Chem. Sci. **10**, 4141 (2019)
18. Wilson, A.J., Marchuk, K., Willets, K.A.: Imaging electrogenerated chemiluminescence at single gold nanowire electrodes. Nano Lett. **15**, 6110 (2015)
19. Chen, M.-M., Xu, C.-H., Zhao, W., Chen, H.-Y., Xu, J.-J.: Observing the structure-dependent electrocatalytic activity of bimetallic Pd–Au nanorods at the single-particle level. Chem. Commun. **56**, 3413 (2020)
20. Zhu, H., Jiang, D., Zhu, J.-J.: High-resolution imaging of catalytic activity of a single graphene sheet using electrochemiluminescence microscopy. Chem. Sci. **12**, 4794 (2021)
21. Fan, F.-R.F., Bard, A.J.: Observing single nanoparticle collisions by electrogenerated chemiluminescence amplification. Nano Lett. **8**, 1746 (2008)
22. Defnet, P.A., Zhang, B.: Detection of transient nanoparticle collision events using electrochemiluminescence on a closed bipolar microelectrode. ChemElectroChem **7**, 252 (2020)
23. Dick, J.E., Renault, C., Kim, B.-K., Bard, A.J.: Simultaneous detection of single attoliter droplet collisions by electrochemical and electrogenerated chemiluminescent responses. Angew. Chem. Int. Ed. **53**, 11859 (2014)
24. Lebègue, E., Anderson, C.M., Dick, J.E., Webb, L.J., Bard, A.J.: Electrochemical detection of single phospholipid vesicle collisions at a Pt ultramicroelectrode. Langmuir **31**, 11734 (2015)
25. Zanut, A., Fiorani, A., Rebeccani, S., Kesarkar, S., Valenti, G.: Electrochemiluminescence as emerging microscopy techniques. Anal. Bioanal. Chem. **411**, 4375 (2019)
26. Rebeccani, S., Zanut, A., Santo, C.I., Valenti, G., Paolucci, F.: A guide inside electrochemiluminescent microscopy mechanisms for analytical performance improvement. Anal. Chem. **94**, 336 (2022)
27. Yan, Y., Ding, L., Ding, J., Zhou, P., Su, B.: Recent advances in electrochemiluminescence visual biosensing and bioimaging. ChemBioChem **25**, e202400389 (2024)
28. Sentic, M., Milutinovic, M., Kanoufi, F., Manojlovic, D., Arbault, S., Sojic, N.: Mapping electrogenerated chemiluminescence reactivity in space: mechanistic insight into model systems used in immunoassays. Chem. Sci. **5**, 2568 (2014)
29. Kerr, E., Knezevic, S., Francis, P.S., Hogan, C.F., Valenti, G., Paolucci, F., Kanoufi, F., Sojic, N.: Electrochemiluminescence amplification in bead-based assays induced by a freely diffusing iridium(III) complex. ACS Sens. **8**, 933 (2023)
30. Knežević, S., Kerr, E., Valenti, G., Paolucci, F., Francis, P.S., Hogan, C.F., Sojic, N., Kanoufi, F.: Electrocatalytic amplification of coreactant electrochemiluminescence using redox mediators. Electrochim. Acta **499**, 144677 (2024)
31. Blom, S.J., Adamson, N.S., Kerr, E., Doeven, E.H., Wenger, O.S., Schaer, R.S., Hayne, D.J., Paolucci, F., Sojic, N., Valenti, G., Francis, P.S.: Redox mediated enhancement and quenching of co-reactant electrochemiluminescence by iridium(III) complexes. Electrochim. Acta **484**, 143957 (2024)
32. Zanut, A., Fiorani, A., Canola, S., Saito, T., Ziebart, N., Rapino, S., Rebeccani, S., Barbon, A., Irie, T., Josel, H.-P., Negri, F., Marcaccio, M., Windfuhr, M., Imai, K., Valenti, G., Paolucci, F.: Insights into the mechanism of coreactant electrochemiluminescence facilitating enhanced bioanalytical performance. Nat. Commun. **11**, 2668 (2020)

33. Adamson, N.S., Blom, S.J., Doeven, E.H., Connell, T.U., Hadden, C., Knežević, S., Sojic, N., Fracassa, A., Valenti, G., Paolucci, F., Ding, J., Wang, Y., Su, B., Hua, C., Francis, P.S.: Electrochemiluminescence enhanced by a non-emissive dual redox mediator. Angew. Chem. Int. Ed. **63**, e202412097 (2024)
34. Arias-Aranda, L.R., Salinas, G., Kuhn, A., Xu, G., Kanoufi, F., Bouffier, L., Sojic, N.: Complex electrochemiluminescence patterns shaped by hydrodynamics at a rotating bipolar electrode. Chem. Sci. **15**, 8723 (2024)
35. Ludvík, J., Volke, J.: Non-uniform current density distribution at a *rde*. Observation by means of photographing electrochemiluminescence. Electrochim. Acta **35**, 1983 (1990)
36. Knežević, S., Totoricaguena-Gorriño, J., Gajjala, R.K.R., Hermenegildo, B., Ruiz-Rubio, L., Vilas-Vilela, J.L., Lanceros-Méndez, S., Sojic, N., Del Campo, F.J.: Enhanced electrochemiluminescence at the gas/liquid interface of bubbles propelled into solution. J. Am. Chem. Soc. **146**, 22724 (2024)
37. Hill, M.L., Layman, B.R., Dick, J.E.: Real-Time visualization of an elusive, strong reducing agent during tris(2,2'-bipyridyl)ruthenium(II) electro-oxidation in water. J. Am. Chem. Soc. **147**, 17701 (2025)
38. Wang, Y., Zhou, S., Zheng, Y., Wang, Y., Hou, Y., Wu, K., Huang, C., Liu, S., Shen, Y., Chen, R., Zhang, Y.: Measurements of local pH gradients for electrocatalysts in the oxygen evolution reaction by electrochemiluminescence. J. Am. Chem. Soc. (2025). https://doi.org/10.1021/jacs.5c04896
39. Wang, Y., Su, B.: Measuring and regulating the reaction layer for enhanced electrochemiluminescence analysis. ACS Electrochem. (2025). https://doi.org/10.1021/acselectrochem.5c00086
40. Belotti, M., El-Tahawy, M.M.T., Yu, L.-J., Russell, I.C., Darwish, N., Coote, M.L., Garavelli, M., Ciampi, S.: Luciferase-free luciferin electrochemiluminescence. Angew. Chem. Int. Ed. **61**, e202209670 (2022)
41. Belotti, M., El-Tahawy, M.M.T., Darwish, N., Garavelli, M., Ciampi, S.: Electrochemically generated luminescence of luminol and luciferin in ionic liquids. ChemElectroChem **10**, e202201033 (2023)
42. Amatore, C., Pebay, C., Servant, L., Sojic, N., Szunerits, S., Thouin, L.: Mapping electrochemiluminescence as generated at double-band microelectrodes by confocal microscopy under steady state. ChemPhysChem **7**, 1322 (2006)
43. Amatore, C., Oleinick, A., Klymenko, O.V., Thouin, L., Servant, L., Svir, I.: Confocal microscopy imaging of electrochemiluminescence at double band microelectrode assemblies: numerical solution of the inverse optical problem. ChemPhysChem **8**, 1664 (2007)
44. Svir, I., Oleinick, A., Klymenko, O.V., Amatore, C.: Strong and unexpected effects of diffusion rates on the generation of electrochemiluminescence by amine/transition-metal(II) systems. ChemElectroChem **2**, 811 (2015)
45. Szunerits, S., Tam, J.M., Thouin, L., Amatore, C., Walt, D.R.: Spatially resolved electrochemiluminescence on an array of electrode tips. Anal. Chem. **75**, 4382 (2003)
46. Wang, Y., Jin, R., Sojic, N., Jiang, D., Chen, H.: Intracellular wireless analysis of single cells by bipolar electrochemiluminescence confined in a nanopipette. Angew. Chem. Int. Ed. **59**, 10416 (2020)
47. Han, J.-H., Lee, E., Park, S., Chang, R., Chung, T.D.: Effect of nanoporous structure on enhanced electrochemical reaction. J. Phys. Chem. C **114**, 9546 (2010)
48. Voci, S., Ismail, A., Pham, P., Yu, J., Maziz, A., Mesnilgrente, F., Reynaud, L., Livache, T., Mailley, P., Buhot, A., Leichle, T., Kuhn, A., Leroy, L., Bouchet-Spinelli, A., Sojic, N.: Wireless enhanced electrochemiluminescence at a bipolar microelectrode in a solid-state micropore. J. Electrochem. Soc. **167**, 137509 (2020)
49. Nikolaou, P., Valenti, G., Paolucci, F.: Nano-structured materials for the electrochemiluminescence signal enhancement. Electrochim. Acta **388**, 138586 (2021)
50. Guo, W., Zhou, P., Sun, L., Ding, H., Su, B.: Microtube electrodes for imaging the electrochemiluminescence layer and deciphering the reaction Mechanism. Angew. Chem. Int. Ed. **60**, 2089 (2021)

51. Liu, H., Hussain, A., Zholudov, Y.T., Snizhko, D.V., Sojic, N., Xu, G.: Self-Powered electrochemiluminescence for imaging the corrosion of protective coating of metal and quantitative analysis. Angew. Chem. Int. Ed. **63**, e202411764 (2024)
52. Voci, S., Vannoy, K.J., Dick, J.E.: Femtoliter oil droplets act as CO_2 micropumps for uninterrupted electrochemiluminescence at the waterloil interface. J. Colloid Interface Sci. **661**, 853 (2024)
53. Vogel, Y.B., Evans, C.W., Belotti, M., Xu, L., Russell, I.C., Yu, L.-J., Fung, A.K.K., Hill, N.S., Darwish, N., Gonçales, V.R., Coote, M.L., Swaminathan Iyer, K., Ciampi, S.: The corona of a surface bubble promotes electrochemical reactions. Nat. Commun. 11, 1 (2020)
54. Layman, B.R., Dick, J.E.: Through-Space electrochemiluminescence reveals bubble forces at remote phase boundaries. J. Am. Chem. Soc. **146**, 707 (2024)
55. Layman, B.R., Dick, J.E.: Phase-resolved electrochemiluminescence with a single luminophore. J. Phys. Chem. Lett. **14**, 8151 (2023)
56. Glasscott, M.W., Dick, J.E.: Visualizing phase boundaries with electrogenerated chemiluminescence. J. Phys. Chem. Lett. **11**, 4803 (2020)
57. Glasscott, M.W., Voci, S., Kauffmann, P.J., Chapoval, A.I., Dick, J.E.: Mapping solvent entrapment in multiphase systems by electrogenerated chemiluminescence. Langmuir **37**, 2907 (2021)
58. Liu, C., Bard, A.J.: Electrostatic electrochemistry at insulators. Nat. Mater. **7**, 505 (2008)
59. Zhang, J., Coote, M.L., Ciampi, S.: Electrostatics and electrochemistry: mechanism and scope of charge-transfer reactions on the surface of tribocharged insulators. J. Am. Chem. Soc. **143**, 3019 (2021)
60. Fu, Y., Xie, B., Liu, M., Hou, S., Zhu, Q., Kuhn, A., Zhang, L., Yang, W., Sojic, N.: Bipolar electrochemiluminescence at the water/organic interface. Chem. Sci. **15**, 19907 (2024)
61. Nami-Ana, S.F., Mehrgardi, M.A., Mofidfar, M., Zare, R.N.: Sustained regeneration of hydrogen peroxide at the water-gas interface of electrogenerated microbubbles on an electrode surface. J. Am. Chem. Soc. **146**, 31945 (2024)
62. Mariani, C., Fracassa, A., Pastore, P., Bogialli, S., Paolucci, F., Valenti, G., Zanut, A.: Singling out the electrochemiluminescence profile in microelectrode arrays. Chem. Biomed. Imaging (2025). https://doi.org/10.1021/cbmi.5c00022
63. Chovin, A., Garrigue, P., Vinatier, P., Sojic, N.: Development of an ordered array of optoelectrochemical individually readable sensors with submicrometer dimensions: application to remote electrochemiluminescence imaging. Anal. Chem. **76**, 357 (2004)
64. Chovin, A., Garrigue, P., Sojic, N.: Electrochemiluminescent detection of hydrogen peroxide with an imaging sensor array. Electrochim. Acta **49**, 3751 (2004)
65. Sentic, M., Virgilio, F., Zanut, A., Manojlovic, D., Arbault, S., Tormen, M., Sojic, N., Ugo, P.: Microscopic imaging and tuning of electrogenerated chemiluminescence with boron-doped diamond nanoelectrode arrays. Anal. Bioanal. Chem. **408**, 7085 (2016)
66. Fiorani, A., Han, D., Jiang, D., Fang, D., Paolucci, F., Sojic, N., Valenti, G.: Spatially resolved electrochemiluminescence through a chemical lens. Chem. Sci. **11**, 10496 (2020)
67. Feng, Y., Wang, C., Zhou, W., Yang, X., Paolucci, F., Valenti, G., Qi, H.: Tomography electrogenerated chemiluminescence imaging from magnetic microbeads. Small 2500804 (2025)
68. Han, D., Jiang, D., Valenti, G., Paolucci, F., Kanoufi, F., Chaumet, P.C., Fang, D., Sojic, N.: Optics determines the electrochemiluminescence signal of bead-based immunoassays. ACS Sens. **8**, 4782 (2023)
69. Han, D., Goudeau, B., Lapeyre, V., Ravaine, V., Jiang, D., Fang, D., Sojic, N.: Enhanced electrochemiluminescence at microgel-functionalized beads. Biosens. Bioelectron. **216**, 114640 (2022)
70. Zhao, W., Chen, H.-Y., Xu, J.-J.: Electrogenerated chemiluminescence detection of single entities. Chem. Sci. **12**, 5720 (2021)
71. Dong, J., Lu, Y., Xu, Y., Chen, F., Yang, J., Chen, Y., Feng, J.: Direct imaging of single-molecule electrochemical reactions in solution. Nature **596**, 244 (2021)

72. Dong, J., Feng, J.: Electrochemiluminescence from single molecule to imaging. Anal. Chem. **95**, 374 (2023)
73. Voci, S., Goudeau, B., Valenti, G., Lesch, A., Jović, M., Rapino, S., Paolucci, F., Arbault, S., Sojic, N.: Surface-Confined electrochemiluminescence microscopy of cell membranes. J. Am. Chem. Soc. **140**, 14753 (2018)
74. Ding, H., Guo, W., Su, B.: Imaging cell-matrix adhesions and collective migration of living cells by electrochemiluminescence microscopy. Angew. Chem. Int. Ed. **59**, 449 (2020)
75. Valenti, G., Scarabino, S., Goudeau, B., Lesch, A., Jović, M., Villani, E., Sentic, M., Rapino, S., Arbault, S., Paolucci, F., Sojic, N.: Single cell electrochemiluminescence imaging: from the proof-of-concept to disposable device-based analysis. J. Am. Chem. Soc. **139**, 16830 (2017)
76. Li, B., Huang, X., Lu, Y., Fan, Z., Li, B., Jiang, D., Sojic, N., Liu, B.: High electrochemiluminescence from Ru(bpy)32^{+} embedded metal-organic frameworks to visualize single molecule movement at the cellular membrane. Adv. Sci. **9**, 2204715 (2022)
77. Chen, Y., Gou, X., Ma, C., Jiang, D., Zhu, J.-J.: A synergistic coreactant for single-cell electrochemiluminescence imaging: guanine-rich ssDNA-Loaded high-index faceted gold nanoflowers. Anal. Chem. **93**, 7682 (2021)
78. Ma, C., Wang, M.-X., Wei, H.-F., Wu, S., Zhang, J.-R., Zhu, J.-J., Chen, Z.: Catalytic route electrochemiluminescence microscopy of cell membranes with nitrogen-doped carbon dots as nano-coreactants. Chem. Commun. **57**, 2168 (2021)
79. Wang, N., Gao, H., Li, Y., Li, G., Chen, W., Jin, Z., Lei, J., Wei, Q., Ju, H.: Dual intramolecular electron transfer for in situ coreactant-embedded electrochemiluminescence microimaging of membrane protein. Angew. Chem. Int. Ed. **60**, 197 (2021)
80. Ma, C., Wu, S., Zhou, Y., Wei, H., Zhang, J., Chen, Z., Zhu, J., Lin, Y., Zhu, W.: Bio-Coreactant-Enhanced electrochemiluminescence microscopy of intracellular structure and transport. Angew. Chem. Int. Ed. **60**, 4907 (2021)
81. Zhou, J., Ma, G., Chen, Y., Fang, D., Jiang, D., Chen, H.: Electrochemiluminescence imaging for parallel single-cell analysis of active membrane cholesterol. Anal. Chem. **87**, 8138 (2015)
82. Chen, M.-M., Xu, C.-H., Zhao, W., Chen, H.-Y., Xu, J.-J.: Super-Resolution electrogenerated chemiluminescence microscopy for single-nanocatalyst imaging. J. Am. Chem. Soc. **143**, 18511 (2021)
83. Xu, L., Zhou, Z., Zhang, C., He, Y., Su, B.: Electrochemiluminescence imaging of latent fingermarks through the immunodetection of secretions in human perspiration. Chem. Commun. **50**, 9097 (2014)
84. Xu, L., Li, Y., Wu, S., Liu, X., Su, B.: Imaging latent fingerprints by electrochemiluminescence. Angew. Chem. Int. Ed. **51**, 8068 (2012)
85. Tan, J., Xu, L., Li, T., Su, B., Wu, J.: Image-Contrast technology based on the electrochemiluminescence of porous silicon and its application in fingerprint visualization. Angew. Chem. Int. Ed. **53**, 9822 (2014)
86. Ding, H., Zhou, P., Fu, W., Ding, L., Guo, W., Su, B.: Spatially selective imaging of cell-matrix and cell-cell junctions by electrochemiluminescence. Angew. Chem. Int. Ed. **60**, 11769 (2021)
87. Ma, Y., Colin, C., Descamps, J., Arbault, S., Sojic, N.: Shadow electrochemiluminescence microscopy of single mitochondria. Angew. Chem. Int. Ed. **60**, 18742 (2021)
88. Zhou, Y., Dong, J., Zhao, P., Zhang, J., Zheng, M., Feng, J.: Imaging of single bacteria with electrochemiluminescence microscopy. J. Am. Chem. Soc. **145**, 8947 (2023)
89. Tønnesen, J., Inavalli, V.V.G.K., Nägerl, U.V.: Super-Resolution imaging of the extracellular space in living brain tissue. Cell **172**, 1108 (2018)
90. Sigal, Y.M., Zhou, R., Zhuang, X.: Visualizing and discovering cellular structures with super-resolution microscopy. Science **361**, 880 (2018)
91. Urban, N.T., Willig, K.I., Hell, S.W., Nägerl, U.V.: STED nanoscopy of actin dynamics in synapses deep inside living brain slices. Biophys. J. **101**, 1277 (2011)
92. Nägerl, U.V., Willig, K.I., Hein, B., Hell, S.W., Bonhoeffer, T.: Live-cell imaging of dendritic spines by STED microscopy. Proc. Natl. Acad. Sci. **105**, 18982 (2008)

93. Deng, Y., He, X., Jin, R., Jiang, D., Fang, D.: Local electrochemiluminescence imaging using scanning electrochemical cell microscopy. Electrochem. Commun. **164**, 107737 (2024)
94. Gao, H., Han, W., Qi, H., Gao, Q., Zhang, C.: Electrochemiluminescence imaging for the morphological and quantitative analysis of living cells under external stimulation. Anal. Chem. **92**, 8278 (2020)
95. Han, D., Yang, M., Feng, Z., Wu, Y., Sojic, N., Jiang, D.: Thickness-Resolved electrochemiluminescence microscopy of extracellular matrix at tumor tissues for rapid cancer diagnosis. ACS Appl. Mater. Interfaces **16**, 32078 (2024)
96. Dong, J., Xu, Y., Zhang, Z., Feng, J.: Operando imaging of chemical activity on gold plates with single-molecule electrochemiluminescence microscopy. Angew. Chem. Int. Ed. **61**, e202200187 (2022)
97. Han, D., Goudeau, B., Jiang, D., Fang, D., Sojic, N.: Electrochemiluminescence microscopy of cells: essential role of surface regeneration. Anal. Chem. **93**, 1652 (2021)
98. Gou, X., Zhang, Y., Xing, Z., Ma, C., Mao, C., Zhu, J.-J.: Site-selective heat boosting electrochemiluminescence for single cell imaging. Chem. Sci. **14**, 9074 (2023)
99. Gou, X., Xing, Z., Zhang, Z., Jin, R., Xu, Q., Sojic, N., Zhu, J.-J., Ma, C.: Designable electrochemiluminescence patterning for renewable and enhanced bioimaging. Angew. Chem. Int. Ed. **64**, e202410825 (2025)
100. Laser, D., Bard, A.J.: Semiconductor electrodes. Photo-induced electrogenerated chemiluminescence and up-conversion at semiconductor electrodes. Chem. Phys. Lett. 34, 605 (1975)
101. Luttmer, J.D., Bard, A.J.: Electrogenerated chemiluminescence: 34. Photo-Induced electrogenerated chemiluminescence and up-conversion at semiconductor electrodes. J. Electrochem. Soc. 126, 414 (1979)
102. Descamps, J., Zhao, Y., Goudeau, B., Manojlovic, D., Loget, G., Sojic, N.: Infrared photoinduced electrochemiluminescence microscopy of single cells. Chem. Sci. **15**, 2055 (2024)
103. Han, D., Goudeau, B., Manojlovic, D., Jiang, D., Fang, D., Sojic, N.: Electrochemiluminescence loss in photobleaching. Angew. Chem. Int. Ed. **60**, 7686 (2021)
104. Zhao, Y., Yu, J., Xu, G., Sojic, N., Loget, G.: Photoinduced electrochemiluminescence at silicon electrodes in water. J. Am. Chem. Soc. **141**, 13013 (2019)
105. Yu, J., Saada, H., Abdallah, R., Loget, G., Sojic, N.: Luminescence amplification at $BiVO_4$ photoanodes by photoinduced electrochemiluminescence. Angew. Chem. Int. Ed. **59**, 15157 (2020)
106. Zhao, Y., Bouffier, L., Xu, G., Loget, G., Sojic, N.: Electrochemiluminescence with semiconductor (nano)materials. Chem. Sci. **13**, 2528 (2022)
107. Zhao, Y., Descamps, J., Léger, Y., Sojic, N., Loget, G.: Light conversion by electrochemiluminescence at semiconductor surfaces. Acc. Chem. Res. **57**, 2144 (2024)
108. Han, S., Lee, H.J., Kim, T., Lim, S.Y., Kim, J.: Flexible and dynamic light-guided electrochemiluminescence for spatiotemporal imaging of photoelectrochemical processes on hematite. Anal. Chem. **96**, 11146 (2024)
109. Xing, Z., Gou, X., Yang, R., Wang, Y., Tang, W., Jiang, L.-P., Zhu, W., Chen, Z., Xu, Q., Zhu, J.-J., Ma, C.: Pre-Label-Free three-dimensional imaging of cellular bottom topography with a charge-lock electrochemiluminescence microscope. Nano Lett. **25**, 7507 (2025)
110. Ding, L., Zhou, P., Yan, Y., Su, B.: Electrochemiluminescence imaging of cellular contact guidance on microfabricated substrates. Chem. Biomed. Imaging **1**, 558 (2023)
111. Ding, L., Ding, H., Zhou, P., Xi, L., Su, B.: Surface-Sensitive imaging analysis of cell-microenvironment interactions by electrochemiluminescence microscopy. Anal. Chem. **94**, 10885 (2022)
112. Yan, Y., Zhou, P., Ding, L., Hu, W., Chen, W., Su, B.: T cell antigen recognition and discrimination by electrochemiluminescence imaging. Angew. Chem. Int. Ed. **62**, e202314588 (2023)
113. Knežević, S., Kerr, E., Goudeau, B., Valenti, G., Paolucci, F., Francis, P.S., Kanoufi, F., Sojic, N.: Bimodal electrochemiluminescence microscopy of single cells. Anal. Chem. **95**, 7372 (2023)

114. Xu, J., Huang, P., Qin, Y., Jiang, D., Chen, H.: Analysis of intracellular glucose at single cells using electrochemiluminescence imaging. Anal. Chem. **88**, 4609 (2016)
115. Zhang, J., Jin, R., Chen, Y., Fang, D., Jiang, D.: Enhanced electrochemiluminescence at single lithium iron phosphate nanoparticles for the local sensing of hydrogen peroxide efflux from single living cell under a low voltage. Sens. Actuators B Chem. **329**, 129208 (2021)
116. Zhang, J., Ding, H., Zhao, S., Jiang, D., Chen, H.-Y.: Confined electrochemiluminescence in vertically ordered silica mesochannels for the imaging of hydrogen peroxide released from single cells. Electrochem. Commun. **98**, 38 (2019)
117. Liu, G., Ma, C., Jin, B.-K., Chen, Z., Zhu, J.-J.: Direct electrochemiluminescence imaging of a single cell on a chitosan film modified electrode. Anal. Chem. **90**, 4801 (2018)
118. Ding, H., Guo, W., Ding, L., Su, B.: Confined electrochemiluminescence at microtube electrode ensembles for local sensing of single cells. Chin. J. Chem. **39**, 2911 (2021)
119. Qin, X., Jin, H.-J., Li, X., Li, J., Pan, J.-B., Wang, K., Liu, S., Xu, J.-J., Xia, X.-H.: Label-Free electrochemiluminescence imaging of single-cell adhesions by using bipolar nanoelectrode array. Eur. J. Chem. **28**, e202103964 (2022)
120. Li, X., Qin, X., Wang, Z., Wu, Y., Wang, K., Xia, X., Liu, S.: In situ imaging of endogenous hydrogen peroxide efflux from living cells via bipolar gold nanoelectrode array and electrochemiluminescence technology. ACS Sens. **7**, 2446 (2022)
121. He, R., Tang, H., Jiang, D., Chen, H.: Electrochemical visualization of intracellular hydrogen peroxide at single cells. Anal. Chem. **88**, 2006 (2016)
122. Zhang, J., Jin, R., Jiang, D., Chen, H.-Y.: Electrochemiluminescence-Based capacitance microscopy for label-free imaging of antigens on the cellular plasma membrane. J. Am. Chem. Soc. **141**, 10294 (2019)
123. Ni, J., Huang, S., Yang, W., Chen, Q., Lin, Z.: Electrochemiluminescence detecting and imaging of yeast metabolism indicated by endogenetic co-reactant. Anal. Chem. **97**, 921 (2025)
124. Cao, J.-T., Wang, Y.-L., Zhang, J.-J., Dong, Y.-X., Liu, F.-R., Ren, S.-W., Liu, Y.-M.: Immuno-Electrochemiluminescent imaging of a single cell based on functional nanoprobes of heterogeneous Ru(bpy)3^{2+}@SiO_2/Au Nanoparticles. Anal. Chem. **90**, 10334 (2018)
125. Yu, S., Hu, X., Pan, J., Lei, J., Ju, H.: Nanoconfined cathodic electrochemiluminescence for self-sensitized bioimaging of membrane protein. Anal. Chem. **95**, 16593 (2023)
126. Wang, N., Ao, H., Xiao, W., Chen, W., Li, G., Wu, J., Ju, H.: Confined electrochemiluminescence imaging microarray for high-throughput biosensing of single cell-released dopamine. Biosens. Bioelectron. **201**, 113959 (2022)
127. Zhang, H., Gao, W., Liu, Y., Sun, Y., Jiang, Y., Zhang, S.: Electrochemiluminescence-Microscopy for microRNA imaging in single cancer cell combined with chemotherapy-photothermal therapy. Anal. Chem. **91**, 12581 (2019)
128. Chen, M.-M., Xu, C.-H., Zhao, W., Chen, H.-Y., Xu, J.-J.: Single cell imaging of electrochemiluminescence-driven photodynamic therapy. Angew. Chem. Int. Ed. **61**, e202117401 (2022)
129. Gao, W., Liu, Y., Zhang, H., Wang, Z.: Electrochemiluminescence biosensor for nucleolin imaging in a single tumor cell combined with synergetic therapy of tumor. ACS Sens. **5**, 1216 (2020)
130. Descamps, J., Colin, C., Tessier, G., Arbault, S., Sojic, N.: Ultrasensitive imaging of cells and sub-cellular entities by electrochemiluminescence. Angew. Chem. Int. Ed. **62**, e202218574 (2023)
131. Hiramoto, K., Ino, K., Komatsu, K., Nashimoto, Y., Shiku, H.: Electrochemiluminescence imaging of respiratory activity of cellular spheroids using sequential potential steps. Biosens. Bioelectron. **181**, 113123 (2021)
132. Hiramoto, K., Villani, E., Iwama, T., Komatsu, K., Inagi, S., Inoue, K.Y., Nashimoto, Y., Ino, K., Shiku, H.: Recent advances in electrochemiluminescence-based systems for mammalian cell analysis. Micromachines (Basel) 11, 530 (2020)
133. Hiramoto, K., Komatsu, K., Shikuwa, R., Konno, A., Sato, Y., Hirano-Iwata, A., Ino, K., Shiku, H.: Evaluation of respiratory and secretory activities of multicellular spheroids *via* electrochemiluminescence imaging. Electrochim. Acta **458**, 142507 (2023)

134. Ino, K., Mashiko, M., Kanno, Y., Tang, Y., Masui, S., Nisisako, T., Hiramoto, K., Abe, H., Shiku, H.: Extended spherical diffusion theory: electrochemiluminescence imaging analysis of diffusive molecules from spherical biosamples. Anal. Chem. **96**, 18967 (2024)
135. Hiramoto, K., Konno, A., Nashimoto, Y., Hirano-Iwata, A., Ino, K., Shiku, H.: Osteogenic differentiation of mesenchymal stem cell spheroids: a microfluidic device and electrochemiluminescence imaging study. Electrochim. Acta **491**, 144291 (2024)
136. Gupta, V., Falciani, F., Layman, B.R., Hill, M.L., Rapino, S., Dick, J.E.: Real-Time visualization of endogenous H_2O_2 production in mammalian spheroids by electrochemiluminescence. Chem. Biomed. Imaging **3**, 310 (2025)
137. Shi, J., Han, D., Feng, Z., Jiang, D., Jiang, D.: Electrochemiluminescence imaging of a membrane carcinoembryonic antigen at single tissue sections. Analyst **148**, 2511 (2023)
138. Fu, C., Han, D., Jiang, D., Jiang, D.: Antigen retrieval assisted electrochemiluminescence imaging of surface antigens at single tissue section. Electrochim. Acta **474**, 143550 (2024)
139. Han, D., Sojic, N., Jiang, D.: Spatial profiling of multiple enzymatic activities at single tissue sections via fenton-promoted electrochemiluminescence. J. Am. Chem. Soc. **147**, 9610 (2025)
140. Wang, H., Zhang, S., Xia, S., Zhou, J., Liu, Y.: In Situ "Confocal" electrochemiluminescence 3D imaging: from cell to tissue section. Angew. Chem. Int. Ed. e202503594 (2025)
141. Liu, S., Yuan, H., Bai, H., Zhang, P., Lv, F., Liu, L., Dai, Z., Bao, J., Wang, S.: Electrochemiluminescence for electric-driven antibacterial therapeutics. J. Am. Chem. Soc. **140**, 2284 (2018)
142. Luy, J., Ameline, D., Thobie-Gautier, C., Boujtita, M., Lebègue, E.: Detection of bacterial rhamnolipid toxin by redox liposome single impact electrochemistry. Angew. Chem. Int. Ed. **61**, e202111416 (2022)
143. Ben Trad, F., Wieczny, V., Delacotte, J., Morel, M., Guille-Collignon, M., Arbault, S., Lemaître, F., Sojic, N., Labbé, E., Buriez, O.: Dynamic electrochemiluminescence imaging of single giant liposome opening at polarized electrodes. Anal. Chem. 94, 1686 (2022)
144. Ben Trad, F., Delacotte, J., Guille-Collignon, M., Lemaître, F., Arbault, S., Sojic, N., Burlina, F., Labbé, E., Buriez, O.: Electrochemiluminescence imaging of liposome permeabilization by an antimicrobial peptide: melittin. Chem. Biomed. Imaging 1, 58 (2023)
145. Hiramoto, K., Hirano-Iwata, A., Ino, K., Shiku, H.: Electrochemiluminescence of $[Ru(bpy)3]^{2+}$/tri-n-propylamine to visualize different lipid compositions in supported lipid membranes. Chem. Commun. **61**, 4495 (2025)
146. Liang, Z., Wang, P., Li, Z., Wang, D., Ma, Q.: Magnetoplasmonic metasurface-modulated electrochemiluminescence strategy for extracellular vesicle detection. Anal. Chem. **96**, 4909 (2024)
147. Liang, Z., Wang, P., Li, Z., Li, W., Ma, Q.: Au Nanorings/TiO_2 NPs@MXene-based metasurfaces with a magnetic mirror-modulated ECL strategy for extracellular vesicle detection. Anal. Chem. **96**, 16443 (2024)
148. Ben Trad, F., Carré, B., Delacotte, J., Lemaître, F., Guille-Collignon, M., Arbault, S., Sojic, N., Labbé, E., Buriez, O.: Electrochemiluminescent imaging of a NADH-based enzymatic reaction confined within giant liposomes. Anal. Bioanal. Chem. 416, 7385 (2024)
149. Ben Trad, F., Delacotte, J., Lemaître, F., Guille-Collignon, M., Arbault, S., Sojic, N., Labbé, E., Buriez, O.: Shadow electrochemiluminescence imaging of giant liposomes opening at polarized electrodes. Analyst 149, 3317 (2024)
150. Zhu, W., Dong, J., Ruan, G., Zhou, Y., Feng, J.: Quantitative single-molecule electrochemiluminescence bioassay. Angew. Chem. Int. Ed. **62**, e202214419 (2023)
151. Liu, Y., Zhang, H., Li, B., Liu, J., Jiang, D., Liu, B., Sojic, N.: Single biomolecule imaging by electrochemiluminescence. J. Am. Chem. Soc. **143**, 17910 (2021)
152. Huang, X., Li, B., Lu, Y., Liu, Y., Wang, S., Sojic, N., Jiang, D., Liu, B.: Direct visualization of nanoconfinement effect on nanoreactor via electrochemiluminescence microscopy. Angew. Chem. Int. Ed. **62**, e202215078 (2023)
153. Lu, Y., Huang, X., Wang, S., Li, B., Liu, B.: Nanoconfinement-Enhanced electrochemiluminescence for in situ imaging of single biomolecules. ACS Nano **17**, 3809 (2023)

154. Lu, Y., Ning, Y., Li, B., Liu, B.: Dual-Signal imaging mode based on fluorescence and electrochemiluminescence for ultrasensitive visualization of SARS-CoV-2 spike protein. Anal. Chem. **96**, 463 (2024)
155. Wang, X.-Y., Wu, S.-T., Lin, Y.-Z., Ding, S.-N., Xu, J.-J., Chen, H.-Y.: Confinement effect enhanced bipolar electrochemistry: structural color coding coupled with wireless electrochemiluminescence imaging technology. Anal. Chem. **96**, 14372 (2024)
156. Li, H., Garrigue, P., Bouffier, L., Arbault, S., Kuhn, A., Sojic, N.: Double remote electrochemical addressing and optical readout of electrochemiluminescence at the tip of an optical fiber. Analyst **141**, 4299 (2016)
157. Li, H., Zhang, T., Zhou, H., Zhang, Z., Liu, M., Wang, C.: Enhanced electrochemiluminescence in a microwell bipolar electrode array prepared with an optical fiber bundle. ChemElectroChem **8**, 1473 (2021)
158. Guo, W., Ding, H., Zhou, P., Wang, Y., Su, B.: Electrochemiluminescence waveguide in single crystalline molecular wires. Angew. Chem. Int. Ed. **59**, 6745 (2020)
159. Xu, Y., Huang, X., Wang, Y., Qu, W., Guo, W., Su, B., Dai, Z.: Controllable and low-loss electrochemiluminescence waveguide supported by a micropipette electrode. J. Am. Chem. Soc. **146**, 5423 (2024)
160. Wang, Y., Jiang, D., Chen, H.-Y.: Wireless electrochemical visualization of intracellular antigens in single cells. CCS Chemistry **4**, 2221 (2021)
161. Chen, T., Dong, B., Chen, K., Zhao, F., Cheng, X., Ma, C., Lee, S., Zhang, P., Kang, S.H., Ha, J.W., Xu, W., Fang, N.: Optical super-resolution imaging of surface reactions. Chem. Rev. **117**, 7510 (2017)
162. Collinson, M.M., Wightman, R.M.: Observation of individual chemical reactions in solution. Science **268**, 1883 (1995)
163. Bentley, C.L., Kang, M., Unwin, P.R.: Nanoscale surface structure-activity in electrochemistry and electrocatalysis. J. Am. Chem. Soc. **141**, 2179 (2019)
164. Pfisterer, J.H.K., Liang, Y., Schneider, O., Bandarenka, A.S.: Direct instrumental identification of catalytically active surface sites. Nature **549**, 74 (2017)
165. Mefford, J.T., Akbashev, A.R., Kang, M., Bentley, C.L., Gent, W.E., Deng, H.D., Alsem, D.H., Yu, Y.-S., Salmon, N.J., Shapiro, D.A., Unwin, P.R., Chueh, W.C.: Correlative operando microscopy of oxygen evolution electrocatalysts. Nature **593**, 67 (2021)
166. Zhu, M.-J., Pan, J.-B., Wu, Z.-Q., Gao, X.-Y., Zhao, W., Xia, X.-H., Xu, J.-J., Chen, H.-Y.: Electrogenerated chemiluminescence imaging of electrocatalysis at a single Au-Pt Janus nanoparticle. Angew. Chem. Int. Ed. **57**, 4010 (2018)
167. Kerr, E., Hayne, D.J., Soulsby, L.C., Bawden, J.C., Blom, S.J., Doeven, E.H., Henderson, L.C., Hogan, C.F., Francis, P.S.: A redox-mediator pathway for enhanced multi-colour electrochemiluminescence in aqueous solution. Chem. Sci. **13**, 469 (2022)
168. Fracassa, A., Santo, C.I., Kerr, E., Knežević, S., Hayne, D.J., Francis, P.S., Kanoufi, F., Sojic, N., Paolucci, F., Valenti, G.: Redox-mediated electrochemiluminescence enhancement for bead-based immunoassay. Chem. Sci. **15**, 1150 (2024)
169. Fracassa, A., Mariani, C., Fiorani, A., Einaga, Y., Hogan, C.F., Paolucci, F., Sojic, N., Francis, P.S., Valenti, G.: Overcoming kinetic barriers of remote electrochemiluminescence on boron-doped diamond via catalytic coreactant oxidation. Chem. Commun. **61**, 3900 (2025)
170. Zhao, Y., Descamps, J., Ababou-Girard, S., Bergamini, J.-F., Santinacci, L., Léger, Y., Sojic, N., Loget, G.: Metal-Insulator-Semiconductor anodes for ultrastable and site-selective upconversion photoinduced electrochemiluminescence. Angew. Chem. Int. Ed. **61**, e202201865 (2022)
171. Vogel, Y.B., Darwish, N., Ciampi, S.: Spatiotemporal control of electrochemiluminescence guided by a visible light stimulus. Cell Rep. Phys. Sci. **1**, 100107 (2020)
172. Xue, J.-W., Xu, C.-H., Zhao, W., Chen, H.-Y., Xu, J.-J.: Photoinduced electrogenerated chemiluminescence imaging of plasmonic photoelectrochemistry at single nanocatalysts. Nano Lett. **23**, 4572 (2023)
173. Zhao, Y., Descamps, J., al Hoda Al Bast, N., Duque, M., Esteve, J., Sepulveda, B., Loget, G., Sojic, N.: All-Optical electrochemiluminescence. J. Am. Chem. Soc. 145, 17420 (2023)

174. Maar, R.R., Zhang, R., Stephens, D.G., Ding, Z., Gilroy, J.B.: Near-Infrared photoluminescence and electrochemiluminescence from a remarkably simple boron difluoride formazanate dye. Angew. Chem. Int. Ed. **58**, 1052 (2019)
175. Valenti, G., Rampazzo, E., Kesarkar, S., Genovese, D., Fiorani, A., Zanut, A., Palomba, F., Marcaccio, M., Paolucci, F., Prodi, L.: Electrogenerated chemiluminescence from metal complexes-based nanoparticles for highly sensitive sensors applications. Coord. Chem. Rev. **367**, 65 (2018)
176. Hesari, M., Ding, Z.: Identifying highly photoelectrochemical active sites of two Au21 nanocluster isomers toward bright near-infrared electrochemiluminescence. J. Am. Chem. Soc. **143**, 19474 (2021)
177. Zhao, P., Zhu, W., Zheng, M., Feng, J.: Deep learning enhanced electrochemiluminescence microscopy. Anal. Chem. **95**, 4803 (2023)
178. Gao, W., Muzyka, K., Ma, X., Lou, B., Xu, G.: A single-electrode electrochemical system for multiplex electrochemiluminescence analysis based on a resistance induced potential difference. Chem. Sci. **9**, 3911 (2018)
179. Ma, X., Qi, L., Gao, W., Yuan, F., Xia, Y., Lou, B., Xu, G.: A portable wireless single-electrode system for electrochemiluminescent analysis. Electrochim. Acta **308**, 20 (2019)
180. Du, F., Dong, Z., Guan, Y., Zeid, A.M., Ma, D., Feng, J., Yang, D., Xu, G.: Single-electrode electrochemical system for the visual and high-throughput electrochemiluminescence immunoassay. Anal. Chem. **94**, 2189 (2022)
181. Dong, Z., Chen, Y., Xia, S., Alboull, A.M.A., Hussain, A., Tian, Y., Xu, G.: Single-electrode flow cell for electrochemiluminescent flow analysis. Anal. Chem. **96**, 18927 (2024)

Electrochemiluminescence Nano-Emitters and Electrochemiluminescence Microscopy-Based Single Nanoparticle Analysis

Zejing Xing, Zhichen Zhang, Jingjuan Xu, and Jun-Jie Zhu

Abstract Nanomaterials have gained burgeoning attention in various fields including biomedicine and energy during last decades. The traditional strategies to research nanomaterials can only record the average characteristics of ensemble materials, which may lead to analysis deviation. The emergence of single-nanoparticle analysis techniques has tackled the problem, but their low analysis speed and sensitivity limit their further applications. Electrochemiluminescence (ECL) is an electrochemical reactions-dependent luminescence process without external light sources. Coupled with a microscopy configuration, a novel imaging methodology ECL microscopy (ECLM) has been developed. ECLM, due to its distinctive imaging mode, holds considerable potential in the field of single-nanoparticle research. This chapter will initially present an overview of some common ECL nanoemitters, including the basic principles of their ECL emission and their current applications. The next section will discuss diverse applications of ECLM in studying single nanoparticles. Finally, we will summarize the challenges and outline the prospects for further development of single-nanoparticle ECLM technique.

Keywords Nanomaterials · Electrochemiluminescence · Microscopy · Single-nanoparticle analysis

The original version of the chapter has been revised. A correction to this chapter can be found at https://doi.org/10.1007/978-3-032-14602-1_14

Z. Xing · Z. Zhang · J. Xu (✉) · J.-J. Zhu (✉)
School of Chemistry and Chemical Engineering, Nanjing University, Nanjing, China
e-mail: xujj@nju.edu.cn

J.-J. Zhu
e-mail: jjzhu@nju.edu.cn

F. Paolucci et al. (eds.), *Electrogenerated Chemiluminescence Microscopy and Imaging*, https://doi.org/10.1007/978-3-032-14602-1_9

1 Introduction

The continuous advancement of scientific and technological analysis has led to the discovery of novel materials, providing deeper insights into scientific development and promoting their application in various fields [1, 2]. Nanomaterials are simply defined as those with a size range of 1–100 nm [3]. Since their discovery, nanomaterials have attracted significant attention from various fields and have developed rapidly [4]. The significant progress and widespread interest in nanomaterials can be attributed to their unique physical and chemical properties, which differ from those of bulk materials [5]. With the structure of small particle size and large specific surface area, the research of nanomaterials had been extended to various areas, including preparation, characterization, physical and chemical properties [6–9]. In recent decades, nanoscale analysis of novel materials has successfully integrated with other fields, involving biomedicine, optics, engineering, and electronics [10–14]. In particular, advanced characterization and analysis tools for nanomaterials have significantly developed, providing strong potential for in-depth research. However, despite advancements in nanoscale analysis, it still remains several challenges including limited resolution of traditional methods at the level of nanoscale, difficulty in evaluating surface and interface effects, and complex interactions with environment. These challenges hinder a clear understanding of nanomaterials and necessitate the development of novel strategies for the controlled preparation of high-performance nanomaterials.

Previous studies have demonstrated the misconception of nanoparticles being considered a homogeneous system, whereas their properties are, in fact, heterogeneous [15]. On the one hand, the heterogeneity arises from the inherent properties of nanomaterials, which possess multiple thermodynamically stable states. Specifically, the uneven distribution of components results in variations in material properties, such as structure, size, and chemical state. On the other hand, variations in synthesis conditions and the resulting particle environment contribute to heterogeneity due to differing thermal fluctuations and mass transport. However, traditional analysis methods only recorded the average characteristics and activities of ensemble particles, omitting specific information of individual particles and causing the deviation of analysis [16]. To understand the exceptional properties and investigate the mechanisms behind the synthesis of high-performance nanomaterials for real-world applications, imaging nanoparticles and evaluating their surrounding environment at the single-particle level are essential. Single-particle analysis provides detailed information about individual particles, enabling further exploration of the conditions required to meet the performance standards for nanomaterials.

Several techniques have been developed for imaging single particles to study their physical and chemical properties at the microscale. Scanning electrochemical microscopy (SECM) is a powerful technique that provides specific information including quantification, spatiotemporal information, and chemical properties at a controllable distance from the substrates [17]. Scanning electrochemical cell microscopy (SECCM) demonstrates alternative high-throughput technique under the use of nanopipette as a scanning probe to reveal characterization of surface structures and visualize surface activity site for providing detailed electrochemical information [18]. Although scanning probe methods achieve nanometer-scale resolution, their analysis speed limits both the resolution of acquired dimensional data and further applications. To meet the need for high-throughput and rapid single-particle analysis, optical imaging techniques and instruments are applied. Bright-field microscopy (BFM) is a fundamental optical technique known for its low cost and ease of control [19]. Additionally, the external light source exerts no perturbation on the chemical environment surrounding the particles, and the compatibility of chemical reaction conditions of BFM minimizes interference to the nanoparticle system, making it suitable for basic single-particle analysis. Building upon BFM, improved methods have been developed for optimal imaging, tailored to the specific properties of individual particles. Dark-field microscopy (DFM) utilizes light scattering from individual nanoparticles with a high signal-to-noise ratio, enabling the extraction of detailed information about the particles, including their composition, morphology, microenvironment, and coupling [20]. This makes DFM a powerful tool for imaging single particles, particularly plasmonic particles. Similarly, surface plasmon resonance microscopy (SPRM) exploits the interaction between a metal phase and a dielectric material to visualize single particles [21]. Raman scattering microscopy (SRSM) and surface-enhanced Raman scattering (SERS) imaging can identify fingerprint information for detecting trace substances with exceptional sensitivity, resolution, and specificity [22]. Although scattering-based optical imaging methods are widely used in fields such as biosensing, the scattering intensity is affected by environmental noise, making it difficult to detect small nanoparticles. Fluorescence microscopy (FLM) is the dominant method for single-particle analysis, requiring an external light source to excite specific fluorescence and provide sufficient intensity for high-resolution imaging [23]. However, photobleaching and photoblinking must be considered to ensure high-resolution single-particle imaging.

Comparing with the above-mentioned single-particle imaging strategies, electrochemiluminescence microscopy (ECLM) offers a superior integration of optical imaging and electrochemical detection. Optical tracking images clearly provide direct information and identification of individual properties at the single-particle level during electrochemical detection. Luminescence is triggered by electrochemical reactions, offering the advantage of well-controlled temporal resolution. Moreover, the absence of an additional light source reduces optical background noise and minimizes other interference, such as phototoxicity. Additionally, ECLM demonstrates high-throughput and excellent spatial resolution, enabling the visualization of local electrochemical features in different particles [24]. Therefore, ECLM is rapidly becoming a powerful approach for analyzing and evaluating single-particle properties, providing deeper insights into nanomaterials research.

In this chapter, we primarily focus on recent advances in ECL imaging analysis at the single-particle level. Based on the properties of nanomaterials, we summarize the research progress in ECL nanoemitters and systematically introduce applications of ECLM in single-nanoparticle electrochemistry research. Finally, we discuss the challenges and prospects for single-particle ECLM technique.

2 Nanomaterials as ECL Emitter

2.1 ECL Nanoemitters

With the growing demand for the expansion of ECL applications, various nanomaterials with distinct luminescent properties have been reported as potential ECL emitters [25]. Based on their composition and structure, various metal clusters have been explored for ECL detection, exhibiting distinct properties such as ECL intensity, spectral characteristics, and electrochemical behavior [26–28]. In addition to the commonly used ECL nanoemitters, other nanomaterials, such as perovskites, carbon quantum dots, and aggregation-induced emission nanomaterials, show promise for future development in ECL analysis [29–31]. Although various ECL emitters have been explored for ECL applications, few luminophores exhibit sufficient quantum efficiency for further ECL imaging analysis.

While enhanced ECL has prompted the possibility for visualization strategies in ECL detection, further exploration is needed to uncover the individual ECL behavior of single nanoparticles. Following the exploration of nanomaterials with ECL emission, it is essential to investigate their mechanisms and properties at the single-particle level to guide the development of ECL nanoemitters. Based on the specific ECL behaviors of nanomaterials, various materials with sufficient ECL efficiency have been used to evaluate their physical and chemical characteristics using ECLM. Due to the limited ECL efficiency of nanoluminophores in aqueous solutions, early ECL imaging analysis of single particles focused more on organic systems. To record the ECL signal of single nanoparticles on the surface of a macroscopic electrode, Barbara's group developed an ECL imaging study using a conjugate polymer (poly(9,9-dioctylfluorene-co-benzothiadiazole, F8BT), approximately 25 nm in size) [32]. With single-molecule electrochemical spectroscopy, they first recorded ECL images of single nanoparticles and investigated the electrochemical heterogeneity of individual particles (Fig. 1a). Although organic polymer semiconductors provided new insights into ECL imaging of single particles, the higher ECL strength in organic systems hindered the application of this technique in aqueous solutions. To enable ECL imaging of single particles in aqueous solutions, other types of Pdots were developed, meeting the required intensity for achieving the necessary ECL imaging resolution. Ju et al. synthesized tertiary amine-embedded polymer dots (TEA-Pdots) as self-enhanced emitters, leveraging efficient electron transfer between short-lived coreactant radicals and luminophores to enhance ECL intensity, where coreactants were incorporated into the emitter backbone [33]. The probe exhibited higher efficiency through dual intramolecular electron transfer involving tertiary amines and the conjugated superstructure of coreactant-embedded Pdots, enabling the evaluation of specific protein expression in living cells. Xu's group developed three-in-one conjugated Pdots for self-enhanced ECL sensors [34]. The development of Pdots applications requires the implementation of appropriate modification strategies. The proposed Pdots incorporated various functions: The commercial conjugated polymer poly was used for its bifunctional properties, including ECL efficiency through the amidoxime-conjugated backbone as the ECL emitter and coreactant group for self-enhanced ECL, as well as a capture group for detecting UO_2^{2+}. UO_2^{2+} caused ECL quenching due to the resonance energy transfer effect on ECL emission, providing new insights for detecting radioactive elements in public water safety.

In addition to polymer nanoparticles, other novel nanomaterials with sufficient light emission efficiency for ECL imaging have been proposed. Due to the unique photoelectric properties of semiconductor quantum dots, Bard first proposed the surface-state hypothesis for the ECL mechanism using silicon nanocrystal quantum dots, pioneering the study of quantum dots (QDs) in ECL [36]. Subsequently, other researchers explored various quantum dot nanomaterials for ECL emission, including CdSe, CdTe, CdTeSe, and others [37–39].

As a non-metallic organic polymer material, graphitic phase carbon nitride (g-C_3N_4) has garnered increasing attention as a novel photoelectric material [40]. Since the initial exploration of g-C_3N_4's ECL emission, the ECL mechanism of g-C_3N_4 and

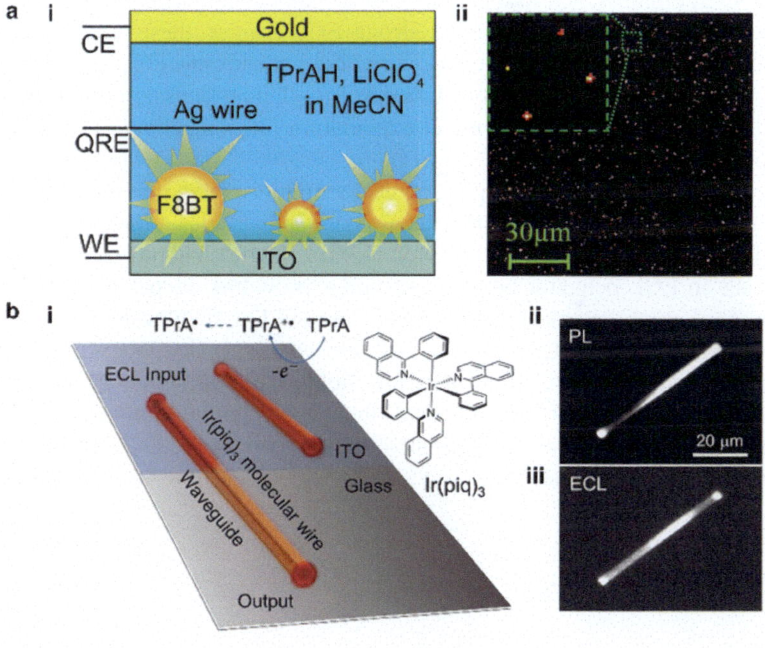

Fig. 1 **a** ECL emission of single-conjugated polymer nanoparticles. (i) Schematic diagram of single-molecule spectroelectrochemistry technique. (ii) Wide-field ECL image (inset: expanded region showing ECL from four particles). The intensity scale (0–200 counts) and integration time (0.5 s) of images c and d are the same. Reprinted with permission from Ref. [32]. Copyright 2008 American Chemical Society. **b** Using single-crystalline molecular wires as ECL emitter. (i) Schematic illustration of ECL waveguide in single molecular wires of $Ir(piq)_3$ on a patterned ITO electrode. (ii) PL and (iii) ECL images of single molecular wires of $Ir(piq)_3$ on the ITO electrode in 0.1 M PB solution (pH 7.4) containing 0.1 M TPrA. ECL images were captured at +1.5 V. The exposure time was 60 s. Reprinted with permission from Ref. [35]. Copyright 2020 Wiley–VCH

its further functional modifications in composites have been extensively discussed in various applications [41]. The ECL mechanism of g-C_3N_4 can be classified into two categories: cathodic and anodic ECL emission. In cathodic route, g-C_3N_4 shows ECL emission with $K_2S_2O_8$ as the coreactant [42]. The reduction of $K_2S_2O_8$ generates the strong oxidant $SO_4^{\bullet-}$ and injected into the valence band of g-C_3N_4. As the potential continues to decrease, an electron is injected into the conduction band for the generation of negative charged g-$C_3N_4^{\bullet-}$. The electron transfer between $SO_4^{\bullet-}$ and g-$C_3N_4^{\bullet-}$ induces the generation of the excited state g-C_3N_4*, which produced light emission subsequently. As for anodic ECL, the oxidized g-C_3N_4 can react with the strong reducing intermediates from electrochemical oxidizing products of coreactants such as TPrA [43]. Exploiting the sufficient photon emission efficiency in cathodic ECL of g-C_3N_4, Zhu's group synthesized hollow carbon nitride nanospheres to image, for the first time, the ECL blinking phenomenon during competitive electrocatalytic hydrogen evolution reactions [44]. Gaining further insight into the ECL mechanism of nanoemitters, Su's group employed single-crystalline molecular wires as ECL emitters, self-assembled from cyclometalated iridium(III) complexes ($Ir(piq)_3$) [35]. In addition to being a strong ECL emitter, these single-crystalline molecular wires also acted as active waveguides, confining ECL generation and propagation over distances of up to 100 mm for efficient emission at the terminals (Fig. 1b). This strategy offers a novel explanation for contactless electrochemical analysis.

The strategy of ECL imaging using nanoemitters opens new possibilities for studying the physical and chemical behaviors of nanoparticles. Imaging single particles as ECL emitters, without the need for additional labeled luminophores, revealed differences in ECL behavior between particles, providing deeper insights into the luminescence mechanisms of excited states and guiding the development of nanomaterials with higher ECL efficiency. Although the proposed label-free strategies demonstrate potential applications in various fields, the properties of ECL nanoemitters limit the exploration of nanomaterials that do not exhibit luminescence. Therefore, it is essential to develop alternative nanomaterials for ECL imaging using appropriate decoration strategies.

2.2 *Dye-Decorated Nanoparticles*

In order to expand the application of single-particle imaging in ECLM, alternative strategies were developed for the single-particle analysis which was disabled in ECL emission. Compared to exploring nanomaterials with inherent ECL efficiency, additional luminophores have been introduced as ECL labels to provide stable luminescence, improving the resolution of single-particle imaging and expanding the application of ECLM. Selecting appropriate luminescent molecules and decoration strategies is crucial for ECL imaging of nanomaterials at the single-particle level.

Many molecules exhibit both satisfactory electrochemical activity and light emission as ECL emitters, which can be classified into inorganic complexes and organic aromatic compounds. Inorganic ECL systems include various coordination compounds that utilize elements such as Ag, Au, Cu, Ir, and Ru as central coordination atoms [45, 46]. Among inorganic ECL systems, Tris-(2,2'-bipyridyl)dichlororuthenium(II) hexahydrate ($Ru(bpy)_3^{2+}$) is widely used due to its superior ECL efficiency, high stability, and unique water solubility [47]. Thanks to the unique structure of the derivative bipyridine ligands in $Ru(bpy)_3^{2+}$, it can be modified with antibodies, nucleic acids, and nanomaterials to satisfy the requirements of ECL detection [48–51]. In organic systems, luminol and its derivatives are also typical luminophores used to establish ECL analysis platforms [52].

Among various luminophores, $Ru(bpy)_3^{2+}$ is commonly used to modify nanomaterials. Decorated nanomaterials for ECL emission can be classified based on the type of substrate material. Polystyrene (PS) microspheres are suitable as substrate materials for surface modification and functionalization due to their excellent chemical stability and controllable surface properties [55, 56]. As a favorable nanocarrier for ECL decoration, Sojic's group prepared 3 μm PS beads modified with three different antibodies for identification through an immuno-sandwich reaction and fixed them onto the surface of a microelectrode array [57]. In this ECL immunoassay imaging system, the secondary antibody labeled with $Ru(bpy)_3^{2+}$ was attached based on the interaction between biotin and streptavidin. To distinguish between different PS beads with varying antibodies at the single-nanoparticle level, they introduced a fluorescence probe at different concentrations to locate the PS beads based on fluorescence images with varying luminescence intensity. The proposed strategy enabled the sensitive detection of different antibodies under an applied potential to induce ECL, and imaging dozens of different nanoparticles simultaneously to meet high-throughput analysis requirements. This immuno-modification strategy can also be applied to other electrodes, which offers simpler instruments and more convenient modification compared to microelectrode arrays. Roda's team improved the imaging technology through building a novel transparent electrochemical system that provides information on single PS beads with an 8 μm diameter [58]. They used a similar modification strategy to label ECL luminophores on the surface of PS beads via biotin–streptavidin interaction. Clear ECL imaging of a single particle was captured on the surface of a fluorine-doped tin oxide electrode, indicating the potential for future analysis in combination with other multiplexed imaging techniques. In addition to the immuno-modification strategy to immobilize ECL luminophores on PS beads, they further investigated a novel approach based on $Ru(bpy)_3^{2+}$-grafted microgels [59]. Due to the aggregation of ECL luminophores and the enhanced ECL reactions, the microgel-functionalized ECL nanoemitters demonstrated significantly higher ECL intensity compared to traditional bead assays. Additionally, top-view and side-view images suggested the extended ECL layer caused by microgels. The improved ECL imaging strategy for single PS beads provides a foundation for future single-entity imaging analysis [60].

Silicon dioxide nanomaterials are another suitable carrier for modification with specific target molecules or materials, demonstrating excellent biocompatibility, and large specific surface areas [61–63]. Owing to the electrostatic interaction between the surface of SiO_2 nanomaterials and ECL luminophores, they can serve as nanocarriers to load various ECL emitters, acting as amplification probes for the construction of highly sensitive ECL sensor platforms [64]. Given the numerous advantages of silica nanoparticles as carriers of luminophores with superior biocompatibility, functionalized ECL nanoprobes based on silica nanoparticles have been identified as essential components for sensitive detection [46, 65]. Among the various strategies for decorating silica nanoparticles with ECL luminophores, the microemulsion method is commonly used to synthesize ECL emitters with sufficient intensity for single-particle imaging [66]. However, the reported synthetic method still has limitations in application of doping nanoparticles with organic or inorganic dyes, primarily due to solubility issues and electrostatic interactions that prevent the incorporation of negatively charged species. Zanut et al. avoided these inhibitions and eliminated the interference of positively charged surfaces by using biotinylated $Ru(bpy)_3^{2+}$-doped silica nanoparticles synthesized via the reverse microemulsion method [53]. Comparing with the commercial ECL-based immunoassay system, the proposed ECL nanoparticles possessed superior ECL properties with enhanced intensity and increased stability (Fig. 2a).

The main factors influencing ECL intensity, which limit imaging resolution, include electrochemical efficiency and radiative recombination efficiency. In single-molecule ECL reactions, the emission of a single photon limits the ECL sensitivity during luminescence. The aforementioned strategy involves utilizing stable nanoparticles including PS and SiO_2 as substrates for the enrichment of ECL emitters, and they are able to improve ECL signal for better imaging resolution at the single-particle level. However, the number of ECL emitters that can be modified with these stable nanomaterials is inherently limited. Therefore, in addition to introducing stable nanomaterials as luminescence carriers, various strategies, such as using appropriate labels and enriching luminophores, have been employed to further amplify ECL intensity and meet the imaging resolution requirements.

In addition to nanomaterials with specific morphologies used as carriers for luminophores, other materials have also shown favorable properties for enhancing luminophore accumulation. Lu et al. developed an ECL nanoreactor to amplify ECL signals by confining the reactants within a narrow region, thereby increasing the possibility of collisions between them to enhance ECL efficiency [67]. This strategy involves a nanoscale-confined region that not only provides a more favorable microenvironment for enhancing ECL emission, but also facilitates the accumulation of intermediate free radicals and electron transfer through a stable porous structure. $Ru(bpy)_3^{2+}$ was enriched in the negatively charged nanopores of zeolite due to electrostatic interactions. The nanoconfinement strategy enhanced ECL and enabled clear imaging of target proteins at the single-molecule level by using zeolite as a nanoreactor for electroactive substances.

Currently, various types of porous nanomaterials are widely used as fundamental structures for loading luminescent molecules. In addition to enriching ECL luminescent molecules, another method for ECL imaging at the single-particle level has been proposed, which focuses on enhancing intramolecular charge transfer by appropriately regulating the nanoparticle structure. Metal–organic frameworks (MOFs) are

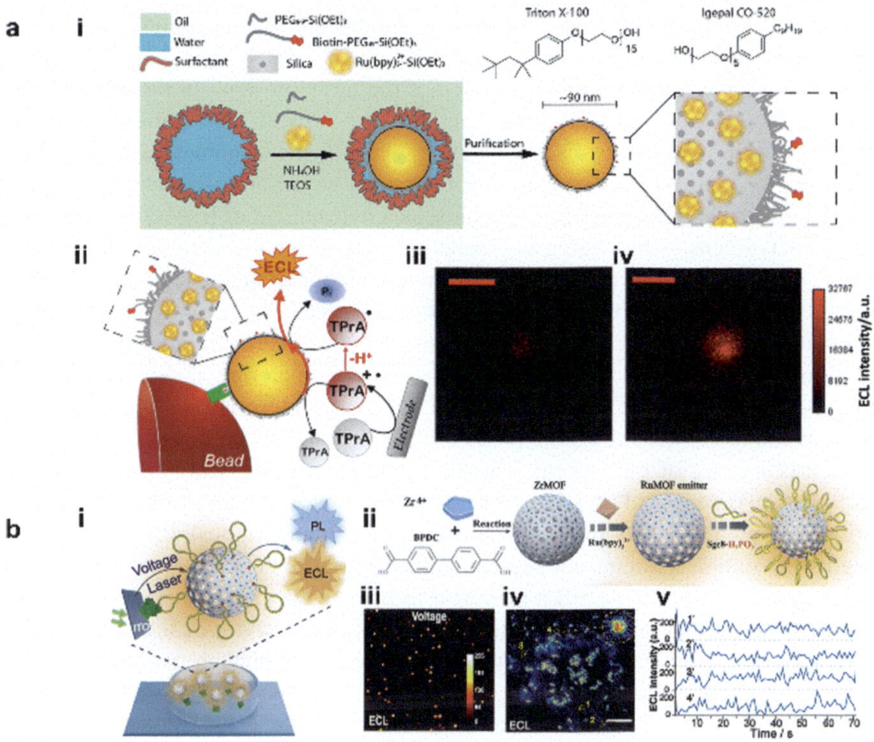

Fig. 2 **a** Dye-doped silica nanoparticles to amplify ECL signal. (i) Schematic representation of the reverse microemulsion synthesis of $Ru(bpy)_3{}^{2+}$-doped silica nanoparticles. (ii) The heterogeneous mechanism for the "oxidative-reduction" coreactant ECL generation obtained using 2.8 mm beads labeled with nanoparticles (yellow sphere) through a streptavidin (green tool)-biotin (red tool) bond. (iii) ECL imaging of single bead labeled with biotinylated $Ru(bpy)_3{}^{2+}$ complex and (iv) with bio-Triton@RuNp. They were obtained by applying a constant potential of 1.4 V (vs. Ag/AgCl) for 4 s in 180 mM TPrA and 0.2 M PB. EMCCD camera coupled with a potentiostat. Scale bar, 5 µm. Reprinted with permission from Ref. [53]. Copyright 2020 Wiley–VCH. **b** $Ru(bpy)_3{}^{2+}$ -embedded MOF for imaging of ECL events at single-molecule level. (i) Imaging mechanism for the ECL visualization. (ii) Synthesis scheme of the RuMOFs via a facile solvothermal reaction. (iii) Single-molecule ECL image of nanoemitters recorded with a voltage or laser. Exposure time is 1 s. (iv) ECL image of CCRF-CEM cells. Scale bars, 3.5 µm. (v) ECL trajectories of single-molecule ECL signal spots marked in ECL image of (v). Reprinted with permission from Ref. [54]. Copyright 2022 Wiley–VCH

crystalline porous materials with periodic network structures, constructed by assembling organic ligands with metal centers [68]. MOFs have the advantages of adjustable porosity and tunable chemistry, in addition to their inherent porous structures [69]. Based on the properties of luminophore enrichment and a confined microenvironment provided by MOFs, Lu et al. fully utilized porous MOF nanoparticles to load luminescent molecules as ECL nanoemitters [70]. To avoid interference from single-output analysis due to low concentrations and complex environments, they further developed a dual-output sensing platform by modifying the nanoemitters and electrode. This platform captured the disappearance of the fluorescence signal and the generation of ECL images in the presence of the target protein. In this dual-model detection, they eliminated false-positive and false-negative signals, demonstrating the potential of ECL imaging for disease diagnosis. Li et al. incorporated $Ru(bpy)_3{}^{2+}$ into a Zr-based MOF as an ECL nanoemitter to prevent aggregation-induced quenching [54]. Additionally, the well-confined nanoscale environment of RuMOF reduces the loss of ECL intensity by decreasing collision probabilities and inhibiting thermal motion, which minimizes energy loss. Due to the higher and more stable ECL intensity of RuMOF, they captured single-molecule ECL events at the imaging pixel level to monitor the behavior of individual proteins on the surface of living cells (Fig. 2b).

Effective modification of luminophores onto nanomaterials is widely used for single-particle imaging. The enhanced ECL intensity is attributed to the combined effects of nanoconfinement, luminophore enrichment, and the improved chemical and physical properties of the system. The amplified ECL signal from single particles highlights the potential of ECLM in revealing critical insights for material characterization, as well as cellular and immune analysis. Although optical diffraction limits imaging resolution and complicates the detection of subtle variations in particles, ECLM of single particle is still a powerful to investigate electrochemical behavior of nanomaterials. The ability to monitor and track the behavior of individual particles makes ECLM to become a valuable technique for advancing materials science, catalysis, and biomedical research.

3 Applications of ECLM in Single-Nanoparticle Analysis

3.1 Studying on Nano-Electrocatalysts at the Single-Particle Level

Originating from the discovery of nanomaterials, they have attracted extensive attention and developed rapidly. The significant progress and interest in nanomaterials can be attributed to their specific physical and chemical properties, which are different from their macroscopic counterparts. Nanomaterials possess tunable particle sizes and large specific surface area, which affords them unique properties that have endowed their use in many different applications, particularly in the field of electrochemistry. However, traditional electrochemical analytical tools struggle to clearly elucidate the electrochemical properties of nanomaterials because of the erroneous assumption that nanoparticles behave as a homogeneous system, thereby obscuring the heterogeneous nature of nanomaterial properties. To elucidate the unique properties and resolve the debated mechanisms for synthesizing high-performance nanomaterials in practical applications, it is crucial to image nanoparticles and evaluate their surrounding environment at the single-particle level. In recent years, ECLM has been widely utilized for evaluating various nano-electrocatalysts and catalytic reactions. We would then systematically discuss these applications.

Noble metal nanoparticles have become the most efficient electrocatalysts for applications in sensing, energy production, and storage owing to their unique features [71, 72]. Among them, gold nanoparticles are particularly notable for their excellent catalytic performance and tunable properties [73]. In such a context, the study of gold particles based on ECL microscopy came out first. Gold nanoparticles can efficiently catalyze ECL reactions, producing a high optical signal contrast with the background [74]. Thus, ECLM is an outstanding technique for in situ study of gold particles. In 2015, Pan et al. synthesized gold nanoparticles ranging from 30 to 300 nm in diameter and demonstrated that ECL generation is positively correlated with particle size and is also regulated by the local chemical environment and charge-transfer dynamics [75]. This was the inaugural instance of ECLM enabling the examination of local redox occurrences of gold particles at the single-nanoparticle level. Moreover, researchers have found that the ECL emission process is also affected by particle size. For small gold particles (<40 nm), the ECL emission is majorly quenched, while large gold particles (>80 nm) enhance their local ECL emission ascribing to the localized surface plasmon resonance (LSPR) [76]. Therefore, to obtain the optimized ECL from gold particles, it is necessary to consider their impact on both the electrochemical redox process and the luminescence process. Variations in ECL activity among individual gold nanoparticles reveal particle-to-particle heterogeneity, and similarly, subparticle heterogeneity can also be observed in 2D gold nanomaterials. Using ECLM, Chen et al. observed a non-uniform ECL distribution on single gold nanoplates with submicron resolution, confirming that corners and edges have more defect sites and exhibit more effective reactivity, while compared to flat facets, their stability is relatively low [77]. Moreover, thanks to its capability in continuous in situ imaging, ECLM allows the elucidation of the real-time spatial variations in the electrocatalytic active sites on single nanoplates. However, ECLM, as an optical microscopy technique, still encounters the diffraction limit, hindering its endeavor to study the catalytic spatial heterogeneity at resolutions below 200 nm. Recently, Chen et al. employed super-resolution radial fluctuations (SRRF) to study the spatial distribution of reactivity on gold plates with a resolution of approximately 100 nm [78] (Fig. 3a). Subsequently, Dong et al. utilized their developed single-molecule ECL microscopy to image electrocatalytic ECL reactions on individual gold plates, achieving single-reaction sensitivity and spatial resolution of several tens of nanometers, while enabling quantitative kinetic analysis at the single-molecule level [79] (Fig. 3b). The above-mentioned super-resolution ECLM techniques are important milestones in the fields of single-particle electrocatalysis research. The unstable surface properties of monometal nanoparticles often cause light emission attenuation for a long time, drawing attention to the electrocatalytic performance of different bimetallic nanomaterials. Janus particles, owing to their structural asymmetry and directionality, have been widely applied in multitarget sensing, drug delivery, and electrochemistry, as they confer distinct functionalities to individual particles [80, 81]. Gold-platinum Janus nanoparticle (Au-Pt JNP) is a novel bimetallic material, and Zhu et al. have found that it exhibits enhanced ECL intensity and stability when compared with the monometal nanoparticles [82] (Fig. 4a). It is further confirmed that the phenomenon is originated from the different electrotransfer rate constants between the Au and Pt interfaces. This creates an osmotic gradient and induces local turbulence, which facilitate redox reactions while limiting surface oxidation. This significant observation shows Janus materials hold promises in electrocatalysis with high efficiency and stability. With the similar strategy, the electrocatalytic activities of bimetallic Pd-Au nanostructures including Pd-tipped, Pd-covered, and anisotropic Pd-Au Janus nanorods were also evaluated at single-particle level [83] (Fig. 4b). The results from ECLM and digital simulations demonstrated that Pd-Au nanorods with different elemental distributions exhibited heterogeneous electrocatalytic reactivity and stability. These studies provided valuable insights for designing high-performance bimetallic electrocatalysts, underscoring the importance of precise structural and compositional control in bimetallic nano-electrocatalysis. Thus, in situ and dynamic monitoring electrocatalytic activity of bimetallic catalysts during their formation at the single-particle level is urgently demanded, but remains challenging. In this context, Xue et al. developed a platform that integrates online synthesis with real-time imaging of Au@Metal nanoparticles based on ECLM [84]. Using this designed platform, they successfully quantified the optimal surface component ratios of Au-M nanoparticles (M = Pt, Pd, and Rh) for maximum catalytic activity, enabling the design of tailored bimetallic nano-electrocatalysts with enhanced efficiency and a comprehensive understanding of catalytic behavior during the deposition process.

Beyond noble metal nanoparticles, the electrocatalytic properties of various other materials can also be assessed using ECLM. QDs, due to their atomic-like band structure, are highly promising candidates for electrocatalysis and have been widely applied in water-splitting catalysis in recent years [85–87]. In this context, Chen et al. employed the ECLM to study the catalytic ability of various QDs including CdSe, CdTe, and CdTeSe [88]. These QDs were initially loaded onto a silica microparticle, which were capable of generating hydroperoxide surface intermediates as coreactants, thereby facilitating ECL emission from the luminol derivative L-012. By recording optical signals, the catalytic activities of water oxidation with different QDs were visualized in situ, revealing that CdTeSe QDs exhibited the strongest ECL emission compared to other QDs. This is attributed to the reduced formation energy of intermediates in CdTeSe QDs, stemming from their higher surface defect density and more favorable valence band position. Subsequently, the same strategy was applied to investigate the electrocatalytic rates of ZnO crystals with different facets. Both experimental and theoretical results demonstrated that the catalytic performance is facet-dependent [89]. Unfortunately, above strategies are only suited to catalytic reactions linked with ECL emission, thereby limiting the study of other electrocatalytic reactions at the single-particle level. To overcome this limitation, Ma et al. synthesized hollow carbon nitride nanospheres (HCNSs), which serve as both efficient ECL nanoemitters and catalysts for the hydrogen evolution reaction (HER) [44]. Under negative potential scanning, ECL emission from HCNSs was triggered. However, this emission was suppressed when H_2 nanobubbles formed on their surface, resulting in an ECL blinking phenomenon. Given the ECL on/off mechanism is closely correlated

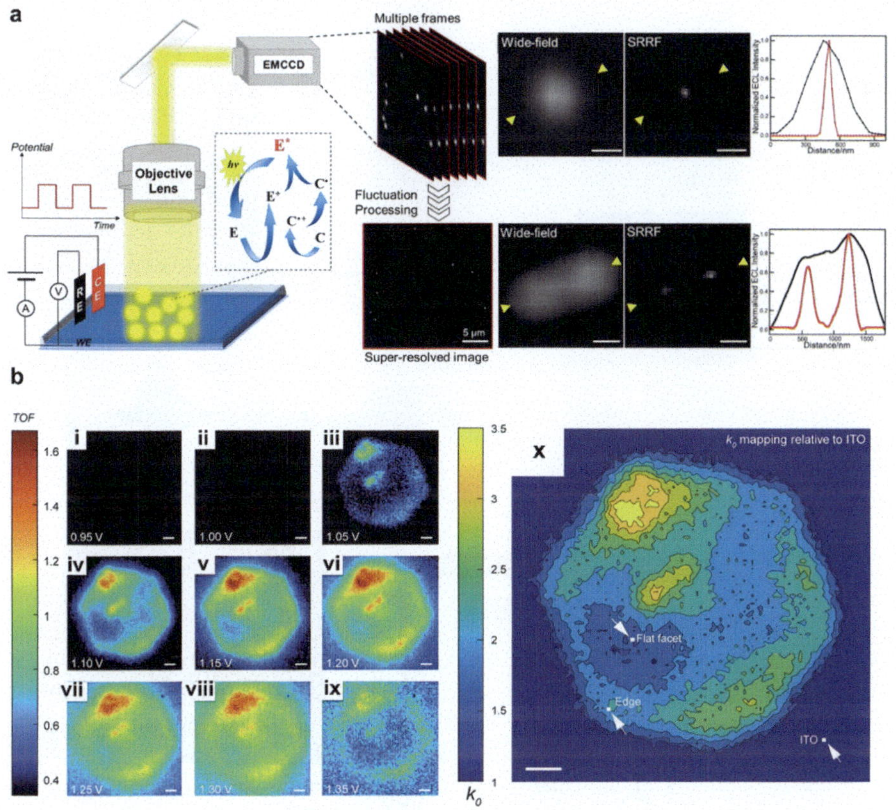

Fig. 3 **a** Basic principle and results of SRRF analysis of Au nanospheres. Reprinted with permission from Ref. [78]. Copyright 2021, American Chemical Society. **b** Mapping of electrochemical reaction kinetics on a single Au plate. (i)–(ix) Turnover frequency (TOF) mapping of the Au plate under different voltages and (x) the spatial distribution of reaction rate constant. Reprinted with permission from Ref. [79]. Copyright 2022, Wiley–VCH

with the generation, growth, and collapse of H_2 nanobubbles, as well as the power-law distribution of the durations of both states, the power-law coefficients during this dynamic process can be utilized to assess HER activities. By loading other catalysts including AuPd, NiS, and Pt on HCNSs, their electrocatalytic performance for HER has been successfully compared. Furthermore, layered nanomaterials have also been widely utilized in electrocatalysis [90], underscoring necessity to elucidate the evolving mechanisms of catalytic active sites and the corresponding structure–property relationships. Zhu et al. designed a high-resolution ECLM strategy for imaging the catalytic activity of a single reduced graphene oxide (rGO) microsheet, which exhibits excellent conductivity and tunable surface properties [91]. To mitigate the impact of diffusion on measurement accuracy, the exposure time was shortened to 0.2 ms and consecutive images were compiled into a graph, reaching a high spatial resolution, which is 400 nm. This approach revealed the presence of discrete optical

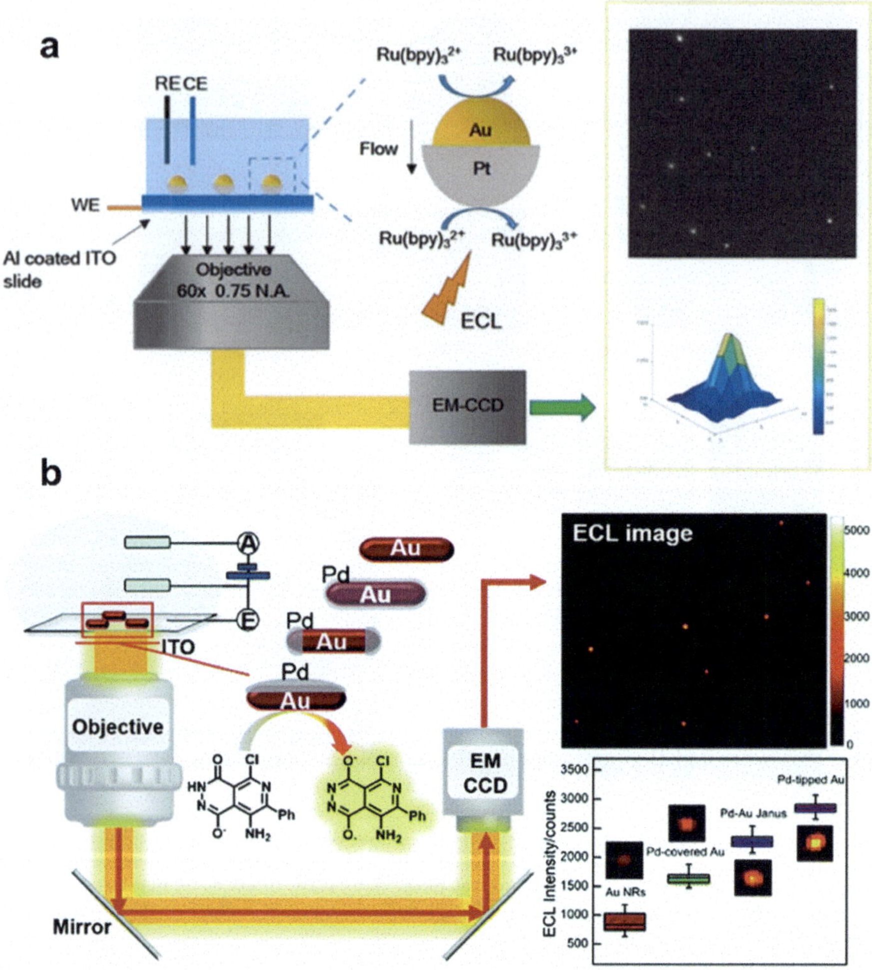

Fig. 4 Schematic illustration of ECL imaging of electrocatalysis at **a** single Au-Pt JNPs and **b** single Pd-Au nanorods with different structures. Reprinted with permission from Ref. [82]. Copyright 2018, Wiley–VCH and reprinted with permission from Ref. [83]. Copyright 2020, Royal Society of Chemistry, respectively

signals at disparate locations on the rGO microsheet, thereby substantiating the existence of spatial heterogeneity in the absorption of luminophores and coreactants. Although ECLM excels at in situ evaluation of catalytic activity at the single- and subparticle levels, obtaining structural information, such as composition and physical/chemical state, remains a significant challenge. This limitation restricts the study of catalytic mechanisms. Therefore, there is an urgent requirement to design an integrated platform that combines ECLM with other techniques. In light of these considerations, Zhao et al. combined ECLM with vis-absorption microscopy to conduct an investigation of the electrocatalytic oxygen evolution reaction (OER) process in situ at individual cobalt-layered hydroxide nanoplates [92]. The spatial and temporal distributions of electrocatalytic activities were unveiled with ECLM, and their corresponding valence state variation is shown by vis-absorption imaging. Ultimately, researchers determined that during the electrocatalytic OER process, the active site originates from the formed high-valence state of cobalt (Co^{4+}). Additionally, they observed facilitated reversible tetrahedral Co phase transition due to the residual interlayer water within the bulk phase transition.

3.2 Exploring Microscopic Mechanism of ECL

In addition to study the electrocatalytic performance of materials, the advantages of ECLM endow them more novel applications. It is well known that ECL reactions involve many steps such as deprotonation, homo-/heterogeneous charge transfer as well as light emission is involved in ECL reactions, and therefore it is critical to understand the complex processes in depth. With its high spatiotemporal resolution, ECLM serves as an efficient tool for studying ECL reaction mechanisms by dynamically monitoring ECL emission processes and mapping the spatial distribution of optical signals. In 2014, Sentic et al. decorated ruthenium labels on polystyrene (PS) microbeads via a sandwich immunoassay and then mapped the ECL reactivity at individual ruthenium-functionalized PS beads by using a 3D imaging approach [60]. The ECL emission from the individual microbead exhibited a non-uniform spatial distribution along the z-axis and occurred only in the 3 μm region near the electrode (Fig. 5a). This finite spatial extent is owing to the short lifetime of TPA cation radicals, highlighting the importance of TPA and its coreactant in regulating the light-emitting layer (Fig. 5b). Researchers investigated strategies to broaden the ECL-emitting range, as the narrow ECL layer significantly limits the development of ECLM. Fiorani et al. have found that there is a negative correlation between the thickness of the ECL-emitting layer and the phosphate buffer (PB) concentration from the side-view ECL images of the ruthenium-decorated PS beads [93] (Fig. 5c, d). This is due to the critical role of PB in the deprotonation of TPA cation radicals, which determines their lifetime and the spatial extent of ECL emission. Another way to extend the ECL-emitting range has been developed by functionalizing a biotinylated carbon nanotube labeled with a ruthenium complex on micron-sized magnetic beads [94]. In conventional bead systems, ECL emission relies solely on the oxidation of TPA at the electrode. However, functionalized carbon nanotubes act as a conductive layer, facilitating the direct oxidation of labeled luminophores, thereby extending the ECL emission range and enhancing the ECL intensity. Revealing the unknown reactions involved in the ECL process is also a research hotspot. Leveraging the previous 3D imaging model and the ECLM technique, the global process could be directly visualized. Zanut et al. unexpectedly identified an additional pathway in the $Ru(bpy)_3^{2+}$-TPA system, involving the formation of a transient, short-lived dipropylamine radical that significantly enhances ECL intensity [95]. This discovery helped to the selection of a more efficient branched amine coreactant, DPIBA, which has remarkably enhanced ECL intensity by 128%. Besides intrinsic reactions of the ECL process, Kerr et al. have discovered that the addition of another metal coordination complex $Ir(sppy)_3^{3-}$ could also be involved in this process [96]. By recording the ECL emission from the $Ru(bpy)_3^{2+}$ labels on a PS bead, a dramatic $Ir(sppy)_3^{3-}$-mediated enhancement of the ECL signal at a relatively lower oxidation potential was demonstrated (70.9-fold improvement at 0.9 V). Additionally, ECLM has revealed other potential factors influencing bead-based ECL imaging and bioassays. For instance, electrode surface passivation can attenuate ECL intensity during electrochemical oxidation [97], and optical effects may lead to varied ECL patterns at different focal planes [98]. Despite the properties and possible regulatory strategies on the bead-based ECL system have been fully discussed, the ECL emission pathway of $Ru(bpy)_3^{2+}$ functionalized on beads differs from that of diffusing $Ru(bpy)_3^{2+}$. To deeply understand the solution-phase ECL reactions, Guo et al. synthesized microtube electrodes with varying sizes and imaged the TEL from $Ru(bpy)_3^{2+}$-coreactants mixed solutions inside the microtube electrodes [99] (Fig. 6). By continuously increasing the concentration of $Ru(bpy)_3^{2+}$, it was observed that the corresponding TEL increased from 3.1 μm to >4.5 μm when TPA acted as a coreactant. This result is originated from the contribution of the so-called catalytic route. In contrast, when DBAE was used, the TEL remained almost constant regardless of the $Ru(bpy)_3^{2+}$ concentration. These phenomena confirmed that the lifetime of DBAE-derived radicals is significantly shorter than that of TPA radicals in solution. From this point of view, the selection of a suitable coreactant is crucial for specific applications.

3.3 Observing Microscopic Electrochemical Behaviors

Since ECL is electrically triggered, this unique feature grants ECLM an exceptional ability to directly observe diverse electrochemical behaviors. Conductivity, a key parameter of electrode materials determines the charge-transfer resistance at the electrode. As an electrochemical reaction, the luminescence generated by the ECL process is dominated by the electron transfer rate of the luminophores. Therefore, the output optical signal serves as an indicator to evaluate the charge-transfer resistance of electrode materials. Due to its unique structure and superior electrical conductivity, graphene is expected to substitute traditional electrode materials [100]. Nevertheless, the underlying relationships between its conductivity and intrinsic defects are unclear. Zhu et al. thus developed a perturbation ECLM to visualize the fluctuation of charge-transfer resistance in single graphene microsheets with redox-induced defects [101] (Fig. 7a, b). Their study revealed that redox-induced defects formed under high potential lead to increased charge-transfer resistance. After that, they investigated transition metal dichalcogenides (TMDCs), another layered material widely used in electronic devices [102]. During electrochemical processes, the performance of TMDCs is primarily altered by spontaneous oxidation, which is attributed to the higher bond

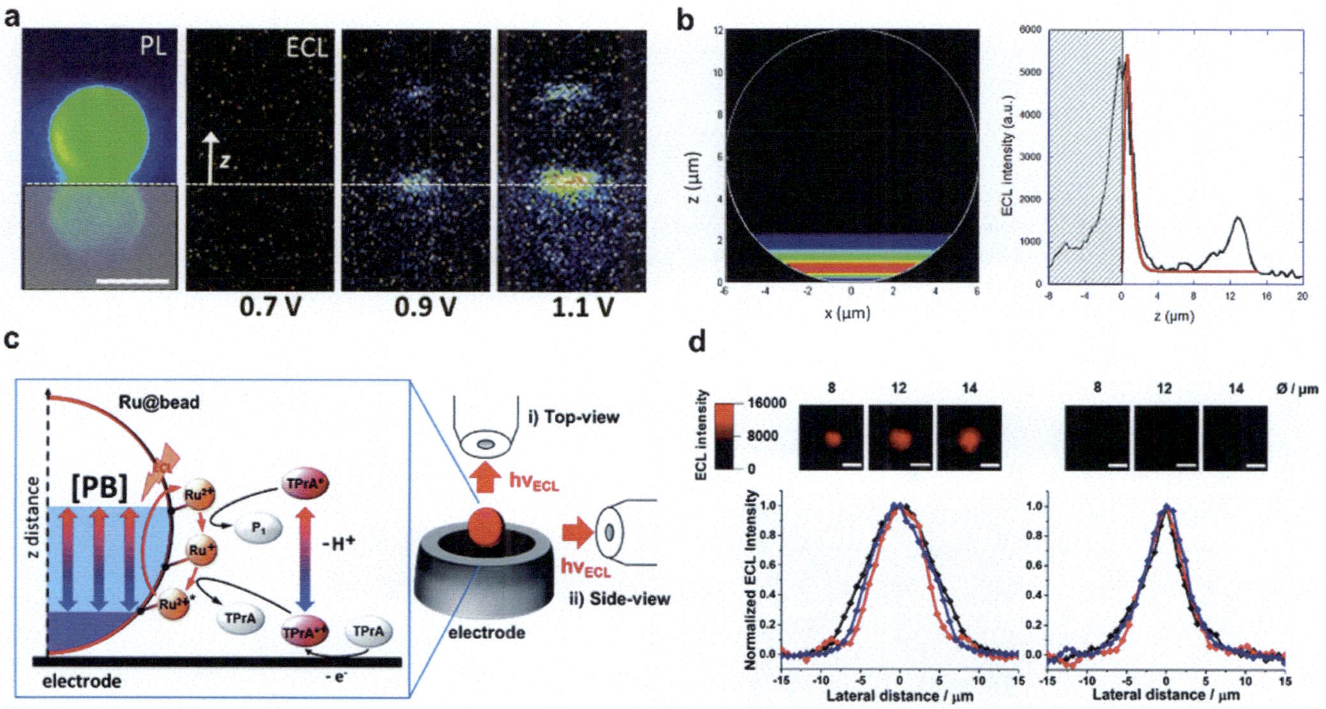

Fig. 5 **a** PL and ECL images of a ruthenium complex-labeled PS bead and **b** the simulated distribution of $Ru(bpy)_3{}^{2+*}$ (i.e., ECL intensity) in side view. Scale bar is 10 μm. Reprinted with permission from Ref. [60]. Copyright 2014, Royal Society of Chemistry. **c** Schematic illustration of the chemical lens strategy and **d** the ECL image of single Ru@beads in different PB concentrations (left: 0.01 M; right: 1 M). Scale bar is 10 μm. Reprinted with permission from Ref. [93]. Copyright 2020, Royal Society of Chemistry

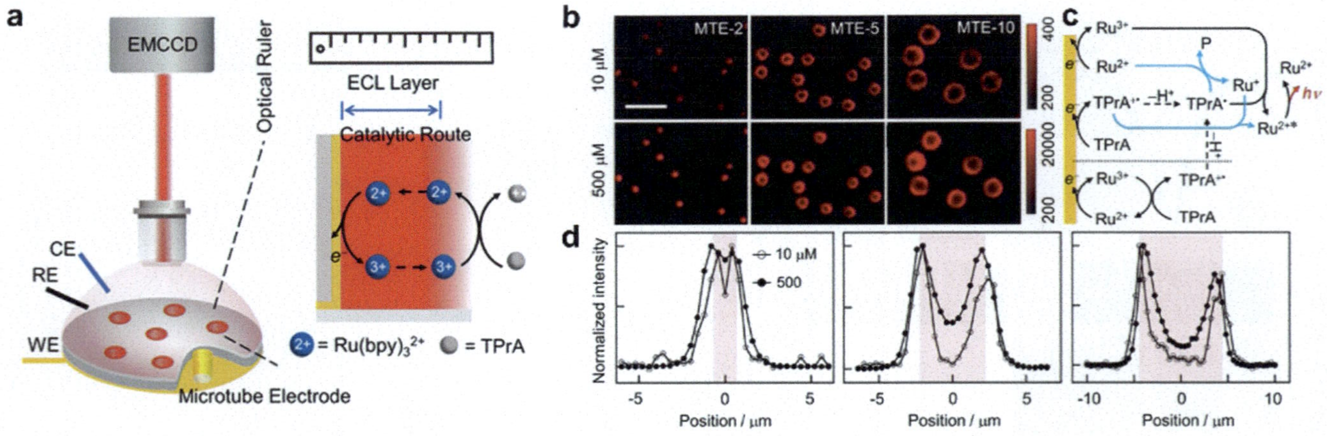

Fig. 6 **a** Schematic illustration of measuring TEL in microtube electrodes by ECLM. **b** ECL images and **d** the ECL intensity variations along the radial direction of MTE-2, MTE-5, and MTE-10 at a low and a high $Ru(bpy)_3^{2+}$ concentration. Scale bar is 20 μm. **c** ECL reaction routes of $Ru(bpy)_3^{2+}$-TPrA system. Reprinted with permission from Ref. [99]. Copyright 2021, Wiley–VCH

strength of M–O compared to M–S [103]. Consequently, the interaction between oxygen and TMDCs has garnered significant attention, necessitating methods for dynamically monitoring this process. Using the ECLM technique, they confirmed the generation of O-atom active sites on a single MoS_2 particle by recording the ECL blinking [104]. According to their observation, a new insight into the process of synergistic oxidation in MoS_2 was revealed that the oxygen doping is an intermittent process rather than a straightforward insertion into sulfur vacancies (Fig. 7c, d). Additionally, the researchers elucidated that the interaction between the edge and basal plane of MoS_2 is conducive to the oxidation process. Nanostructure electrodes are becoming increasingly prevalent in electrochemical research, particularly in the field of single-molecule electrochemistry due to their nanoscale dimensions and distinctive characteristics. To assess the functionality of the gold nanowire electrode, Wilson et al. directly imaged ECL emission from $Ru(bpy)_3{}^{2+}$ at individual electrodes [105]. The findings indicate that an inexpensive polymer layer PEDOT: PSS-PVA coated on gold nanowires can protect them from electrochemical deterioration, irrespective of the coating thickness. It is anticipated that this approach will provide insights into future investigations into the impact of nanocrystal electrode morphology, dimensions, and composition on their electrochemical characteristics. The governing processes of electrochemical reactions are not limited to charge transfer at the electrode surface; diffusion of reactants also plays a significant role. In order to gain deeper insight into diffusion behaviors, researchers further deployed the ECLM technique, which offers fast time-resolution capabilities. Contact between reagents and electrodes is a prerequisite for electrochemical reactions, but detailed information obtained through ultramicroelectrode techniques is limited. As a result, Ma et al. synthesized nanoemitters $Ru(bpy)_3{}^{2+}$-doping silica nanoparticles (RuDSNs) and visualized their stochastic collision electrochemistry at single-particle level with ECLM, which effectively avoids the interference from current and external light sources [106]. The designed strategy permitted the tracing of the collision trajectory of multiple nanoparticles and the categorization of their collisions as either spike or staircase types (Fig. 8a). Meanwhile, the correlation between the collision behaviors and particles natures has also been revealed, highlighting its great potential in electrochemical analysis. Liposomes are a common type of vesicle that are used in number of fields, including drug delivery and the creation of artificial cells [107, 108]. Consequently, there has been a growing interest in the mechanisms of rupture and opening of these vesicles. By employing liposome-encapsulated luminophores, Trad et al. effectively integrated ECLM with amperometry and epifluorescence imaging, enabling real-time monitoring of liposome collisions and the rupture of liposomes on electrode surface [109] (Fig. 8b). In light of the fact that electron transfer is not possible across a lipid bilayer, this strategy offers a robust method for studying collision electrochemistry, extra-vesicular release processes, and content diffusion behaviors with an improved signal-to-noise ratio.

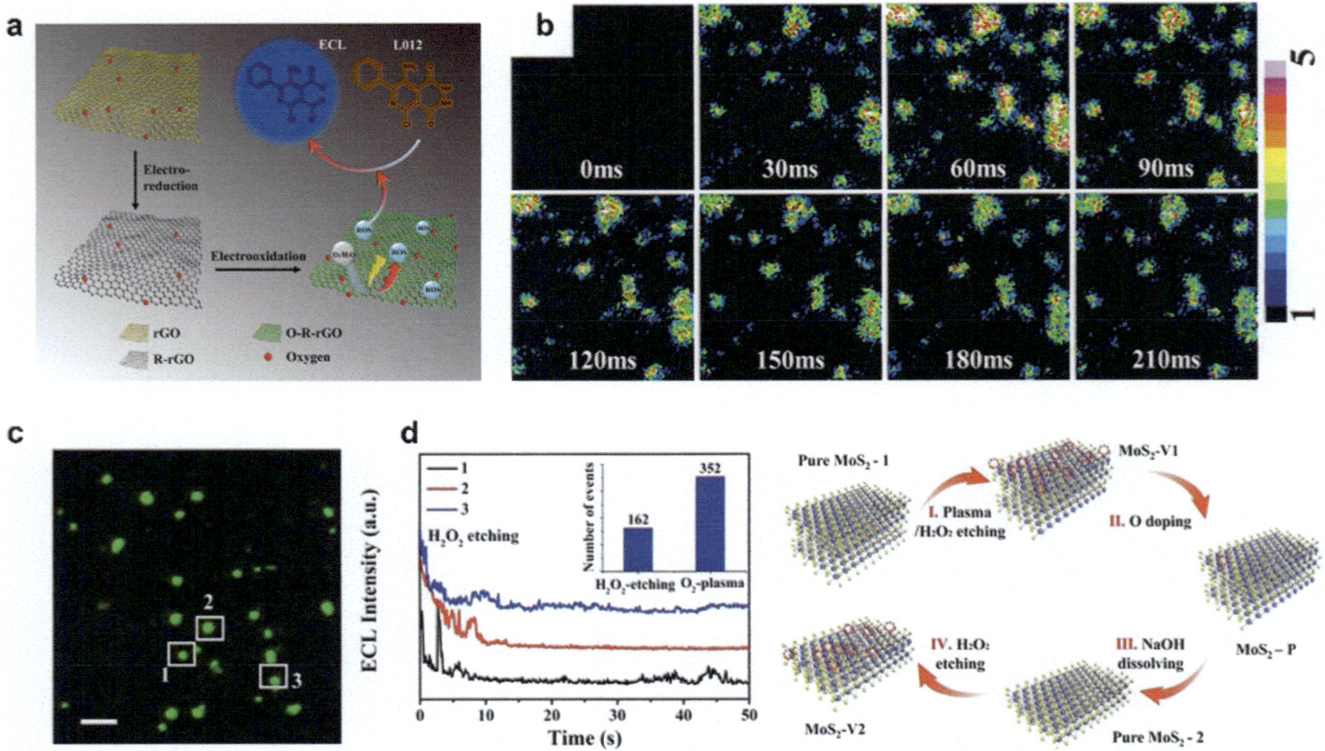

Fig. 7 **a** Schematic illustration of the perturbation ECL imaging in individual rGO microsheets to monitor the fluctuation of charge-transfer resistance. **b** ECL imaging of individual rGO microsheets at different time. Reprinted with permission from Ref. [101]. Copyright 2019, American Chemical Society. **c** ECL image of MoS_2–P particles and the ECL intensity variations of three selected particles. **d** Illustration of the mechanism of the electrocatalytic activity burst from MoS_2. Reprinted with permission from Ref. [104]. Copyright 2021, Wiley–VCH

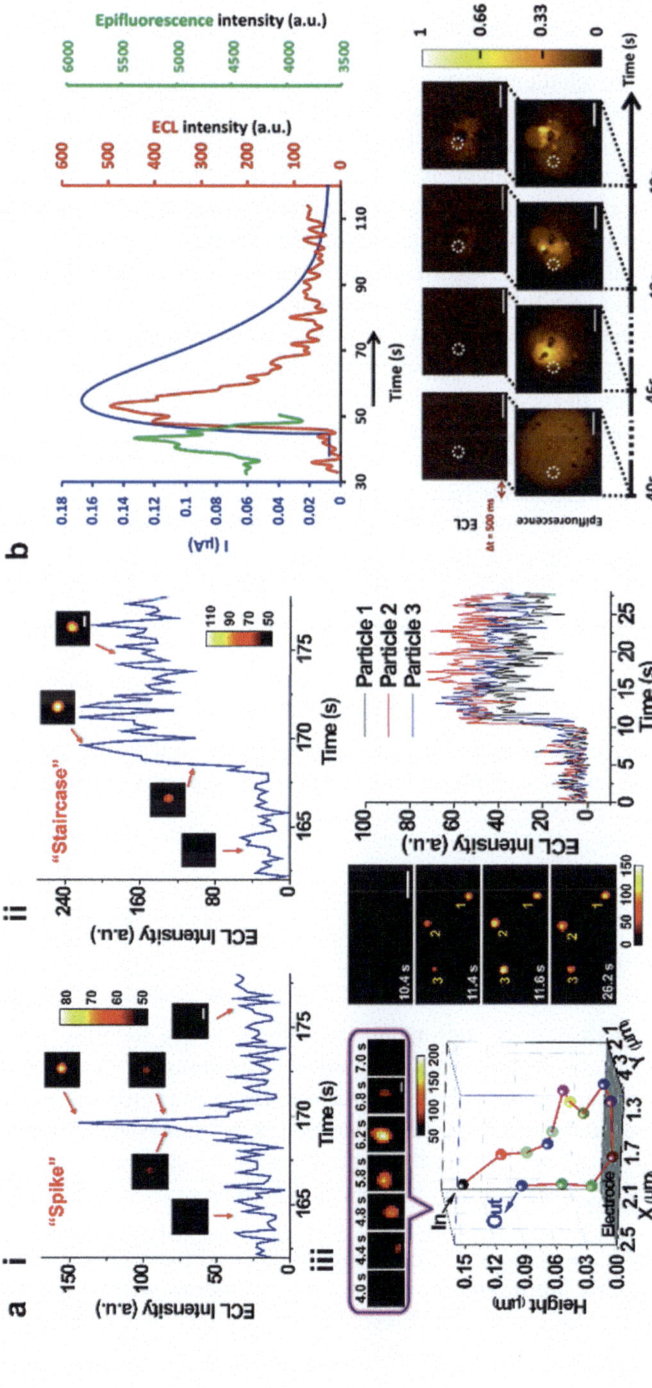

Fig. 8 **a** ECL imaging of individual nanoparticles during (i) "spike" and (ii) "staircase" collision processes. Scale bar is 2 μm. (iii) ECL imaging of three RuDSNs during an elastic collision with Brownian motion. Scale bar is 5 μm. Reprinted with permission from Ref. [106]. Copyright 2018, Royal Society of Chemistry. **b** The current, PL, and ECL intensity variations as a function of time when a single liposome opening at electrode surface. Scale bar is 200 μm. Reprinted with permission from Ref. [109]. Copyright 2022, American Chemical Society

3.4 Studying on Photoelectrochemical Property of Nanomaterials

In order to address the challenges of energy and environmental sustainability, researchers have devoted a significant amount of effort to the efficient conversion and storage of solar energy. In this context, photoelectrochemistry has been the subject of considerable research and development as a means of directly storing solar energy in chemical fuels. Briefly, photoelectrochemistry is a process that occurs at the interface of semiconductors during illumination. The produced photogenerated carriers serve to decrease the potential, thereby triggering electrochemical reactions. The imposition of high potential is typically required for the majority of ECL reactions, which significantly impedes their applicability in biosensing and cell imaging. It is therefore proposed that the concept of photoelectrochemistry may thus provide a pathway to overcome this challenge. In recent years, researchers have investigated a multitude of n-type semiconductor photoanode materials including n-Si/SiO_x/metal [110, 111] and $BiVO_4$ [112], with the objective of developing a photoelectrochemiluminescence (PECL) system capable of producing light from photogenerated h^+. These pioneering works prompted researchers to couple the PECL system with a microscopy technique for the study optical–electrical conversion at micro/nanointerfaces. For example, the photocatalytic process on a single gold nanoparticle on TiO_2 was monitored with photoinduced electrochemiluminescence microscopy (PECLM) [113]. In the presence of laser illumination, photogenerated carriers are produced and subsequently separated by an electric field, leaving holes at the surface of gold nanoparticles to drive ECL reactions. Accordingly, the PECL intensity is found to be in close correlation with the number of photogenerated carriers. Experimental results revealed that the photoelectrochemical conversion efficiency is significantly influenced by the particle size. This study demonstrated that PECLM provides a novel perspective for understanding photocatalysis.

3.5 Nano-Electrochemical Bioassays

The early diagnosis of disease is contingent upon the sensitive identification of biomarkers. In recent years, a plethora of immunoassay techniques have emerged for the detection of various biomarkers. Nevertheless, the rising costs of analysis, the complexity of the procedures, and the interference of signals have impeded the advancement of these techniques. With the combination of electrochemical and optical approaches, ECL represents many superiorities including excellent spatiotemporal controllability, high sensitivity, and a wide dynamic range. These characteristics expand the scope of ECL applications in the analysis of diverse biomarkers, including proteins and nucleic acids. Furthermore, the ECL-based immunoassay techniques have become the gold standard for clinical diagnosis. The advent of imaging methods has enabled the determination of multiple analytes at the micro-/nanoscale, thereby inaugurating a new era for nano-electrochemical bioassays. In 2009, Deiss et al. firstly demonstrated that ECL imaging could be used as an alternative strategy for detecting multiple antigens in a microarray format [57]. Briefly, they adapted a conventional sandwich immunoassay using polystyrene beads, where ECL emission occurred only in the presence of analytes, originating from streptavidin-modified $Ru(bpy)_3^{2+}$ linked to biotinylated antibodies. By imaging different individual beads, a method for the simultaneous detection of multiplex targets was developed and successfully validated. This innovative analytical approach has inspired further optimization of ECL immunoassays (ECLIAs). Two primary objectives of developing ECLIAs are to enhance detection sensitivity and to improve multiplexing capacity. Single-molecule ECL imaging represented a revolutionary technique, pushing traditional ECLM to achieve ultimate single-photon sensitivity and suboptical diffraction resolution [114]. This advanced technique was further integrated with conventional ECLIAs to develop a single-molecule ECL bioassay, which enables direct quantification of single biomolecules [115]. The proposed single-molecule ECL bioassay exhibited an ultralow detection limit of 67 attomoles, approximately 10,000 times greater than that of traditional ECLIAs (Fig. 9). Yang et al. put forth an enhanced ECLIA methodology for the detection of a greater number of targets [116]. They developed a size-encoded multiplex ECLIA by substituting magnetic microbeads with gold-coated ones. Because these gold-coated microbeads are capable of activating a proximity LOP route, the ECL emission layer is no longer constrained by the generated short-lived coreactant radicals and microbeads of varying diameters can be employed in size-encoded multiplex immunoassays. Benefiting from its high spatial resolution, researchers further attempted to extend ECLM-based bioassay techniques to the detection of single biomolecules. Unlike traditional ensemble measurements, single-biomolecule analytical techniques offer insights into heterogeneities and stochastic processes in biological systems. Similarly, as with bead-based ECLIAs, Liu et al. employed monoclonal antibody-coated electrodes to capture individual proteins, followed by the introduction of antibodies labeled with ECL nanoemitters (RuDSN/AuNPs) [117]. The ECL image of RuDSN/AuNPs permitted direct visualization of these single biomolecules. In comparison to photoluminescence (PL), the ECL emission demonstrated a near-to-zero background signal, thereby a high signal-to-noise (S/N) ratio for the observation of single proteins. Inspired by the ECLM-based single biomolecules detection strategy, the Ru-doped metal–organic framework or other nanoporous nanoparticles were used instead of RuDSN as ECL nanoemitters. Furthermore, monoclonal antibodies were replaced with cost-effective aptamer DNA for specific binding to target biomolecules. These modified strategies enabled the successful imaging of single biomolecules, including the SARS-CoV-2 spike protein and EpCAM protein [67, 70]. The high performance of these single-biomolecule detection approaches enables the direct quantification of biomarker proteins on cell membranes, marking a significant advance in nano-electrochemical analysis for biological applications. Another notable application of ECLM in the field of nanobiology is the in situ investigation of protein corona formation kinetics on individual nanomedicine surfaces. The protein corona plays a predominant role in the functionality of nanomedicines. It can not only disrupt the targeting ability of

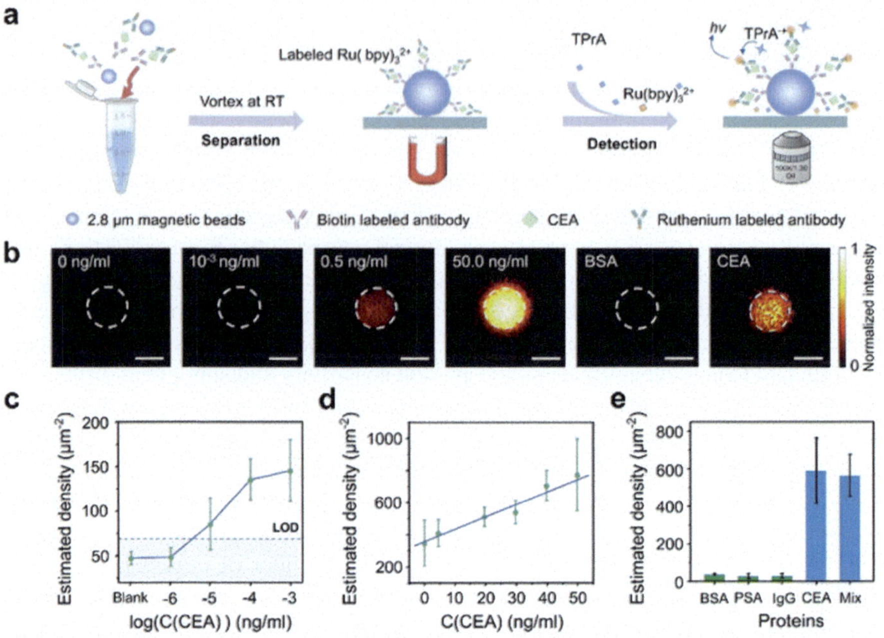

Fig. 9 **a** Schematic illustration of ECLM-based immunoassay. **b** ECL images of labeled beads at different biomarker concentrations and detection targets. Scale bar: 2 μm. **c** Sensitivity, **d** linear, and **e** selectivity analysis of the technique. Reprinted with permission from Ref. [115]. Copyright 2023, Wiley–VCH

nanomedicines, but also can protect cells from the injury of nanomaterials [118, 119]. Therefore, it is crucial to gain insight into the mechanisms and regulatory pathways involved in the formation of the protein corona. Xing and colleagues developed a self-regulated ECLM based on the electrostatic interaction between luminophores and the nanoparticle surface [120]. Due to the hindrance effect of proteins, the ECL signals from nanoparticles progressively decreased during protein corona formation. The high spatiotemporal resolution of ECLM allows for the real-time monitoring of the dynamic process of protein corona formation on single nanomedicines, thereby paving the way for advancements in bio-nano interface exploration, drug delivery, and precision cancer treatment.

4 Conclusion

The aim of this chapter is to present the recent developments in ECL nanoemitters and the application of ECLM in single-nanoparticle analysis. Operando imaging electrochemical processes and detecting individual entities in an electrochemical situation at the nanometer levels are gradually becoming mainstream. A significant challenge remains the improvement of throughput and analysis speed. To address this challenge, the development of novel wide-field imaging strategies with high spatiotemporal resolution is required. The distinctive attributes of ECL, including its minimal background and direct correlation with electron transfer reactions, render ECL microscopy tremendous potential in the domain of nano-electrochemical research. To date, ECL microscopy has enabled the observation of particle-to-particle and subparticle electrocatalytic heterogeneity, as well as the study of microscopic electrochemical interfacial behaviors and single-molecule electrochemical sensing.

Nevertheless, considerable obstacles remain to be overcome before ECL microscopy can be developed further and made more widely applicable. A significant challenge lies in the dearth of highly efficient ECL emitters. Despite the numerous ECL emitters that have been reported in the literature, only $Ru(bpy)_3^{2+}$ and luminol are utilized for ECL microscopy. This limitation arises from the low ECL efficiency of alternative emitters and the restricted photon collection capability of current cameras. Recently, certain nanoemitters, such as HCNS, have been demonstrated to be efficacious for ECL microscopy. However, their substantial dimensions and inadequate diffusivity may diminish the spatiotemporal resolution. Therefore, it is imperative that researchers continue to prioritize the development of small-molecule ECL emitters. In the future, quantum chemical calculations may be employed to predict potential ECL-active molecules derived from existing fluorescent luminophores. Subsequently, the ECL efficiency of the selected molecules will be evaluated. Following this, the researchers will then screen the desired ECL-active molecules and construct a molecular library. The combination of machine learning methods will facilitate the discovery of additional small molecular ECL emitters. It is noteworthy that water solubility is a crucial factor in the application of electrochemical bioassays. Another issue that requires attention is necessity for the application of hyphenated techniques. Although ECL microscopy has made a significant contribution to the study of single-particle electrocatalysis, it currently only provides information on catalytic activity heterogeneity. This significantly impedes our ability to elucidate the fundamental mechanisms underlying catalytic processes, thereby motivating researchers to develop ECL microscopy-based coupling technologies. For example, the combination of ECL microscopy with in situ infrared absorption or Raman spectroscopy represents a promising strategy. Infrared absorption or Raman spectroscopy can be employed to elucidate chemical heterogeneity in nanomaterials and identify produced intermediates, providing direct evidence of catalytic mechanisms. Another potential approach is coupling ECL microscopy with scanning probe techniques, such as AFM, SECCM, and SICM. By linking the differences in physical properties with catalytic activity heterogeneity within individual materials, researchers can gain deeper insights into the structure–function relationship.

Current ECL systems have been overwhelmingly used in the field of single-nanoparticle electrochemical analysis, but we posit that there is significant opportunity to expand the scope of ECL microscopy applications. Firstly, ECL microscopy could be employed to investigate additional electrocatalytic reactions. So far, OER and ORR have been sufficiently discussed with the aid of luminol-based ECL systems, while ECL microscopy has yet to be applied in the investigation of other reactions. Amine compounds are crucial coreactants for the ECL emission of $Ru(bpy)_3^{2+}$, and they can be derived from nitrate. It is therefore evident that $Ru(bpy)_3^{2+}$-amine ECL systems represent a promising avenue for the study of nitrate reduction reactions (NO_3RRs). Similarly, oxalate, another common coreactant for $Ru(bpy)_3^{2+}$-based ECL, is closely linked to carbon dioxide generation, enabling the study of CO_2 reduction reactions (CO_2RR) using ECL techniques. Secondly, ECL microscopy may serve as a platform for operando imaging of the microscopic alterations in electrode materials. Lithium-ion batteries have revolutionized portable electronics, with the majority of studies concentrating on the characterization of the microscopic morphology and component variation on the material level. This is because it has a significant impact on the energy density and safety of batteries. Voltage-triggered light emission provides a reliable signal for the dynamic monitoring of material variations during battery operation. Thirdly, the combination of ECL microscopy with nanopore methods has the potential to facilitate the development of next-generation sequencing technologies. Despite the considerable potential of nanopore electrochemistry as a tool for DNA sequencing, the inherent background noise limits the sensitivity of detection. In comparison to conventional electrochemical methodologies, ECL produces a multidimensional output signal, comprising intensity, spectrum, onset voltage, and peak voltage, which collectively facilitate single-base detection. Through collaboration between ECL researchers and scientists from other fields, we believe ECL microscopy holds great promise for advancing next-generation nano-electrochemical analysis.

References

1. Roh, K.H., Martin, D.C., Lahann, J.: Biphasic Janus particles with nanoscale anisotropy. Nat. Mater. **4**, 759–763 (2005). https://doi.org/10.1038/nmat1486
2. Liu, H.L., Nosheen, F., Wang, X.: Noble metal alloy complex nanostructures: controllable synthesis and their electrochemical property. Chem. Soc. Rev. **44**, 3056–3078 (2015). https://doi.org/10.1039/c4cs00478g
3. Borm, P.J., Robbins, D., Haubold, S., et al.: The potential risks of nanomaterials: a review carried out for ECETOC. Part. Fibre Toxicol. **3**, 1–35 (2006). https://doi.org/10.1186/1743-8977-3-11
4. Zhang, Q., Uchaker, E., Candelaria, S.L., et al.: Nanomaterials for energy conversion and storage. Chem. Soc. Rev. **42**, 3127–3171 (2013). https://doi.org/10.1039/c3cs00009e
5. Roduner, E.: Size matters: why nanomaterials are different. Chem. Soc. Rev. **35**, 583–592 (2006). https://doi.org/10.1039/b502142c
6. Zhang, H., Watanabe, T., Okumura, M., et al.: Catalytically highly active top gold atom on palladium nanocluster. Nat. Mater. **11**, 49–52 (2011). https://doi.org/10.1038/nmat3143
7. Valden, M., Lai, X., Goodman, D.W.: Onset of catalytic activity of gold clusters on titania with the appearance of nonmetallic properties. Science **281**, 1647–1650 (1998). https://doi.org/10.1126/science.281.5383.1647
8. Hvolbæk, B., Janssens, T.V.W., Clausen, B.S., et al.: Catalytic activity of Au nanoparticles. Nano Today **2**, 14–18 (2007). https://doi.org/10.1016/s1748-0132(07)70113-5
9. Fujita, T., Guan, P., McKenna, K., et al.: Atomic origins of the high catalytic activity of nanoporous gold. Nat. Mater. **11**, 775–780 (2012). https://doi.org/10.1038/nmat3391

10. Xiang, G., Wang, Y.-G.: Exploring electronic-level principles how size reduction enhances nanomaterial surface reactivity through experimental probing and mathematical modeling. Nano Res. **15**, 3812–3817 (2021). https://doi.org/10.1007/s12274-021-3910-1
11. Zhou, L., Zhuang, Z., Zhao, H., et al.: Intricate hollow structures: controlled synthesis and applications in energy storage and conversion. Adv. Mater. 29 (2017). https://doi.org/10.1002/adma.201602914
12. Li, Y., Shi, J.: Hollow-structured mesoporous materials: chemical synthesis, functionalization and applications. Adv. Mater. **26**, 3176–3205 (2014). https://doi.org/10.1002/adma.201305319
13. Li, Y., Li, Z., Chi, C., et al.: Plasmonics of 2D nanomaterials: properties and applications. Adv. Sci. **4**, 1600430 (2017). https://doi.org/10.1002/advs.201600430
14. Alfieri, A., Anantharaman, S.B., Zhang, H., et al.: Nanomaterials for quantum information science and engineering. Adv. Mater. **35**, e2109621 (2023). https://doi.org/10.1002/adma.202109621
15. Chen, Q.: Beyond snowflakes: heterogeneity in nanomaterials. Nano Lett. **22**, 3–5 (2022). https://doi.org/10.1021/acs.nanolett.1c03400
16. Wang, W.: Imaging the chemical activity of single nanoparticles with optical microscopy. Chem. Soc. Rev. **47**, 2485–2508 (2018). https://doi.org/10.1039/C7CS00451F
17. Sun, T., Yu, Y., Zacher, B.J., et al.: Scanning electrochemical microscopy of individual catalytic nanoparticles. Angew. Chem. Int. Ed. **53**, 14120–14123 (2014)
18. Takahashi, Y., Kumatani, A., Munakata, H., et al.: Nanoscale visualization of redox activity at lithium-ion battery cathodes. Nat. Commun. **5**, 5450 (2014)
19. Suran, S., Bharadwaj, K., Raghavan, S., et al.: Bright-field nanoscopy: visualizing nanostructures with localized optical contrast using a conventional microscope. Sci. Rep. **6**, 25011 (2016)
20. Huang, W., Yu, L., Zhu, Y., et al.: Single-Particle imaging of anion exchange reactions in cuprous oxide. ACS Nano **15**, 6481–6488 (2021). https://doi.org/10.1021/acsnano.0c09094
21. Jiang, D., Jiang, Y., Li, Z., et al.: Optical imaging of phase transition and Li-ion diffusion kinetics of single $LiCoO_2$ nanoparticles during electrochemical cycling. J. Am. Chem. Soc. **139**, 186–192 (2017)
22. Gao, X., Lang, X., El Khoury, E., et al.: Quantitative label-free chemical imaging of PLGA nanoparticles in cells and tissues with single-particle sensitivity. Nano Lett. **24**, 1024–1033 (2024). https://doi.org/10.1021/acs.nanolett.3c04463
23. Xie, R.C., Gao, J., Wang, S.C., et al.: Optically imaging in situ effects of electrochemical cycling on single nanoparticle electrocatalysis. Anal. Chem. **96**, 2455–2463 (2024). https://doi.org/10.1021/acs.analchem.3c04425
24. Liu, Y., Guo, W., Su, B.: Recent advances in electrochemiluminescence imaging analysis based on nanomaterials and micro-/nanostructures. Chin. Chem. Lett. **30**, 1593–1599 (2019)
25. Jin, R., Zeng, C., Zhou, M., et al.: Atomically precise colloidal metal nanoclusters and nanoparticles: fundamentals and opportunities. Chem. Rev. **116**, 10346–10413 (2016). https://doi.org/10.1021/acs.chemrev.5b00703
26. Chen, D., Li, B., Cai, S., et al.: Dual targeting luminescent gold nanoclusters for tumor imaging and deep tissue therapy. Biomaterials **100**, 1–16 (2016)
27. Antonello, S., Hesari, M., Polo, F., et al.: Electron transfer catalysis with monolayer protected Au 25 clusters. Nanoscale **4**, 5333–5342 (2012)
28. Wu, M.-S., Shi, H.-W., He, L.-J., et al.: Microchip device with 64-site electrode array for multiplexed immunoassay of cell surface antigens based on electrochemiluminescence resonance energy transfer. Anal. Chem. **84**, 4207–4213 (2012)
29. Xue, J., Zhang, Z., Zheng, F., et al.: Efficient solid-state electrochemiluminescence from high-quality perovskite quantum dot films. Anal. Chem. **89**, 8212–8216 (2017)
30. Zheng, L., Chi, Y., Dong, Y., et al.: Electrochemiluminescence of water-soluble carbon nanocrystals released electrochemically from graphite. J. Am. Chem. Soc. **131**, 4564–4565 (2009)

31. Carrara, S., Aliprandi, A., Hogan, C.F., et al.: Aggregation-induced electrochemiluminescence of platinum (II) complexes. J. Am. Chem. Soc. **139**, 14605–14610 (2017)
32. Chang, Y., Palacios, R.E., Fan, F.R., et al.: Electrogenerated chemiluminescence of single conjugated polymer nanoparticles. J. Am. Chem. Soc. **130**, 8906–8907 (2008). https://doi.org/10.1021/ja803454x
33. Wang, N., Gao, H., Li, Y., et al.: Dual intramolecular electron transfer for in situ coreactant-embedded electrochemiluminescence microimaging of membrane protein. Angew. Chem. Int. Ed. **60**, 197–201 (2021)
34. Wang, Z., Gao, H., Liu, P., et al.: Visualized uranium rapid monitoring system based on self-enhanced electrochemiluminescence-imaging of amidoxime functionalized polymer nanoparticles. Chin. Chem. Lett. **33**, 3456–3460 (2022). https://doi.org/10.1016/j.cclet.2021.11.019
35. Guo, W., Ding, H., Zhou, P., et al.: Electrochemiluminescence waveguide in single crystalline molecular wires. Angew. Chem. Int. Ed. **59**, 6745–6749 (2020). https://doi.org/10.1002/anie.201915984
36. Ding, Z., Quinn, B.M., Haram, S.K., et al.: Electrochemistry and electrogenerated chemiluminescence from silicon nanocrystal quantum dots. Science **296**, 1293–1297 (2002)
37. Myung, N., Ding, Z., Bard, A.J.: Electrogenerated chemiluminescence of CdSe nanocrystals. Nano Lett. **2**, 1315–1319 (2002)
38. Bae, Y., Myung, N., Bard, A.J.: Electrochemistry and electrogenerated chemiluminescence of CdTe nanoparticles. Nano Lett. **4**, 1153–1161 (2004)
39. Sun, L., Bao, L., Hyun, B.-R., et al.: Electrogenerated chemiluminescence from PbS quantum dots. Nano Lett. **9**, 789–793 (2009)
40. Zhou, Z., Zhang, Y., Shen, Y., et al.: Molecular engineering of polymeric carbon nitride: advancing applications from photocatalysis to biosensing and more. Chem. Soc. Rev. **47**, 2298–2321 (2018)
41. Cheng, C., Huang, Y., Tian, X., et al.: Electrogenerated chemiluminescence behavior of graphite-like carbon nitride and its application in selective sensing Cu^{2+}. Anal. Chem. **84**, 4754–4759 (2012)
42. Chen, L., Zeng, X., Si, P., et al.: Gold nanoparticle-graphite-like C3N4 nanosheet nanohybrids used for electrochemiluminescent immunosensor. Anal. Chem. **86**, 4188–4195 (2014)
43. Cheng, C., Huang, Y., Wang, J., et al.: Anodic electrogenerated chemiluminescence behavior of graphite-like carbon nitride and its sensing for rutin. Anal. Chem. **85**, 2601–2605 (2013)
44. Ma, C., Wei, H.-F., Wang, M.-X., et al.: Hydrogen evolution reaction monitored by electrochemiluminescence blinking at single-nanoparticle level. Nano Lett. **20**, 5008–5016 (2020). https://doi.org/10.1021/acs.nanolett.0c01129
45. Ritchie, E.L., Pastore, P., Wightman, R.M.: Free energy control of reaction pathways in electrogenerated chemiluminescence. J. Am. Chem. Soc. **119**, 11920–11925 (1997)
46. Dong, Y.-P., Chen, G., Zhou, Y., et al.: Electrochemiluminescent sensing for caspase-3 activity based on Ru (bpy) $_3{}^{2+}$-doped silica nanoprobe. Anal. Chem. **88**, 1922–1929 (2016)
47. Xu, Y., Bai, H., Lu, G., et al.: Flexible graphene films via the filtration of water-soluble noncovalent functionalized graphene sheets. J. Am. Chem. Soc. **130**, 5856–5857 (2008)
48. Zhou, X., Zhu, D., Liao, Y., et al.: Synthesis, labeling and bioanalytical applications of a tris (2, 2′-bipyridyl) ruthenium (II)-based electrochemiluminescence probe. Nature Protocols **9**, 1146–1159 (2014)
49. Zhang, J., Qi, H., Li, Y., et al.: Electrogenerated chemiluminescence DNA biosensor based on hairpin DNA probe labeled with ruthenium complex. Anal. Chem. **80**, 2888–2894 (2008)
50. Miao, W., Bard, A.J.: Electrogenerated chemiluminescence. 80. C-reactive protein determination at high amplification with $[Ru(bpy)_3]^{2+}$-containing microspheres. Anal. Chem. **76**, 7109–7113 (2004)
51. Miao, W., Bard, A.J.: Electrogenerated chemiluminescence. 77. DNA hybridization detection at high amplification with $[Ru(bpy)_3]^{2+}$-containing microspheres. Anal. Chem. **76**, 5379–5386 (2004)

52. Kuwana, T., Epstein, B., Seo, E.T.: Electrochemical generation of solution luminescence. J. Phys. Chem. **67**, 2243–2244 (1963). https://doi.org/10.1021/j100804a516
53. Zanut, A., Palomba, F., Rossi Scota, M., et al.: Dye-Doped silica nanoparticles for enhanced ECL-based immunoassay analytical performance. Angew. Chem. Int. Ed. **59**, 21858–21863 (2020). https://doi.org/10.1002/anie.202009544
54. Li, B., Huang, X., Lu, Y., et al.: High electrochemiluminescence from $Ru(bpy)_3{}^{2+}$ embedded metal-organic frameworks to visualize single molecule movement at the cellular membrane. Adv. Sci. **9**, e2204715 (2022). https://doi.org/10.1002/advs.202204715
55. Chen, L., Huang, J., Shi, Y., et al.: Polystyrene-based nanospheres with controllable microstructures for exceptional solid phase microextraction of organic pollutants. Chem. Eng. J. **428**, 132527 (2022)
56. Hao, C., Chen, J., Chen, Y., et al.: Preparation of silver-coated polystyrene microspheres intermediated with polyaniline and their application in array-patterned anisotropic conductive films. Adv. Compos. Hybrid Mater. **8**, 67 (2025)
57. Deiss, F., LaFratta, C.N., Symer, M., et al.: Multiplexed sandwich immunoassays using electrochemiluminescence imaging resolved at the single bead level. J. Am. Chem. Soc. **131**, 6088–6089 (2009). https://doi.org/10.1021/ja901876z
58. Chovin, A., Garrigue, P., Vinatier, P., et al.: Development of an ordered array of optoelectrochemical individually readable sensors with submicrometer dimensions: application to remote electrochemiluminescence imaging. Anal. Chem. **76**, 357–364 (2003). https://doi.org/10.1021/ac034974w
59. Han, D., Goudeau, B., Lapeyre, V., et al.: Enhanced electrochemiluminescence at microgel-functionalized beads. Biosens. Bioelectron. **216**, 114640 (2022). https://doi.org/10.1016/j.bios.2022.114640
60. Sentic, M., Milutinovic, M., Kanoufi, F., et al.: Mapping electrogenerated chemiluminescence reactivity in space: mechanistic insight into model systems used in immunoassays. Chem. Sci. **5**, 2568–2572 (2014). https://doi.org/10.1039/c4sc00312h
61. Wen, J., Yang, K., Liu, F., et al.: Diverse gatekeepers for mesoporous silica nanoparticle based drug delivery systems. Chem. Soc. Rev. **46**, 6024–6045 (2017)
62. Vallet-Regí, M., Balas, F., Arcos, D.: Mesoporous materials for drug delivery. Angew. Chem. Int. Ed. **46**, 7548–7558 (2007)
63. Duan, H., Dong, J., Gu, X., et al.: Hydrodeoxygenation of water-insoluble bio-oil to alkanes using a highly dispersed Pd–Mo catalyst. Nat. Commun. **8**, 591 (2017)
64. Zanarini, S., Rampazzo, E., Ciana, L.D., et al.: Ru $(bpy)_3$ covalently doped silica nanoparticles as multicenter tunable structures for electrochemiluminescence amplification. J. Am. Chem. Soc. **131**, 2260–2267 (2009)
65. Hoffmann, F., Cornelius, M., Morell, J., et al.: Silica-based mesoporous organic–inorganic hybrid materials. Angew. Chem. Int. Ed. **45**, 3216–3251 (2006)
66. Zhang, L., Dong, S.: Electrogenerated chemiluminescence sensors using $Ru(bpy)_3{}^{2+}$ doped in silica nanoparticles. Anal. Chem. **78**, 5119–5123 (2006)
67. Lu, Y., Huang, X., Wang, S., et al.: Nanoconfinement-Enhanced electrochemiluminescence for in situ imaging of single biomolecules. ACS Nano **17**, 3809–3817 (2023). https://doi.org/10.1021/acsnano.2c11934
68. Li, X., Yang, X., Xue, H., et al.: Metal–organic frameworks as a platform for clean energy applications. EnergyChem **2**, 100027 (2020)
69. Yang, W., Li, X., Li, Y., et al.: Applications of metal-organic-framework-derived carbon materials. Adv. Mater. **31**, e1804740 (2019). https://doi.org/10.1002/adma.201804740
70. Lu, Y., Ning, Y., Li, B., et al.: Dual-Signal imaging mode based on fluorescence and electrochemiluminescence for ultrasensitive visualization of SARS-CoV-2 spike protein. Anal. Chem. **96**, 463–470 (2024). https://doi.org/10.1021/acs.analchem.3c04535
71. Cui, L., Wu, J., Ju, H.: Electrochemical sensing of heavy metal ions with inorganic, organic and bio-materials. Biosens. Bioelectron. **63**, 276–286 (2015). https://doi.org/10.1016/j.bios.2014.07.052

72. Li, Y., Sun, Y., Qin, Y., et al.: Recent advances on water-splitting electrocatalysis mediated by noble-metal-based nanostructured materials. Adv. Energy Mater. **10**, 1903120 (2020). https://doi.org/10.1002/aenm.201903120
73. Guo, S., Wang, E.: Synthesis and electrochemical applications of gold nanoparticles. Anal. Chim. Acta **598**, 181–192 (2007). https://doi.org/10.1016/j.aca.2007.07.054
74. Li, X., Wu, D., Ma, H., et al.: Ultrasensitive amyloid-β proteins detection based on curcumin conjugated ZnO nanoparticles quenching electrochemiluminescence behavior of luminol immobilized on Au@MoS2/Bi2S3 nanorods. Biosens. Bioelectron. **131**, 136–142 (2019). https://doi.org/10.1016/j.bios.2019.01.066
75. Pan, S., Liu, J., Hill, C.M.: Observation of local redox events at individual Au nanoparticles using electrogenerated chemiluminescence microscopy. J. Phys. Chem. C **119**, 27095–27103 (2015). https://doi.org/10.1021/acs.jpcc.5b06829
76. Wei, Y., Zhang, Y., Pan, J., et al.: Plasmon-enhanced electrochemiluminescence at the single-nanoparticle level. Angew. Chem. Int. Ed. **62**, e202214103 (2023). https://doi.org/10.1002/anie.202214103
77. Chen, M.M., Zhao, W., Zhu, M.J., et al.: Spatiotemporal imaging of electrocatalytic activity on single 2D gold nanoplates via electrogenerated chemiluminescence microscopy. Chem. Sci. **10**, 4141–4147 (2019). https://doi.org/10.1039/c9sc00889f
78. Chen, M.-M., Xu, C.-H., Zhao, W., et al.: Super-Resolution electrogenerated chemiluminescence microscopy for single-nanocatalyst imaging. J. Am. Chem. Soc. **143**, 18511–18518 (2021). https://doi.org/10.1021/jacs.1c07827
79. Dong, J., Xu, Y., Zhang, Z., et al.: Operando imaging of chemical activity on gold plates with single-molecule electrochemiluminescence microscopy. Angew. Chem. Int. Ed. **61**, e202200187 (2022). https://doi.org/10.1002/anie.202200187
80. Walther, A., Müller, A.H.E.: Janus particles: synthesis, self-assembly, physical properties, and applications. Chem. Rev. **113**, 5194–5261 (2013). https://doi.org/10.1021/cr300089t
81. Loget, G., Zigah, D., Bouffier, L., et al.: Bipolar electrochemistry: from materials science to motion and beyond. Acc. Chem. Res. **46**, 2513–2523 (2013). https://doi.org/10.1021/ar400039k
82. Zhu, M.J., Pan, J.B., Wu, Z.Q., et al.: Electrogenerated chemiluminescence imaging of electrocatalysis at a single Au-Pt janus nanoparticle. Angew. Chem. Int. Ed. **57**, 4010–4014 (2018). https://doi.org/10.1002/anie.201800706
83. Chen, M.-M., Xu, C.-H., Zhao, W., et al.: Observing the structure-dependent electrocatalytic activity of bimetallic Pd-Au nanorods at the single-particle level. Chem. Commun. **56**, 3413–3416 (2020). https://doi.org/10.1039/d0cc00185f
84. Xue, J.W., Xu, C.H., Zhao, W., et al.: Unveiling the dynamic electrocatalytic activity of online synthesized bimetallic nanocatalysts via electrochemiluminescence microscopy. Nano Lett. **24**, 4665–4671 (2024). https://doi.org/10.1021/acs.nanolett.4c01026
85. Wu, T., Liu, X., Liu, Y., et al.: Application of QD-MOF composites for photocatalysis: energy production and environmental remediation. Coord. Chem. Rev. **403**, 213097 (2020). https://doi.org/10.1016/j.ccr.2019.213097
86. Han, Z., Eisenberg, R.: Fuel from water: the photochemical generation of hydrogen from water. Acc. Chem. Res. **47**, 2537–2544 (2014). https://doi.org/10.1021/ar5001605
87. Fernando, K.A.S., Sahu, S., Liu, Y., et al.: Carbon quantum dots and applications in photocatalytic energy conversion. ACS Appl. Mater. Interfaces **7**, 8363–8376 (2015). https://doi.org/10.1021/acsami.5b00448
88. Chen, Y., Fu, J., Cui, C., et al.: In Situ visualization of electrocatalytic reaction activity at quantum dots for water oxidation. Anal. Chem. **90**, 8635–8641 (2018). https://doi.org/10.1021/acs.analchem.8b01935
89. Chen, Y., Zhao, D., Fu, J., et al.: In Situ imaging facet-induced spatial heterogeneity of electrocatalytic reaction activity at the subparticle level via electrochemiluminescence microscopy. Anal. Chem. **91**, 6829–6835 (2019). https://doi.org/10.1021/acs.analchem.9b01044
90. Wang, M., Langer, M., Altieri, R., et al.: Two-Dimensional layered heterojunctions for photoelectrocatalysis. ACS Nano **18**, 9245–9284 (2024). https://doi.org/10.1021/acsnano.3c12274

91. Zhu, H., Jiang, D., Zhu, J.-J.: High-resolution imaging of catalytic activity of a single graphene sheet using electrochemiluminescence microscopy. Chem. Sci. **12**, 4794–4799 (2021). https://doi.org/10.1039/d0sc06967a
92. Zhao, X., Li, Y., Cui, Y., et al.: Spatiotemporally and chemically resolved imaging of electrocatalytic oxygen evolution on single nanoplates of cobalt-layered hydroxide. J. Am. Chem. Soc. **145**, 20897–20906 (2023). https://doi.org/10.1021/jacs.3c06062
93. Fiorani, A., Han, D., Jiang, D., et al.: Spatially resolved electrochemiluminescence through a chemical lens. Chem. Sci. **11**, 10496–10500 (2020). https://doi.org/10.1039/D0SC04210B
94. Rebeccani, S., Wetzl, C., Zamolo, V.A., et al.: Electrochemiluminescent immunoassay enhancement driven by carbon nanotubes. Chem. Commun. (Camb.) **57**, 9672–9675 (2021). https://doi.org/10.1039/D1CC03457J
95. Zanut, A., Fiorani, A., Canola, S., et al.: Insights into the mechanism of coreactant electrochemiluminescence facilitating enhanced bioanalytical performance. Nat. Commun. **11**, 2668 (2020). https://doi.org/10.1038/s41467-020-16476-2
96. Kerr, E., Knezevic, S., Francis, P.S., et al.: Electrochemiluminescence amplification in bead-based assays induced by a freely diffusing Iridium(III) complex. ACS Sens **8**, 933–939 (2023). https://doi.org/10.1021/acssensors.2c02697
97. Dutta, P., Han, D., Goudeau, B., et al.: Reactivity mapping of luminescence in space: insights into heterogeneous electrochemiluminescence bioassays. Biosens. Bioelectron. **165**, 112372 (2020). https://doi.org/10.1016/j.bios.2020.112372
98. Han, D., Jiang, D., Valenti, G., et al.: Optics determines the electrochemiluminescence signal of bead-based immunoassays. ACS Sens **8**, 4782–4791 (2023). https://doi.org/10.1021/acssensors.3c01878
99. Guo, W., Zhou, P., Sun, L., et al.: Microtube electrodes for imaging the electrochemiluminescence layer and deciphering the reaction mechanism. Angew. Chem. Int. Ed. **60**, 2089–2093 (2021). https://doi.org/10.1002/anie.202012340
100. Pumera, M.: Electrochemistry of graphene: new horizons for sensing and energy storage. Chem. Rec. **9**, 211–223 (2009). https://doi.org/10.1002/tcr.200900008
101. Zhu, H., Jin, R., Jiang, D., et al.: Perturbation electrochemiluminescence imaging to observe the fluctuation of charge-transfer resistance in individual graphene microsheets with redox-induced defects. ACS Appl. Mater. Interfaces **11**, 46666–46670 (2019). https://doi.org/10.1021/acsami.9b14017
102. Wang, Q.H., Kalantar-Zadeh, K., Kis, A., et al.: Electronics and optoelectronics of two-dimensional transition metal dichalcogenides. Nat. Nanotechnol. **7**, 699–712 (2012). https://doi.org/10.1038/nnano.2012.193
103. Martincová, J., Otyepka, M., Lazar, P.: Is single layer MoS2 stable in the air? Chem. Eur. J. **23**, 13233–13239 (2017). https://doi.org/10.1002/chem.201702860
104. Zhu, H., Jin, R., Chang, Y.C., et al.: Understanding the synergistic oxidation in dichalcogenides through electrochemiluminescence blinking at millisecond resolution. Adv. Mater. **33**, e2105039 (2021). https://doi.org/10.1002/adma.202105039
105. Wilson, A.J., Marchuk, K., Willets, K.A.: Imaging electrogenerated chemiluminescence at single gold nanowire electrodes. Nano Lett. **15**, 6110–6115 (2015). https://doi.org/10.1021/acs.nanolett.5b02383
106. Ma, C., Wu, W., Li, L., et al.: Dynamically imaging collision electrochemistry of single electrochemiluminescence nano-emitters. Chem. Sci. **9**, 6167–6175 (2018). https://doi.org/10.1039/c8sc02251h
107. Bulbake, U., Doppalapudi, S., Kommineni, N., et al.: Liposomal formulations in clinical use: an updated review. In Pharmaceutics (2017)
108. Sercombe, L., Veerati, T., Moheimani, F., et al.: Advances and challenges of liposome assisted drug delivery. Front. Pharmacol. **6** (2015)
109. Ben Trad, F., Wieczny, V., Delacotte, J., et al.: Dynamic electrochemiluminescence imaging of single giant liposome opening at polarized electrodes. Anal. Chem. **94**, 1686–1696 (2022). https://doi.org/10.1021/acs.analchem.1c04238

110. Zhao, Y., Descamps, J., Ababou-Girard, S., et al.: Metal-Insulator-Semiconductor anodes for ultrastable and site-selective upconversion photoinduced electrochemiluminescence. Angew. Chem. Int. Ed. **61**, e202201865 (2022). https://doi.org/10.1002/anie.202201865
111. Zhao, Y., Yu, J., Xu, G., et al.: Photoinduced electrochemiluminescence at silicon electrodes in water. J. Am. Chem. Soc. **141**, 13013–13016 (2019). https://doi.org/10.1021/jacs.9b06743
112. Yu, J., Saada, H., Abdallah, R., et al.: Luminescence amplification at $BiVO_4$ photoanodes by photoinduced electrochemiluminescence. Angew. Chem. Int. Ed. **59**, 15157–15160 (2020). https://doi.org/10.1002/anie.202004634
113. Xue, J.W., Xu, C.H., Zhao, W., et al.: Photoinduced electrogenerated chemiluminescence imaging of plasmonic photoelectrochemistry at single nanocatalysts. Nano Lett. **23**, 4572–4578 (2023). https://doi.org/10.1021/acs.nanolett.3c01028
114. Dong, J., Lu, Y., Xu, Y., et al.: Direct imaging of single-molecule electrochemical reactions in solution. Nature **596**, 244–249 (2021). https://doi.org/10.1038/s41586-021-03715-9
115. Zhu, W., Dong, J., Ruan, G., et al.: Quantitative single-molecule electrochemiluminescence bioassay. Angew. Chem. Int. Ed. **62**, e202214419 (2023). https://doi.org/10.1002/anie.202214419
116. Yang, X., Hang, J., Qu, W., et al.: Gold microbeads enabled proximity electrochemiluminescence for highly sensitive and size-encoded multiplex immunoassays. J. Am. Chem. Soc. **145**, 16026–16036 (2023). https://doi.org/10.1021/jacs.3c04250
117. Liu, Y., Zhang, H., Li, B., et al.: Single biomolecule imaging by electrochemiluminescence. J. Am. Chem. Soc. **143**, 17910–17914 (2021). https://doi.org/10.1021/jacs.1c06673
118. Monopoli, M.P., Åberg, C., Salvati, A., et al.: Biomolecular coronas provide the biological identity of nanosized materials. Nat. Nanotechnol. **7**, 779–786 (2012). https://doi.org/10.1038/nnano.2012.207
119. Lesniak, A., Fenaroli, F., Monopoli, M.P., et al.: Effects of the presence or absence of a protein corona on silica nanoparticle uptake and impact on cells. ACS Nano **6**, 5845–5857 (2012). https://doi.org/10.1021/nn300223w
120. Xing, Z., Gou, X., Jiang, L.-P., et al.: An in situ investigation of the protein corona formation kinetics of single nanomedicine carriers by self-regulated electrochemiluminescence microscopy. Angew. Chem. Int. Ed. **62**, e202308950 (2023). https://doi.org/10.1002/anie.202308950

Exploring Multiphase Environments with Electrochemiluminescence

Brady R. Layman, Megan L. Hill, Daniel M. Carrel, and Jeffrey E. Dick

Abstract When examining multiphase systems, primarily ones with liquid|liquid and liquid|gas interfaces, many interesting physicochemical phenomena such as reaction rate acceleration and surprising reaction pathways are present. To probe these chemistries, Multiphase Electrochemiluminescence Microscopy can be used to visualize both the statics and dynamics of these systems. Here, we will describe the progress of the field from its beginning, before the inclusion of microscopy, to current advancements. Furthermore, how multiphase ECL microscopy has elucidated exciting physicochemical phenomena will be discussed. Finally, we will look toward the future to examine research gaps in this field.

Keywords Multiphase electrochemistry · Liquid–liquid interface · Bubbles · Droplets · Stochastic electrochemistry · Droplet electrochemistry

1 Introduction

In the past two decades, multiphase systems have captured the interest of the broad scientific community due to their prevalence in nature, usefulness in anthropogenic applications, and their interesting physicochemical properties. These unique properties include spontaneous formation of emulsions at the interface and reaction rate enhancement [1, 2]. Multiphase systems are often comprised of liquid|liquid and/or liquid|liquid|gas phase boundaries between immiscible (or sometimes partially miscible [3, 4]) liquids such as 1,2-dichloroethane (1,2-DCE) and water. Moreover, gases such as carbon dioxide (CO_2) and hydrogen (H_2) are commonly present in these systems and are involved in the curious reactions.

While a nice model system for reaction enhancement and curious chemistry, there are many places in nature where liquid|liquid phase boundaries are important. For example, biological cells take advantage of multiphase systems, namely liquid|liquid interfaces, for fundamental purposes in reaction rate enhancement through P Granule

B. R. Layman · M. L. Hill · D. M. Carrel · J. E. Dick (✉)
Department of Chemistry, Purdue University, West Lafayette, USA
e-mail: jdick@purdue.edu

F. Paolucci et al. (eds.), *Electrogenerated Chemiluminescence Microscopy and Imaging*,
https://doi.org/10.1007/978-3-032-14602-1_10

localization through controlled dissolution and demixing [5], mitosis, demixing [6], forming liquid|liquid boundaries [6–8], and increasing enzymatic reaction rate [9]. Other exciting applications of multiphase systems include organic synthesis such as "on water catalysis [10]," nanomaterial deposition [11–13], unexpected new methods of greenhouse gas remediation [14], ammonia synthesis [15], and the formation of strong reductive agents [16]. Large, complex questions that humans have pondered for millennia, such as the origin of life, have also implicated multiphase systems, in this case: sea spray [17].

There are a few established techniques that are used to study these liquid|liquid and liquid|gas interfaces. First, mass spectrometry is the instrument of choice for much of the community due to its sensitivity and selectivity [2, 18, 19]. However, it lacks spatial information when it comes to heterogeneities inside microdroplets. Other techniques include vibrational sum frequency generation spectroscopy [20–26]. Overall, ultrafast spectroscopy techniques have large potential for elucidating the strange and curious chemistry at these interfaces; however, they often have problems pushing into the solvent (namely the water phase) window [27].

Although these before-mentioned techniques all have their strengths, multiphase Electrochemiluminescence (ECL) Microscopy has the greatest potential to address the major research gap: how do we visualize heterogeneities in space in multiphase systems? Generally, this technique uses a radical coreactant-mediated ECL scheme. ECL has two special characteristics that are taken advantage of in these multiphase systems. First, the radical lifetime of the coreactant confines the emission based on its diffusion [28]. Additionally, the detector noise dictates the signal-to-noise ratio, since there are no excitation photons, such as in fluorescence, the signal acquisition can be quite low if an ultrasensitive camera is used [29].

Prior to the integration of microscopy in multiphase ECL, there was work performed by Dick and colleagues with oil-in-water emulsions: a common model system used in multiphase chemistry, as well as single entity- and stochastic electrochemistry [30–32]. In these experiments, oil-phase droplets were loaded with the luminophore (Rubrene) and coreactant (Tri-*N*-Propyl Amine (TPrA)) which was then suspended via sonication in an aqueous bulk phase. The ECL emission process is confined to the oil-phase droplets (Fig. 1a). On macroelectrodes, droplets were adsorbed onto the electrode, and the ECL intensity was correlated with both cyclic voltammetry (Fig. 1b) and alternating 0 and + 1.0 V versus Ag/AgCl steps with a square-wave potential form (Fig. 1d). Current transients were also measured for the 0 and + 1.0 V versus Ag/AgCl pulses (Fig. 1c). Additionally, on an ultramicroelectrode, collisions of these droplets and their corresponding current transients (Fig. 1e, black trace) were correlated with the intensity of the ECL signal (Fig. 1e, red trace).

Even without microscopy, these results offer a large step forward by performing ECL experiments under nanoconfinement. These early experiments laid the foundation for stochastic electrochemistry experiments correlated with the acquisition of ECL photons in confined environments.

This movement toward microscopy in the ECL multiphase community to include optical imaging, also parallels the ECL community at large. In the past *ca.* decade ECL techniques have been coupled with optical microscopy [33, 34] to offer spatial

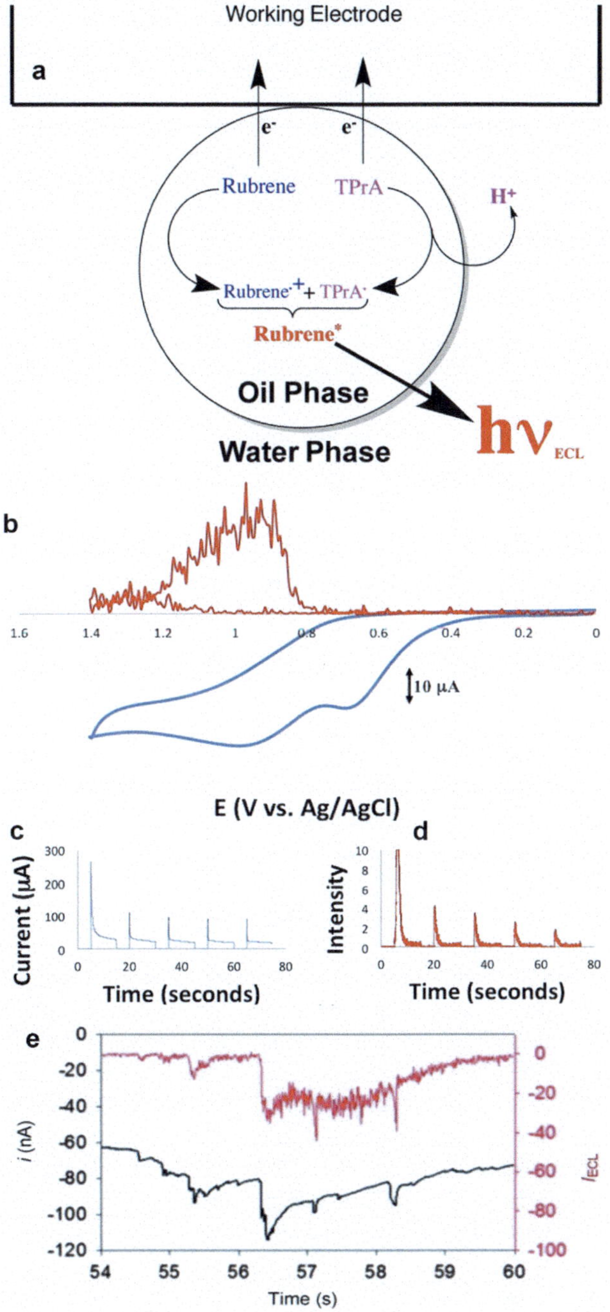

Working Electrode
a
e-
e-
Rubrene
TPrA
H+
Rubrene·+ + TPrA·
Rubrene*
Oil Phase
Water Phase
hν ECL
b
1.6
1.4
1.2
1
0.8
0.6
0.4
0.2
0
10 μA
E (V vs. Ag/AgCl)
c
Current (μA)
300
200
100
0
0
20
40
60
80
Time (seconds)
d
Intensity
10
8
6
4
2
0
0
20
40
60
80
Time (seconds)
e
i (nA)
0
-20
-40
-60
-80
-100
-120
54
55
56
57
58
59
60
Time (s)
I ECL
0
-20
-40
-60
-80
-100

◂**Fig. 1** **a** Schematic and proposed oxidative coreactant ECL mechanism for Rubrene and Tri-*N*-Propyl Amine (TPrA) in a Toluene droplet adsorbed to an ultramicroelectrode. The Rubrene and TPrA undergo co-oxidation; then the oxidized species undergo a homogeneous charge-transfer reaction which produces an excited state of Rubrene that can emit a photon. **b** Cyclic voltammetry of Rubrene and TPrA organic droplets adsorbed onto a glassy carbon macroelectrode (blue curve) correlated with the ECL emission (red trace). **c** Current versus time traces as potential is modulated from 0 to + 1.0 V. **d** Correlated with **c**. ECL intensity versus time traces. **e** Collision events of same rubrene system from **a** to **d** The black trace is current versus time and the red trace is the ECL intensity versus time. Data from adapted from Ref. [30] and [31]. (**a**–**d**) Adapted with permission from 30 Copyright 2014 American Chemical Society. **e** reproduced from Ref. [31] with permission from John Wiley and Sons, Copyright © 2014 WILEY–VCH Verlag GmbH & Co. KGaA, Weinheim

chemical reactivity information confined to the surface of the electrode (due to the short radical lifetimes) with applications in nanomaterial catalysis [35], forensics [36], and biology [37].

Here, we will outline the experimental instrumentation and work done so far by the ECL multiphase community, and end with a perspective toward the future.

2 Multiphase ECL Microscopy Instrumentation

In many experimental procedures discussed, the setup used for data acquisition was very similar, so a general schematic will be outlined (Fig. 2) [38–42]. By utilizing a micro-positioner, one can view the bottom of the working electrode (usually a glassy carbon inlaid disk macroelectrode) and adjust the position to best fit with the goals of the trial. In conjunction with this, a reference electrode with an agarose salt bridge and counter electrode, usually a glassy carbon rod, is also included in the system.

Due to the wide variety of potential avenues of inquiry in these electrochemical processes, different configurations can be used to best tailor acquisition to the goal of the experiment. For example, the working electrode can be placed upright, so the inlaid disk of the electrode is viewed "face-on." The electrode can also be placed in a "profile" configuration, allowing for droplets to be viewed from the side and the ECL emission layer accessible for imaging [38, 41]. Additionally, an optically transparent yet electrically conductive Indium-Tin-Oxide (ITO) glass electrode can be used in place of the glass coverslip at the bottom of the electrochemical cell [39]. This allows for the bottom of the droplets and ECL emission layer to be imaged. The figure below depicts the varying setups described in which the aqueous bulk phase containing Tris(bipyridine)ruthenium(II) ($[Ru(bpy)_3]^{2+}$) is colored light orange, with organic inclusions adsorbed to each working electrode surface.

Depending on what information is needed, different objectives or cameras can be used. Usually a 4–6 × objective is used, as one can easily view the entire electrode surface, but 40 × or oil-immersed 63 × objectives have also been used. The camera used for acquisition is more dependent on the process at hand than the objective. There are two main options for imaging: an electron-multiplying charge capture device (EMCCD) and complementary metal oxide semiconductor (CMOS) camera.

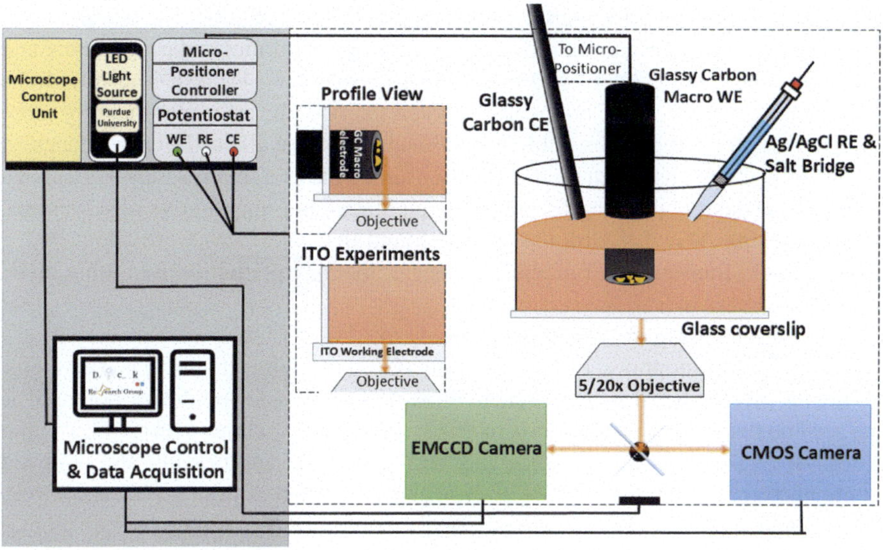

Fig. 2 Schematic depicting the inverted microscopy setups most often used. The electrochemical cells have three configurations, each offering a unique perspective on the system. A glassy carbon inlaid disk electrode is used in the "glass coverslip" and "profile view" experiments. An Ag/AgCl electrode with agarose salt bridge is used as the reference electrode, and the counter electrode is a glassy carbon rod. In the "ITO Experiments" panel, a panel of Indium-Tin-Oxide(ITO) coated glass is used in place of the coverslip at the bottom of the cell, and the ITO is used as the working electrode (reference and counter electrodes are kept the same). In all schematics, the organic/oil inclusions are colored yellow and the emission layer of the ECL is colored red orange. The propagation of light is represented by orange arrows, where the light is then directed by a mirror to one of the two cameras after traveling through the objective. All components in the schematic are drawn for clarity, not to scale. Adapted with permission from [30]. Copyright 2022 American Chemical Society

EMCCD cameras are best used in time-resolved situations, in which exposure time needs to be less than half a second. They also excel in low-intensity ECL reactions, as they have single-photon sensitivity in principle, though their spatial resolution lacks slightly when compared to the CMOS. Alternatively, although they have improved spatial resolution, CMOS cameras require a longer exposure time for imaging, ranging from one to ten seconds per image.

3 Proof-of-Concept: Measuring the Thickness of Interfaces

Since 1992, the interface between 1,2-DCE and water (1,2-DCE|H_2O) has been extensively studied, initially by Ilan Benjamin, who used nonlinear spectroscopy and molecular dynamics simulations to predict the characteristics of the interface [43]. This work led Benjamin to the conclusion that the boundary was "sharp" in nature, containing spike-like capillaries as long as 0.8 nm. In 2004, Dave Walker and colleagues coupled surface vibrational spectroscopy with molecular dynamics and concluded that the 1,2-DCE|H_2O interface is anisotropic [44, 45]. These findings revealed that the molecular boundary between the two phases is significantly more disorganized than other immiscible liquid boundaries like alkane|H_2O interfaces.

This information has been previously used alongside ECL microscopy to visualize the phase boundary, allowing for not only qualitative but quantitative observations. After analyzing $N = 128$ droplets, it was found that interfaces were about 9 ± 3 microns thick; four orders of magnitude greater than the prediction Benjamin made in 1992 [46].

A common reaction mechanism used in these ECL systems includes a $[Ru(bpy)_3]^{2+}$ luminophore and an oxalate $(C_2O_4)^{2-}$ coreactant [47]. The light-producing mechanism for these reactions is as follows. An oxidative potential is first applied to the system, which co-oxidizes $[Ru(bpy)_3]^{2+}$ along with $(C_2O_4)^{2-}$ into $[Ru(bpy)_3]^{3+}$ and $(C_2O_4)^{\cdot-}$, respectively (Eq. 1 and 2). Additionally, the oxidized luminophore can potentially act as a strong oxidizing agent in this system as well, which can also oxidize the coreactant (Eq. 3). After this, the oxalate radical spontaneously (an often cocerted step) decomposes into CO_2 and $CO_2^{\cdot-}$, and this resulting radical acts as a strong reducing agent toward the previously oxidized luminophore (Eqs. 4 and 5, respectively). The now reduced +2-oxidaition state luminophore is left in an electronically excited state, which radiatively relaxes, and emits a photon at about 620 nm wavelength (Eq. 6) (Fig. 3e).

$$\left[Ru(bpy)_3\right]^{2+} - e^- \rightarrow \left[Ru(bpy)_3\right]^{3+} \tag{1}$$

$$(C_2O_4)^{2-} - e^- \rightleftharpoons (C_2O_4)^{\cdot-} \tag{2}$$

$$(C_2O_4)^{2-} + \left[Ru(bpy)_3\right]^{3+} \rightarrow (C_2O_4)^{\cdot-} + \left[Ru(bpy)_3\right]^{2+} \tag{3}$$

$$(C_2O_4)^{\cdot-} \rightarrow CO_2^{\cdot-} + CO_2 \tag{4}$$

$$CO_2^{\cdot-} + \left[Ru(bpy)_3\right]^{3+} \rightarrow \left[Ru(bpy)_3\right]^{2+*} + CO_2 \tag{5}$$

$$\left[Ru(bpy)_3\right]^{2+*} \rightarrow \left[Ru(bpy)_3\right]^{2+} + h\nu(620\,\text{nm}) \tag{6}$$

To image the two-dimensional contact between the two phases, one can take advantage of solubility behaviors of a selected luminophore and coreactant to have the reaction proceed selectively in the aqueous phase (Fig. 3d). The ECL signal produced by the co-oxidation of $[Ru(bpy)_3]^{2+}$ and $C_2O_4{}^{2-}$ at + 1.3 V provides a significant contrast between the aqueous and oil phases. When this is combined with optical microscopy, the sizes of the relevant droplets can be quantified (Fig. 3f). The

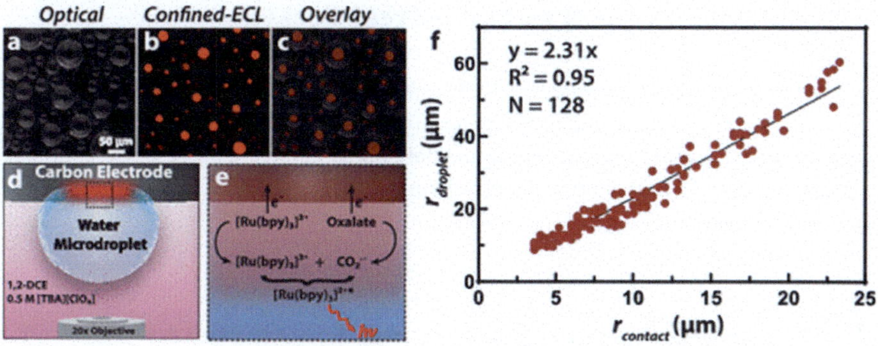

Fig. 3 **a** Brightfield micrograph showing adsorbed water microdroplets that contain the $[Ru(bpy)_3]^{2+}$ luminophore and the oxalate coreactant on a glassy carbon inlaid disk macroelectrode. Scale bar is 50 μm. **b** ECL micrograph of the same system shown in (**a**). **c** Overlay of (**a**) and (**b**). **d** Schematic where ECL signal is confined to the microdroplet. **e** ECL reaction between the luminophore and coreactant. **f** Linear trend revealed when plotting the radius of the microdroplets as a function of their respective contact radii. Adapted with permission from Ref. [46]. Copyright 2020 American Chemical Society

ECL micrograph (Fig. 3b) and its corresponding brightfield image (Fig. 3a) can be overlaid (Fig. 3c) for further analysis of the droplets and their boundaries.

When examining these small systems, it is crucial to estimate the length of time light acquisition will last. To approximate this lifetime, the average length of bulk electrolysis can be used. Assuming 99% electrolysis, (Eq. 7) can be used to relate the two times:

$$i(t) = i_0 e^{-\frac{m_0 A}{V} t} = e^{-\frac{4Dr}{V} t} \tag{7}$$

Given that $i(t)$ is the current at time t, i_0 is the initial current, m_0 is the mass transfer coefficient, A is the average contact area of the microdroplet, r is the contact radius, D is the diffusion coefficient of the $[Ru(bpy)_3]^{2+}$, and V is the average droplet volume, estimated from the contact area.

For the droplets used in this study, electrolyzing about 99% of the material in the droplet took about 30 s, and for proper analysis, a 10 s exposure on a CMOS camera was used to ensure presentable visual contrast and light collection.

To measure the droplet radius, the ECL signal resulting from this study was consistent with the brightfield/optical image, and in terms of resolution, the ECL micrograph proved to be more effective. This was particularly useful when studying the contact radii for much smaller droplets.

4 An Interesting Observation: Solvent Pockets

After establishing the usefulness of ECL microscopy in determining the thickness of the 1,2-DCE|H_2O interface experimentally, other observations led to even more questions. Firstly, some droplets exhibited dark areas in an otherwise homogenously emitting droplet. The dynamic physicochemical phenomena of solvent inclusions was next to be investigated.

4.1 Characterization and Formation of Solvent Pockets

To map these solvent pockets (or sometimes called "entrapments"), droplets were adsorbed to the surface of the glassy carbon working electrode on an inverted optical microscope (Fig. 4a) [48]. As time passes, the emulsion settles and separates, allowing one to take brightfield images [Fig. 4a(B) and a(E)] of the surface of the electrode without scattering caused by droplets suspended in bulk solution. Then, ECL microscopy [Fig. 4a(C)] was performed, and the droplets light up while the solvent entrapments appear dark, since the $(C_2O_4)^{2-}$ cannot partition into the 1,2-DCE. Line intensity analysis (4a(F) and a(H), respectively) was performed for a single selected droplet to compare the brightfield (Fig. 4a(E)) and ECL micrographs (Fig. 4a(G)), respectively. The ECL image displays a large intensity decrease with high spatial resolution, compared to no difference present in the brightfield line intensity plot. This set of experiments demonstrates how the heterogeneous phase boundaries present in these complex multiphase systems can be fully visualized with this imaging technique at the various triple phase boundaries.

ECL microscopy also has the ability to visualize the formation of these solvent inclusions. To accomplish this, Voci and colleagues utilized the same oxidative oxalate coreactant system to watch neighboring water droplets coalesce [42]. The droplets, as described above, were adsorbed onto the glassy carbon inlaid disk macroelectrode. ECL imaging could take place once the emulsion settled. Since the turbidity of the solution and rate of coalescence is inversely related, the probability of capturing a coalescence event decreases with time. This is an experimental trade-off in these trials that must often be accounted for.

Figure 4b shows a series of ECL micrographs that map the formation of solvent inclusions during the merging of two water droplets. To begin, the two water droplets exhibited a discrete interface, showing a reflective ECL signal (Fig. 4b(A)). To mark the original locations of the water droplets, the red dashed circles provide this information in each panel. Then, small pools of solvent are formed as the droplets begin to coalesce (Fig. 4b(A) to b(C)). As the series of images continue, the amorphous solvent inclusions appear dark, while the water droplet phase emits light (Fig. 4b(E) to b(K)).

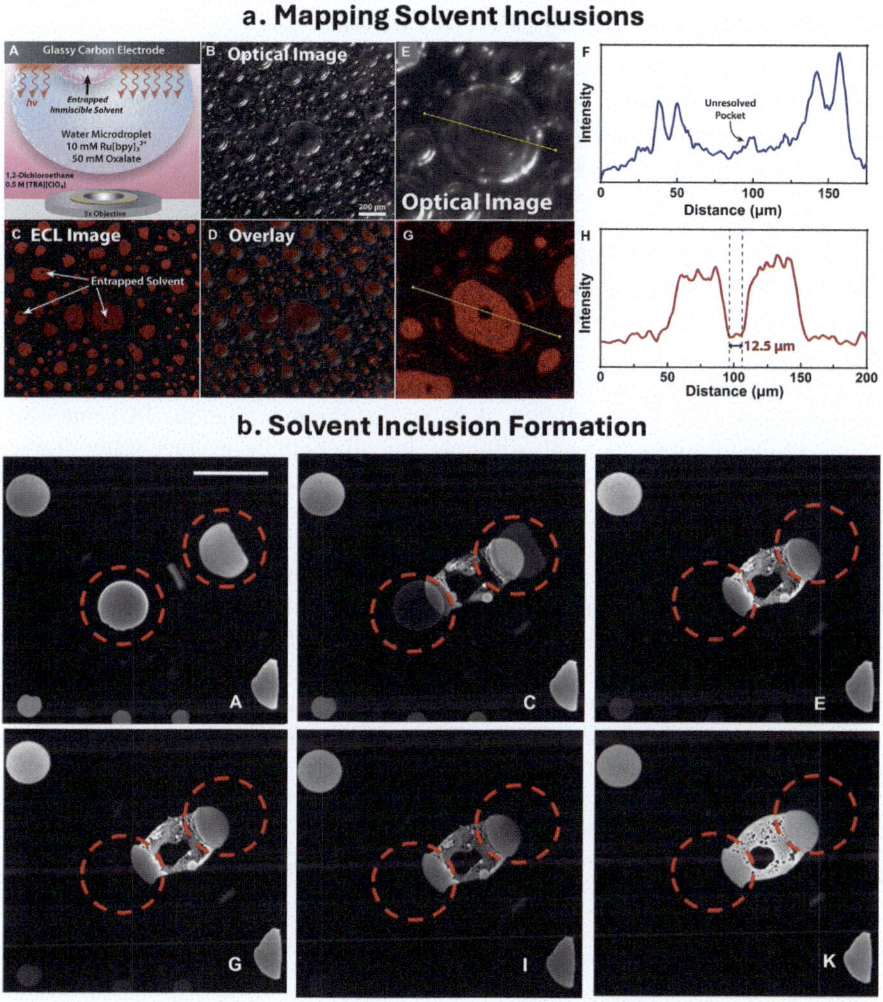

Fig. 4 **a** Mapping Solvent Inclusions (**A**) Schematic of droplets **adsorbed** to a glassy carbon macroelectrode with the luminophore and coreactant confined to the droplet phase. The red arrows depict the ECL emission. (**B**) Optical brightfield micrograph of droplets adsorbed to the working electrode. Scale bar is 200 µm. (**C**) ECL micrograph corresponding to (**B**) where the solvent entrapments appear as black spots within the red emission regions. (**D**) Overlay of (**B**) and (**C**). (**E**) Brightfield micrograph of a selected adsorbed droplet. (**F**) Line scan corresponding to the (**E**). (**G**) ECL micrograph of the same selected droplet as in (**E**). (**H**) Corresponding line scan to (**G**). **b** Solvent Inclusion Formation (**A**) two droplets that are adsorbed to a glassy carbon macroelectrode. (**C–I**) The water droplets merge and solvent pockets form. (**K**) The droplets have fully coalesced. A 10 s. exposure on a CMOS camera was used to acquire these images. (**a**) Adapted with permission from Ref. [48] Copyright 2021 American Chemical Society. (**b**) Reproduced from Ref. [42] with permission from the Royal Society of Chemistry

4.2 Physical Processes of Solvent Pockets

In Fig. 4b(K) the boundaries of the solvent inclusions seemed to be brighter than the rest of the emission region. In order to investigate this observation, ECL microscopy was once again carried out to determine mechanistically what was occurring to cause this differential in ECL intensity [41].

Aqueous droplets containing $[Ru(bpy)_3]^{2+}$ and $(C_2O_4)^{2-}$ were adsorbed to a glassy carbon macroelectrode. Contrasted to previous experiments, the emulsion was allowed to settle to the point that there was a single large water droplet covering the electrode in its entirety. Patterned across the electrode's surface were solvent inclusions.

The ECL intensity profile (Fig. 5a) of the experiment at different locations on the macroelectrode (Fig. 5b) is offered, showing that through time, the ECL intensity profiles of the locations are vastly different. Initially both the average water droplet and 1,2-DCE|water interface ECL intensity was similar; however, by 400 s. the former had decreased to the camera's noise level. Moreover, at the interface there was sustained ECL emission for more than 600 s.

CO_2 has a three order of magnitude increase in solubility in 1,2-DCE compared to water. Therefore, gaseous or solvated CO_2 will selectively partition into the organic solvent inclusion phase preferentially compared to the aqueous droplet phase. Furthermore, the flux profiles will then favorably promote the exit of the CO_2 from the apex of the interface (Fig. 5c). This increased mass transport, from the solvent inclusions acting essentially as micropumps, at the 1,2-DCE|Water interface is responsible for the sustained ECL present at the interface.

5 Imaging Both Sides of a Single Interface

In all the aforementioned work, the ECL imaging was limited to one phase of the multiphase system. Specifically, for multiphase systems consisting of liquid|liquid boundaries, ECL emission was restricted to either the aqueous phase, or the organic phase, but never both. By limiting the ECL emission to one phase, more detailed studies of these multiphase systems, such as probing the time-dependent dynamics of phase boundaries, can be difficult to achieve. However, Layman and co-workers overcame this restriction by generating light on both sides of a liquid|liquid interface by simply playing with the solubility of the luminophore [49]. This technique, coined "phase-resolved electrochemiluminescence," made use of the slight solubility (*ca. 10%)* of $Ru(bpy)_3Cl_2$ in 1,2-DCE as compared to water to enable $[Ru(bpy)_3]^{2+}$ to act as the luminophore in both phases. Moreover, by carefully choosing coreactants that have negligible solubility in the opposite phase, namely Sodium Oxalate [41, 46, 48] ($Na_2C_2O_4$) and Benzoyl Peroxide [50] (BPO), for the aqueous and organic phases, respectively, one can enable ECL imaging of both phases with use of a single luminophore. Temporal control over when each phase emitted photons was

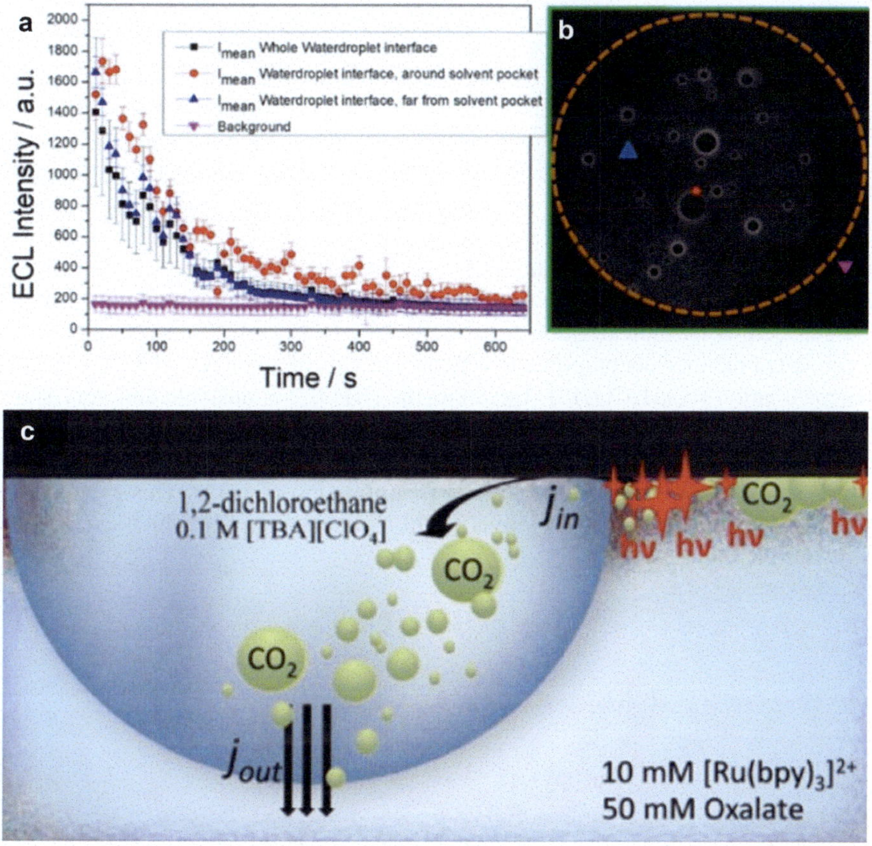

Fig. 5 **a** ECL intensity-time curves for different locations in the multiphase environment. **b** ECL micrograph of the macroelectrode with sample locations given in (**a**). **c** Schematic of the CO_2 fate inside the aqueous droplets where the CO_2 is electrogenerated from the oxidation of $(C_2O_4)^{2-}$ in the droplet aqueous phase. The CO_2 then partitions into the solvent inclusion and exits at the apex interface. Reprinted from [41]: Journal of Colloid and Interface Science, 661, Silvia Voci, Kathryn J. Vannoy, Jeffrey E. Dick, Femtoliter oil droplets act as CO_2 micropumps for uninterrupted electrochemiluminescence at the water|oil interface, 853–860, Copyright 2024, with permission from Elsevier

attained by selecting coreactants that undergo ECL in opposite regimes. The water-soluble coreactant, oxalate, emits light in the oxidative (anodic) regime, while the 1,2-DCE-soluble coreactant, BPO, undergoes ECL in the reductive (cathodic) regime. By alternating the potential between the two different ECL regimes, temporally controlled ECL imaging was carried out on both sides of a liquid|liquid interface.

In these experiments, the aqueous phase contained 10 mM of $[Ru(bpy)_3]^{2+}$ and 50 mM of $(C_2O_4)^{2-}$ within 100 mM KCl, while the organic phase contained only 50 mM of BPO within 100 mM Tetra-Butyl Ammonium Perchlorate (TBAP) in 1,2-DCE, initially. After the organic phase and the aqueous phase came into contact via the emulsification of 300 μL of the aqueous phase and 5 mL of the organic phase, the $[Ru(bpy)_3]^{2+}$ from the aqueous phase partitioned into the organic phase over the 45 min waiting period before ECL was performed to ultimately enable the phase-resolved analysis. Chronoamperometry with a square waveform was performed, pulsing in increments of 5 s, between + 1.5 V versus Ag/AgCl and 1.5 V for 50 cycles. The anodic ECL was the aqueous oxalate/$[Ru(bpy)_3]^{2+}$ ECL reaction (with the mechanism shown in Eq. 1 through Eq. 6), which has a mass-transfer limited current for the oxidation of oxalate occurring at + 1.5 V that was verified with cyclic voltammetry. The cathodic ECL was the BPO/$[Ru(bpy)_3]^{2+}$ ECL reaction, which has its mass-transfer limited potential at 1.5 V. In this reaction, the BPO is first reduced to its radical anion form ($(BPO)^{\cdot -}$) (Eq. 8) that subsequently (often in concerted manner)decomposes to a Benzoate radical anion ($(C_6H_5CO_2)^{\cdot -}$) and a Benzoate anion ($(C_6H_5CO_2)^{-}$) (Eq. 9) as the $(BPO)^{\cdot -}$ is unstable and very short-lived. Concurrently, the luminophore, $[Ru(bpy)_3]^{2+}$, is also reduced to $[Ru(bpy)_3]^{+}$ (Eq. 10) which undergoes a homogeneous charge-transfer reaction with the Benzoate radical anion to form the excited state of $[Ru(bpy)_3]^{2+*}$ (Eq. 11). This excited state of the luminophore can then radiatively emit a photon (Eq. 12), which is measured as part of the ECL emission.

$$\mathrm{BPO} + e^{-} \rightarrow (\mathrm{BPO})^{\cdot -} \tag{8}$$

$$(\mathrm{BPO})^{\cdot -} \rightarrow (C_6H_5CO_2)^{\cdot -} + (C_6H_5CO_2)^{-} \tag{9}$$

$$\left[\mathrm{Ru(bpy)_3}\right]^{2+} + e^{-} \rightarrow \left[\mathrm{Ru(bpy)_3}\right]^{+} \tag{10}$$

$$\left[\mathrm{Ru(bpy)_3}\right]^{+} + (C_6H_5CO_2)^{\cdot -} \rightarrow \left[\mathrm{Ru(bpy)_3}\right]^{2+*} + (C_6H_5CO_2)^{-} \tag{11}$$

$$\left[\mathrm{Ru(bpy)_3}\right]^{2+*} \rightarrow \left[\mathrm{Ru(bpy)_3}\right]^{2+} + h\nu(620\,\mathrm{nm}) \tag{12}$$

As these reactions are in the opposite potential regime, with their max intensities at + 1.5 V and – 1.5 V, for the aqueous and organic ECl reactions, respectively, ECl emission was observed on both sides of the liquid|liquid interface in alternating 5 s pulses. It is important to note that prior to the initial pulsing between the anodic and cathodic regimes, a 5 s quiet time at 0 V was held.

To set up the experiments, the emulsion of the droplet, aqueous phase and the bulk, organic phase was placed into an electrochemical cell with a glassy carbon inlaid disk electrode in it, and the emulsion was left to settle for 45 min. Before ECL was performed, a brightfield image (Fig. 6a) and a fluorescence micrograph (Fig. 6b) were taken of the settled droplets on the electrode. The brightfield is advantageous because it allows for the imaging of solvent inclusions within the droplets, while fluorescence microscopy provides information on the prevalence of the luminophore within the phases. Then, ECL was carried out. As shown in Fig. 6c, the aqueous phase emitted light in the first 5 s (Fig. 6ci), followed by the organic phase for the next 5 s

(Fig. 6cii). The white dashed line seen in Fig. 6cii highlights an optical effect that is unique to this system. In this box, there appear to be three stripes of higher intensity ECL emission. The stripes to the left and the right are reflections of photons coming off the nearby droplets, while the center stripe is the junction between the electrode surface and the droplets where the light is thought to originate. This optical artifact reveals that, by reflecting light from objects far from the electrode, ECL can be taken to the third dimension to yield more information about the system of interest.

Additionally, an interesting physical phenomenon elucidated by this ECL technique is the mapping of dynamic events such as water droplet coalescence from both sides of the interface in Fig. 6e. In Fig. 6d and Fig. 6e, two water droplets coalesce over time, which was visualized via the phase-resolved ECL. In Fig. 6d(i), which is an overlay of the ECL micrographs of the aqueous, droplet phase and bulk, organic phase, at time 26.964 s, the white circle brings attention to an aqueous droplet that at time 56.817 s coalesces with the larger, nearby aqueous droplet [Fig. 6d(ii)]. This coalescence event is evidenced by the disappearance of the aqueous, anodic ECL and the appearance of the organic, cathodic ECL emission. Moreover, this area seen in Fig. 6d is probed more deeply in Fig. 6e. At time 0 s. in Fig. 6e, two separate aqueous droplets are seen, in the anodic regime, separated by an interface that is highlighted with the white dashed line. Then, at 1.605 s to 3.531 s, in the cathodic regime, the aqueous droplets slowly start to merge as evidenced at 4.173 s where the organic phase is imaged exhibiting ECL where the water droplet once was, indicating that the movement of the water droplet away from that area freed up space on the electrode surface to drive the BPO/$[Ru(bpy)_3]^{2+}$ ECL. This emission in the cathodic regime grows in 4.494 and 4.815 s. At 6.420 and 7.383 s, the anodic regime supports the droplet merging event as the droplet ECL emission has been significantly reduced. By 13.805 s, in the cathodic regime, approximately half of the droplet has merged with the other aqueous droplet, as demonstrated by the area of the dashed white circle, which represents the original droplet area, is now displaying the organic phase cathodic ECL emission.

Another method to achieve multiphase ECL imaging in which both phases emit light is to employ multiple luminophores within the system. Previously, Voci and colleagues demonstrated the use of various luminophores to attain multicolor ECL emission [51]. They then expanded upon this work to develop the innovative "bipolar electrochemiluminescence" in which an ECL processes can occur independently from the opposite reaction. Bipolar ECL is an extension of bipolar electrochemistry, which is advantageous due to its ability to both drive electrochemical oxidation and reduction on a single electrode and because it is intrinsically "wireless." [52–56] Fu and co-workers developed bipolar ECL by placing a Janus microbead (an amphiphilic glassy carbon microbead) at a liquid|liquid interface so that once it was polarized by an electric field, it triggered two different ECL reactions, one in the aqueous phase and one in the organic phase [57]. Moreover, the ECL reactions from the water phase and the organic phase result in different color emissions, blue (425 nm) and red (620 nm), respectively, allowing for further phase resolution, an advancement from Layman and colleagues' previous work.

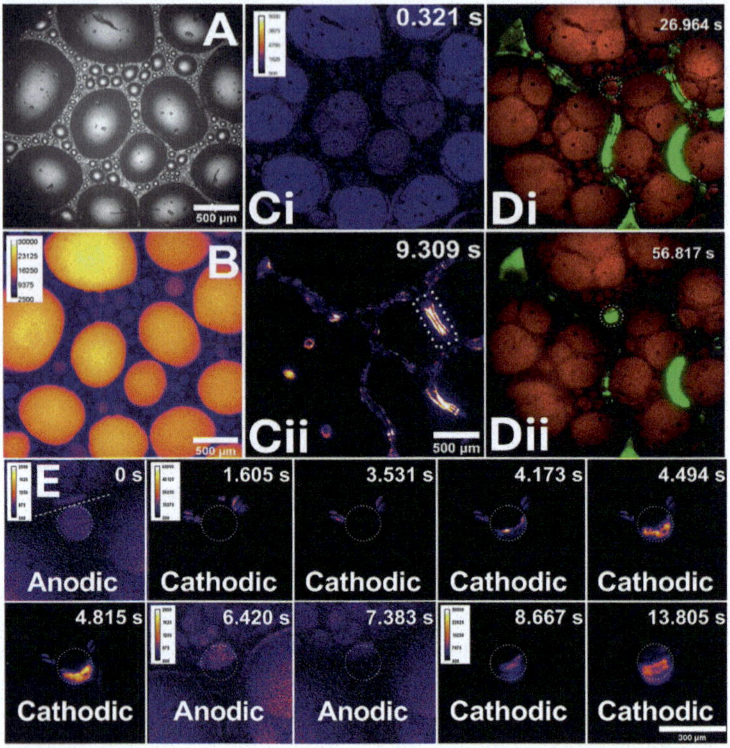

Fig. 6 **a** Brightfield micrograph of the water droplets settled onto the glassy carbon inlaid disk electrode within the bulk organic phase. **b** Fluorescence of the water droplets with a provided calibration bar. The more intense the fluorescence, the higher $[Ru(bpy)_3]^{2+}$ concentration. **ci** Anodic ECL emission from the aqueous, droplet phase at $t = 0.321$ s into the chronoamperometry. This ECL image was taken at + 1.5 V versus Ag/AgCl within the first pulse (5 s) of the 50 s experiment. **cii** Cathodic ECL emission from the organic, bulk phase at $t = 9.309$ s occurs during the second pulse (between 5 and 10 s) at 1.5 V. The white dashed box highlights the interesting ECL emission where the central line emits the ECL and the lines next to it are interfaces from which the light reflected off. **d** Aqueous, droplet phase (red) and bulk, organic phase (green) overlaid ECL signals taken during a chronoamperometry experiment. As each pulse is 5 s, and, therefore, each pulse collects a total of 15 images, for consistency, the 7th image of each pulse was represented with the time indicating the point within the entire experiment the organic phase image was taken. **di** At time 26.964 s into the experiment, a small water droplet can be seen next to a larger one, as pointed out with the white dashed circle around it. **dii** Between time 26.964 and 56.817 s, this small water droplet merged with the larger one next to it. At $t = 56.817$ s in this image, the white dashed circle where the water droplet once was is now the bulk, organic phase. **e** A series of ECL micrographs showing the coalescence of two water droplets. All panels labeled "Anodic" had an applied potential of + 1.5 V and showed ECL emission from the aqueous droplets. All "Cathodic" panels were taken at a 1.5 V potential and depicted the bulk, organic phase ECL emission. The calibration scales demonstrate the intensity of the emissions, and the white dashed circle that remains constant in each image represents the original location of the small droplet that coalescence. The straight, dashed, white line in the first panel ($t = 0$ s) highlights the interface between the small and large droplets that merged. Adapted with permission from Ref. [49] Copyright 2023 American Chemical Society

In these experiments, the amphiphilic Janus microbeads were placed at an aqueous|organic interface. The Janus microbeads are capable of being stabilized at a liquid|liquid interface due to them having both hydrophobic (the portion of the bead within the organic phase) and hydrophilic (the part of the bead within the aqueous phase) regions, the ratio of which can be changed during the preparation of the beads. The aqueous phase of these experiments was composed of 5 mM Luminol [58, 59], 50 mM H_2O_2, and 0.1 M NaOH, for which this ECL reaction is driven by the oxidation of both the luminophore, Luminol, and the coreactant, H_2O_2. Briefly, in this reaction, first, the hydrogen peroxide, H_2O_2, is electrochemically oxidized into a superoxide radical ($O_2^{\cdot-}$) (Eq. 13), while luminol (LH) is also electrochemically oxidized into a reactive intermediate ($L^{\cdot-}$) (Eq. 14) that immediately reacts with the $O_2^{\cdot-}$ to form the excited-state 3-Aminophthalate Dianion (AP^{2-*}) (Eq. 15). This dianion then emits a photon at 425 nm (blue light) when it relaxes back down to the ground state (Eq. 16).

$$H_2O_2 - e^- - 2H^+ \rightarrow O_2^{\cdot-} \tag{13}$$

$$LH - e^- - 2H^+ \rightarrow L^{\cdot-} \tag{14}$$

$$L^{\cdot-} + O_2^{\cdot-} \rightarrow AP^{2-*} \tag{15}$$

$$AP^{2-*} \rightarrow AP^{2-} + h\nu(425\,\text{nm}) \tag{16}$$

The organic phase was comprised of 1 mM $Ru(bpy)_3(PF_6)_2$, 20 mM BPO, and 20 mM Tetra-Butyl Ammonium Hexafluorophospate ($TBAPF_6$) in 1,2-DCE. This ECL emission emits photons at 620 nm (red light), and its mechanism is described in Eqs. 8–12.

Initially, in this study, the ratio of the hydrophilic and hydrophobic parts of the bead was kept equal, stabilizing the bead on the liquid|liquid interface so that the bead area is nearly equal in each phase, as shown in Fig. 7b. Then, an electric field was applied to polarize the microbead, which then initiates the ECL by triggering the electrochemical oxidation and reduction reactions in the aqueous and the organic phases, respectively. In Fig. 7a, the ECL reaction scheme is shown for an electric field that is applied parallel to the liquid|liquid interface, with the ECL micrographs with increasing electric field strength depicted in Fig. 7c with the white, dashed line indicating the location of the bead and the liquid|liquid interface. Separately, Fig. 7d conveys the scheme for when the electric field is applied perpendicular to the interface, with the corresponding ECL images, again with increasing electric field strength, shown in Fig. 7e. It is important to note that over the course of these experiments, the bead is stabilized at the interface so that there is no change in bead position over the course of the experiments. For both Fig. 7c and e, the same trend was observed where under lower electric field strengths (3.8 V/cm and 4.3 V/cm, respectively), ECL is only seen in the extremity of the bead that is located within the bulk, aqueous solution. As the luminol/H_2O_2 ECL reaction occurs at a lower, anodic potential than the cathodic, BPO/$[Ru(bpy)_3]^{2+}$ ECL reaction, it follows that the emission from the organic phase only is seen at higher electric field strengths. Similar to the aqueous ECL reaction, the organic ECL emission is initially only seen at the ends of the bead farthest from the interface. For both the aqueous and organic ECL emissions in Fig. 7e, a trend can be noted in which ECL emission moves closer to the interface as the electric field strength increases, likely due to the correlated increase in the polarization region of the bead itself. Additionally, in the aqueous phase of both Fig. 7c and e, bubble nucleation can be seen as a result of water oxidation being driven at the higher electric field potentials.

An interesting difference between the ECL emissions shown in Fig. 7c and e is that in Fig. 7e, the black region in the center of the bead can be clearly seen, but in Fig. 7c, it entirely disappears. This black region is hypothesized to be the "interfacial width" of the liquid|liquid interface, which is thought to arise due to nano- and microemulsions being formed at the interface as a result of the high, perpendicular electric field inducing ion migration across the phase boundary. These emulsions are theorized to inhibit ECL reactions, manifesting optically as the black region. However, when an electric field is applied parallel to the interface, as in Fig. 7c, the ion transfer across the interface may be significantly reduced if not suppressed entirely, which could be why the black region is no longer seen.

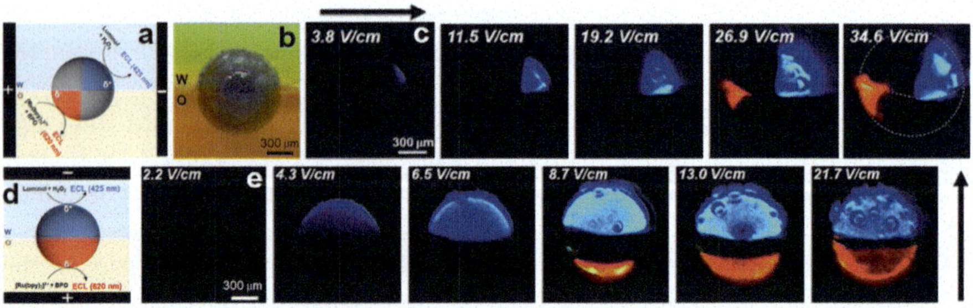

Fig. 7 **a** Schematic of the bipolar ECL emission when the electric field is parallel to the water|organic interface. The amphiphilic Janus microbead ($r = 450$ μm) is stabilized at this interface. The black rectangles represent the feeder electrodes, while the "*W*" denotes the aqueous phase, and the "*O*" symbolizes the organic phase. **b** White-light image of the amphiphilic Janus microbead at the liquid|liquid interface. **c** Panel of ECL images of the microbead under increasingly higher electric field strengths, indicated in the top right corner of the image, where the electric field is parallel to the liquid|liquid interface. The arrow indicates the direction of the electric field. **d** Schematic of the bipolar ECL emission when the electric field is perpendicular to the water|organic interface. **e** Panel of ECL images of the microbead, oriented as seen in the schematic in this figure **d**, with the electric field increasing in magnitude from 2.2 to 21.7 V/cm (denoted in the upper, right corner of the image). The arrow indicates the direction of the electric field, which is perpendicular to the liquid|liquid interface. Reproduced from Ref. [56] with permission from the Royal Society of Chemistry

6 More Phases, More ECL: Introducing the Gas Phase into the Mix

Throughout this chapter, ECL has been described in detail for multiphase systems comprised of liquid|liquid and liquid|solid interfaces. Liquid|gas interfaces have only been mentioned in passing, primarily in the context of bubble nucleation that occurs within specific ECL reactions and, do not contribute directly to the ECL emission. Nevertheless, liquid|gas interfaces are important to study, as they are found commonly in nature [60–63], such as in sea spray, and have a variety of applications [2, 13, 17, 64, 65]. Liquid|gas interfaces are also critical to investigate as they can provide a gateway to understanding certain curious chemical phenomena, such as reaction acceleration reported in microdroplets [66–70]. In the context of electrochemistry, liquid|gas interfaces, specifically bubble nucleation on electrodes, are often a competitive reaction that hinders the electrochemical reaction of interest by acting as a blocking agent, taking up valuable space on the electrode that could otherwise be used to carry out the intended electron-transfer reaction. These electrogenerated bubbles, hence, cause the efficiency of important electrochemical reactions to be significantly reduced and can make electroanalysis difficult. In this section, however, liquid|gas interfaces within multiphase ECL will be discussed as an asset to the system, capable of enhancing ECL in space and time and elucidating physiochemical information such as interfacial tension of adsorbed droplets and the buoyancy force of bubbles [39, 71, 72].

A fundamental limitation of ECL is that it is a surface-confined process, which occurs because it is restricted by the lifetimes of the electrogenerated radicals and excited-state species. This results in ECL emission being bound to only a few microns from the surface of the electrode. To overcome this ECL limitation, Layman and Dick developed a new method, using the reflectance of light off CO_2 bubbles that nucleated and grew within 1,2-DCE, to enable the imaging of phase boundaries 100 s of μm from the surface of the electrode [39]. This method, called "*through-space* ECL," earned its moniker from its ability to image ECL through space and in a three-dimensional capacity. Beyond extending ECL to the third dimension, this method can also be used to measure specific physiochemical properties of multiphase systems, including the buoyancy force of the trapped CO_2 bubble.

In the experiments conducted by Dick and colleagues, they used a multiphase system comprised of a bulk, aqueous phase containing 10 mM of $[Ru(bpy)_3]^{2+}$ and 50 mM of $[C_2O_4]^{2+}$ within 100 mM KCl, while the droplet, organic phase was composed of 100 mM TBAP in 1,2-DCE. In the experiment for which the results are shown in Fig. 8a, a 15 μL droplet of the organic phase was pipetted onto ITO (setup similar to the schematic in Fig. 2, except the working electrode is ITO, not a glassy carbon inlaid disk electrode), and subsequently submerged in the aqueous phase. Then, amperometry was performed by applying + 1.5 V versus Ag/AgCl, which is the potential at which the highest ECL intensity is achieved for the $[Ru(bpy)_3]^{2+}$/$[C_2O_4]^{2+}$ ECL reaction (the mechanism and details of how this potential was determined to have the highest ECL intensity is described in the previous section of this chapter). In Fig. 8a, by 142.20 s, it is seen that a bubble has nucleated within the 1,2-DCE and had become trapped at the apex of the liquid|liquid interface, as evidenced by the small, central circle of a higher ECL intensity that comes from reflectance of the bulk, aqueous ECL emission being reflected off of the bubble. In both 199.34 and 251.02 s, the circle of greater intensity ECL emission continues to get bigger, supporting that the bubble continues to grow throughout the application of + 1.5 V. Episcopic brightfield images (not shown) also reinforce the hypothesis that the CO_2 bubble gets trapped at the liquid|liquid interface, because in comparing the images taken of the 1,2-DCE droplet before and after ECL, the after image has a dark circle in the center, which was confirmed to be a bubble with a side-profile brightfield image. In the side-profile brightfield image (not included), the droplet can be seen to be located at the apex of the liquid|liquid interface *ca.* 300 mm above the electrode's surface.

In Fig. 8b, an emulsion consisting of 300 μL of the organic phase and 5 mL of the aqueous phase was pipetted into an ECL cell with ITO as the working electrode, and ECL was immediately performed. At 495.95 s, no bubble is visible; however, quickly after at 496.59 and 496.90 s, the bubble is seen and appears to grow, as shown by the increasing intensity and size of the ring of light that indicates the presence of a bubble. At 497.23 s, the 1,2-DCE droplet undergoes a coalescence event with a nearby 1,2-DCE droplet, enabling the bubble to continue to grow, as indicated by the much larger ring seen in 522.59 s. Fluorescence microscopy was performed on the system simultaneously with the ECL reaction and can be seen at 523.87 s, which shows that the CO_2 bubble makes up most of the larger DCE droplet. The immediate performance of ECL after adding the emulsion to the ECL cells in these experiments highlights an important advantage of this system—the removal of the waiting period often needed for multiphase ECL systems. By changing the ECL emission layer to be microns away from the ITO (located at the bottom of the ECL cell), there is no longer the requirement to wait for the emulsion to settle on the electrode surface and, therefore, allows for the dynamic multiphase microenvironment to be imaged in real time.

The proposed mechanism for this phenomenon is exhibited in Fig. 8c, which depicts how the CO_2 generated in the aqueous phase can enter the organic phase and result in bubble nucleation and growth. By using the higher level of solubility of CO_2 gas in 1,2-DCE as opposed to water, the organic droplets on the electrode essentially act as a reservoir for the CO_2 gas produced by the bulk aqueous phase, demonstrating a carbon capture capability for small volumes. Furthermore, from the ECL images, the buoyancy force (F_B) acting upon the bubble by the liquid|liquid interface can be calculated. Specifically, by measuring the diameter of the ring of light, which is approximated to be equivalent to the diameter of the bubble, and assuming a spherical shape, the F_B can be calculated using Eq. 17.

$$F_B = \frac{1}{24} g \pi d^3 (\rho_1 + 3\rho_2) \tag{17}$$

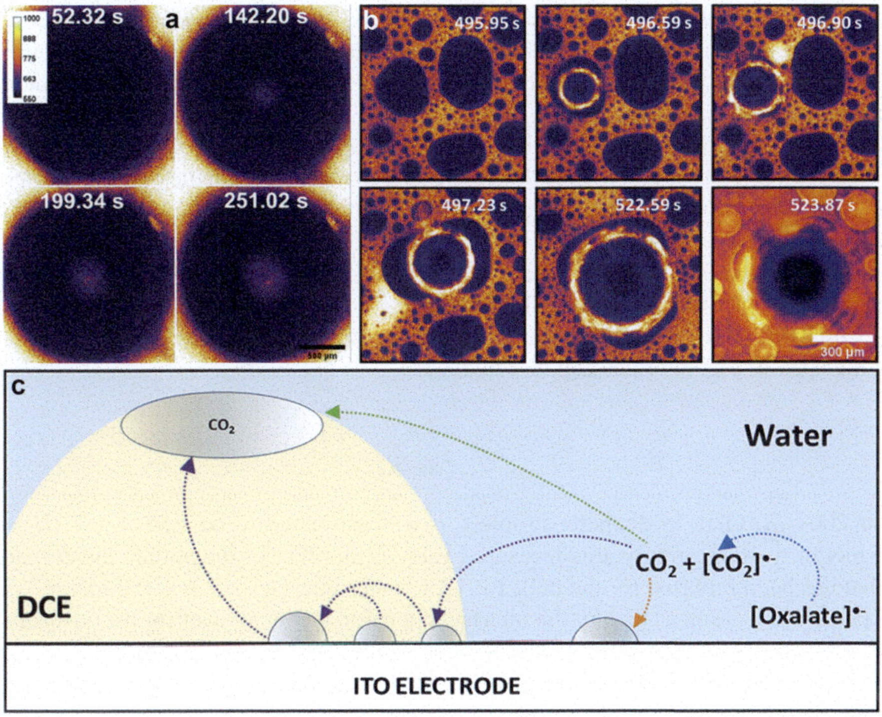

Fig. 8 **a** ECL images showing the progression of CO_2 bubble nucleation and growth within a large droplet of 1,2-DCE. The droplet was pipetted onto an ITO electrode and was submerged within a bulk, aqueous phase. At time $t = 52.32$ s, no bubble was seen. However, at time $t = 142.20$ s, a small ECL emission is seen in the center of the droplet, indicating bubble nucleation. It continues to grow throughout the rest of the experiment, as seen when comparing the bubble size in $t = 199.34$ s and $t = 251.02$ s. **b** CO_2 bubble nucleation and growth within a 1,2-DCE droplet after an emulsion of a droplet phase of 1,2-DCE and an aqueous bulk phase was left to settle on an ITO electrode. At time $t = 495.95$ s, no bubble nucleation is seen; however, at $t = 496.59$ s, the bubble is visible, as evidenced by the glowing ring of light. This ring continues to grow, as depicted in the increasing radius of the ring from $t = 496.90$ s to $t = 522.59$. The growth occurs as the droplet itself increases in size due to merging with nearby 1,2-DCE droplets. Fluorescence is shown for $t = 523.87$ s. **c** Proposed mechanism for CO_2 bubble formation with the corresponding schematic for this system. In this mechanism, the oxidation of the oxalate on the ITO surface generates the radical oxalate anion. The blue arrow shows the subsequent decomposition of the radical oxalate anion into CO_2 and its radical anion. After, multiple pathways are possible for the CO_2. The orange arrow represents the nucleation and growth of the CO_2 bubble within the aqueous phase while adhered to the ITO. The purple arrow shows the CO_2 radicals transport into the 1,2-DCE, where the bubbles can nucleate, grow, merge, and then, once meeting a critical size, detach due to the buoyancy force, and get trapped at the liquid|liquid interface. Bubbles in the aqueous phase will also detach once they are sufficiently large enough (once again, due to the buoyancy force), and can also join the trapped bubble at the liquid|liquid interface (green arrow). Adapted with permission from Ref. [39]. Copyright 2023 American Chemical Society

In Eq. 17, g is the gravitational constant (9.81 ms^{-2}), d is the measured diameter of the bubble, ρ_1 is the first phase density, and ρ_2 is the second phase's density. Moreover, in a separate study conducted by Dick and colleagues using the same system, these calculations were then extended to also calculate the interfacial tension (g) of the liquid|liquid interface of the adsorbed 1,2-DCE droplet within a continuous aqueous phase [71]. It is important to note that, to be able to measure both the buoyancy force (F_B) and the interfacial tension (g) using this through-space ECL methodology, it can only be done once the CO_2 bubble leaves the adsorbed 1,2-DCE droplet on the electrode surface as that is the point the interfacial tension (g) exhibited by the adsorbed droplet is overcome by the buoyancy force (F_B).

From Eq. 18, in which r_{droplet} is the radius of the droplet, the interfacial tension of the apex of this interface can be calculated.

$$\gamma = \frac{F_B}{r_{\text{droplet}}} \tag{18}$$

For 200—700 μm in diameter droplets, it was estimated to be 1.56 ± 0.7 ($N = 7$) mN m^{-1}. Interestingly, this measured value is an order of magnitude lower than what has been reported for the bulk 1,2-DCE|water interface (32.6 ± 0.8 mN m^{-1}). A postulated reason as to why the interfacial tension is much lower for the adsorbed 1,2-DCE droplets is that the reorganization within the droplet makes it favorable for the bubble to leave the droplet only when the bubble has reached a diameter that is equivalent to the droplet's diameter. This observation further exemplifies both the capability of and future potential of multiphase systems to reveal curious, new chemical phenomena.

Outside of the oxalate/$[Ru(bpy)_3]^{2+}$ ECL reaction that nucleates CO_2 bubbles, another common ECL reaction that also results in bubble nucleation is luminol/ H_2O_2. This system, which is one of the most popular ECL reactions in the literature, nucleates O_2 and N_2 bubbles. Vogel et al. used the bubbles nucleated from the luminol/ H_2O_2 ECL reaction to drive the oxidation of water-soluble species, from which they learned that the interface between the electrode, solution, and the bubbles was more effective at the oxidation of these species than only the bulk liquid|electrode interface [73]. In their study, Vogel and colleagues postulated that, at the solid|gas|liquid interface, there is an excess of hydroxide ions (OH–) that, when destabilized due to electrostatic repulsion from each other, oxidize into hydroxyls that can then be harnessed to drive other redox reactions. They validated this hypothesis by coupling the luminol/H_2O_2 ECL to a step polymerization, in which both processes are initiated by reactive oxygen species (ROS) and are oxidative processes. The luminol/ H_2O_2 ECL acted as a light source that was then quenched by the polymer; hence, the coupling of these systems acted as a real-time 2D map of the polymer growth. They observed that the polymer filaments did not grow radially from the bubble but rather grew in a way that linked neighboring bubbles. This behavior allowed them to monitor the concentration of ROS species away from the bubble. Moreover, they learned that the luminol/H_2O_2 ECL emission is greater at the solid|gas|liquid interface than the bulk liquid|solid interface, which they attributed to a higher concentration of hydroxyl radicals present at the solid|gas|liquid interface, further supporting their hypothesis.

Nami-Ana and co-workers also conducted luminol/H_2O_2 ECL to explore the sustained generation of H_2O_2 on the surface of bubbles [74]. They did so by placing a copper electrode within a 5 mM luminol and pH 9 0.1 M phosphate buffer solution, using a split closed bipolar electrode setup to separate the anodic and cathodic chambers. This was done so that the cathodic and anodic processes could be studied separately. Then, bubbles nucleated on the electrode surface via the luminol/H_2O_2 ECL reaction, varying the applied potential from +2 up to +30 V using a DC power supply, and took images of the ECL emission at these different potentials and at various time points. They observed that at higher applied potentials, the intensity of the ECL emission increased, along with the number of microbubbles nucleated. Afterglow ECL was also seen, with the sustained ECL emission after the application of potential ended correlated to the continued generation of H_2O_2 at the microbubble surface [38, 75]. Upon removal of microbubbles from the electrode surface, the afterglow ended immediately, further evidencing that H_2O_2 is produced at microbubble surfaces and contributing to the afterglow phenomenon. It is important to highlight this observation of the afterglow ECL as it is exciting because it enables the extension of the duration of ECL emission, which is typically limited by the often-short lifetimes of the radical and excited-state species generated at the electrode.

The duration of ECL was extended in space and time, as well as its signal amplified, by Knežević and co-workers through their development of bubble-enhanced ECL [72]. Using an ECL reaction between H_2O_2 and a luminol derivative, 8-amino-5-chloro-2,3-dihydro-7-phenyl-pyrido[3,4-d]pyridazine-1,4-dione (L-012) as the luminophore, they demonstrated that the nucleation of bubbles predominantly composed of chlorine gas (Cl_2) on an electrode surface can produce enhanced and long-lasting ECL on the bubble's surface. The production of hypochlorite (ClO^-) from the reaction between the chlorine gas bubbles and the aqueous solution oxidizes the L-012 while, at the same time, the concentration gradient of hydroxyl radicals, hypothesized by Vogel and described earlier in this section, generates ROS species at the interface between the bubble, electrode, and solution to further push forward the ECL reaction.

In this study, a glass ECL cell was filled with an aqueous solution that contained 0.42 mM L-012, 0.5 M H_2O_2, and 0.05 M KCl or KNO_3 within 0.2 M phosphate buffer at pH 8. A gold inlaid disk microelectrode ($d = 25\ \mu m$) was then submerged in this solution and used to perform both L-012/H_2O_2 ECL (seen in Fig. 9a) and chloride oxidation. The L012/H_2O_2 reaction follows the proposed reaction mechanism outlined in Eqs. 13–16, and the ECL emission intensity is, therefore, linearly dependent on the H_2O_2 concentration. Chlorine oxidation was carried out simultaneously with the L-012/H_2O_2 ECL by either conducting cyclic voltammograms from +0 to +3 V versus Ag/AgCl/1 M or by holding the potential at +0.7, +2.4, or +3 V. The chlorine oxidation results in the nucleation of the chlorine gas bubble on the electrode surface, seen in the two different schematics of the L-012/ H_2O_2 ECL emission process seen in Fig. 9b, c. To investigate how the chemical species of the supporting electrolyte influences this process, both 0.5 mM of KNO_3 and KCl were used in this system, separately, and studied. It was determined that in KNO_3, the primary process at high anodic potentials is that of the oxygen evolution reaction (OER), while in KCl, the chlorine evolution reaction (CER) is dominant over OER. Moreover, it was found that, in the presence of chloride, low-potential anodic L-012 ECL experiences an enhancement in the ECL emission, which was attributed to the generation of intermediates from OER and CER that facilitate the homogeneous oxidation of the ECL species. This enhancement in the presence of chloride is depicted in Fig. 9d, where at high anodic potentials (+2.4 and +3 V), the ECL emission is much brighter in KCl than in KNO_3. Although OER produces ROS as intermediates that can react with L-012 to produce light, the hypochlorite (ClO^-) formed from the reaction between the electrogenerated Cl_2 gas and water results in a much stronger ECL emission because this species can promote the oxidation of L-012. This demonstrates that, even with the nucleation of O_2 bubbles on the electrode in KNO_3, the composition of the bubble plays an important role in this process.

Using this bubble-enhanced ECL process, the signal intensity was not only amplified by the Cl_2 gas bubble, but it also was extended in both space and time. Neso and colleagues achieved this by electrochemically generating a Cl_2 bubble on a glassy carbon inlaid disk electrode and imaging it at various time points after the bubble detached from the electrode, which is shown in Fig. 9e. It was discovered that, through this process, the bubble's path can be traced from the diffusion pattern of the dissolved chlorine as well as the changes in the bubble size and velocity as the buoyancy force propels it upwards from the electrode surface can also be probed with the ECL emission. Furthermore, it was hypothesized that the dissolution of gas trapped inside the bubble can result in the delayed formation of the reactive radical species, which are then implicated in extending the ECL emission in space and time. Figure 9f depicts this effect, in that it was observed that the ECL imaging of the propagation of the Cl_2 gas bubble up and away from the electrode surface could be spatially resolved up to 5 mm away from the electrode surface. Interestingly, the maximum ECL intensity was measured to be in regions far from the electrode surface. It was hypothesized that in these regions, "fresh" reactants are present that can reactant with the chlorine and sustain the reaction, until ultimately the dissolution of the chlorine causes the bubble to collapse, leaving with a sharp, strong emission that quickly vanishes.

In addition to the spatial extension, the temporal extension, or the afterglow component of the system, was observed to last up to 145 s after the end of the application of the potential to the bubble. However, this afterglow component was observed to exponentially decrease over time, immediately following the end of the application of the potential. It is important to note that the size of the bubble within this bubble-enhanced ECL process influences both the extent of signal amplification and the distance the reaction extends from the electrode surface. As bubbles are nucleated on the electrode and grow, the signal intensity and distance from the electrode where the ECL emission occurs continue to increase until bubble size is reached where, for O_2 bubbles, the elevated electrochemical reactivity no longer accounts for the electrode surface area that is covered by the bubble and the ECL emission diminishes. While, at this critical bubble size for Cl_2 bubbles, the ECL emission reaches a plateau where there is no longer an advantage to increasing the

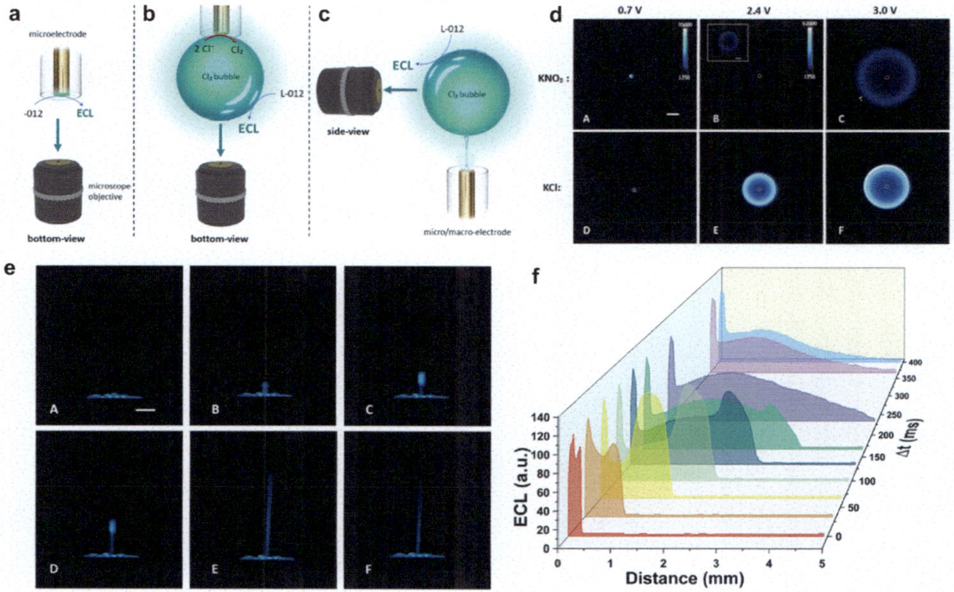

Fig. 9 **a** Schematic demonstrating the L-012 ECL emission process at the surface of a gold microelectrode in the bottom-view configuration. **b** Illustration of the L-012 ECL emission process at the gaslliquid interface in the bottom-view configuration. The gas phase was composed of Cl_2, which was electrogenerated by the microelectrode, and the liquid was the bulk aqueous solution. **c** Schematic of the L-012 ECL emission process after the detachment of the Cl_2 bubble from the microelectrode from the side-profile configuration. **d** ECL emission in the bottom-view configuration shown for two different electrolytes, KNO_3 (**a–c**) and KCl (**d–f**), at potentials of +0.7, +2.4, and +3.0 V. Nucleation and growth shown for an O_2 bubble (**a–c**) and a Cl_2 bubble. **e** ECL emission in the side-profile view configuration of the nucleation, growth, and detachment of a Cl_2 bubble on a glassy carbon macroelectrode over different time points (A-F: 0, 33, 67, 100, 233, and 367 ms, respectively). The ECL emission shows the trajectory of the single electrogenerated Cl_2 bubble. **f** ECL emission extended spatially with the Cl_2 bubble, as evidenced by the ECL intensity versusdistance from the electrode with the correlated time points. ECL was observed up to 5 mm from the electrode surface. Adapted with permission from Ref. [72] Copyright 2024 American Chemical Society

bubble size, as the electrogenerated chlorine species mediates the ECL reaction and sustains the emission of light even without the presence of ROS. Moreover, this same logic follows to explain why afterglow was seen for the Cl_2 bubbles, and not the O_2 bubbles.

As a final note, this system is an important advancement in ECL for a multitude of reasons. Not only does this system enhance the ECL emission of the L-012/H_2O_2 ECL reaction and extend it in space and time, but it also allows for the study and visualization of fundamental physiochemical processes of interest. These processes include diffusion of chemical species, dissolution of gases, and reactions in general that occur both in bulk and at phase interfaces. Additionally, this bubble-enhanced ECL system also has promise for biosensing applications. In the same study, Neso and co-workers extended this system to detect glucose. Specifically, by performing the ECL in the presence of glucose oxidase and in both the presence and absence of glucose, ECL emission was 5.2 times higher at +2.4 V than at +0.7 V that was only seen in the presence of glucose (the emission intensity was lower and nearly identical at both potentials in the absence of glucose). This preliminary study introduces this bubble-enhanced ECL system as a potential technology for the enzymatic detection of glucose. Evidently, this study is an important example of how the gas|liquid interface, which is typically viewed as a problem within electrochemistry, can be utilized to make advances not only in ECL, but also in measurement science as a whole.

7 More Complex Multiphase Measurement's: Toward Biologics

As mentioned in the introduction, biological cells often exhibit complex multiphase environments either through covariate droplets, or membranes. Moreover, the ECL community has been largely interested in applying this technique to these biology systems to visualize phase boundaries, primarily membrane|electrode, membrane|membrane, and membrane|extracellular matrix. Cellular function is often dependent on liposomal transport, where liposomes act as transport vehicles for various molecules. To start, Bois and colleagues' work where they used luminophore-grafted microgels to image the 2D interface between oil and water phases was the first hint at what could be accomplished with more complex multiphase systems [76].

In continuing the study of these more complex systems and their associated phenomena, Trad and colleagues have published a series of four papers using the giant liposome as a model system [77–80].

Here, giant liposomes are synthesized and collided with an optically transparent ITO electrode. This collision induces both the optical and electrochemical signal (Fig. 10a). This shows the power of packaging (a.k.a. encapsulating) the luminophore ($[Ru(bpy)_3]^{2+}$) and the coreactant (tri-n-propyl amine (TPrA)) in this confined volume to visualize the liposome opening (Fig. 10b). In Fig. 10b, there is a liposome opening in the 3rd ECL frame which is marked by the start of the ECL emission. This is contrasted to the epifluorescence emission shown below which depicts sustained emission due to the lack of surface confinement. This shows the importance of using ECL to study these biophysical processes since it provides real-time dynamical information. Furthermore, ECL emission is sustained and provides information about the diffusion of the luminophore after the collision (Fig. 10c).

Moreover, they employ different types of ECL microscopy [77]. When the liposome is left empty and the luminophore and coreactant are confined to the solution, shadow ECL takes place around the liposome to create a negative image of the process. When the liposomes collide with the ITO electrode, the ECL signal turns off locally, and as diffusion takes place over the electrodes surface the dark region will spread.

Additionally, one can introduce membrane permeabilization agents into the solutions when liposomes are placed onto the surface of the ITO [78]. Before the introduction of the Melittin, the luminophore and coreactant could not interact with the surface of the electrode. However, as the liposome breaks apart from the introduction of the Melittin, there is a positive ECL signal (Fig. 10d). This provides real-time information regarding membrane permeability, something that the epifluorescence micrographs do not (Fig. 10e).

Additionally, enzymatic reactions can be monitored using this technique by pairing a coreactant, such as NADH, which forms from the enzymatic activity of glucose dehydrogenase. The NADH and $[Ru(bpy)_3]^{2+}$ then proceed in an ECL reaction to emit light [78].

While giant liposomes are an excellent model system for cellular transport, another liquid|liquid phase boundary occurs at the biological cell|extracellular matrix. It is outside the scope of this chapter to review the excellent work performed by the biological ECL imaging community; however, it is important to note this perspective.

This budding community includes Gupta and colleagues' recent work using endogenously produced peroxide as a coreactant to the luminol luminophore, which showed the boundaries of the tumor model at the electrode surface [81]. Additionally, others have imaged other biological entities such as tissue sections [82], biological cells [37], and mitochondria [83] to visualize their membrane boundaries. Overall, the field of liquid|liquid and liquid|gas interfacial ECL microscopy provides an excellent fundamental basis for pursuing these biological avenues of inquiry.

8 Concluding Remarks

Over the past *ca.* decade, the field of multiphase ECL has seen its impact by informing us about peculiar chemistry under confinement. From the original studies without microscopy, to expanding multiphase ECL "*through-space*," "*bubble enhancing*," and use of giant liposomes, these measurements have revealed new physicochemical phenomena present in these environments. Such phenomena include the formation of solvent inclusions through droplet coalescence, micropumping, CO_2 entrapment, and chlorine gas evolution [72].

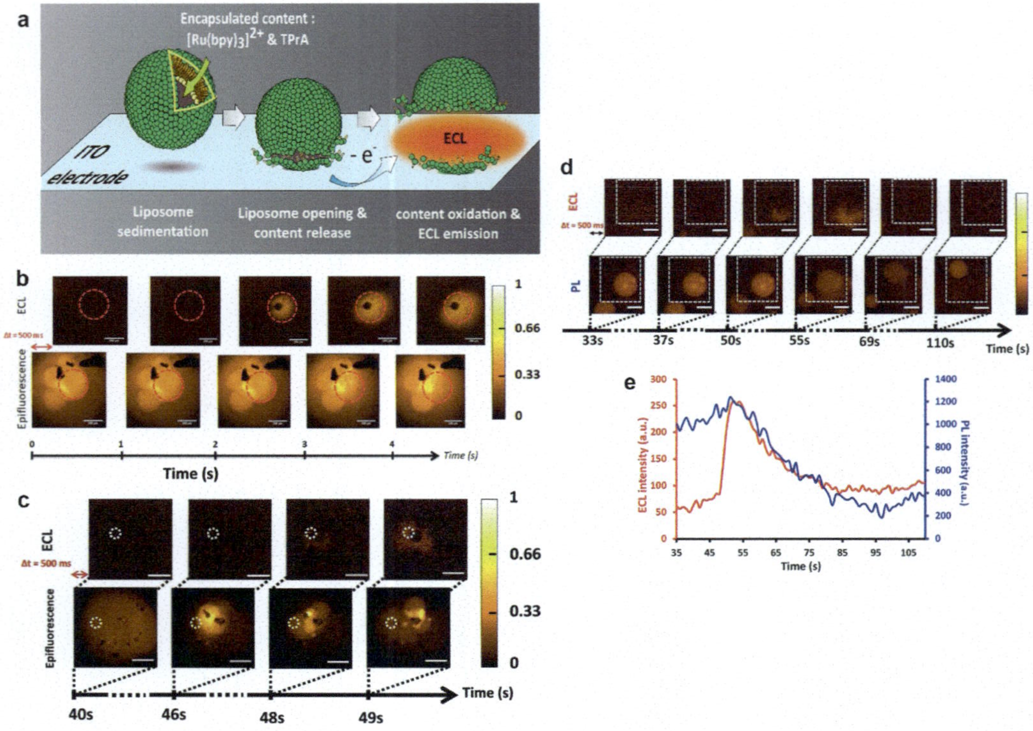

Fig. 10 **a** Schematic of giant liposome that encapsulates $[Ru(bpy)_3]^{2+}$ and TPrA. As the liposome impacts, the ITO surface breaks open and releases the ECL reagents. **b** ECL and epifluorescence micrographs of the collisions of the giant liposomes on the ITO electrode surface. **c** Diffusion of $[Ru(Bpy)_3]^{2+}$ from the collided giant liposome across the electrode surface. **d** ECL and photoluminescence (PL) micrographs of the effect of permeabilization of the liposome by Melittin. **e** ECL and PL intensity with time corresponding to (**f**). All scale bars are 200 µm. (**a–c**) Adapted with permission from Ref. [80] Copyright 2022 American Chemical Society. (**d–e**) Adapted with permission from Ref. [79] Copyright © 2023 The Authors. Co-published by Nanjing University and American Chemical Society

Looking forward, there is still much to discover about multiphase environments through ECL microscopy. Effects such as enhanced mass transfer and kinetics, precipitation, gas bubble nucleation, other organic phases, and relevant co-electrochemical reactions such as the oxygen evolution and hydrogen evolution reactions need to be studied [84]. Furthermore, the inclusion of novel electrodes and electrochemical cell, such as nanopipettes and bubble walls [85], is of promise in these systems. As chemists we *mostly* don't do science in a "vacuum" so to speak. therefore, we should also be taking inspiration from other fields such as lasing, emulsion science, cellular biology, and environmental science to generate new avenues of inquiry for both fundamentals and applications.

Another field that the ECL Multiphase Microscopist should be partnering with is the reaction acceleration community. By decoupling the stimulus and signal from standard electrochemistry (where the reaction is controlled and measured by a potentiostat), the ECL community has an incredibly powerful tool for studying the peculiar and curious chemistry that happens under confinement. The preliminary work performed by Trad and colleagues is an excellent springboard into the concept of enhanced and confined biocatalysis, particularly how biological cells use these confined spaces to speed up and change the dynamics of these relevant reactions [77–80].

Additionally, we would also like to point out the opportunity for ECL in multiphase microscopy to be used with students to visualize these ideas [84]. There are less than a handful of articles centered around the ability for ECL in general to visualize electrochemistry for students [84, 86–89], and there are many avenues of inquiry for these exciting events as far as creativity will take the experimenter. Furthermore, there are no, to our knowledge, multiphase ECL experiments for students to explore chemical reactivity at interfaces in a teaching laboratory setting. With luminol and certain other reactions being very high efficiency (*bright*), it is only a matter of time till these concepts can be taken to students in an exciting way.

Overall, ECL multiphase microscopy has rapidly advanced our understanding of these complex environments by allowing us to view physicochemical phenomena through coupled excited-state electrochemistry and microscopy. Broader impacts range from CO_2 sequestration to cellular and cancer cell biology. This is still a field that is only around a decade old, so just beginning its adolescence, so to speak. The authors very much encourage more ECL microscopists to allow their creativity to find new truths about nature using this technique and hope that their before-mentioned suggestions of "jumping off points" are of use.

Acknowledgements Brady R. Layman is currently supported by the Department of Defense (DoD) through the National Defense Science and Engineering Graduate (NDSEG) Fellowship. Megan L. Hill is currently supported by the Department of Energy (DOE) through the University Nuclear Leadership Program (UNLP). For previous work we would like to acknowledge Matthew Glascott (Army Research Lab, DEVCOM), Silvia Voci (University of Paris), and Kathryn Vannoy (Leiden University). Brady would like to thank Silvia Voci (University of Paris) for her mentorship through his first year. For helpful discussions we would like to thank all current and past Dick Research Group Members along with, Christophe Renault (Layola University Chicago), Neso Sojic (University of Bordeaux), and countless others. For funding, we acknowledge the National Institutes of General

Medical Science (NIGMS) under grant R35GM138133. The MIRA program has provided us with the flexibility to explore the fundamental principles of multiphase chemical systems, which directly inform the advancement of enabling biotechnologies.

Competing Financial Interests: The authors declare no competing financial interests.

References

1. Colón-Quintana, G.S., Clarke, T.B., Dick, J.E.: Interfacial solute flux promotes emulsification at the water|oil interface. Nat. Commun. **14**(1) (2023). https://doi.org/10.1038/s41467-023-35964-9
2. Yan, X., Bain, R.M., Cooks, R.G.: Organic reactions in microdroplets: reaction acceleration revealed by mass spectrometry. Angew. Chem. Int. Ed. (2016). https://doi.org/10.1002/anie.201602270
3. Nguyen, J.H., Rana, A., Dick, J.E.: Amplifying the electrochemical footprint of < 1000 molecules in a dissolving microdroplet. Analyst **149**(16), 4222–4229 (2024). https://doi.org/10.1039/D4AN00504J
4. Rana, A., Clarke, T.B., Nguyen, J.H., Dick, J.E.: Adsorbed microdroplets are mobile at the nanoscale. Proc. Natl. Acad. Sci. **121**(47), e2412148121 (2024). https://doi.org/10.1073/pnas.2412148121
5. Brangwynne, C.P., Eckmann, C.R., Courson, D.S., Rybarska, A., Hoege, C., Gharakhani, J., Jülicher, F., Hyman, A.A.: Germline P granules are liquid droplets that localize by controlled dissolution/condensation. Science (1979) **324**(5935) (2009). https://doi.org/10.1126/science.1172046
6. Hyman, A.A., Weber, C.A., Jülicher, F.: Liquid-liquid phase separation in biology. Annu. Rev. Cell Dev. Biol. (2014). https://doi.org/10.1146/annurev-cellbio-100913-013325
7. Zwicker, D., Seyboldt, R., Weber, C.A., Hyman, A.A., Jülicher, F.: Growth and division of active droplets provides a model for protocells. Nat. Phys. **13**(4) (2017). https://doi.org/10.1038/nphys3984
8. Stroberg, W., Schnell, S.: On the origin of non-membrane-bound organelles, and their physiological function. J. Theor. Biol. **434** (2017). https://doi.org/10.1016/j.jtbi.2017.04.006
9. Vannoy, K.J., Lee, I., Sode, K., Dick, J.E.: Electrochemical quantification of accelerated FADGDH rates in aqueous nanodroplets. Proc. Natl. Acad. Sci. USA **118**(25) (2021). https://doi.org/10.1073/pnas.2025726118
10. Jung, Y., Marcus, R.A.: On the theory of organic catalysis "on Water." J. Am. Chem. Soc. **129**(17) (2007). https://doi.org/10.1021/ja068120f
11. Lee, J.K., Samanta, D., Nam, H.G., Zare, R.N.: Spontaneous formation of gold nanostructures in aqueous microdroplets. Nat. Commun. **9**(1) (2018). https://doi.org/10.1038/s41467-018-04023-z
12. Glasscott, M.W., Pendergast, A.D., Goines, S., Bishop, A.R., Hoang, A.T., Renault, C., Dick, J.E.: Electrosynthesis of high-entropy metallic glass nanoparticles for designer, multi-functional electrocatalysis. Nat. Commun. **10**(1) (2019). https://doi.org/10.1038/s41467-019-10303-z
13. Spoorthi, B.K., Debnath, K., Basuri, P., Nagar, A., Waghmare, U.V., Pradeep, T.: Spontaneous weathering of natural minerals in charged water microdroplets forms nanomaterials. Science (1979) **384**(6699), 1012–1017 (2024). https://doi.org/10.1126/science.adl3364
14. Song, X., Meng, Y., Zare, R.N.: Spraying water microdroplets containing 1,2,3-triazole converts carbon dioxide into formic acid. J. Am. Chem. Soc. **144**(37) (2022). https://doi.org/10.1021/jacs.2c07779
15. Song, X., Basheer, C., Zare, R.N.: Making ammonia from nitrogen and water microdroplets. Proc. Natl. Acad. Sci. USA **120**(16) (2023). https://doi.org/10.1073/pnas.2301206120

16. Lee, J.K., Samanta, D., Nam, H.G., Zare, R.N.: Micrometer-sized water droplets induce spontaneous reduction. J. Am. Chem. Soc. **141**(27) (2019). https://doi.org/10.1021/jacs.9b03227
17. Holden, D.T., Morato, N.M., Cooks, R.G.: Aqueous microdroplets enable abiotic synthesis and chain extension of unique peptide isomers from free amino acids. Proc. Natl. Acad. Sci. U S A **119**(42) (2022). https://doi.org/10.1073/pnas.2212642119
18. Qiu, L., Wei, Z., Nie, H., Cooks, R.G.: Reaction acceleration promoted by partial solvation at the gas/solution interface. Chempluschem **86**(10) (2021). https://doi.org/10.1002/cplu.202100373
19. Li, Y., Liu, Y., Gao, H., Helmy, R., Wuelfing, W.P., Welch, C.J., Cooks, R.G.: Accelerated forced degradation of pharmaceuticals in levitated microdroplet reactors. Chem. Eur. J. **24**(29) (2018). https://doi.org/10.1002/chem.201801176
20. Matsuzaki, K., Kusaka, R., Nihonyanagi, S., Yamaguchi, S., Nagata, T., Tahara, T.: Partially hydrated electrons at the air/water interface observed by UV-excited time-resolved heterodyne-detected vibrational sum frequency generation spectroscopy. J. Am. Chem. Soc. **138**(24) (2016). https://doi.org/10.1021/jacs.6b02171
21. Mondal, J.A., Nihonyanagi, S., Yamaguchi, S., Tahara, T.: Three distinct water structures at a zwitterionic lipid/water interface revealed by heterodyne-detected vibrational sum frequency generation. J. Am. Chem. Soc. **134**(18) (2012). https://doi.org/10.1021/ja300658h
22. Nihonyanagi, S., Yamaguchi, S., Tahara, T.: Ultrafast dynamics at water interfaces studied by vibrational sum frequency generation spectroscopy. Chem. Rev. (2017). https://doi.org/10.1021/acs.chemrev.6b00728
23. Nihonyanagi, S., Mondal, J.A., Yamaguchi, S., Tahara, T.: Structure and dynamics of interfacial water studied by heterodyne-detected vibrational sum-frequency generation. Annu. Rev. Phys. Chem. **64** (2013). https://doi.org/10.1146/annurev-physchem-040412-110138
24. Wen, Y.C., Zha, S., Liu, X., Yang, S., Guo, P., Shi, G., Fang, H., Shen, Y.R., Tian, C.: Unveiling microscopic structures of charged water interfaces by surface-specific vibrational spectroscopy. Phys. Rev. Lett. **116**(1) (2016). https://doi.org/10.1103/PhysRevLett.116.016101
25. Tang, F., Ohto, T., Sun, S., Rouxel, J.R., Imoto, S., Backus, E.H.G., Mukamel, S., Bonn, M., Nagata, Y.: Molecular structure and modeling of water-air and ice-air interfaces monitored by sum-frequency generation. Chem. Rev. (2020). https://doi.org/10.1021/acs.chemrev.9b00512
26. Nagata, Y., Mukamel, S.: Vibrational sum-frequency generation spectroscopy at the water/lipid interface: molecular dynamics simulation study. J. Am. Chem. Soc. **132**(18) (2010). https://doi.org/10.1021/ja100508n
27. Liu, H., Klein, I.M., Michelsen, J.M., Cushing, S.K.: Element-specific electronic and structural dynamics using transient XUV and soft X-ray spectroscopy. Chem **7**(10), 2569–2584 (2021). https://doi.org/10.1016/j.chempr.2021.09.005
28. Ding, H., Zhou, P., Fu, W., Ding, L., Guo, W., Su, B.: Spatially selective imaging of cell–matrix and cell–cell junctions by electrochemiluminescence. Angew. Chem. Int. Ed. **60**(21) (2021). https://doi.org/10.1002/anie.202101467
29. Gou, X., Xing, Z., Ma, C., Zhu, J.J.: A close look at mechanism, application, and opportunities of electrochemiluminescence microscopy. Chem. Biomed. Imaging (2023). https://doi.org/10.1021/cbmi.2c00007
30. Dick, J.E., Renault, C., Kim, B.K., Bard, A.J.: Simultaneous Detection of single attoliter droplet collisions by electrochemical and electrogenerated chemiluminescent responses. Angew. Chem. Int. Ed. **53**(44) (2014). https://doi.org/10.1002/anie.201407937
31. Dick, J.E., Renault, C., Kim, B.K., Bard, A.J.: Electrogenerated chemiluminescence of common organic luminophores in water using an emulsion system. J. Am. Chem. Soc. **136**(39) (2014). https://doi.org/10.1021/ja507198r
32. Dick, J.E., Renault, C.: Chapter 11: single entity electrogenerated chemiluminescence. In: RSC Detection Science, vol. 2020-January (2020). https://doi.org/10.1039/9781788015776-00309
33. Sentic, M., Virgilio, F., Zanut, A., Manojlovic, D., Arbault, S., Tormen, M., Sojic, N., Ugo, P.: Microscopic imaging and tuning of electrogenerated chemiluminescence with boron-doped diamond nanoelectrode arrays. Anal. Bioanal. Chem. **408**(25) (2016). https://doi.org/10.1007/s00216-016-9504-1

34. Meng, C., Knežević, S., Du, F., Guan, Y., Kanoufi, F., Sojic, N., Xu, G.: Recent advances in electrochemiluminescence imaging analysis. eScience (2022). https://doi.org/10.1016/j.esci.2022.10.004
35. Ma, C., Wu, W., Li, L., Wu, S., Zhang, J., Chen, Z., Zhu, J.J.: Dynamically imaging collision electrochemistry of single electrochemiluminescence nano-emitters. Chem. Sci. **9**(29) (2018). https://doi.org/10.1039/c8sc02251h
36. Xu, L., Li, Y., Wu, S., Liu, X., Su, B.: Imaging latent fingerprints by electrochemiluminescence. Angew. Chem. Int. Ed. **51**(32) (2012). https://doi.org/10.1002/anie.201203815
37. Voci, S., Goudeau, B., Valenti, G., Lesch, A., Jović, M., Rapino, S., Paolucci, F., Arbault, S., Sojic, N.: Surface-confined electrochemiluminescence microscopy of cell membranes. J. Am. Chem. Soc. **140**(44) (2018). https://doi.org/10.1021/jacs.8b08080
38. Layman, B.R., Dick, J.E.: Electroprecipitating the sulfate radical anion amplifies electrochemiluminescence in space and time. J. Am. Chem. Soc. **146**(38), 26216–26222 (2024). https://doi.org/10.1021/jacs.4c07852
39. Layman, B.R., Dick, J.E.: Through-space electrochemiluminescence reveals bubble forces at remote phase boundaries. J. Am. Chem. Soc. **146**(1), 707–713 (2024). https://doi.org/10.1021/jacs.3c10505
40. Herchenbach, P.J., Layman, B.R., Dick, J.E.: Quantifying the interfacial tension of adsorbed droplets on electrified interfaces. J. Colloid Interface Sci. (2024)
41. Voci, S., Vannoy, K.J., Dick, J.E.: Femtoliter oil droplets act as CO_2 micropumps for uninterrupted electrochemiluminescence at the water|oil interface. J. Colloid Interface Sci. (2024). https://doi.org/10.1016/j.jcis.2024.01.100
42. Voci, S., Clarke, T.B., Dick, J.E.: Abiotic microcompartments form when neighbouring droplets fuse: an electrochemiluminescence investigation. Chem. Sci. **14**(9) (2022). https://doi.org/10.1039/d2sc06553c
43. Benjamin, I.: Theoretical study of the water/1,2-dichloroethane interface: structure, dynamics, and conformational equilibria at the liquid-liquid interface. J. Chem. Phys. **97**(2) (1992). https://doi.org/10.1063/1.463219
44. Walker, D.S., Brown, M.G., McFearin, C.L., Richmond, G.L.: Evidence for a diffuse interfacial region at the dichloroethane/water interface. J. Phys. Chem. B (2004). https://doi.org/10.1021/jp031147q
45. Walker, D.S., Richmond, G.L.: Depth profiling of water molecules at the liquid-liquid interface using a combined surface vibrational spectroscopy and molecular dynamics approach. J. Am. Chem. Soc. **129**(30) (2007). https://doi.org/10.1021/ja071740b
46. Glasscott, M.W., Dick, J.E., Dick, J.E.: Visualizing phase boundaries with electrogenerated chemiluminescence. J. Phys. Chem. Lett. **11**(12) (2020). https://doi.org/10.1021/acs.jpclett.0c01207
47. Rubinstein, I., Bard, A.J.: Electrogenerated Chemiluminescence. 37. Aqueous Eel Systems Based on Ru(2,2′-Bipyridine)$^{32+}$ and Oxalate or Organic Acids, vol. 103 (1981). https://pubs.acs.org/sharingguidelines
48. Glasscott, M.W., Voci, S., Kauffmann, P.J., Chapoval, A.I., Dick, J.E.: Mapping solvent entrapment in multiphase systems by electrogenerated chemiluminescence. Langmuir **37**(9) (2021). https://doi.org/10.1021/acs.langmuir.0c03445
49. Layman, B.R., Dick, J.E.: Phase-resolved electrochemiluminescence with a single luminophore. J. Phys. Chem. Lett. **14**(36), 8151–8156 (2023). https://doi.org/10.1021/acs.jpclett.3c01993
50. Yuan, Y., Li, J., Xu, G.: Chapter 4: electrochemiluminescence coreactants. In: RSC Detection Science, vol. 2020-January (2020). https://doi.org/10.1039/9781788015776-00092
51. Voci, S., Duwald, R., Grass, S., Hayne, D.J., Bouffier, L., Francis, P.S., Lacour, J., Sojic, N.: Self-enhanced multicolor electrochemiluminescence by competitive electron-transfer processes. Chem. Sci. **11**(17) (2020). https://doi.org/10.1039/d0sc00853b
52. Fu, Y., Xie, B., Liu, M., Hou, S., Zhu, Q., Kuhn, A., Zhang, L., Yang, W., Sojic, N.: Bipolar electrochemiluminescence at the water/organic interface. Chem. Sci. **15**(47), 19907–19912 (2024). https://doi.org/10.1039/D4SC06103A

53. Arias-Aranda, L.R., Salinas, G., Kuhn, A., Xu, G., Kanoufi, F., Bouffier, L., Sojic, N.: Complex electrochemiluminescence patterns shaped by hydrodynamics at a rotating bipolar electrode. Chem. Sci. **15**(23), 8723–8730 (2024). https://doi.org/10.1039/D4SC02528H
54. Sentic, M., Arbault, S., Goudeau, B., Manojlovic, D., Kuhn, A., Bouffier, L., Sojic, N.: Electrochemiluminescent swimmers for dynamic enzymatic sensing. Chem. Commun. **50**(71) (2014). https://doi.org/10.1039/c4cc04105d
55. Sentic, M., Loget, G., Manojlovic, D., Kuhn, A., Sojic, N.: Light-emitting electrochemical "swimmers." Angew. Chem. Int. Ed. **51**(45) (2012). https://doi.org/10.1002/anie.201206227
56. Bouffier, L., Zigah, D., Adam, C., Sentic, M., Fattah, Z., Manojlovic, D., Kuhn, A., Sojic, N.: Lighting up redox propulsion with luminol electrogenerated chemiluminescence. ChemElectroChem **1**(1) (2014). https://doi.org/10.1002/celc.201300042
57. Fu, Y., Xie, B., Liu, M., Hou, S., Zhu, Q., Kuhn, A., Zhang, L., Yang, W., Sojic, N.: Bipolar electrochemiluminescence at the water/organic interface. Chem. Sci. (2024). https://doi.org/10.1039/D4SC06103A
58. Feng, M., Dauphin, A.L., Bouffier, L., Zhang, F., Wang, Z., Sojic, N.: Enhanced cathodic electrochemiluminescence of luminol on iron electrodes. Anal. Chem. **93**(49), 16425–16431 (2021). https://doi.org/10.1021/acs.analchem.1c03139
59. Xia, H., Zheng, X., Li, J., Wang, L., Xue, Y., Peng, C., Han, Y., Wang, Y., Guo, S., Wang, J., Wang, E.: Identifying luminol electrochemiluminescence at the cathode via single-atom catalysts tuned oxygen reduction reaction. J. Am. Chem. Soc. **144**(17) (2022). https://doi.org/10.1021/jacs.2c00865
60. Ruiz-Lopez, M.F., Francisco, J.S., Martins-Costa, M.T.C., Anglada, J.M.: Molecular reactions at aqueous interfaces. Nat. Rev. Chem. (2020). https://doi.org/10.1038/s41570-020-0203-2
61. Deal, A.M., Rapf, R.J., Vaida, V.: Water-air interfaces as environments to address the water paradox in prebiotic chemistry: a physical chemistry perspective. J. Phys. Chem. A (2021). https://doi.org/10.1021/acs.jpca.1c02864
62. Ellison, G.B., Tuck, A.F., Vaida, V.: Atmospheric processing of organic aerosols. J. Geophys. Res. Atmos. **104**(D9) (1999). https://doi.org/10.1029/1999JD900073
63. Griffith, E.C., Vaida, V.: In situ observation of peptide bond formation at the water-air interface. Proc. Natl. Acad. Sci. USA **109**(39) (2012). https://doi.org/10.1073/pnas.1210029109
64. Wang, W., Qiao, L., He, J., Ju, Y., Yu, K., Kan, G., Guo, C., Zhang, H., Jiang, J.: Water microdroplets allow spontaneously abiotic production of peptides. J. Phys. Chem. Lett. **12** (2021). https://doi.org/10.1021/acs.jpclett.1c01083
65. Nam, I., Lee, J.K., Nam, H.G., Zare, R.N.: Abiotic production of sugar phosphates and uridine ribonucleoside in aqueous microdroplets. Proc. Natl. Acad. Sci. USA **114**(47) (2017). https://doi.org/10.1073/pnas.1714896114
66. Vannoy, K.J., Edwards, M.Q., Renault, C., Dick, J.E.: An electrochemical perspective on reaction acceleration in microdroplets. Ann. Rev. Anal. Chem. **17**(Volume 17, 2024), 149–171 (2024). https://doi.org/10.1146/annurev-anchem-061622-030919
67. Lee, J.K., Banerjee, S., Nam, H.G., Zare, R.N.: Acceleration of reaction in charged microdroplets. Q. Rev. Biophys. **48**(4) (2015). https://doi.org/10.1017/S0033583515000086
68. Ruiz-López, M.F., Martins-Costa, M.T.C.: Disentangling reaction rate acceleration in microdroplets. Phys. Chem. Chem. Phys. **24**(48) (2022). https://doi.org/10.1039/d2cp04998h
69. Banerjee, S., Gnanamani, E., Yan, X., Zare, R.N.: Can all bulk-phase reactions be accelerated in microdroplets? Analyst (2017). https://doi.org/10.1039/c6an02225a
70. Wei, Z., Li, Y., Cooks, R.G., Yan, X.: Accelerated reaction kinetics in microdroplets: overview and recent developments. Annu. Rev. Phys. Chem. (2020). https://doi.org/10.1146/annurev-physchem-121319-110654
71. Herchenbach, P.J., Layman, B.R., Dick, J.E.: Quantifying the interfacial tension of adsorbed droplets on electrified interfaces. J. Colloid Interface Sci. **674**, 474–481 (2024). https://doi.org/10.1016/j.jcis.2024.06.141
72. Knežević, S., Totoricaguena-Gorriño, J., Gajjala, R.K.R., Hermenegildo, B., Ruiz-Rubio, L., Vilas-Vilela, J.L., Lanceros-Méndez, S., Sojic, N., Del Campo, F.J.: Enhanced electrochemiluminescence at the gas/liquid interface of bubbles propelled into solution. J. Am. Chem. Soc. **146**(32), 22724–22735 (2024). https://doi.org/10.1021/jacs.4c07566

73. Vogel, Y.B., Evans, C.W., Belotti, M., Xu, L., Russell, I.C., Yu, L.J., Fung, A.K.K., Hill, N.S., Darwish, N., Gonçales, V.R., Coote, M.L., Swaminathan Iyer, K., Ciampi, S.: The corona of a surface bubble promotes electrochemical reactions. Nat. Commun. **11**(1) (2020). https://doi.org/10.1038/s41467-020-20186-0
74. Nami-Ana, S.F., Mehrgardi, M.A., Mofidfar, M., Zare, R.N.: Sustained regeneration of hydrogen peroxide at the water-gas interface of electrogenerated microbubbles on an electrode surface. J. Am. Chem. Soc. **146**(46), 31945–31949 (2024). https://doi.org/10.1021/jacs.4c11422
75. Chen, L., Zhu, X., Wei, J., Tian, L., Hu, C., Xiang, X., Zhou, S.-F.: Afterglow electrochemiluminescence from nitrogen-deficient graphitic carbon nitride. Anal. Chem. **95**(5), 2917–2924 (2023). https://doi.org/10.1021/acs.analchem.2c04566
76. Bois, R., Scarabino, S., Ravaine, V., Sojic, N.: Two-dimensional electrochemiluminescence: light emission confined at the oil-water interface in emulsions stabilized by luminophore-grafted microgels. Langmuir **33**(29) (2017). https://doi.org/10.1021/acs.langmuir.7b01585
77. Ben Trad, F., Delacotte, J., Lemaître, F., Guille-Collignon, M., Arbault, S., Sojic, N., Labbé, E., Buriez, O.: Shadow electrochemiluminescence imaging of giant liposomes opening at polarized electrodes. Analyst **149**(12), 3317–3324 (2024). https://doi.org/10.1039/D4AN00470A
78. Ben Trad, F., Carré, B., Delacotte, J., Lemaître, F., Guille-Collignon, M., Arbault, S., Sojic, N., Labbé, E., Buriez, O.: Electrochemiluminescent imaging of a NADH-based enzymatic reaction confined within giant liposomes. Anal. Bioanal. Chem. (2024). https://doi.org/10.1007/s00216-024-05133-y
79. Ben Trad, F., Delacotte, J., Guille-Collignon, M., Lemaître, F., Arbault, S., Sojic, N., Burlina, F., Labbé, E., Buriez, O.: Electrochemiluminescence imaging of liposome permeabilization by an antimicrobial peptide: melittin. Chem. Biomed. Imaging **1**(1) (2023). https://doi.org/10.1021/cbmi.3c00003
80. Ben Trad, F., Wieczny, V., Delacotte, J., Morel, M., Guille-Collignon, M., Arbault, S., Lemaître, F., Sojic, N., Labbé, E., Buriez, O.: Dynamic electrochemiluminescence imaging of single giant liposome opening at polarized electrodes. Anal. Chem. **94**(3) (2022). https://doi.org/10.1021/acs.analchem.1c04238
81. Gupta, V., Falciani, F., Layman, B.R., Hill, M.L., Rapino, S., Dick, J.E.: Real-time visualization of endogenous H_2O_2 production in mammalian spheroids by electrochemiluminescence. Chem. Biomed. Imaging (2025). https://doi.org/10.1021/cbmi.4c00105
82. Han, D., Sojic, N., Jiang, D.: Spatial profiling of multiple enzymatic activities at single tissue sections via fenton-promoted electrochemiluminescence. J. Am. Chem. Soc. **147**(11), 9610–9619 (2025). https://doi.org/10.1021/jacs.4c17749
83. Ma, Y., Colin, C., Descamps, J., Arbault, S., Sojic, N.: Shadow electrochemiluminescence microscopy of single mitochondria. Angew. Chem. Int. Ed. 60(34) (2021). https://doi.org/10.1002/anie.202105867
84. Layman, B.R., Krushinski, L.E., Dick, J.E.: Visualizing competing charge-transfer reactions with electrochemiluminescence. J. Chem. Educ. **102**(1), 437–443 (2025). https://doi.org/10.1021/acs.jchemed.4c00778
85. Beladi-Mousavi, S.M., Salinas, G., Bouffier, L., Sojic, N., Kuhn, A.: Wireless electrochemical light emission in ultrathin 2D nanoconfinements. Chem. Sci. **13**(48) (2022). https://doi.org/10.1039/d2sc04670a
86. Ibanez, J.G., Zavala-Araiza, D., Sotomayor-Martinez Barranco, B., Torres-Perez, J., Camacho-Zuniga, C., Bohrmann-Linde, C., Tausch, M.W.: A demonstration of simultaneous electrochemiluminescence. J. Chem. Educ. **90**(4) (2013). https://doi.org/10.1021/ed300701a
87. Bolton, E., Richter, M.M.: Light emission at electrodes: an electrochemiluminescence demonstration. J. Chem. Educ. **78**(5), 641 (2001). https://doi.org/10.1021/ed078p641
88. Han, Z., Zhang, L., Lu, X.: Sensitive detection of trace hydrogen peroxide via dual-emissive electrochemiluminescence from a luminol/porphyrin system: comprehensive innovative experiments on analytical instruments for undergraduates. J. Chem. Educ. **101**(3), 1248–1256 (2024). https://doi.org/10.1021/acs.jchemed.3c01240

89. De Poulpiquet, A., Sojic, N., Bouffier, L., Kuhn, A., Zigah, D.: Wireless electronic light emission: an introduction to bipolar electrochemistry. J. Chem. Educ. **100**(2) (2023). https://doi.org/10.1021/acs.jchemed.2c00573

Bipolar Electrochemiluminescence Imaging for Electroanalysis

Elena Villani, Kosuke Ino, Shinsuke Inagi, and Hitoshi Shiku

Abstract The coupling between bipolar electrochemistry and electrochemiluminescence (ECL) technologies has resulted in a fascinating field of the academic research, because the advantages of both techniques are often emphasized in their combined form. For instance, the light emission resulting from an electrochemical reaction typical of the ECL process can be exploited for reporting redox events occurring on a wireless bipolar electrode (BPE). In this chapter, we summarized recent trends of bipolar ECL imaging for electroanalytical purposes. These applications include the wireless detection and quantification of biological analytes exploiting ECL as the reporting technology, the measurement of potential and current gradients on BPEs with ECL, and the use of conducting polymers as electrode materials to trigger the ECL reaction.

Keywords Bipolar electrochemistry · Electrochemiluminescence · Cellular activity · Immunosensors · Potential mapping · Conducting polymers

1 Introduction and First Historical Contributions to Bipolar ECL

Bipolar electrochemistry was first demonstrated in the late 1960s through the application of fluidized bed electrodes and capillary gap cells for the industrial-scale electrochemical production of key commodity chemicals [1]. Subsequently, in the mid-1980s, Fleischmann and colleagues pioneered the use of bipolar electrochemistry to induce redox reactions on wireless, micrometer-scale conductive particles [2]. Despite these pioneering advances, the field remained largely overlooked until the

E. Villani (✉) · S. Inagi
Department of Chemical Science and Engineering, School of Materials and Chemical Technology, Institute of Science Tokyo, Yokohama, Japan
e-mail: villani.e@mct.isct.ac.jp

K. Ino · H. Shiku
Graduate School of Engineering, Tohoku University, Sendai, Japan

F. Paolucci et al. (eds.), *Electrogenerated Chemiluminescence Microscopy and Imaging*,
https://doi.org/10.1007/978-3-032-14602-1_11

early 2000s, when it underwent a significant revival within the academic community. Today, bipolar electrochemistry is employed across diverse fields, including materials synthesis [3–6], flow reactors [7–9], corrosion science [10], actuators [11], electroanalytical chemistry [12], and beyond.

Over two decades ago, two research groups independently published the first studies on the integration of bipolar electrochemistry with electrochemiluminescence (ECL). In one report, Manz and coworkers introduced a wireless ECL detector featuring a microfabricated "U"-shaped floating platinum (Pt) electrode positioned across the separation channel of a microfluidic device [13]. This configuration enabled the micellar electrokinetic chromatographic separation and ECL-based detection of two ruthenium complexes and three natural amino acids. In a parallel effort, the Crooks group developed a comparable microfluidics platform that combined electrochemical sensing with ECL signaling, achieving, for the first time, the chemical decoupling between the sensing reaction and the ECL emission using benzyl viologen as the target analyte [14]. Surprisingly, these pioneering works were not promptly built upon; it was not until the late 2000s that Crooks and colleagues further advanced the field by demonstrating an electrochemical DNA microarray sensor operated with just two driving electrodes, in which the faradaic current through an array of wireless bipolar electrodes was visualized via an ECL emission [15].

Subsequently, this approach was scaled up to an array of one thousand individual wireless electrodes, all controllable using only a pair of driving electrodes and a basic power supply [16]. Notably, Crooks and coworkers made substantial contributions to both the theoretical and experimental foundations of ECL at bipolar electrodes (BPEs), establishing a direct relationship between ECL emission intensity and the current passing through the BPE [17]. In a related development, the same group introduced an electroanalytical method that integrates BPEs and ECL to capture optical voltammograms in a single micrograph, with all data encoded spatially rather than temporally, as in traditional voltammetry [18].

These foundational investigations have greatly advanced the comprehension of bipolar ECL, highlighting the versatility and promise of this combined technique [19, 20], and have inspired numerous research groups worldwide to pursue analogous strategies across a wide range of applications [21].

2 Principles of Bipolar ECL for Analytical Applications

To explain the concept of bipolar electrochemistry and of a BPE one should consider a conductive material placed in an electrolytic solution between two driving electrodes that are connected through an external power supply (Fig. 1). When a potential difference is applied between the driving electrodes (E_{tot}), an electric field is generated in the electrolyte, which gradually decreases due to the resistance of the solution. When electroactive species, red_1 and ox_2, are introduced into the electrolyte, electrochemical reactions take place at opposite ends of the BPE, as described by Eqs. (1) and (2). Here, n_1 and n_2 represent the number of electrons transferred in each process, red_1/ox_1 and ox_2/red_2 denote the respective redox couples, and E_1° and E_2° are the standard potentials of the two reactions:

$$\mathrm{red}_1 \rightarrow \mathrm{ox}_1 + n_1\mathrm{e}^- \ (E_1^\circ) \tag{1}$$

$$\mathrm{ox}_2 + n_2\mathrm{e}^- \rightarrow \mathrm{red}_2 \ (E_2^\circ) \tag{2}$$

To trigger an electrochemical reaction on the BPE, the minimum applied potential difference (ΔV_{min}) must theoretically be at least equal to the difference between the standard potentials of the two redox couples [22], as expressed in Eq. (3):

$$\Delta V_{min} \geq E_1^\circ - E_2^\circ \tag{3}$$

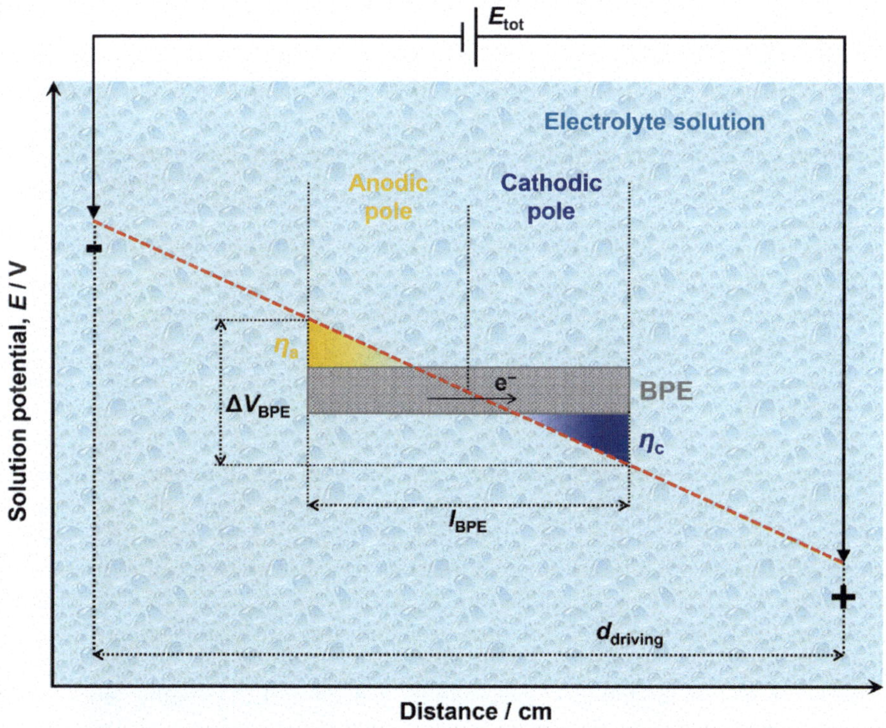

Fig. 1 Schematic representation showing the principle of bipolar electrochemistry in an open configuration: the gray rectangle represents the BPE, E_{tot} is the total applied voltage between the driving electrodes, l_{BPE} is the length of the BPE, $d_{driving}$ represents the distance between the two driving electrodes, ΔV_{BPE} represents the fraction of E_{tot} dropped along the BPE, and η_a and η_c represent the anodic and cathodic overpotentials, respectively, established on the BPE. The red dashed line represents the solution potential drop across the electrolyte [23]

To preserve the electroneutrality of the system, the anodic and cathodic redox processes are electrically coupled through the BPE and must proceed simultaneously at its opposite ends with equal rates. Such a case is rather different compared to conventional electrochemistry, where a working electrode is directly connected to a power supply and it is used to drive the faradaic reaction of interest [3].

The potential drop across the BPE (ΔV_{BPE}), representing the portion of E_{tot} experienced by the BPE, can be approximated using Eq. (4), under the assumption that the solution resistance is significantly lower than the overall faradaic resistance of the BPE [4, 24, 25]:

$$\Delta V_{BPE} = \left(E_{tot} / d_{driving} \right) l_{BPE}, \tag{4}$$

where E_{tot} is the total applied voltage, $d_{driving}$ represents the distance between the driving electrodes, and l_{BPE} is the length of the BPE. This implies that, for a given electric field ($E_{tot}/d_{driving}$), longer BPEs gain high values of ΔV_{BPE} and generate larger currents flowing through the BPE.

The foregoing theoretical description of bipolar electrochemistry applies to an "open" bipolar configuration, in which the BPE is fully immersed in the electrolyte and does not fully partition the cell, thereby permitting ionic continuity between the two driving electrodes. In contrast, a "closed" configuration physically isolates the solutions in contact with the anodic and cathodic poles of the BPE, for example, by embedding the BPE within an insulating layer that blocks the communication between the two half-cells. This arrangement enables the spatial separation of reducing and oxidizing agents, which is advantageous when reaction products or reactants must be isolated for compatibility reasons, or when one pole of the BPE is employed as a reporter for electrochemical processes occurring at the opposite pole. Nevertheless, ECL imaging applications can be easily integrated with both bipolar configurations.

The contactless nature of faradaic reactions occurring at the BPE, although fascinating and relatively simple to exploit in the most diverse domains, additionally implicates the difficulty to precisely determine the magnitude of both current and potential established on this type of electrodes during the progress of the reaction. Even though several experimental methods have been developed to overcome this limitation [23, 26], the coupling of bipolar electrochemistry with other optical readout methods, such as ECL, is a successful and straightforward strategy to visualize redox processes occurring at the BPE level. In addition, the ease of operation of the combined methodologies, generally by generating ECL via the "coreactant pathway" represents a simple way to achieve analytical detection. Generally, three approaches can be exploited for the bipolar ECL detection [19]:

(i) The target analyte can be either the coreactant or the luminophore involved in the ECL process (Fig. 2, upper sketch),
(ii) The target analyte is directly involved in the ECL process, causing quenching or enhancement of the light emission (Fig. 2, upper sketch),

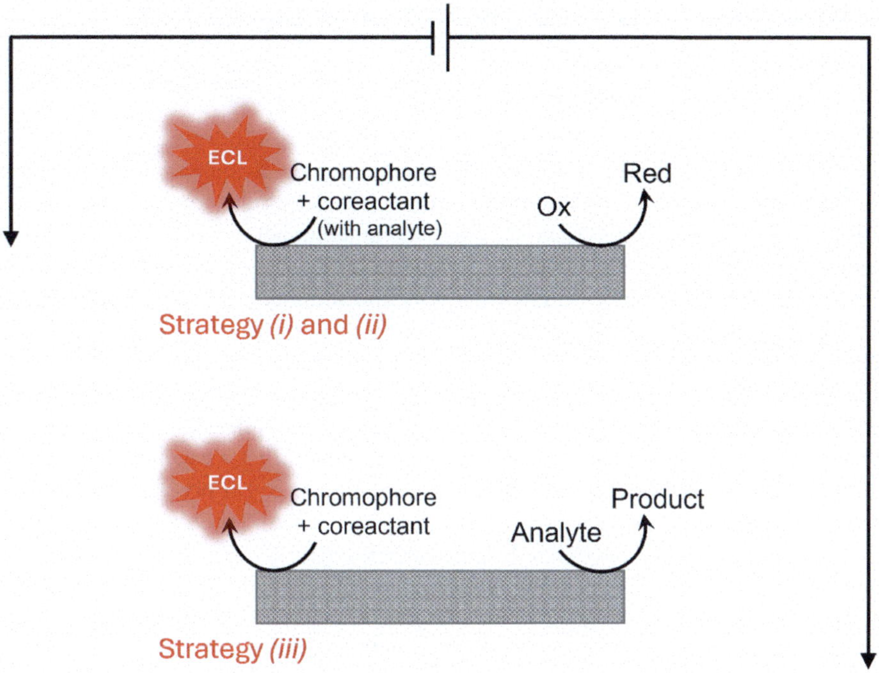

Fig. 2 Coupling of bipolar electrochemistry and ECL for analytical sensing resulting in the direct method for analyte detection (strategies (i) and (ii), upper sketch), and in the indirect method (strategy (iii), lower sketch) [19]

(iii) The electrochemical reaction of the target analyte at one edge of the BPE is electrically coupled with an ECL reaction occurring on the other BPE edge, offering visual readout of the analyte reaction progress (Fig. 2, lower sketch).

While the first two methods provide a direct bipolar ECL readout because the target analyte is directly involved in the ECL process, the third approach can be considered as an indirect bipolar reporting method, because the sensing and reporting functions are physically separated [14, 27]. However, by exploiting the last approach, the detection of a larger pool of analytes can be extended to molecules that are not directly involved in the ECL emission process and the concentration of the analyte can be directly correlated to the ECL intensity [17].

3 Examples of Analytical Applications of Bipolar ECL

3.1 Bipolar ECL Analysis of Cellular Activity

The analysis of cell culture is crucial for regenerative medicine and in vitro assays for drug screening [28, 29]. For these assays, it is important to evaluate cellular activity in a high-throughput manner due to the inherent heterogeneity of cells [30]. Electrochemical sensors are widely used to evaluate cellular activities because of their excellent features such as high sensitivity and simplicity [31]. For example, the oxygen consumption rate of cultured cells can be precisely evaluated using scanning electrochemical microscopy (SECM) [32]. In addition to single-cell analysis, three-dimensional (3D) cultured cells, including spheroids [33], have also been investigated. Recently, ECL imaging has gained widespread application in cell analysis because of its unique advantages [34, 35]. For example, ECL reagents were added to the measurement solution with spheroids, and the oxygen concentration near the spheroids was clearly visualized in ECL images [36]. As the oxygen consumption rate reflects cellular metabolic activity, this approach is useful for assessing the quality of cells. However, cells are assayed in the presence of ECL reagents, which may affect the results of drug screening. To eliminate the influence of these reagents, closed bipolar ECL devices are proposed as they allow the separation of assay and reporter chambers.

Previously, BPE-ECL devices have been reported to measure cellular respiratory activity (Fig. 3) [37]. Closed BPE arrays were designed to physically separate the assay and reporter chambers (Fig. 3a) to address the aforementioned issue. The electrodes were fabricated using Pt sputtering. The reporter chamber was filled with $[Ru(bpy)_3]^{2+}$ and tri-*n*-propylamine (TPrA). In this study, a bias of 1.5 V was applied between the driving electrodes. The electrons generated at the anodic poles of the BPEs in the reporter chambers were transferred to oxygen (O_2) in the assay chambers, and the O_2 was reduced to water (H_2O) at the Pt BPEs. The ECL emission from $[Ru(bpy)_3]^{2+}$ depends on the O_2 reduction at the cathodic poles; therefore, the oxygen concentration can be determined by monitoring ECL signals. When cancer spheroids were placed on the cathodic poles (Fig. 3b), they consumed O_2 because of their metabolic activity. Consequently, the ECL signals decreased in the presence of the spheroids. In addition to spheroid analysis, this concept has also been applied to evaluate the respiratory activity of *Escherichia coli* and *Pseudomonas aeruginosa* [38]. These BPE-ECL devices are advantageous over conventional electrochemical imaging obtained using SECM, because they do not require probe scanning. However, the spatial resolution of the obtained images is inferior to that using SECM. To address this issue, massive BPE arrays have been reported, as described in the next section.

In addition to the analysis of oxygen consumption rate, reactive oxygen species (ROS) production from cells was assessed using BPE-ECL devices. Hydrogen peroxide (H_2O_2) is one of the ROS molecules and an important molecule for cellular signaling. A BPE-ECL device with Pt driving electrodes and a gold (Au) BPE was used to detect H_2O_2 released from leukemia cells [39]. In this study, the cells were

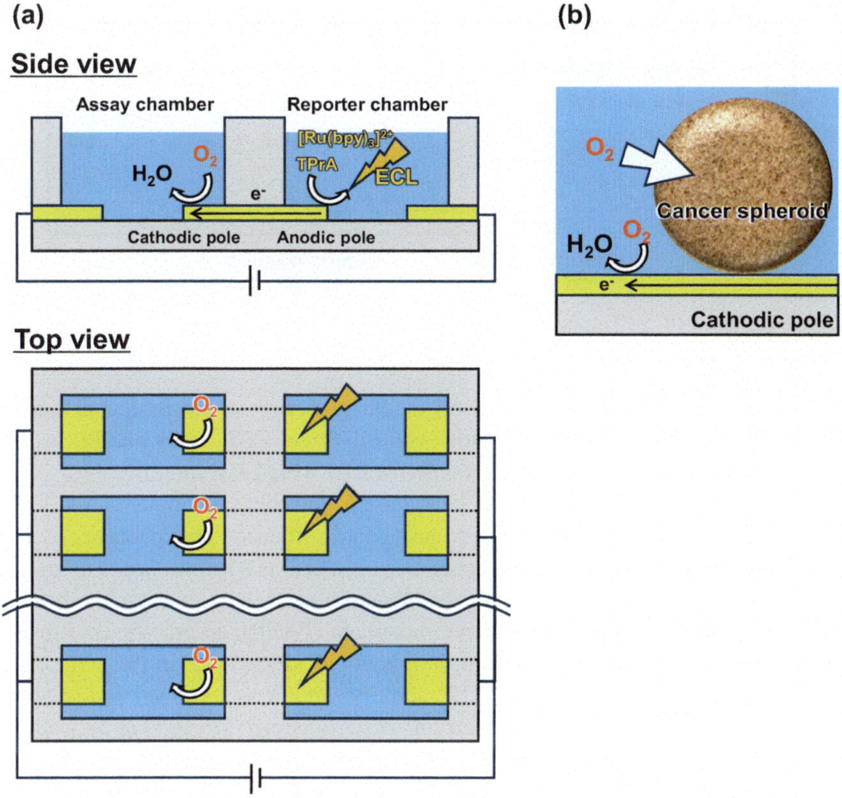

Fig. 3 Strategy of BPE-ECL assay for measuring the respiratory activity of cells. **a** Schematic of the device and electrochemical reactions for O_2 detection (cells are not depicted). **b** Schematic for spheroid analysis

introduced at the cathodic pole in the assay chamber and stimulated with ascorbic acid to induce H_2O_2 production. Then, the released H_2O_2 was reduced at the cathodic pole. The reporter chamber was filled with luminol and H_2O_2 to generate luminol-ECL emission (Fig. 4). The limit of detection (LOD) of the device is 1.8 nM (12 cells), which is highly sensitive compared to amperometric biosensors for cell analysis. As H_2O_2 is a product of oxidase activity, enzymatic substrates and oxidase activity can be evaluated by monitoring H_2O_2 with BPE-ECL devices. For example, glucose concentration can be determined using this approach [40]. In this study, a Prussian blue-based glucose sensor was set at the cathodic pole to allow the catalytic reduction of H_2O_2, which is enzymatically produced through the reaction between O_2 and glucose at glucose oxidase. At the anodic pole, luminol was oxidized to produce ECL emission. Although this study did not focus on cell analysis, this concept can be applied to such applications because glucose is one of the most important molecules for cellular metabolism. Even without the use of bipolar electrochemistry, H_2O_2 released from cells can be visualized in luminol-ECL images

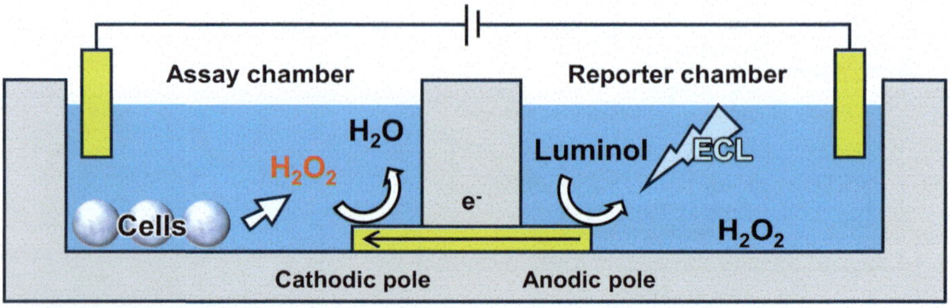

Fig. 4 Strategy of BPE-ECL assay for measuring H_2O_2 released from cells

[41]. In this system, cells were assayed in the presence of luminol, and luminol reacts directly with released H_2O_2. However, the direct ECL assay conditions may affect cell functions, which could be a drawback of BPE-ECL imaging.

BPE-ECL devices are also used to evaluate cellular enzymatic activity. For example, a device was applied to detect endogenous alkaline phosphatase (ALP) in single HepG2 cells, liver cancer cells (Fig. 5) [42]. ALP is a biomarker to assess cell differentiation and cell quality. HepG2 cells were placed in the assay chamber. The chamber was filled with *p*-aminophenyl phosphate (PAPP), an enzymatic substrate for ALP. PAPP was hydrolyzed to *p*-aminophenol (PAP) at ALP in the cells; therefore, the amount of PAP indicates ALP activity. PAP was oxidized at the anodic pole of the BPE, and the resulting electrons were transferred to H_2O at the cathodic pole to reduce H_2O in the reporter chamber. Furthermore, the reporter chamber was filled with $[Ru(bpy)_3]^{2+}$ and TPrA, and these reagents were oxidized at the driving electrode for ECL emission. Thus, the place of ECL emission differs, allowing the reductant, such as PAP, to be measured using the BPE-ECL system.

PAP is measured using redox cycling with an interdigitated array (IDA) electrode [43], and an ECL assay using an IDA electrode has been reported (Fig. 6) [44]. In the assay chamber, the IDA electrodes consisting of the driving electrode (+) and cathodic pole were set. The reporter chamber contained the anodic pole and the

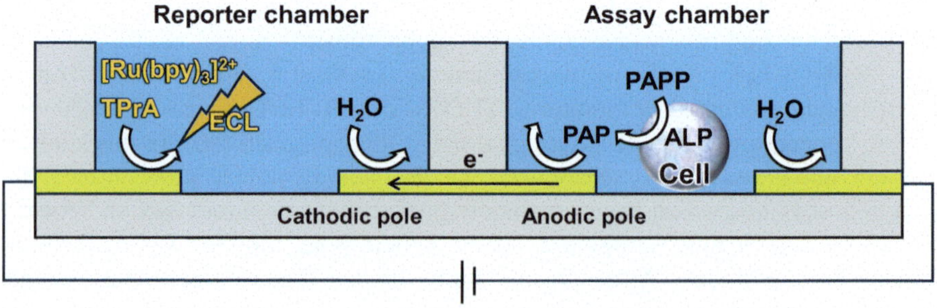

Fig. 5 Strategy of BPE-ECL assays to measure endogenous ALP activity in cells

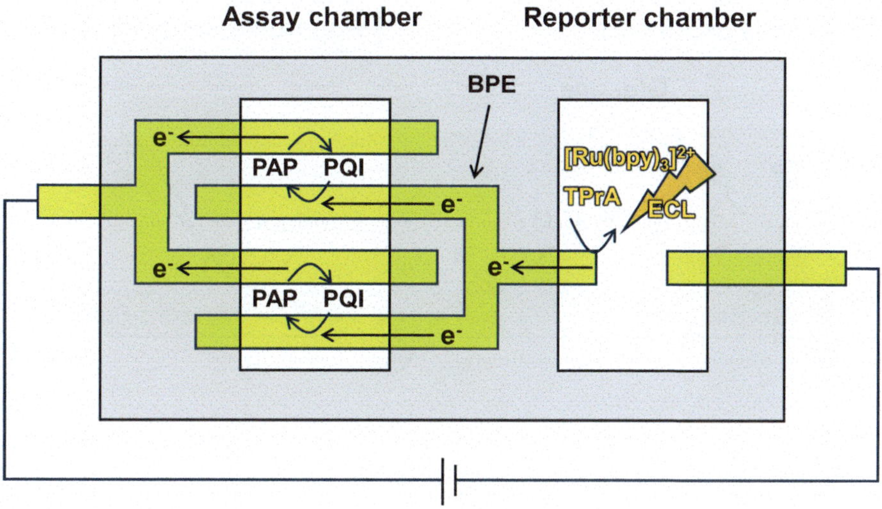

Fig. 6 Strategy of BPE-ECL assays based on redox cycling to measure ALP activity. PAPP is hydrolyzed by ALP in cells to generate PAP. Cells are not illustrated to simplify the scheme

driving electrode (−). In the presence of ALP-labeled cells, the resulting PAP was oxidized to *p*-quinoneimine (PQI) at the driving electrode (+). As the redox cycling was induced, PQI was reduced at the cathodic pole. Therefore, $[Ru(bpy)_3]^{2+}$ and TPrA were oxidized at the anodic pole to generate the ECL emission. In general, redox cycling can improve sensitivity because signals can be amplified.

3.2 *Bipolar ECL in Immunosensors and Aptasensors*

Electrochemical biosensors have been widely used in diagnostics, food analysis, and environmental assessments [45–47]. These biosensors consist of transducers, such as electrodes, and bioreceptors, such as antibodies and nucleic acids. In electrochemical biosensor-based immunoassays, antibodies are immobilized on electrodes to capture target analytes. In a typical sandwich immunoassay, the captured analytes are further modified with secondary antibodies conjugated to tags, such as enzymes and nanomaterials. These tags provide signals at the electrodes.

By placing these biosensor elements at the cathodic poles of closed BPEs, ECL-based immunoassays can be conducted. In an immunoassay using a BPE-ECL device (Fig. 7) [48], glucose oxidase (GO*x*) was used as a tag. The GO*x* generated H_2O_2 in the presence of prostate-specific antigen (PSA). In this study, a Fe-single-atom catalyst with catalase and peroxidase-like activity was immobilized on the cathodic pole to reduce H_2O_2 to water. Catalyzing the decomposition of H_2O_2 enhanced the luminol-ECL emission at the anodic pole. The ECL intensity was recorded with a potential scan from 0 to 2.6 V. The LOD is 0.62 pg/mL. In another immunosensor

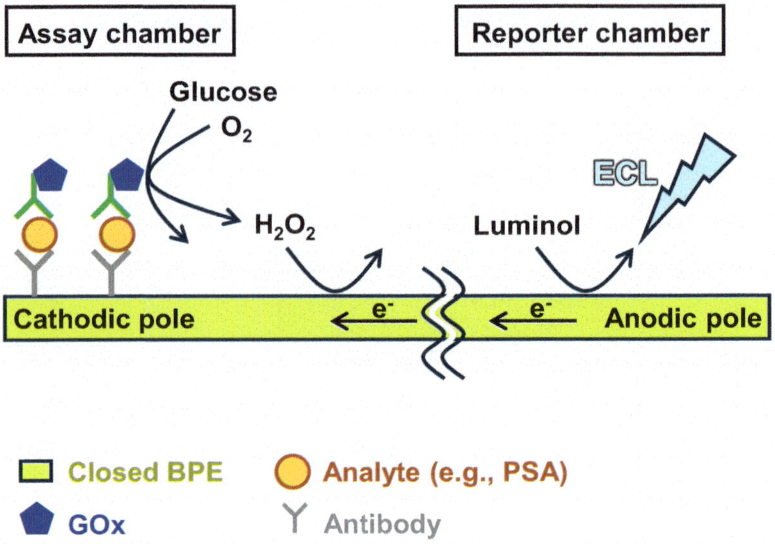

Fig. 7 Strategy for BPE-ECL immunoassays using GO*x* as a tag

using a closed BPE device [49], thionine-doped silica nanoparticles were conjugated with antibodies and used as electrochemical tags. An immunoreaction was induced at cathodic poles to capture cancer biomarkers. By applying a suitable potential between the driving electrodes, thionine (Ox) was reduced at the cathodic poles, and $[Ru(bpy)_3]^{2+}$ and TPrA were oxidized at the anodic poles in a reporter chamber for ECL emission. Thus, some tags have been proposed for BPE-ECL-based immunoassays.

In addition to antibodies, aptamers have been widely used as bioreceptors in electrochemical biosensors [50]. Aptamers are oligomers or peptides that bind tightly to a broad range of targets. DNA aptamers have been extensively utilized in BPE-ECL assays. For instance, an aptamer was employed to PSA in a BPE-ECL system [51, 52]. A probe DNA was immobilized at the cathodic pole where capture DNA was present, because of the hybridization in the presence of the aptamer. DNA exhibits remarkable properties beyond binding like an aptamer; it can also possess enzymatic functions like a DNAzyme [53] and form nanostructures [54] by properly designing its sequence. Given these advantages, the development of DNA-based sensors is expected to progress further in the future.

3.3 Massive Arrays of BPEs Embedded into Thin Films for ECL Imaging with High Spatial Resolutions

The aforementioned devices are useful for bioanalysis; however, an issue with electrode density was identified because both cathodic and anodic poles need to be arranged two-dimensionally. To improve electrode density, vertical arrangements of cathodic and anodic poles in BPE arrays have been reported (Fig. 8) [55–58]. For fabrication, Au was electrodeposited into the pores of a track-etched membrane with electroless plating, resulting in a 41-μm interval BPE array [57]. In another process, SU-8 was patterned with lithography and pyrolyzed to prepare arrays of 8-μm diameter carbon electrodes. Then, Parylene C was deposited to obtain a thin film. Finally, the Parylene film embedding BPEs was collected [56]. Anodic aluminum oxide (AAO) templates were also used for device fabrication. Briefly, Au was electrodeposited inside the AAO pores [58]. Using this concept, the sensor pitch was shortened to less than 1 μm, resulting in high sensor density. In addition to these devices, commercially available anisotropic conductive films can also be used as BPE films for ECL imaging [55].

One of these devices has been applied to evaluate the respiratory activity of spheroids (Fig. 8) [57]. This type of device can visualize the spatial distribution of target analytes, which is advantageous compared to the aforementioned device. In addition to oxygen consumption, the cell adhesion region was successfully visualized at the single-cell level via oxygen reduction at the cathodic poles [58].

This type of device containing nano BPE arrays has also been applied to evaluate ALP activity as a label [59]. In this study, epidermal growth factor receptor (EGFR) in cancer cells, a biomarker of diseases, is placed on the anodic poles in the assay

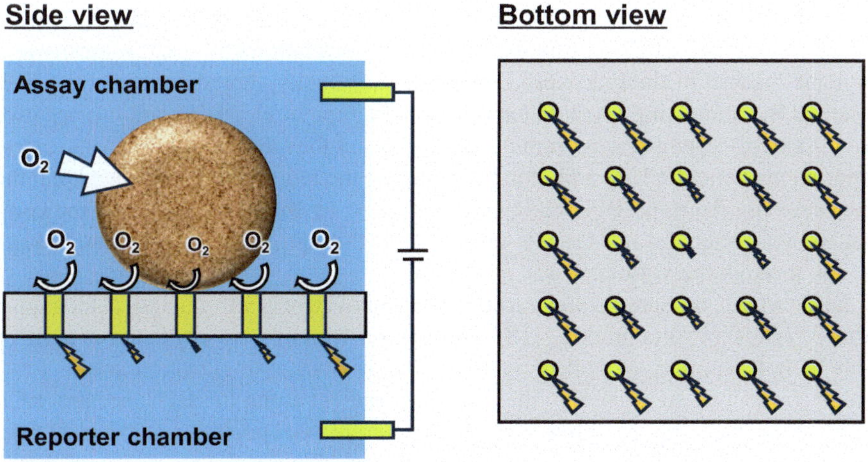

Fig. 8 Massive arrays of BPEs embedded into thin films for ECL imaging with high spatial resolutions to measure the respiratory activity of cell spheroids

chamber. The cells were labeled with antibodies conjugated with ALP. PAPP was used as an enzymatic substrate, and PAP was oxidized at the anodic poles of the AAO membrane where BPEs were embedded. At the cathodic poles in the reporter chamber, cadmium selenide quantum dots (CdSe QDs) were immobilized for ECL emission. $S_2O_8^{2-}$ was used as a coreactant to improve the ECL intensity. Thus, the QDs are useful for cathodic ECL emission, and the nanoelectrode arrays provide high-resolution images.

This type of device has also been applied to BPE-fluorescence imaging as well as BPE-ECL imaging [60]. In the reporter chamber of this study, resazurin was reduced to resorufin, and resorufin can be measured by fluorescence. As such, there are many potential luminophores, and the search for them is ongoing [61].

3.4 ECL Imaging as a Readout Method for Wireless Electrochemical Reactions

Because BPEs are contactless electrodes and electrochemical reactions occur without the need of an ohmic contact, potential and current density distributions cannot be precisely measured, since the contactless nature of this type of electrodes hampers the direct monitoring of these parameters [23]. On the other hand, ECL imaging is a reliable strategy to overcome this limitation because the optical readout, i.e., the light emission, is a function of the potential and it can be related to the current that flows through the wireless electrode.

Indeed, by employing a split-BPE configuration, in which two conductive elements immersed in the electrolyte are electrically linked via an ammeter allowing direct measurement of current flow upon application of a constant voltage between the driving electrodes [62], Crooks and coworkers successfully quantified the current induced across the BPE and established a direct correlation with ECL intensity as a photonic readout of the redox process [17]. Additionally, they determined the BPE potential by connecting a voltmeter between the bipolar electrode and the negative driving electrode, which was grounded and served as the reference ($E = 0$ V). Control experiments using ECL as a photonic reporter of the redox reaction showed that the light intensity from the BPE remained unaffected by the voltmeter. This indicates that the voltmeter did not impede current flow through the BPE and that leakage current through the voltmeter was negligible.

Furthermore, the same group introduced an innovative electroanalytical technique termed "snapshot voltammetry" [18]. This approach blends two key features of BPEs with ECL to capture an optical voltammogram from a single micrograph. First, faradaic reactions are driven by the potential gradient along the BPE; second, ECL serves as a visual reporter of reaction rate at different axial positions on the electrode. By analyzing the ECL intensity profile across axial locations in one micrograph, the researchers determined the interfacial potential as a function of position and directly correlated ECL intensity with the potential (Fig. 9).

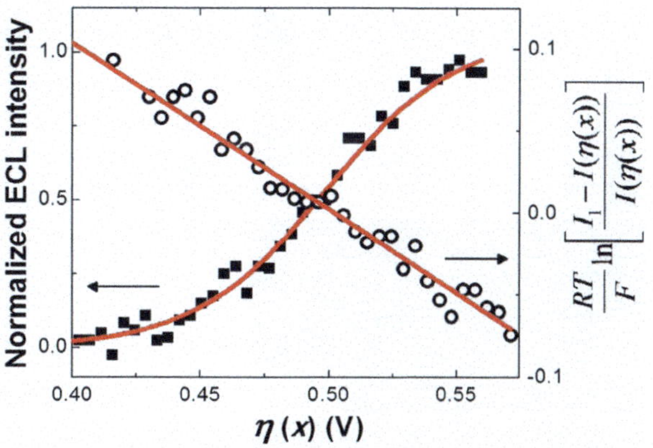

Fig. 9 Correlation between an ECL intensity profile (left *y*-axis) with data measured by snapshot voltammetry (right *y*-axis) proposed by Crooks and coworkers. Reprinted with permission from ref. [18]. Copyright 2010, American Chemical Society

Inspired by these previous contributions, our group developed an analytical tool for mapping the potential difference established over a BPE by imaging the location where the potential is suitable to trigger an ECL reaction and by monitoring its evolution as a function of E_{tot} in real time (Fig. 10a) [63]. Since the ECL intensity is only dependent on the axial location parallel to the electric field [18], the key advantage of this approach is that it enables the correlation of the interfacial potential difference across a BPE (ΔE) with the distance (Δd) by measuring the ECL emission intensity (Fig. 10b and c). This generates a spatial map of the anodic BPE potential across multiple bipolar electrochemical cells with varying geometries and driving electrode configurations.

By fully utilizing the anodic polarization zone, Hogan, Sojic, and coworkers showed that the potential gradient along a BPE can simultaneously resolve multiple ECL emissions from luminophores with different oxidation potentials [64]. Precise control over the size and position of ECL-emitting regions was achieved by adjusting the applied electric field, enabling detailed analysis of multicolor emission to demonstrate both spatial and spectral separation of multiple emitters in solution.

In a separate study, Suzuki and coworkers examined the impact of incorporating an Ag/AgCl reference electrode into the BPE structure on the interfacial potential distribution across its surface [65]. In an open bipolar setup, oxygen reduction at the cathodic pole was paired with an ECL reaction at the anodic pole. Their results showed that ECL intensity varied with both the position of the Ag/AgCl electrode and the chloride ion concentration, enabling precise control of the interfacial potential on the BPE surface.

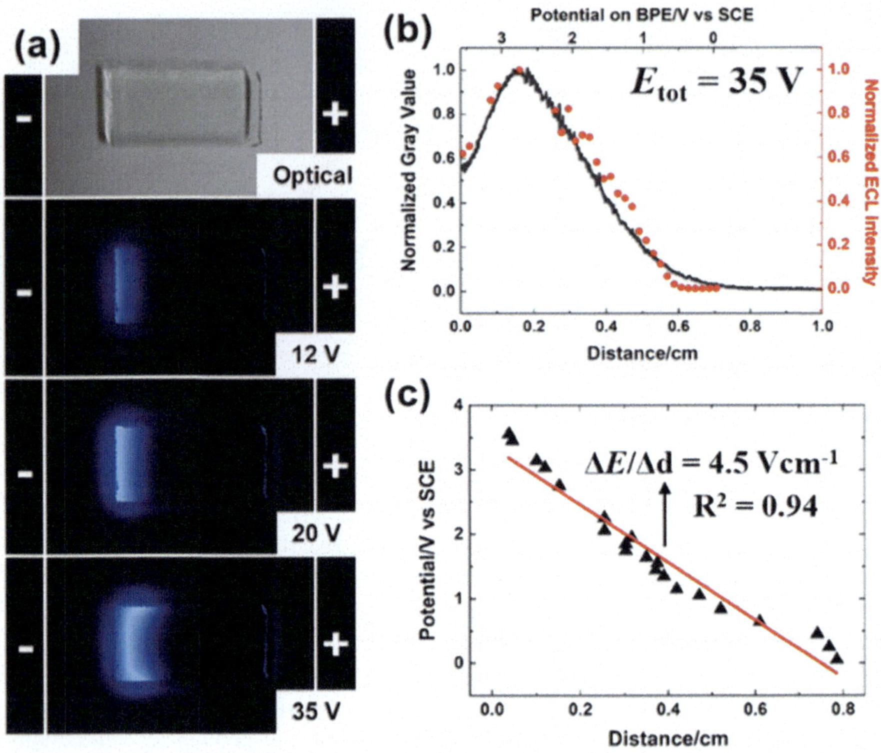

Fig. 10 **a** ECL photographs of an ITO BPE at different applied voltages; **b** correlation between a two-dimensional ECL intensity profile at 35 V (black curve) and the calibration curve developed in this study (red dots), which was then used to generate the plot in **c** reporting the potential difference at the BPE with respect to the distance starting from the anodic BPE edge where ECL emission is generated. Reprinted with permission from ref. [63]. Copyright 2021, American Chemical Society

4 Conducting Polymers as BPEs for ECL Imaging

The excellent conductivity of various conducting polymers (CPs) is frequently harnessed in ECL sensing applications, as it catalyzes the electrode reactions that produce light, thereby boosting ECL intensity [66, 67]. However, the direct use of pristine CPs as electrode materials for ECL generation remains rarely documented in the literature.

We have thus investigated the feasibility of generating ECL directly on the surface of pristine conducting polymer (CP) films, such as poly(3,4-ethylenedioxythiophene) (PEDOT) and polypyrrole (PPy), using bipolar electrochemistry [68]. Dense, uniform CP films were electrodeposited onto stainless-steel wires via constant-current electrolysis and subsequently detached by immersion in methanol. This yielded free-standing, rigid films that served as BPEs to wirelessly drive the electrochemical oxidation of luminol and H_2O_2. The resulting ECL emission was visible to the naked eye and captured with a standard digital camera in a darkened setting. Electrochemical and ECL imaging analysis were employed to correlate the ECL emission patterns on PEDOT and PPy films with E_{tot} and film conductivity.

In our subsequent work, we extended this study to PEDOT thin films and two distinct PEDOT-based composite films produced via alternating current (AC)-bipolar electropolymerization [69], a technique that imparts anisotropic electrical properties to the resulting CP films [3]. Initially, an Au wire placed on a non-conductive substrate (e.g., a glass slide) served as a BPE to enable the in-plane growth of a PEDOT thin film through AC-bipolar electropolymerization of the EDOT monomer. In a following AC-bipolar electrolysis step, these PEDOT films were reused as BPEs to drive either the cathodic electrodeposition of copper particles, yielding PEDOT-Cu composite films, or the anodic electropolymerization of thiophene, producing PEDOT-polythiophene (PEDOT-PT) composite films. All three films (PEDOT, PEDOT-Cu, and PEDOT-PT) were then employed as BPEs to initiate ECL emission from the luminol/H_2O_2 system. By spatially mapping the light distribution and intensity, we assessed their ECL performance via imaging analysis as a function of E_{tot}. Our findings revealed that all three polymer films functioned as suitable ECL electrode materials, though with distinct behaviors influenced by their composition.

We further explored the application of CP films as electrode materials for the electrochemical fabrication of patterned electrodes featuring photoluminescent and electrochemiluminescent colored patterns. This was achieved via a straightforward electrochemical approach that enables precise and facile control over the deposition sites of aggregation-induced emission (AIE) molecules [70]. The method utilizes micelles formed by a redox-active surfactant, within which the target AIE molecule is encapsulated and solubilized in aqueous media. Upon electrochemical stimulation (i.e., surfactant oxidation), the micelles rupture, releasing and depositing the AIE molecule onto the electrode surface. This mechanism, termed the electrolytic micelle disruption (EMD) method [71], has also been integrated with bipolar electrochemistry to wirelessly produce gradient and patterned organic thin films [72]. We initiated our study of this electrochemical method using Au-decorated working electrodes in a standard electrolytic setup, where "AIE" and "TIT" patterns were created on a glass substrate via sputtering. Several types of AIE molecules were incorporated into the micellar structure of α-(11-ferrocenylundecyl)-ω-hydroxypoly(ethyleneglycol) (FPEG) surfactant, and after electrochemical oxidation, the cleavage of the FPEG micelles resulted in the release of the entrapped AIE molecules only on the electroactive area where the EMD process occurred, giving to the interested region a perceptible and homogeneous luminescence under ultraviolet (UV) light irradiation.

5 Conclusions and Future Perspectives

The marriage between bipolar electrochemistry and ECL imaging, that is bipolar ECL imaging, continues to fascinate the electrochemical and analytical research communities, since the ECL emission can be exploited to observe the wireless reaction mode typical of the bipolar electrochemical systems, or can be used to detect and quantify the target analyte of interest. Indeed, in this chapter, we have seen that this light emission can be exploited to map the potential distribution established in bipolar electrochemical systems having different shapes, or used to quantify the current flowing through the BPE. On the other hand, we have also reported some examples of BPE-ECL devices that have been applied for the analysis of single cells and spheroids. In the near future, they are expected to be further utilized for the analysis of more complex cultured cells such as organoids [73] and microphysiological systems [74], and applied for drug screening. Thus, biosensors have been incorporated into BPE-ECL systems for bioanalysis and it is also expected that these systems will be used as wearable devices [75], although they still require fine-tuning. We are confident that BPE-ECL imaging systems will continue to flourish in many applications and developing technologies in the next future.

Acknowledgements The authors acknowledge financial support from Kakenhi Grant-in-Aid (No. JP23K26694) from the Japan Society from the Promotion of Science (JSPS).

References

1. Villani, E., Inagi, S.: Bipolar electrochemistry for synthesis in science of synthesis: electrochemistry in organic synthesis (Ed. Ackermann L), vol. 1, pp. 153–171. Thieme, Stuttgart (2021)
2. Fleischmann, M., Ghoroghchian, J., Rolison, D., Pons, S.: Electrochemical behavior of dispersions of spherical ultramicroelectrodes. J. Phys. Chem. **90**, 6392–6400 (1986)
3. Shida, N., Zhou, Y., Inagi, S.: Bipolar electrochemistry: a powerful tool for electrifying functional material synthesis. Acc. Chem. Res. **52**, 2598–2608 (2019)
4. Loget, G., Zigah, D., Bouffier, L., Sojic, N., Kuhn, A.: Bipolar electrochemistry: from materials science to motion and beyond. Acc. Chem. Res. **46**, 2513–2523 (2013)
5. Shi, Y., Villani, E., Chen, Y., Zhou, Y., Chen, Z., Hussain, A., Xu, G., Inagi, S.: High-throughput electrosynthesis of gradient polypyrrole film using a single-electrode electrochemical system. Anal. Chem. **95**, 1532–1540 (2023)
6. Chen, Z., Zhou, Y., Villani, E., Shida, N., Tomita, I., Inagi, S.: AC-bipolar electropolymerization of 3,4-ethylenedioxythiophene in ionic liquids. Langmuir **39**, 4450–4455 (2023)
7. Kirino, R., Iwai, S., Sato, K., Villani, E., Hasegawa, G., Inagi, S.: Comprehensive study of shape of channels filled with porous materials in streaming potential cells. Electrochemistry **93**, 077003 (2025)
8. Iwai, S., Suzuki, T., Sakagami, H., Miyamoto, K., Chen, Z., Konishi, M., Villani, E., Shida, N., Tomita, I., Inagi, S.: Electropolymerization without an electric power supply. Commun. Chem. **5**, 66 (2022)
9. Suzuki, R., Iwai, S., Kirino, R., Sato, K., Konishi, M., Hasegawa, G., Ishizuka, N., Matsukawa, K., Tanaka, K., Villani, E., Inagi, S.: An electrochemiluminescence device powered by

streaming potential for the detection of amines in flowing solution. Nat. Commun. **16**, 8217 (2025)

10. Zhou, Y., Mahmood, S., Engelberg, D.L.: A novel high throughput electrochemistry corrosion test method: bipolar electrochemistry. Curr. Opin. Electrochem. **39**, 101263 (2023)
11. Salinas, G., Arnaboldi, S., Bouffier, L., Kuhn, A.: Recent advances in bipolar electrochemistry with conducting polymers. ChemElectroChem **9**, e202101234 (2022)
12. Rahn, K.L., Anand, R.K.: Recent advancements in bipolar electrochemical methods of analysis. Anal. Chem. **93**, 103–123 (2021)
13. Arora, A., Eijkel, J.C.T., Morf, W.E., Manz, A.: A wireless electrochemiluminescence detector applied to direct and indirect detection for electrophoresis on a microfabricated glass device. Anal. Chem. **73**, 3282–3288 (2001)
14. Zhan, W., Alvarez, J., Crooks, R.M.: Electrochemical sensing in microfluidic systems using electrogenerated chemiluminescence as a photonic reporter of redox reactions. J. Am. Chem. Soc. **124**, 13265–13270 (2002)
15. Chow, K.F., Mavré, F., Crooks, R.M.: Wireless electrochemical DNA microarray sensor. J. Am. Chem. Soc. **130**, 7544–7545 (2008)
16. Chow, K.F., Mavré, F., Crooks, J.A., Chang, B.Y., Crooks, R.M.: A large-scale, wireless electrochemical bipolar electrode microarray. J. Am. Chem. Soc. **131**, 8364–8365 (2009)
17. Mavré, F., Chow, K.F., Sheridan, E., Chang, B.Y., Crooks, J.A., Crooks, R.M.: A theoretical and experimental framework for understanding electrogenerated chemiluminescence (ECL) emission at bipolar electrodes. Anal. Chem. **81**, 6218–6225 (2009)
18. Chang, B.Y., Mavré, F., Chow, K.F., Crooks, J.A., Crooks, R.M.: Snapshot voltammetry using a triangular bipolar microelectrode. Anal. Chem. **82**, 5317–5322 (2010)
19. Bouffier, L., Arbault, S., Kuhn, A., Sojic, N.: Generation of electrochemiluminescence at bipolar electrodes: concepts and applications. Anal. Bioanal. Chem. **408**, 7003–7011 (2016)
20. Bouffier, L., Manojlovic, D., Kuhn, A., Sojic, N.: Advances in bipolar electrochemiluminescence for the detection of biorelevant molecular targets. Curr. Opin. Electrochem. **16**, 28–34 (2019)
21. Villani, E.: Imaging wireless electrode reactions by bipolar electrochemiluminescence. Electrochemistry **92**, 101007 (2024)
22. Fosdick, S.E., Knust, K.N., Scida, K., Crooks, R.M.: Bipolar electrochemistry. Angew. Chem. Int. Ed. **52**, 10438–10456 (2013)
23. Villani, E., Inagi, S.: Experimental methods for measuring potential and current density distributions at bipolar electrodes. Anal. Chem. **97**, 5837–5846 (2025)
24. Mavré, F., Anand, R.K., Laws, D.R., Chow, K.F., Chang, B.Y., Crooks, J.A., Crooks, R.M.: Bipolar electrodes: a useful tool for concentration, separation, and detection of analytes in microelectrochemical systems. Anal. Chem. **82**, 8766–8774 (2010)
25. Bouffier, L., Zigah, D., Sojic, N., Kuhn, A.: Recent advances in bipolar electrochemistry in Electroanalytical Chemistry—A Series of Advances, vol. 27, pp. 27–117. In: Bard, A.J., Zoski, C.G. (eds.), CRC Press, Boca Raton
26. Fiorani, A., Eβmann, V., Santana Santos, C., Schuhmann, W.: Enhancing electrogenerated chemiluminescence on platinum electrodes through surface modification. ChemElectroChem **7**, 1256–1260 (2020)
27. Fiorani, A., Valenti, G., Irkham, P.F., Einaga, Y.: Quantification of electrogenerated chemiluminescence from tris(bipyridine)ruthenium(II) and hydroxyl ions. Phys. Chem. Chem. Phys. **22**, 15413–15417 (2020)
28. Wei, F., Wang, S., Gou, X.: A review for cell-based screening methods in drug discovery. Biophys. Rep. **7**, 504–516 (2021)
29. Nierode, G., Kwon, P.S., Dordick, J.S., Kwon, S.J.: Cell-based assay design for high-content screening of drug candidates. J. Microbiol. Biotechnol. **26**, 213–225 (2016)
30. Stuart, T., Satija, R.: Integrative single-cell analysis. Nat. Rev. Genet. **20**, 257–272 (2019)
31. Huang, K., Wang, Y.H., Zhang, H., Wang, T.Y., Liu, X.H., Liu, L., Jiang, H., Wang, X.M.: Application and outlook of electrochemical technology in single-cell analysis. Biosens. Bioelectron. **242**, 115741 (2023)

32. Shiku, H., Shiraishi, T., Ohya, H., Matsue, T., Abe, H., Hoshi, H., Kobayashi, M.: Oxygen consumption of single bovine embryos probed by scanning electrochemical microscopy. Anal. Chem. **73**, 3751–3758 (2001)
33. Friedrich, J., Seidel, C., Ebner, R., Kunz-Schughart, L.A.: Spheroid-based drug screen: considerations and practical approach. Nat. Protoc. **4**, 309–324 (2009)
34. Hiramoto, K., Villani, E., Iwama, T., Komatsu, K., Inagi, S., Inoue, K.Y., Nashimoto, Y., Ino, K., Shiku, H.: Recent advances in electrochemiluminescence-based systems for mammalian cell analysis. Micromachines **11**, 530 (2020)
35. Valenti, G., Scarabino, S., Goudeau, B., Lesch, A., Jović, M., Villani, E., Sentic, M., Rapino, S., Arbault, S., Paolucci, F., Sojic, N.: Single cell electrochemiluminescence imaging: from the proof-of-concept to disposable device-based analysis. J. Am. Chem. Soc. **139**, 16830–16837 (2017)
36. Hiramoto, K., Ino, K., Komatsu, K., Nashimoto, Y., Shiku, H.: Electrochemiluminescence imaging of respiratory activity of cellular spheroids using sequential potential steps. Biosens. Bioelectron. **181**, 113123 (2021)
37. Ino, K., Yaegaki, R., Hiramoto, K., Nashimoto, Y., Shiku, H.: Closed bipolar electrode array for on-chip analysis of cellular respiration by cell aggregates. ACS Sens. **5**, 740–745 (2020)
38. Shirato, Y., Hsueh, A.J., Ab Mutalib, N.A., Deng, Y., Suematsu, R., Kato, A., Kearney, B.M., Kinoshita, M., Suzuki, H.: Bipolar Clark-type oxygen electrode arrays for imaging and multiplexed measurements of the respiratory activity of cells. ACS Omega **9**, 10825–10833 (2024)
39. Shokouhi, M., Mehrgardi, M.A.: Cancer cell detection-based on released hydrogen peroxide using a non-modified closed bipolar electrochemical system. ChemElectroChem **7**, 3439–3444 (2020)
40. Eßmann, V., Jambrec, D., Kuhn, A., Schuhmann, W.: Linking glucose oxidation to luminol-based electrochemiluminescence using bipolar electrochemistry. Electrochem. commun. **50**, 77–80 (2015)
41. He, X., Deng, Y., Jiang, D., Fang, D.: Electrochemiluminescence detection and imaging of biomolecules at the single-cell level. Chemosensors **11**, 538 (2023)
42. Yang, X.Y., Bai, Y.Y., Huangfu, Y.Y., Guo, W.J., Yang, Y.J., Pang, D.W., Zhang, Z.L.: Ultrasensitive electrochemiluminescence biosensor based on closed bipolar electrode for alkaline phosphatase detection in single liver cancer cell. Anal. Chem. **93**, 1757–1763 (2021)
43. Niwa, O., Morita, M., Tabei, H.: Electrochemical behavior of reversible redox species at interdigitated array electrodes with different geometries: consideration of redox cycling and collection efficiency. Anal. Chem. **62**, 447–452 (1990)
44. Borchers, J.S., Clark, M.J., Van Scoy, S.B., Anand, R.K.: Array of interdigitated bipolar electrodes for selective capture and analysis of melanoma cells. ChemElectroChem **11**, e202300345 (2024)
45. Jayanthi, S.A., Das, A.B., Saxena, U.: Recent advances in biosensor development for the detection of cancer biomarkers. Biosens. Bioelectron. **91**, 15–23 (2017)
46. Cesewski, E., Johnson, B.N.: Electrochemical biosensors for pathogen detection. Biosens. Bioelectron. **159**, 112214 (2020)
47. Vigneshvar, S., Sudhakumari, C.C., Senthilkumaran, B., Prakash, H.: Recent advances in biosensor technology for potential applications—an overview. Front. Bioeng. Biotechnol. **4**, 11 (2016)
48. Chen, X., Xv, H., Li, C., Kong, L., Li, C., Li, F.: Fe-single-atom catalysts boosting electrochemiluminescence via bipolar electrode integrated with its peroxidase-like activity for bioanalysis. Biosens. Bioelectron. **258**, 116351 (2024)
49. Wu, M.S., Liu, Z., Shi, H.W., Chen, H.Y., Xu, J.J.: Visual electrochemiluminescence detection of cancer biomarkers on a closed bipolar electrode array chip. Anal. Chem. **87**, 530–537 (2015)
50. Song, S., Wang, L., Li, J., Fan, C., Zhao, J.: Aptamer-based biosensors. Trends Anal. Chem. **27**, 108–117 (2008)
51. Lu, H.J., Zhao, W., Xu, J.J., Chen, H.Y.: Visual electrochemiluminescence ratiometry on bipolar electrode for bioanalysis. Biosens. Bioelectron. **102**, 624–630 (2018)

52. Wang, Y.Z., Zhao, W., Dai, P.P., Lu, H.J., Xu, J.J., Pan, J., Chen, H.Y.: Spatial-resolved electrochemiluminescence ratiometry based on bipolar electrode for bioanalysis. Biosens. Bioelectron. **86**, 683–689 (2016)
53. Willner, I., Shlyahovsky, B., Zayats, M., Willner, B.: DNAzymes for sensing, nanobiotechnology and logic gate applications. Chem. Soc. Rev. **37**, 1153–1165 (2008)
54. Hu, Q., Li, H., Wang, L., Gu, H., Fan, C.: DNA nanotechnology-enabled drug delivery systems. Chem. Rev. **119**, 6459–6506 (2019)
55. Akasaka, R., Ino, K., Iwama, T., Inoue, K.Y., Nashimoto, Y., Shiku, H.: Electrochemiluminescence imaging based on bipolar electrochemistry using commercially available anisotropic conductive films. Sens. Mater. **34**, 3113–3122 (2022)
56. Anderson, T.J., Defnet, P.A., Zhang, B.: Electrochemiluminescence (ECL)-based electrochemical imaging using a massive array of bipolar ultramicroelectrodes. Anal. Chem. **92**, 6748–6755 (2020)
57. Iwama, T., Inoue, K.Y., Abe, H., Matsue, T., Shiku, H.: Bioimaging using bipolar electrochemical microscopy with improved spatial resolution. Analyst **145**, 6895–6900 (2020)
58. Qin, X., Jin, H.J., Li, X., Li, J., Pan, J.B., Wang, K., Liu, S., Xu, J.J., Xia, X.H.: Label-free electrochemiluminescence imaging of single-cell adhesions by using bipolar nanoelectrode array. Chem. Eur. J. **28**, e202103964 (2022)
59. Gao, J., Jin, H.J., Wei, X., Ding, X.L., Li, Z.Q., Wang, K., Xia, X.H.: Closed bipolar nanoelectrode array for ultra-sensitive detection of alkaline phosphatase and two-dimensional imaging of epidermal growth factor receptors. ACS Sens. **9**, 3754–3762 (2024)
60. Guerrette, J.P., Percival, S.J., Zhang, B.: Fluorescence coupling for direct imaging of electrocatalytic heterogeneity. J. Am. Chem. Soc. **135**, 855–861 (2013)
61. Villani, E., Sakanoue, K., Einaga, Y., Inagi, S., Fiorani, A.: Photophysics and electrochemistry of ruthenium complexes for electrogenerated chemiluminescence. J. Electroanal. Chem. **921**, 116677 (2022)
62. Villani, E., Inagi, S.: Electrosynthesis with split-bipolar electrodes. Curr. Opin. Electrochem. **44**, 101443 (2024)
63. Villani, E., Inagi, S.: Mapping the distribution of potential gradient in bipolar electrochemical systems through luminol electrochemiluminescence imaging. Anal. Chem. **93**, 8152–8160 (2021)
64. Li, H., Bouffier, L., Arbault, S., Kuhn, A., Hogan, C.F., Sojic, N.: Spatially-resolved multicolor bipolar electrochemiluminescence. Electrochem. Commun. **77**, 10–13 (2017)
65. Ab Mutalib, N.A., Deng, Y., Hsueh, A.J., Kariya, K., Kurihara, T., Suzuki, H.: Control of interfacial potentials and redox reactions on bipolar electrodes using Ag/AgCl. Electroanalysis **33**, 2123–2127 (2021)
66. Dennany, L., O'Reilly, E.J., Innis, P.C., Wallace, G.G., Forster, R.J.: The influence of poly(2-methoxyanaline-5-sulfonic acid) on the electrochemical and photochemical properties of a highly luminescent ruthenium complex. Electrochim. Acta **53**, 4599–4605 (2008)
67. Chen, Z., Villani, E., Inagi, S.: Recent progress in bipolar electropolymerization methods toward one-dimensional conducting polymer structures. Curr. Opin. Electrochem. **28**, 100702 (2021)
68. Villani, E., Shida, N., Inagi, S.: Electrogenerated chemiluminescence of luminol on wireless conducting polymer films. Electrochim. Acta **389**, 138718 (2021)
69. Villani, E., Zhang, Y., Chen, Z., Zhou, Y., Konishi, M., Tomita, I., Inagi, S.: AC-bipolar electrosynthesis and luminol electrochemiluminescence imaging of poly(3,4-ethylenedioxythiophene) and its composite films. ACS Appl. Polym. Mater. **5**, 6186–6198 (2023)
70. Zhou, Y., Villani, E., Kurioka, T., Zhang, Y., Tomita, I., Inagi, S.: Fabrication of luminescent patterns using aggregation-induced emission molecules by an electrolytic micelle disruption approach. Aggregate **4**, e202 (2023)
71. Saji, T., Hoshino, K., Aoyagui, S.: Reversible formation and disruption of micelles by control of the redox state of the head group. J. Am. Chem. Soc. **107**, 6865–6868 (1985)

72. Zhou, Y., Shida, N., Tomita, I., Inagi, S.: Fabrication of gradient and patterned organic thin films by bipolar electrolytic micelle disruption using redox-active surfactants. Angew. Chem. Int. Ed. **60**, 14620–14629 (2021)
73. Clevers, H.: Modeling development and disease with organoids. Cell **165**, 1586–1597 (2016)
74. Abaci, H.E., Shuler, M.L.: Human-on-a-chip design strategies and principles for physiologically based pharmacokinetics/pharmacodynamics modeling. Integr. Biol. **7**, 383–391 (2015)
75. Kim, J., Campbell, A.S., Esteban-Fernández de Ávila, B., Wang, J.: Wearable biosensors for healthcare monitoring. Nat. Biotechnol. **37**, 389–406 (2019)

Electrogenerated Chemiluminescence Devices

Ryoichi Ishimatsu, Takashi Kasahara, and Jun Mizuno

Abstract Electrochemiluminescence or electrogenerated chemiluminescence (ECL) is a light-emission phenomenon that involves redox reactions at the electrodes, electron transfer between the electrochemically generated redox species, the formation of excited states of luminescent molecules, and the radiative electronic transition of the luminescent molecules. Light-emitting devices based on ECL, which generally consist of two electrodes and a phase containing one or several types of luminescent molecules, can be very easily fabricated. In this chapter, the principle of ECL and the main components of ECL devices (ECLDs), such as electrodes, solvents, and gels, are briefly described. Furthermore, characteristics including the device performance of several ECLDs based on organic solvents, ionic liquids, liquid luminescent molecules, and gel are introduced.

Keywords Electrogenerated chemiluminescence · Electroluminescence · Liquid organic molecules · Ionic liquids · Gels · Thin-layer cells · Mixed ECL systems

---

R. Ishimatsu (✉)
Department of Applied Physics, University of Fukui, Fukui, Japan
e-mail: ishima2@u-fukui.ac.jp

Research Center for Fibers and Materials, University of Fukui, Fukui, Japan

T. Kasahara (✉)
Department of Electrical and Electronic Engineering, Hosei University, Tokyo, Japan
e-mail: t.kasahara@hosei.ac.jp

J. Mizuno (✉)
Academy of Innovative Semiconductor and Sustainable Manufacturing, National Cheng Kung University, Tainan City, Taiwan
e-mail: junmizuno@gs.ncku.edu.tw

F. Paolucci et al. (eds.), *Electrogenerated Chemiluminescence Microscopy and Imaging*,
https://doi.org/10.1007/978-3-032-14602-1_12

1 Introduction

Electrochemiluminescence or electrogenerated chemiluminescence (ECL) is a phenomenon that converts electrical energy into light energy. When oxidized ($R^{\cdot+}$) and reduced forms ($R^{\cdot-}$) of luminescent molecules (R) in a solution are electrochemically formed, electron transfer (ion annihilation) occurs between $R^{\cdot-}$ and $R^{\cdot+}$, forming an excited state of R (R^*) [1, 2]. The annihilation ECL is then observed through the radiative deactivation of the emissive excited state (the S_1 and T_1 states for fluorescent and phosphorescent materials, respectively). To elucidate the principles of ECL, a three-electrode cell using a working electrode, a reference electrode, and a counter electrode has normally been used. ECL can be measured by applying a square-wave voltage that generates $R^{\cdot-}$ and $R^{\cdot+}$ at the working electrode. The three-electrode cell is suitable for analyzing current–voltage curves because the potential of the working electrode can be controlled well. A rotating ring-disk electrode has also been employed for ECL analysis [3, 4], where constant voltages can be used to generate radical ions, e.g., $R^{\cdot+}$ at the ring part and $R^{\cdot-}$ at the disk part. R that has the following three characteristics tends to show high ECL efficiency for single ECL systems.

(1) R shows a high photoluminescent quantum yield.
(2) $R^{\cdot-}$ and $R^{\cdot+}$ are stable.
(3) The energy of the pair of $R^{\cdot-}$ and $R^{\cdot+}$ is sufficient to form the emissive excited states.

The emissive state is normally the S_1 state for organic molecules and the T_1 state for metal complexes. According to the spin-statistics rule, the ratio of the S_1 state to the T_1 state formed by the electron transfer is 1:3 [1, 5]. Therefore, metal complexes tend to have higher ECL efficiency than fluorescent molecules. In addition to the three properties mentioned above, the following was recently discovered. The ECL efficiency can be enhanced by using thermally activated delayed fluorescence (TADF) molecules. When the energy gap between the S_1 and T_1 states is small, thermal spin-up conversion from the T_1 to S_1 states can occur [6]. Although TADF molecules were developed for organic light-emitting diodes (OLEDs) [7, 8], several TADF materials have been applied to ECL [9–12], and are attracting attention because of their highly efficient ECL.

On the other hand, when the energy gap between the S_1 and T_1 states becomes very large, the production of the T_1 state by electron transfer is kinetically suppressed (Marcus inversion region [13]). Highly efficient ECL of BODIPY analogues utilizing the mechanism based on lowering the kinetics for the formation of the T_1 state by the electron transfer has also been reported [14].

Although the three-electrode cell is advantageous for analyzing the current–voltage curve and elucidating the ECL mechanism, it is not preferable when constructing light-emitting devices based on ECL, and two-electrode cells are favorable for fabricating ECL devices (ECLDs).

ECLDs using two-electrode cells can be fabricated very easily because it is only necessary to fill the solution containing R between the two electrodes. The direct

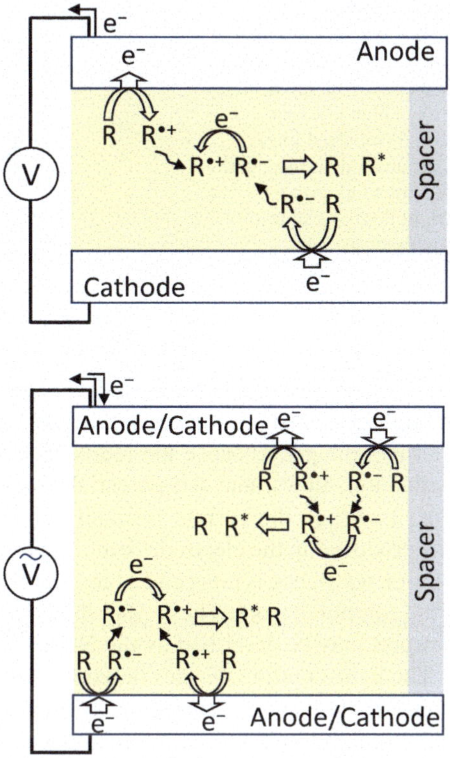

Fig. 1 Illustration of ECL with two-electrode system operated with (top) DC and (bottom) AC voltages

current (DC) and alternating current (AC) voltage can be used for ECLDs. The illustration of ECLDs operated with DC and AC voltage is shown in Fig. 1. In the case of DC voltage, $R^{\cdot -}$ and $R^{\cdot +}$ will be produced at the cathode and anode, respectively. The ions move toward the other electrode, and R^* will be formed through the electron transfer between $R^{\cdot -}$ and $R^{\cdot +}$. When thin-layer cells are used, the distance that $R^{\cdot -}$ and $R^{\cdot +}$ have to travel can be shortened, hence the collision of the radical ions or the electron transfer reaction occurs efficiently, i.e., ECL is generated efficiently. In the case of AC voltage, R^* will be formed near both of the two electrodes. Thus, the electrode gap is not an important factor for the efficient generation of the ECL (see also Sect. 3).

2 ECLD and Light-Emitting Electrochemical Cell

In electrochemistry, a supporting electrolyte is added to the solvent. The addition of the supporting electrolyte reduces the solution resistance and leads to the potential gradient at the electrode interface steeper due to the formation of the electrical double layer. When a sufficient amount of supporting electrolyte is added, the potential

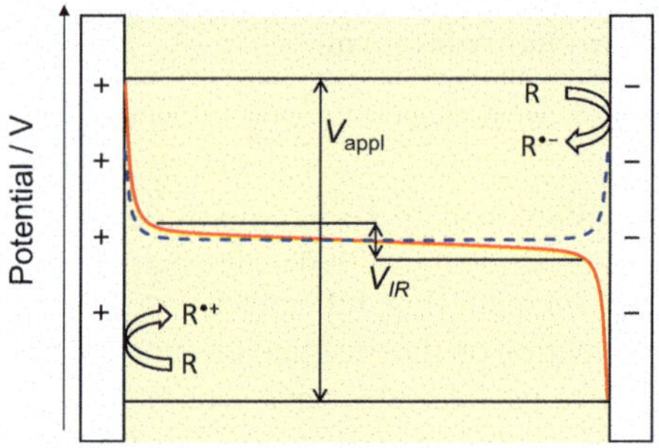

Fig. 2 Schematic representation of a potential profile in the ECLDs when the external voltage is off (broken blue line) and applied (solid red line). V_{appl}: applied voltage, and V_{IR}: potential drop by current flow and solution resistance

profile in the electrochemical cell changes dramatically at the electrode interface and remains constant in the solution phase when no current flows. When sufficient external voltages to cause the redox reactions are applied, current, which is ionic conduction, flows, and the potential gradient because of an IR drop (V_{IR}) arises (Fig. 2). When the current is small and the solution resistance is also small, the ions produced in the electrode reaction move mostly by diffusion, whereas when the solution resistance is huge and a large current flows, the potential gradient or IR drop in the solution becomes large, and the effect of the ion migration increases. The ion transport may be described by the Nernst-Plank equation.

The light-emitting devices based on ionic conduction, with a solution containing sufficient amounts of supporting electrolytes, are often called ECLDs or ECL cells since the light-emission mechanism is similar to that of ECL, i.e., electrochemically generated radical ions move, and the electron transfer reactions occur when $R^{\cdot-}$ and $R^{\cdot+}$ are brought into proximity.

OLEDs are normally solid-state devices, and supporting electrolytes are usually not added. Therefore, a potential gradient is generated in the solid phases [15]. Electrons and holes injected from the electrode move along this gradient. Some OLEDs contain supporting salt. In such devices, the electrical double layer will be formed at the interfaces of the phase containing the supporting electrolyte, and the driving voltage can be somewhat lower than that of solid OLEDs. These kinds of devices and light-emitting devices with larger amounts of R than that of the solvent and supporting electrolytes are sometimes called light-emitting electrochemical cells (LECs) [16].

For OLEDs and LECs, the current flowing through the device is mainly due to electronic conduction. That is, holes and electrons hop between the neutral and reduced or oxidized state and from the anode and cathode, respectively, toward the light-emitting layer, and R^* is formed by the recombination of electrons and holes. Therefore, they are generally driven by DC voltage. On the other hand, for ECLDs, the solvent or electrolyte is present in greater amounts than R, so that the electrochemically generated $R^{\cdot-}$ and $R^{\cdot+}$ collide by mainly diffusion, and R^* is formed by electron transfer (ion annihilation). Light emission is possible with both DC and AC voltages.

3 DC and AC Operation

Unlike OLEDs and LECs devices, which are operated with DC voltages, ECLDs can emit light by applying DC and AC voltages. This is because *R* exists near the anode and cathode, and the redox species generated by the electrode reaction can diffuse (and also migrate). In the case of under DC voltages, undesired reactions except for the electron transfer reaction that forms *R** occur, and the products of these reactions will accumulate at the anode or cathode, resulting in suppressing the electrode reaction of *R* (leading to lowering the ECL intensity). Such accumulation at the electrodes can be reduced by applying AC voltages. In addition, AC voltages are advantageous when the distance between the anode and cathode is large, and when the diffusion coefficient of *R* is small due to using viscous ionic liquids (ILs) or gels.

4 Thin-Layer Electrode Cells

In order to transmit ECL to the outside of the cell, it is necessary to employ a transparent electrode such as an indium-tin oxide (ITO) and F-doped tin oxide (FTO) electrode for one or both of the two electrodes. The ITO electrode is composed of a glass substrate and a thin ITO layer of normally several tens of nm to 100 nm. As the ITO layer becomes thicker, the electrode resistance decreases. The electrical resistance of an ITO film of 100 nm is about 15 Ω/□. A lower resistance is preferable because it can reduce the IR drop and operation voltage, but a thick ITO electrode lowers the transmittance in the visible light region.

For OLEDs, ITO is used for the anode and a thin metal film such as Mg or Mg/Ag is used for the cathode. This is because the work function of Mg is small at about 3.6 eV (the work function of ITO is about 4.7 eV), so the barrier to electron injection is small. Cs_2CO_3 (2.3 eV) deposited on an ITO cathode to suppress the electron injection barrier for liquid light-emitting electrode [17]. The smaller and larger work functions for the cathode and anode, respectively, are advantageous for the systems based on electronic conduction.

For ECLDs and two-electrode electrochemical systems, because the electrochemical oxidation and reduction at the anode and cathode, respectively, determined the current flow, the amount of the electrochemically generated oxidant and reductant should be balanced. For example, in the case of the cyclic voltammograms of a redox-active IL having a ferrocene and viologen moiety (the moiety can be oxidized and reduced, respectively) using two ITO electrodes, the viologen moiety was being reduced at the cathode while the ferrocene moiety was being oxidized at the anode at potentials where positive current flowed [18, 19]. For ECLDs, it is desirable that the ionic species related to the formation of *R** ($R^{\cdot-}$ and $R^{\cdot+}$ in single ECL systems) are equally generated at the cathode and anode. When decomposition of the solvent, dissolution of the electrode, oxidation and reduction of impurities, or differences in the stability of $R^{\cdot-}$ and $R^{\cdot+}$ occur, either $R^{\cdot-}$ and $R^{\cdot+}$ in the cell will become excessive, and ECL will not occur efficiently.

Overall, combinations of electrodes and a solvent that provides a wide polarized potential region are preferred since they allow the use of a wide variety of luminescent molecules. The polarized potential window of the electrode|solvent interface can be estimated by cyclic voltammetry measurement. For example, with a Pt electrode and an organic salt composed of Mg^{2+} dissolved in dimethyl ether, Mg^{2+} is reduced around − 3 V versus SHE and Mg metal formed on the Pt electrode is oxidized around − 0.4 V versus SHE [20]. Hence, using Mg as an electrode may cause an undesired reaction for ECL. The reduction current of the ITO electrode can be observed from around − 2 V when acetonitrile (MeCN) is used as a solvent [21]. When an aqueous solution is used, the oxidation current arises at a potential exceeding 2 V [22]. Although ITO electrodes are frequently used for ECLDs, they could not be suitable for R which is reduced at extremely negative potentials.

5 Spacer

To fabricate a thin-layer electrode cell for ECLDs, it is necessary to sandwich a spacer of several micrometers to several tens of micrometers between the electrodes. The distance between the electrodes is controlled by the size or thickness of the spacer. Glass beads, Ag spattered film, polyimide film, thermal tape, photoresist, etc., which are stable to the solvent, can be used as a spacer.

An epoxy-based negative photoresist (SU-8) can be used as a spacer. The SU-8 layer is mainly formed by spin coating, and the thickness when used as a spacer can be controlled by the concentration of the SU-8 solution used in spin coating and the rotation speed of the spin coater. The thickness can also be reduced by subsequent UV light (365 nm) treatment.

6 Solvents and Polymer Matrices

For ECLDs, solvents normally used in electrochemical measurements, such as MeCN, dichloromethane (DCM), dimethylformamide (DMF), and propylene carbonate (PC), can be used. The relative permittivity of these solvents is relatively high, and supporting electrolytes can be easily dissolved in these solvents. The addition of supporting electrolytes can reduce the solution resistance and lower the operating voltage of the ECLDs. Solvents that are not easily electrolyzed, i.e., solvents showing a wide polarized potential window, are advantageous in electrochemical measurements. The polarized potential window of MeCN is wide on both the positive and negative potential sides (− 2.5 ~ 2.5 V vs. SCE) and that for DCM is narrower on both sides (− 1.8 ~ 1.8 V vs. SCE). In the case of DMF (− 2.7 ~ 1.5 V vs. SCE) and PC (− 2.5 ~ 2.1 V vs. SCE), their positive branches are narrower than that of MeCN. The relative permittivity of MeCN, DMF, and PC are high, and water is easily contaminated. Water reacts with $R^{\cdot+}$, which is disadvantageous for ECL. When R is difficult to dissolve in these polar organic solvents or when aiming for intense ECL using a high-concentration solution of R, mixed solvents with benzene, toluene, and ortho-dichlorobenzene (*o*-DCB) are frequently used.

The evaporation of the organic solvents for ECL leads to a decrease in device performance. Therefore, ILs, which have low vapor pressures, are used as solvents and supporting electrolytes for ECLDs. Because the ECL processes involve the diffusion of R, $R^{\cdot-}$, and $R^{\cdot+}$, and ion migration, ILs with low viscosity are favorable. In addition, electrochemically stable ILs, which lead to wide polarized potential windows, are advantageous. ILs that have the properties are 1-ethyl-3-methylimidazolium bis(fluorosulfonyl)imide ([EMIM][FSI]), 1-ethyl-3-methylimidazolium bis(trifluoromethylsulfonyl)imide ([EMIM][TFSI]), and triethyl(methoxymethyl)phosphonium bis(trifluoromethylsulfonyl)imide ([TEMP][TFSI]). The viscosities of [EMIM][TFSI], [EMIM][FSI], and [TEMP][TFSI] are 18 [23], 34 [24], and 35 mPa s [25], respectively. The polarized potential window of the interfaces between [EMIM][FSI], [EMIM][TFSI], and [TEMP][TFSI] and glassy carbon are − 2.5 ~ 2.0 [23], − 2.5 ~ 2.0 [23], and − 3.0 ~ 2.3 V versus Fc/Fc^{+25}, respectively.

Although it is possible to fabricate flexible ECLDs using organic solvents or ILs, the mechanical strength of ECLDs can be significantly improved by gelling the solution. Various inorganic and polymeric materials have been used as dispersoids. So far, TiO_2, CeO_2, Y_2O_3, and ZnO were used as inorganic dispersoids. As polymer dispersoids, a triblock copolymer, polystyrene-*block*-poly(methylmethacrylate)-*block*-polystyrene, polymethyl methacrylate (PS-b-PMMA-b-PS), polyethylene glycol (PEG), and poly(vinylidene fluoride-*co*-hexafluoropropylene) (P(VDF-HEP)) [26] were employed for gel-based ECLDs. Since P(VDF-HEP) can form free-standing gel with ILs (Fig. 3), it is attractive for flexible EDCLs.

7 Organic Solvent- and IL-Based ECLDs

7.1 DC Operation

Pioneering works on ECLD using organic solvents were done by Nishimura et al. [27]. They dissolved rubrene in a mixed solvent of *o*-DCB and MeCN, and by injecting the solution between two ITO electrodes (electrode gap of 8 μm), they fabricated an ECLD with a luminance of 0.3 cd/m^2 at the applied voltage of about 8 V. Furthermore, they showed that the maximum luminance increased to 183 cd/m^2 by adding 180 mM 1,2-diphenoxyethane (DPE) to this solution (Fig. 4). They pointed out that the role of DPE in the solution without a supporting electrolyte is to stabilize the radical ions generated by the electrode reaction. (It is noted that the stabilization by DPE could not occur when a supporting electrolyte is added

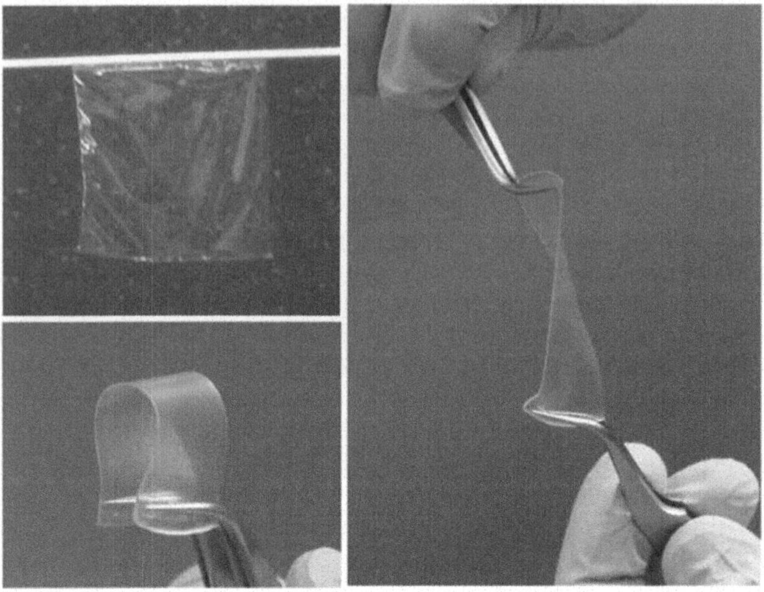

Fig. 3 Optical images of the free-standing gel based on P(VDF-HFP) and [EMIM][TFSA] (1: 4 wt). Reproduced from Ref. [26] by permission of John Wiley & Sons Ltd

to the solution because the redox potential of some luminophores was independent of the presence or absence of DPE [28]). They also reported that when a mixed solvent of *o*-DCB and toluene was used, the driving voltage was high at 80 V, but the luminance reached 986 cd/m^2 with an electrode gap of 3 μm. Moreover, using 9,10-diphenylanthracene (DPA) as blue, 5,12-diphenylnaphtacene as green, rubrene as yellow, and dibenzotetraphenylperiflanthene as red, multicolor ECLDs were fabricated (Fig. 5). The electrode gap is an important factor that affects the maximum luminance. For ECLDs using a 125 mM solution of tris(2,2′-bipyridine)ruthenium(II) bis(hexafluorophosphate) ($Ru(bpy)_3(PF_6)_2$) in PC, the maximum luminance was 175 and 240 cd/m^2 when the electrode distance was 5 and 2 μm, respectively [29].

In case a supporting electrolyte is included or an ionic luminescent species (e.g., $Ru(bpy)_3(PF_6)_2$) is used, the driving voltage is significantly lowered compared to the cases without adding electrolyte. When $LiClO_4$ (50–700 mM) is added to PC, the driving voltage is about 2.5 V [30]. In the case of 125 mM $Ru(bpy)_3(PF_6)_2$ dissolved in PC, the driving voltage is about 3.5 V [29].

When the radical ions generated at the electrode are unstable, the ECL intensity will be low. In such cases, it is possible to increase the ECL intensity by adding a parent molecule that generates a stable radical ion (mixed ECL systems). Kasahara et al. investigated the ECLD behavior of such multi-component systems. For example, when ECL was generated using bis(2-phenylbenzothiazolato)acetylacetonate)iridium(III) ($Ir(BT)_2(acac)$) (whose radical

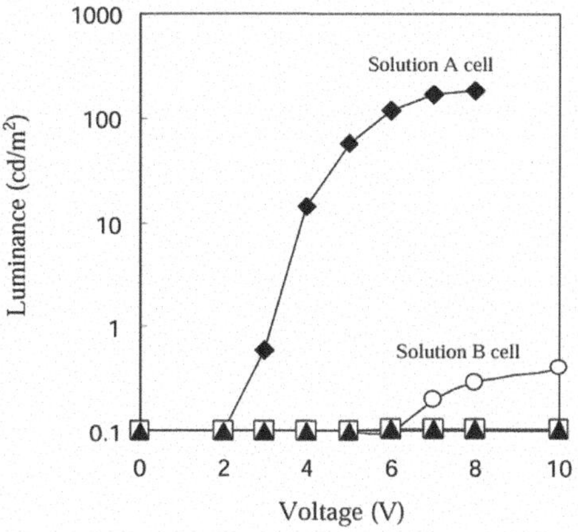

Fig. 4 Luminance–voltage characteristics of four cells. Solvent: *o*-DCB: MeCN = 2: 1 (v/v). 10 mM rubrene + 180 mM DPE (◆), 10 mM rubrene (○), 0 mM rubrene + 180 mM DPE (▲), and 0 mM rubrene + 0 mM DPE (□). © The Japan Society of Applied Physics. Reproduced from Ref. [27] by permission of IOP Publishing Ltd.

anion and cation are unstable and stable, respectively) and 4,5-bis(carbazol-9-yl)-1,2-dicyanobenzene (2CzPN, whose radical anion and cation are stable and unstable, respectively) in a mixed solvent of *o*-DCB and MeCN as judged from cyclic voltammograms shown in Fig. 6, ECL of Ir(BT)$_2$(acac) was observed due to an electron transfer reaction between the radical cation of Ir(BT)$_2$(acac) and the radical anion of 2CzPN [31]. The maximum luminance in this case was about 100 cd/m^2. On the other hand, the maximum luminance of the ECLD when Ir(BT)$_2$(acac) or 2CzPN was used alone was about 0.1 cd/m^2, and it can be seen that the luminance was greatly increased by mixing Ir(BT)$_2$(acac) and 2CzPN. The structure of the electrochemical cell, the maximum luminance, the maximum ECL wavelength (λ_{ECL}), and the electrode gap of several ECLDs using organic solvents operated with DC voltage are summarized in Table 1.

One of the features of liquid-based ECLD is the fluidity of the luminescence solution. Flow ECLDs using microchannels were also developed. Kasahara et al. fabricated a microfluidic ECLD using a photolithography process [32, 33]. Yellow ECL luminescence material flowed from one channel and red ECL luminescence material flowed from the other channel in a Y-shaped ECLD, and these two types of solutions are mixed at the rear of the channel to realize orange ECL [34] (Fig. 7). The characteristics of IL-based ECLDs using AC operation mode have been reported (see below).

So far, various organic solvent-based ECLDs have been developed, and high brightness and multi-color have been realized. However, their operating lifetimes are still short. For example, when a cell with an electrode gap of 5 μm was filled with a mixture of *o*-DCB and MeCN in which rubrene, DPAVB, and DPE were dissolved, and a DC voltage of 4.0 V was applied, the luminance decreased by 50% in about 150 s [37].

Fig. 5 Photographs of multicolor ECLDs using DPA (blue), 5,12-diphenylnaphtacene (green), rubrene (yellow), and dibenzotetraphenylperiflanthene (red).

Fig. 6 CV characteristics of 1 mM $Ir(BT)_2(acac)$ and 1 mM 2CzPN. The inset is an energy diagram of neutral molecules. Reproduced from Ref. [31] with permission of The Electrochemical Society of Japan

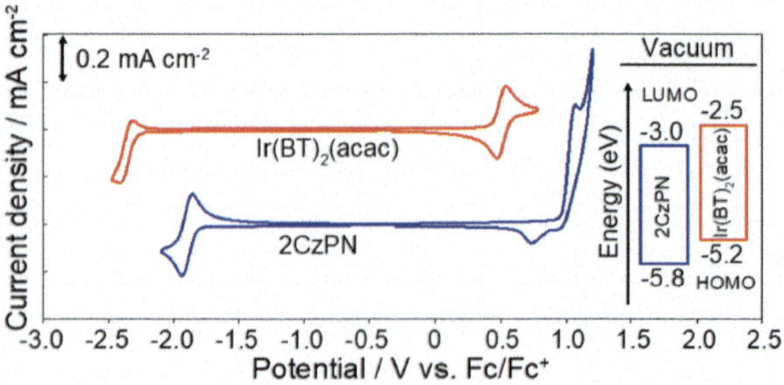

Table 1 Organic solvent-based ECLDs operated with DC voltage

Cell #	Anode		Cathode	λ_{ECL}/nm	Max. lumin./cd m^{-2}	Gap/ µm	Ref.
(1–1)	ITO	10 mM rubrene, 180 mM DPE in *o*-DCB: MeCN = 2: 1 (vol)	ITO		183	8	[27]
(1–2)	ITO	10 mM rubrene in *o*-DCB: MeCN = 2: 1 (vol)	ITO	563	0.3	8	
(1–3)	ITO	10 mM rubrene, 180 mM DPE in *o*-DCB: toluene = 2: 1 (vol)	ITO		983	3	
(2)	ITO	10 mM rubrene, 0.3 mM DBP, 180 mM DPE in *o*-DCB: MeCN = 2: 1 (vol)	ITO	600	11.6	6	[32]
(3)	ITO	1 mg/mL $(pq)_2Ir(pico)$ in PC with $LiClO_4$	ITO	(Yellow)	1.3	60	[30]
(4)	ITO	125 mM $Ru(bpy)_3(PF_6)_2$ in PC	TiO_2, FTO	~ 625	93.5	5	[35]
(5)	FTO	7 mM TBADN, 1.5 mM TMDBQA, 1 mM BDAVBi, 180 mM DPE in *o*-DCB: MeCN = 2: 1 (vol)	ITO	~ 530	40.9	5	[36]
(6)	ITO	125 mM $Ru(bpy)_3(PF_6)_2$ in PC	Al-ZnO, FTO	~ 625	240	2	[29]
(7)	ITO	7 mM TBADN, 1 mM TPBA, 180 mM DPE in *o*-DCB: MeCN = 2: 1 (vol)	FTO	~ 500	7.7	5	[28]
(8)	ITO	10 mM 2CzPN, 4 mM $Ir(BT)_2(acac)$ in *o*-DCB: MeCN = 2: 1 (vol)	FTO	~ 560	113	5	[31]

(continued)

Table 1 (continued)

Cell				λ_{ECL}/nm	Max. lumin./cd m^{-2}	Gap/ μm	Ref.
#	Anode		Cathode				
(9)	ITO	10 mM rubrene 4 mM DPAVB, 180 mM DPE in *o*-DCB: MeCN = 2: 1 (vol)	FTO	~ 560	292	5	[37]
(10)	FTO	70 mg/mL SiQD in *o*-DCB: MeCN = 2: 1 (vol)	ITO	~ 700	1.5	8	[38]

$(pq)_2Ir(pico)$: bis(2-phenylquinoline) iridium (III) picolinic acid, TBADN: 2-*tert*-butyl-9,10-di(naphth-2-yl)anthracene, TMDBQA: 5,12-dibutyl-1,3,8,10-tetramethylquinacridone, BDAVBi: 4,4′-bis[4-(diphenylamino)styryl]biphenyl, DPAVB: 4-(di-p-tolylamino)-4′-[(di-p-tolylamino)styryl]stilbene, SiQD: silicon quantum dot

7.2 AC Operation

In the case of AC operation, the radical ions generated in the electrode reaction do not need to move to the other electrode, so the distance between the electrodes can be longer (thus, the volume of the solution can be larger) than that in DC-operated ECLDs. Even when the gap between the electrodes is large for DC operation, efficient ECL can be seen with AC operation. For example, in an ECLD with an electrode gap of about 70 μm, the luminance of $Ru(bpy)_3^{2+}$ is about 1.4 and 46 cd/m^2 with DC and AC operation, respectively [39]. AC operation is also advantageous when using ILs, which have a higher viscosity than organic solvents. Unlike DC operation, which provides constant light emission, AC driving causes the luminance to vary over time, so the luminance is sometimes reported as the root mean square (RMS) value [40]. The structure of the electrochemical cell, the maximum luminance, λ_{ECL}, and the operation frequency of several ECLDs using organic solvents and ILs operated with AC voltage are summarized in Table 2. Even when using an IL with a higher viscosity than organic solvents, a high luminance of about 700 cd/m^2 has been achieved with AC driving [41].

The highest luminance reported so far for ECLD is 3790 cd/m^2, which was demonstrated by Gather et al. [42]. They focused on the fact that 1,1-bis[(di-4-tolylamino)phenyl]cyclohexane (TAPC) and 2,2′,2″-(1,3,5-benzinetriyl)-tris(1-phenyl-1-H-benzimidazole) (TPBi) form a relatively long-lived exciplex in a mixture of toluene and MeCN (the exciplex exhibited thermally activated delayed fluorescence, i.e., it undergoes thermal reverse intersystem crossing from T_1 to S_1 [43]. The fluorescence lifetimes of the prompted and delayed components were 2.45 and 27.4 ns, respectively, and resulted in an efficient transfer of the excited state energy of the exciplex to yellow-emissive 2,8-di-tert-butyl-5,11-bis(4-tert-butylphenyl)-6,12-diphenyltetracene (TBRb). Moreover, the spin-up conversion is favorable to enhance the ECL efficiency because the T_1 state, which will be formed by the electron transfer

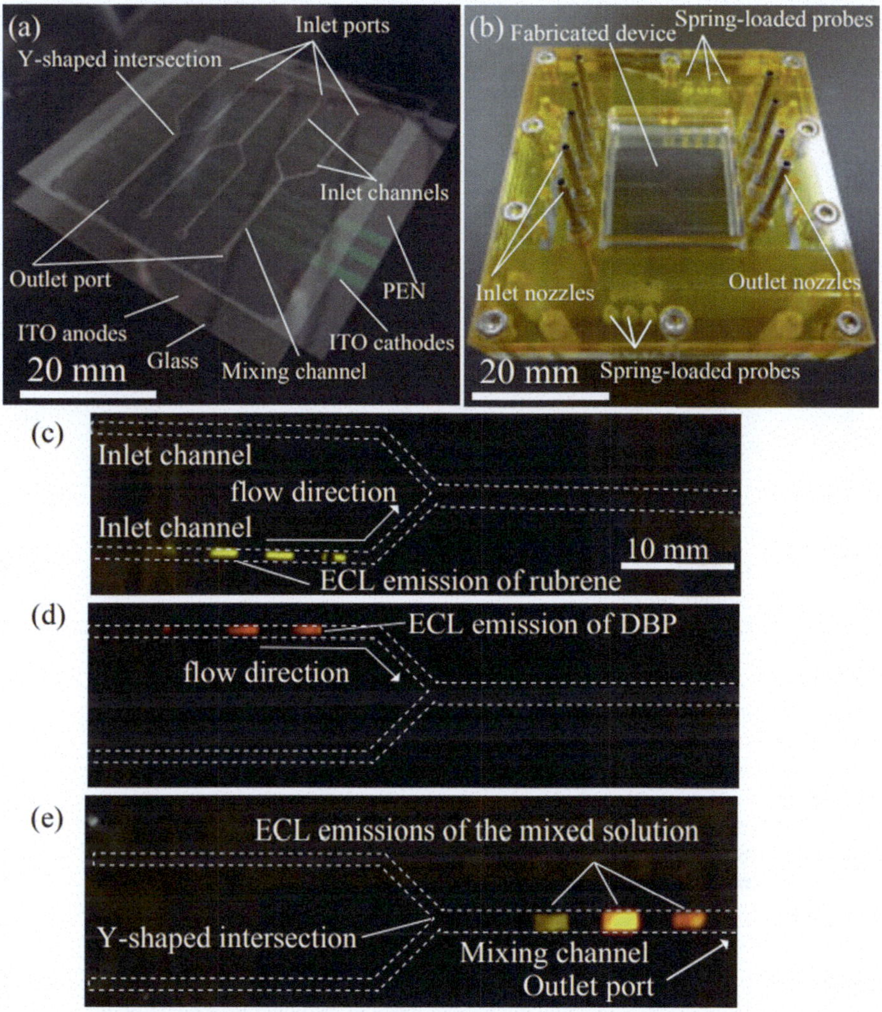

Fig. 7 **a** Fabricated color-tunable microfluidic ECL cells. **b** Experimental set-up for the microfluidic ECL cells. ECL emissions of **c** 10 mM rubrene solution, **d** DBP (0.3 mM)-doped rubrene solution, and **e** mixed solution (microchannel in dash line). © The Japan Society of Applied Physics. Reproduced from Ref. [34] by permission of IOP Publishing Ltd. All rights reserved

reaction between the radical species, can be turned into the S_1 state, which is the emissive state for fluorescent molecules. Overall, the high luminance was achieved by the relatively long lifetime of the exciplex and thermal spin-up conversion.

When the ECLD based on the excimer formation was operated with a voltage to obtain 300 cd/m^2 at 300 Hz, it took about 300 s for the ECL intensity to decrease to 50% of the initial intensity. Kang et al. extended the operation lifetime with a bipolar electrode. When the initial ECL intensity of about 110 cd/m^2 decreased to

Table 2 Organic solvent- and ionic liquid-based ECLDs operated with AC voltage

ki#	Cell			λ_{ECL}/nm	Max. lumin./cd m^{-2}	Freq./Hz	Ref.
(1)	ITO	10 mM $Ru(bpy)_3(PF_6)_2$ in PC with $TBAClO_4$	ITO	620	45.6	50	[39]
(2)	ITO	1 mg/mL $(pq)_2Ir(pico)$ in PC with $LiClO_4$	ITO	(Yellow)	1.2	1	[30]
(3)	ITO	10 wt% $Ru(bpy)_3(PF_6)_2$ in [EMIM][TFSI]	Au	(Orange)	~ 700	500	[45]
(4–1)	ITO	50 mM $Ru(bpy)_3(PF_6)_2$ in [EMIM][TFSI]	ITO	(Orange)	~ 180	500	[41]
(4–2)	ITO	50 mM $Ru(bpy)_3(PF_6)_2$ in [BMIM][TFSI]	ITO	(Orange)	~ 110	300	[41]
(5)	ITO	8 wt% $Ru(bpy)_3(PF_6)_2$ in [EMIM][TFSI]	ITO	(Orange)	207	150	[46]
(6)	ITO	50 mM $Ru(bpy)_3(PF_6)_2$ in [EMIM][TFSI]	Ag	(Orange)	350	300	[47]
(7)	ITO	3 wt% ($Ir(ppy)_2(bpy)PF_6$: $Ir(diFppy)_2(bpy)PF_6$ = 40: 60 mol%) in [EMIM][TFSI]	ITO	~ 597	~ 27	500	[48]
(8)	ITO	30 mM TBRb, 30 mM TAPC, 30 mM TPBi in Toluene: MeCN = 2:1 (vol) with $TBAPF_6$	ITO	575	3790	30	[42]
(9)	ITO	50 mM $Ru(bpy)_3(PF_6)_2$ in [EMIM][TFSI]	AuCr	(Orange)	~ 350	100	[44]

($Ir(ppy)_2(bpy)PF_6$: (2,2′-bipyridine-κN^1,κ$N^{1\prime}$)bis[2-(2-pyridinyl-κ*N*)phenyl-κ*C*]iridium(III) hexafluorophosphate

50% in 1 min. with a sandwiched cell at 3 V with 100 Hz, the lifetime was extended to 5 min. with a bipolar electrode, and further, to 80 min. by applying an AC current with 100 Hz [44]. For practical use, the problem of the short operating lifetime must be improved in the future.

8 Liquid Organic Molecule-Based ECLDs

The first light-emitting device using a liquid organic molecule was reported by Adachi et al. [17]. A thin silver film (thickness of 80 nm) was evaporated between two ITO electrodes as a spacer, and rubrene dissolved in 9-(2-ethylhexyl)carbazole (EHCz), which is a liquid fluorescent molecule, was introduced between the two electrodes to fabricate a light-emitting device (the ITO was partially etched to prevent from shorting between the electrodes). The surface of the ITO anode was covered with poly(3,4-ethylenedioxythiophene): poly(4-styrenesulfonate) (PEDOT:PSS), and the surface of the ITO cathode was coated with Cs_2CO_3. The device emitted light originated mainly from rubrene at $\lambda_{ECL} = 555$ nm and a maximum luminance of about 0.35 cd/m^2. The EHCz system was further optimized. A device using EHCz was used as the liquid host and 6-{5-[3-methyl-4-(methyl-octyl-amino)-phenyl]-thiophen-2-yl}-naphthalene-2-carboxylic acid hexyl ester (BAPTNCE) instead of rubrene as a emitter and $TBAPF_6$ as a supporting electrolyte showed $\lambda_{ECL} = 531$ nm and a maximum luminance of 100 cd/m^2 [49].

So far, EHCz, 9-{2-[2-(2-methoxyethoxy)ethoxy]ethyl}-9H-carbazole (TEGCz) [50], 9,9′-2-[2-(2-methoxyethoxy)ethoxy]ethyl-3,3′-bis(9H-carbazole) ($(TEGCz)_2$) [51], 1-pyrenebutyricacid 2-ethylhexylester (PLQ) [52], and 1-naphthaleneacetic acid 2-ethylhexyl ester (NLQ) [53] have been developed and applied as organic liquid luminescent materials to ECLDs (the molecular structure of the liquid fluorescence molecules are shown in Fig. 8). It was demonstrated that the ECL of PLQ could be observed with a flexible ECLD of about 10 $\times$ 10 cm size with 64 pixels when it was flat and bent [54]. Since PLQ and NLQ have pyrene or naphthalene skeletons, excimer emission can occur. In the case of photoexcitation, excimer formation prevails as the concentration increases however, in the case of ECL, excimer emission becomes more pronounced than in the case of photoexcitation because R exists in the vicinity of R^* generated by electron transfer [55, 56].

Since the concentration of neat organic liquid molecules is a few M, they can be used at a higher concentration than in systems using organic solvents or ILs. This means that a high current density can be achieved, and intense ECL is expected. Because the organic liquid fluorescent molecules except for $(TEGCz)_2$ produce unstable radical cations, the operation lifetime of the devices was short and the luminance was relatively low. The luminance of the device with TEGCz quickly decreased in 50 s, whereas that with $(TEGCz)_2$, which produces stable radical cation and dication, was stable for more than 1000 s [51].

Table 3 summarizes the characteristics of devices using several liquid organic light emitters. Figure 9 shows the design of a microfluidic ECLD, the electrode-gap-dependent luminance-voltage curve with neat PLQ, and multi-color ECL using PLQ and rubrene and DBP [57]. Organic liquid molecules at high concentrations will be oxidized and reduced to form an excited state, and then the excitation energy is transferred to the low-concentration luminescent molecules. Therefore, the PL spectrum of the organic liquid molecules and the absorption spectrum of the low concentration of luminescent molecules must overlap to transfer the energy. For the

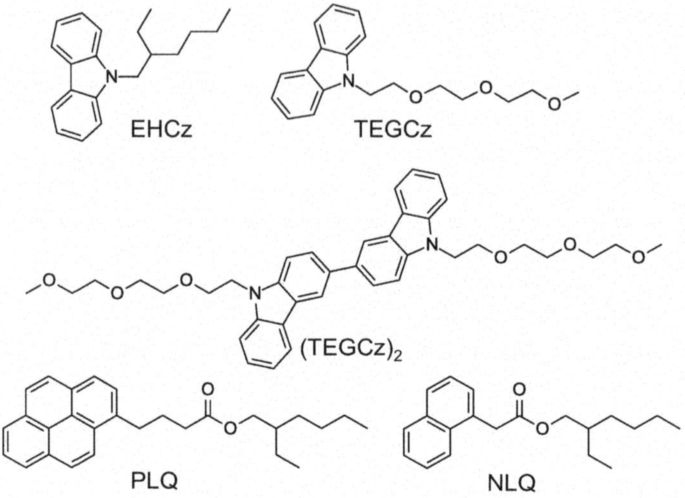

Fig. 8 Molecular structure of liquid fluorescent molecules

system, the absorption spectrum of rubrene and DBP partially overlaps with the PL spectrum of PLQ, resulting in the realization of the multicolor ECL.

9 Gel-Based ECLDs

For gel-based ECLDs, since the diffusion and migration of radical ion species are significantly suppressed, AC operation is suitable. Itoh used TiO_2, CeO_2, Y_2O_3, and ZnO as gel matrices and mixed them with several ILs and $Ru(bpy)_3{}^{2+}$ [59] or deuterated $Ru(bpy)_3{}^{2+}$ ($Ru(d_8\text{-}bpy)_3{}^{2+}$) [40] to fabricate inorganic gel-based ECLDs. He reported the ECL properties at an AC voltage of 60 to 900 Hz. The maximum luminance of ~ 400 cd/m^2 was achieved with $Ru(bpy)_3{}^{2+}$, TiO_2, and [TMPA][TFSI] at 700 Hz [59]. The initial luminance of the gel-based ECLD decreased to 50% in about 30 min [59].

When a pseudo-solid gel is used for the light-emitting layer, mechanically stronger and flexible ECLDs can be fabricated. Moon et al. used a triblock copolymer, (PS-b-PMMA-b-PS) with [EMIM][TFSI] and $Ru(bpy)_3{}^{2+}$ as a luminophore [60]. The block copolymer containing [EMIM][TFSI] with a thickness of about 30 μm was so soft and flexible. The flexible ECLD using the gel and ITO-coated polyethylene terephthalate (PET) electrodes was operated by applying AC voltages. They also demonstrated that by mixing $Ru(bpy)_3{}^{2+}$ with 2,2′-bipyridylbis[2-(2′,4′-difluorophenyl)pyridine]iridium(III) ($Ir(diFppy)_2(bpy)^+$), the ECL from $Ru(bpy)_3{}^{2+}$ was greatly enhanced because of the presence of the pathway of electron transfer

Table 3 Liquid organic molecule-based ECLDs operated with DC voltage

#	Cell					λ_{ECL}/ nm	Max. lumin./ cd m^{-2}	Gap/ µm	Ref.
	Anode				Cathode				
(1)	ITO	PEDOT:PSS	1 wt% rubrene in EHCz	Cs_2CO_3	ITO	555	0.35	0.08	[17]
(2)	ITO	PEDOT:PSS	16.7 wt% BAPTNCE, 0.1 wt% $TBAPF_6$ in EHCz	TiO_2	ITO	531	100	1.1	[49]
(3–1)	ITO	PEDOT:PSS	25 wt% PLQ, 0.25 wt% [TMP][TFSI] in TEGCz	TiO_2	ITO	~ 525	35	1.1	[51]
(3–2)	ITO	PEDOT:PSS	25 wt% BMTB, 0.25 wt% [TMP][TFSI] in TEGCz	TiO_2	ITO	~ 490	44	1.2	[51]
(4)	ITO	PEDOT:PSS	1 wt% rubrene in TEGCz	Cs_2CO_3	ITO	555	~ 0.8	0.68	[50]
(5)	ITO		0.25 wt% [TMP][TFSI] in PLQ	ITO		~ 505	5.6	4.5	[54]
(6)	ITO		0.25 wt% [TMP][TFSI] in PLQ	ITO		~ 505	26	2.5	[57]
(7–1)	ITO		0.25 wt% [TMP][TFSI] in PLQ	ITO		~ 505	2.63	6	[58]
(7–2)	ITO		2 wt% TBRb, 0.25 wt% [TMP][TFSI] in PLQ	ITO		~ 560	5.51	6	[58]
(8)	ITO		2 wt% DPA, 0.25 wt% [TMP][TFSI] in NLQ	ITO		~ 450	4.73	6	[53]

[TMP][TFSI]: tributylmethylphosphonium bis(trifluoromethylsulfonyl)imide, BMTB: 3,3′-{2-[(2-ethylhexyl)oxy]-5-methoxy-1,4-phenylenevinylene]bis(9-ethyl-9*H*-carbazole)

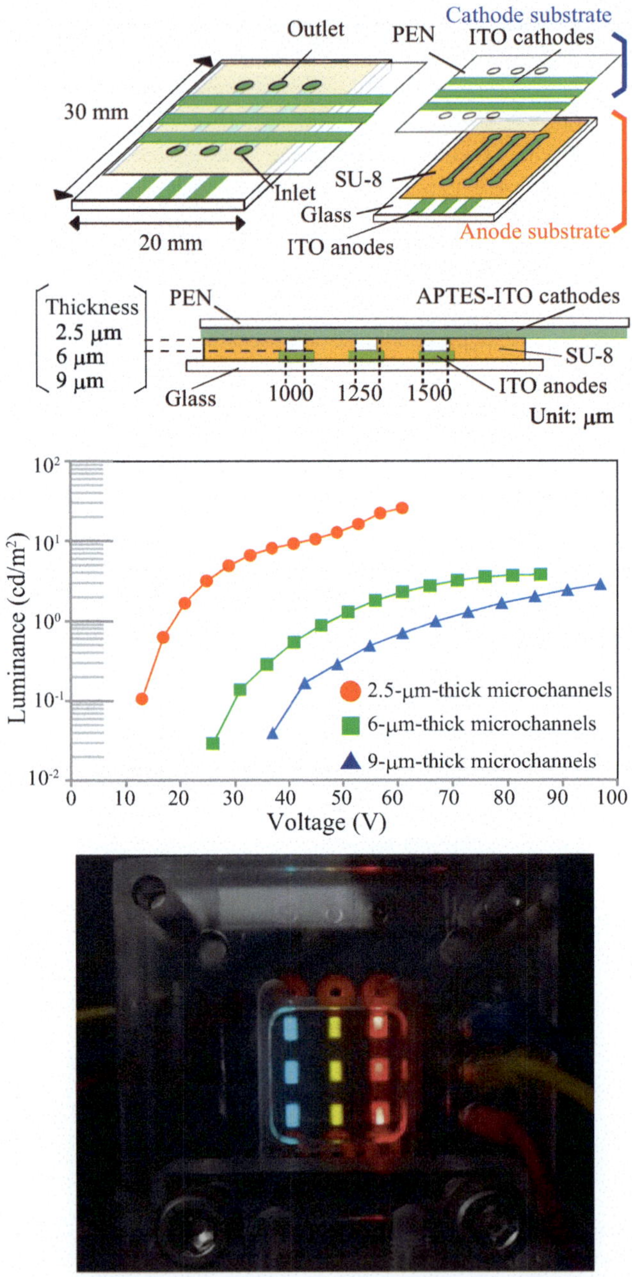

Fig. 9 (Top) Design of microfluidic ECLD. SU-8 microchannels were sandwiched between the ITO anode and cathode. APTES: 3-aminopropyltriethoxysilane. (Middle) Electrode-gap dependent luminance-voltage curves with neat PLQ. (Bottom) PLQ (left column), rubrene-doped PLQ (center column), and DBP-doped PLQ (right column). Reprinted from Ref. [57]. Copyright 2015, with permission from Elsevier

from the excited state of $Ir(diFppy)_2(bpy)^+$ to $Ru(bpy)_3{}^{3+}$ to produce the excited state of $Ru(bpy)_3{}^{2+}$.

Hong et al. reported sticker-type gel-based ECLDs with $Ru(bpy)_3{}^{2+}$, [EMIM][TFSI], and P(VDF-HEP) as a gel phase [61]. They repeated peeling off and sticking the gel phase five times and showed that after five successive peeling off and sticking processes, the ECLD was still active (Fig. 10), although the maximum luminance was dropped somewhat (from about 100 to 60 cd/m^2).

The effect of adding ILs to P(VDF-HEP) on the device characteristics was investigated by Moon et al. [62]. They used [EMIM], [BMIM], [HMIM], and [DMIM] as counter cations of [TFSI]. They showed that the maximum luminance decreased with increasing viscosity of the IL, and the delay in the ECL response under AC voltage application increased from ~ 0.98 to ~ 1.4 ms compared to the current response. Soulsby et al. demonstrated mixed ECL of $Ru(bpy)_3{}^{2+}$ and $Ir(diFppy)_2(bpy)^+$ in a gel composed of P(VDF-HEP) and [BMIM][TFSI] [63]. The ECL color changed from red to green by increasing the ratio of $Ir(diFppy)_2(bpy)^+$.

Kuon and Myoung employed a gel of polymethyl methacrylate, polyethylene glycol, and [EMIM][TFSI] with $Ru(bpy)_3{}^{2+}$, bis[2-(2,4-difluorophenyl)pyridinate](2,2′-dimethyl-4,4′-bipyridine)iridium(III) ($Ir(Fppy)_2(dmb)^+$), or bis[2-(2,4-difluorophenyl)pyridinate](3-methylpyridine-2-carboxylate)iridium(III) ($Ir(Fppy)_2(Mepic)$) as red, green, and blue ECL luminophores [64]. They fabricated flexible ECLD with the gels and ITO-coated PET electrodes and performed mechanical bending tests at a bending radius of 10 mm (Fig. 11). After the bending test of 5000 cycles, the maximum luminance decreased by about 20%. They also fabricated a wearable pressure-sensitive ECLD based on the gel. The ECL luminance of the red, green, and blue colors decreased by about 20% in 60 min [65].

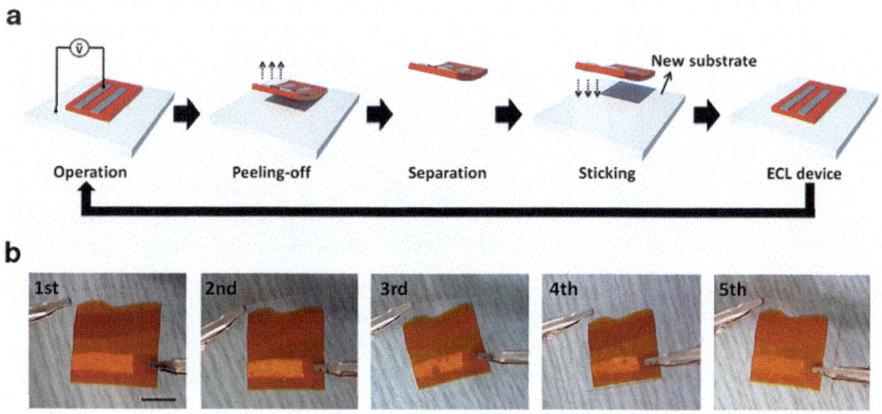

Fig. 10 **a** Schematic of a repeatability test for sticker ECLDs. **b** Photographs taken in each repeatability test step. Scale bar: 1 cm. Reprinted from Ref. [61] with permission from Springer Nature, copyright 2016

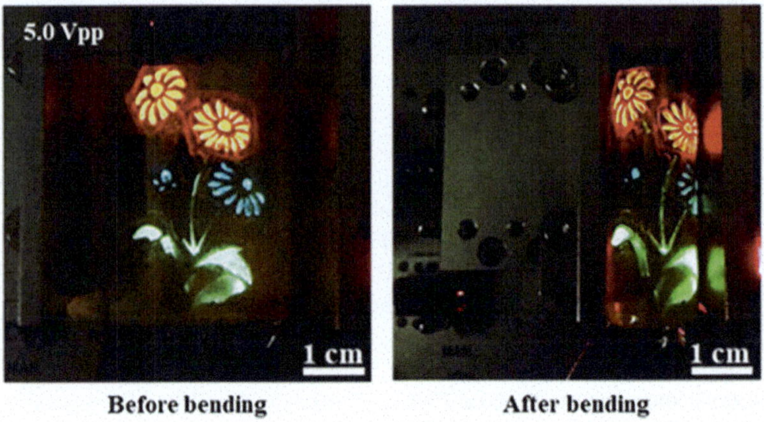

Fig. 11 ECL from the flexible ECLD before and after bending at a bending radius of 10 mm. V_{pp}: peak-to-peak voltage. Reprinted from Ref. [64]. Copyright 2020, with permission from Elsevier

Lee et al. fabricated flexible fiber ECLDs on a Ag wire by dip coating into a mixture of P(VDF-HEP), [EMIM][TFSI], and $Ru(bpy)_3^{2+}$, and a Au foil electrode [66] (Fig. 12). They performed bending tests 100 times with a bending radius of 5 mm and showed that the luminance decreased by 18%.

Table 4 summarizes the cell structures and their characteristics of gel-based ECLDs. Gel-based ECLDs could achieve relatively high luminance with AC voltage. Taking advantage of their gel properties, various types of flexible ECLDs have been fabricated.

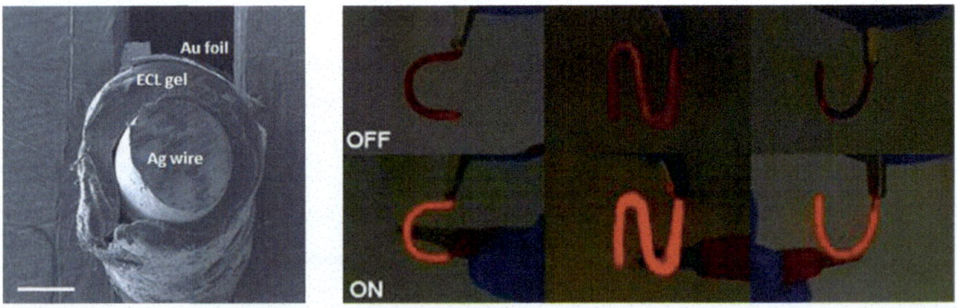

Fig. 12 (Left) Cross-sectional SEM image of the ECL fiber. Scale bar: 500 μm. (Right) Photograph of "CNU" letters of the fiber ECLD. Reprinted with the permission from Ref. [66]. Copyright 2022 American Chemical Society

Table 4 Gel-based ECLDs operated with AC voltage

	Cell				λ_{ECL}/ nm	Max. lumin./ cd m^{-2}	Freq./ Hz	Ref.
(1)		ITO	93 wt% (5 wtl% $Ru(d_8\text{-}bpy)_3(PF_6)_2$ in [ABIM][TFSI]), 7 wt% fumed silica	ITO	(Red)	~ 220	60	[40]
(1)		ITO	55 wt% (1 mol% $Ru(bpy)_3(PF_6)_2$, 99 mol% [TMPA][TFSI]), 45 wt% TiO_2	ITO	(Red)	400	700	[59]
(2–1)		ITO	$Ru(bpy)_3Cl_2$: [EMIM][TFSI]: PS-b-PMMA-b-PS = 1: 16: 4 (wt)	Ag	610	~ 65	60	[60]
(2–2)		ITO	(8 wt% $Ir(diFppy)_2(bpy)PF_6$ in [EMIM][TFSI]), PS-b-PMMA-b-PS	Ag	540	~ 80	N.D	[60]
(3–1)		ITO	10 wt% $Ru(bpy)_3Cl_2$, ([EMIM][TFSI]: P(VDF-HFP): acetone = 4:1:7 (wt)	Ag	630	90	1	[61]
(3–2)	Au	PEDOT: PSS	10 wt% $Ru(bpy)_3Cl_2$, ([EMIM][TFSI]: P(VDF-HFP): acetone = 4:1:7 (wt)	Ag	630	60	1	[61]
(3–3)		Ag	10 wt% $Ru(bpy)_3Cl_2$, ([EMIM][TFSI]: P(VDF-HFP): acetone = 4:1:7 (wt)	Ag	630	30	1	[61]
(4–1)		ITO	$Ir(diFppy)_2(bpy)$: [EMIM][TFSI]: P(VDF-HFP) = 1: 15: 3 (wt)	Ag	(Green)	~ 50	60	[62]
(4–2)		ITO	$Ir(diFppy)_2(bpy)$: [BMIM][TFSI]: P(VDF-HFP) = 1: 15: 3 (wt)	Ag	(Green)	~ 45	60	[62]
(4–3)		ITO	$Ir(diFppy)_2(bpy)$: [HMIM][TFSI]: P(VDF-HFP) = 1: 15: 3 (wt)	Ag	(Green)	~ 35	60	[62]
(4–4)		ITO	$Ir(diFppy)_2(bpy)$: [DMIM][TFSI]: P(VDF-HFP) = 1: 15: 3 (wt)	Ag	(Green)	~ 18	60	[62]
(5–1)		ITO	9 wt% $Ru(bpy)_3(PF_6)_2$ in CH_2Cl_2: 5 wt% PEG400 in [EMIM][TFSI]: 60 mg/mL PMMA in CH_2Cl_2 = 2: 1: 0.1 (vol)	ITO	616	77.4	60	[64]

(continued)

Table 4 (continued)

	Cell				λ_{ECL}/ nm	Max. lumin./ cd m^{-2}	Freq./ Hz	Ref.
(5–2)		ITO	5 wt% $Ir(Fppy)_2(dmb)PF_6$ in CH_2Cl_2: 5 wt% PEG400 in [EMIM][TFSI]: 60 mg/mL PMMA in CH_2Cl_2 = 2: 1: 0.1 (vol)	ITO	532	92.4	60	[64]
(5–3)		ITO	7 wt% $Ir(Fppy)_2$(Mepic) in CH_2Cl_2: 5 wt% PEG400 in [EMIM][TFSI]: 60 mg/mL PMMA in CH_2Cl_2 = 2: 1: 0.1 (vol)	ITO	495	13.2	60	[64]
(6)		ITO	7 mg/mL DPA, 0.8 mg/mL $LiCF_3SO_3$ in xylene: 60 mg/ mL PMMA in CH_2Cl_2 = 2: 1 (vol)	CNT	430	35.4	300	[65]
(7)		Au	1.8 wt% $Ru(bpy)_3Cl_2$, ([EMIM][TFSI],P(VDF-HFP), acetone = 4: 1: 10 (wt)	Ag	(Red)	47	1	[66]

PMMA: Polymethyl methacrylate

10 Summary

ECLDs are very easy to fabricate, and so they are attracting attention as a new light-emitting device instead of OLEDs. Compared to OLEDs, ECLDs have the disadvantages of weaker emission intensity and a shorter device lifetime, but recent research and development has led to ECLDs with a luminance of over 1000 cd/m^2 and longer device lifetime, approaching the performance of OLEDs. Furthermore, flexible ECLDs have also been fabricated, taking advantage of the flexible properties of liquids and gels. It is expected that further research and development will further improve the performance of ECLDs in the future.

References

1. Bard, A.J.: Electrogenerated Chemiluminescence. CRC Press (2004)
2. Richter, M.M.: Electrochemiluminescence (ECL). Chem. Rev. **104**, 3003–3036 (2004)
3. Maloy, J.T., Bard, A.J.: Electrogenerated chemiluminescence. VI. Studies of the efficiency and mechanisms of 9,10-diphenylanthracene, rubrene, and pyrene systems at a rotating-ring-disk electrode. J. Am. Chem. Soc. **93**, 5968–5981 (1971)
4. Wallace, W.L., Bard, A.J.: Electrogenerated chemiluminescence. 35. Temperature dependence of the ECL efficiency of $Ru(bpy)_3{}^{2+}$ in acetonitrile and evidence for very high excited state yields from electron transfer reactions. J. Phys. Chem. **83**, 1350–1357 (1979)

5. Armstrong, N.R., Wightman, R.M., Gross, E.M.: Light-emitting electrochemical processes. Annu. Rev. Phys. Chem. **52**, 391–422 (2001)
6. Parker, C.A., Parker, C.: Photoluminescence of Solutions: With Applications to Photochemistry and Analytical Chemistry. Elsevier Amsterdam (1968)
7. Endo, A., et al.: Thermally activated delayed fluorescence from Sn^{4+}–porphyrin complexes and their application to organic light emitting diodes—a novel mechanism for electroluminescence. Adv. Mater. **21**, 4802–4806 (2009)
8. Uoyama, H., Goushi, K., Shizu, K., Nomura, H., Adachi, C.: Highly efficient organic light-emitting diodes from delayed fluorescence. Nature **492**, 234–238 (2012)
9. Ishimatsu, R., Edura, T., Adachi, C., Nakano, K., Imato, T.: Photophysical properties and efficient, stable, electrogenerated chemiluminescence of donor-acceptor molecules exhibiting thermal spin upconversion. Chem. Eur. J. **22**, 4889–4898 (2016)
10. Ishimatsu, R., et al.: Electrogenerated chemiluminescence of donor-acceptor molecules with thermally activated delayed fluorescence. Angew. Chem. Int. Ed. **53**, 6993–6996 (2014)
11. Huang, P., et al.: Polymer electrochemiluminescence featuring thermally activated delayed fluorescence. ChemPhysChem **22**, 726–732 (2021)
12. Wang, C., Wu, J., Huang, H., Xu, Q., Ju, H.: Electrochemiluminescence of polymer dots featuring thermally activated delayed fluorescence for sensitive DNA methylation detection. Anal. Chem. **94**, 15695–15702 (2022)
13. Marcus, R.A., Sutin, N.: Electron transfers in chemistry and biology. BBA Rev. Bioenerg. **811**, 265–322 (1985)
14. Ishimatsu, R., et al.: Efficient electrogenerated chemiluminescence of pyrrolopyrrole aza-BODIPYs in the near-infrared region with tripropylamine: involving formation of S_2 and T_2 states. J. Am. Chem. Soc. **141**, 11791–11795 (2019)
15. Sasaki, Y., Yamamoto, K., Anada, S., Hirayama, T., Yoshimoto, N.: Direct visualization of electric potential distribution in organic light emitting diode by phase-shifting electron holography. Appl. Phys. Express **14**, 075007 (2021)
16. Meier, S.B., et al.: Light-emitting electrochemical cells: recent progress and future prospects. Mater. Today **17**, 217–223 (2014)
17. Xu, D., Adachi, C.: Organic light-emitting diode with liquid emitting layer. Appl. Phys. Lett. **95** (2009)
18. Tahara, H., Baba, R., Iwanaga, K., Sagara, T., Murakami, H.: Electrochromism of a bipolar reversible redox-active ferrocene–viologen linked ionic liquid. Chem. Commun. **53**, 2455–2458 (2017)
19. Tahara, H., et al.: Electrochromism of ferrocene-and viologen-based redox-active ionic liquids composite. ACS Appl. Mater. Interfaces **11**, 1–6 (2018)
20. Kopač Lautar, A., et al.: Electrolyte reactivity in the double layer in mg batteries: an interface potential-dependent DFT study. J. Am. Chem. Soc. **142**, 5146–5153 (2020)
21. Acevedo-Peña, P., Baray-Calderón, A., Hu, H., González, I., Ugalde-Saldivar, V.M.: Measurements of HOMO-LUMO levels of poly (3-hexylthiophene) thin films by a simple electrochemical method. J. Solid State Electrochem. **21**, 2407–2414 (2017)
22. Benck, J.D., Pinaud, B.A., Gorlin, Y., Jaramillo, T.F.: Substrate selection for fundamental studies of electrocatalysts and photoelectrodes: inert potential windows in acidic, neutral, and basic electrolyte. PLoS ONE **9**, e107942 (2014)
23. Matsumoto, H., et al.: Fast cycling of $Li/LiCoO_2$ cell with low-viscosity ionic liquids based on bis (fluorosulfonyl) imide $[FSI]^-$. J. Power. Sources **160**, 1308–1313 (2006)
24. Klein, J.M., Panichi, E., Gurkan, B.: Potential dependent capacitance of [EMIM][TFSI],$[N_{1114}]$[TFSI] and $[PYR_{13}]$[TFSI] ionic liquids on glassy carbon. Phys. Chem. Chem. Phys. **21**, 3712–3720 (2019)
25. Tsunashima, K., Sugiya, M.: Physical and electrochemical properties of low-viscosity phosphonium ionic liquids as potential electrolytes. Electrochem. Commun. **9**, 2353–2358 (2007)
26. Lee, K.H., et al.: ' Cut and stick' rubbery ion gels as high capacitance gate dielectrics. Adv. Mater. Deerfield Beach Fla **24**, 4457–4462 (2012)

27. Nishimura, K., et al.: Solution electrochemiluminescent cell with a high luminance using an ion conductive assistant dopant. Jpn. J. Appl. Phys. **40**, L1323 (2001)
28. Kato, E., Ishimatsu, R., Koinuma, Y., Mizuno, J., Kasahara, T.: Sky-blue electrogenerated chemiluminescence using anthracene derivatives as host and guest molecules. Jpn. J. Appl. Phys. **61**, 060903 (2022)
29. Okada, K., Ishimatsu, R., Mizuno, J., Kasahara, T.: Microfluidic electrogenerated chemiluminescence cells using aluminum-doped zinc oxide nanoparticles as an electron injection layer. Sens. Actuators Phys. **334**, 113329 (2022)
30. Ko, E.-S., et al.: Pulsed driving methods for enhancing the stability of electrochemiluminescence devices. ACS Photonics **5**, 3723–3730 (2018)
31. Ichinohe, N., et al.: Yellow phosphorescent electrogenerated chemiluminescence cell based on a cyclometalated iridium complex with a redox mediator. Electrochemistry **92**, 027004–027004 (2024)
32. Kasahara, T., et al.: Multi-color microfluidic electrochemiluminescence cells. Sens. Actuators Phys. **214**, 225–229 (2014)
33. Koinuma, Y., et al.: White electrogenerated chemiluminescence using an anthracene derivative host and fluorescent dopants for microfluidic self-emissive displays. Sens. Actuators Phys. **306**, 111966 (2020)
34. Kasahara, T., Ishimatsu, R., Kuwae, H., Shoji, S., Mizuno, J.: Color-tunable microfluidic electrogenerated chemiluminescence cells using Y-shaped micromixer. Jpn. J. Appl. Phys. **57**, 128001 (2018)
35. Okada, K., Ishimatsu, R., Mizuno, J., Kasahara, T.: Fabrication of microfluidic electrogenerated chemiluminescence cells incorporated with titanium dioxide nanoparticles to improve luminescent performances. Appl. Phys. Express **13**, 107001 (2020)
36. Koinuma, Y., Ishimatsu, R., Kato, E., Mizuno, J., Kasahara, T.: Green electrogenerated chemiluminescence using a quinacridone derivative as a guest molecule. Electrochem. Commun. **127**, 107047 (2021)
37. Kato, E., Ishimatsu, R., Mizuno, J., Kasahara, T.: Bright yellow electrogenerated chemiluminescence cell using a rubrene solution doped with an emitting assist dopant. Electrochemistry **91**, 047002–047002 (2023)
38. Sugaya, R., et al.: Multicolor band-edge electrochemiluminescence of colloidal silicon quantum dots in a thin-layer cell. ACS Appl. Opt. Mater. (2025)
39. Nobeshima, T., Morimoto, T., Nakamura, K., Kobayashi, N.: Advantage of an AC-driven electrochemiluminescent cell containing a Ru $(bpy)_3^{2+}$ complex for quick response and high efficiency. J. Mater. Chem. **20**, 10630–10633 (2010)
40. Itoh, N.: Electrochemical light-emitting gel made by using an ionic liquid as the electrolyte. J. Electrochem. Soc. **156**, J37 (2008)
41. Lee, J.I., et al.: Dynamic interplay between transport and reaction kinetics of luminophores on the operation of AC-driven electrochemiluminescence devices. ACS Appl. Mater. Interfaces **10**, 41562–41569 (2018)
42. Moon, C., Butscher, J.F., Gather, M.C.: An exciplex-based light-emission pathway for solution-state electrochemiluminescent devices. Adv. Mater. **35**, 2302544 (2023)
43. Goushi, K., Yoshida, K., Sato, K., Adachi, C.: Organic light-emitting diodes employing efficient reverse intersystem crossing for triplet-to-singlet state conversion. Nat. Photonics **6**, 253–258 (2012)
44. Yee, H., et al.: Extending the operational lifetime of electrochemiluminescence devices by installing a floating bipolar electrode. Small **20**, 2307190 (2024)
45. Kong, S.H., Lee, J.I., Kim, S., Kang, M.S.: Light-emitting devices based on electrochemiluminescence: comparison to traditional light-emitting electrochemical cells. ACS Photonics **5**, 267–277 (2018)
46. Oh, H., Kim, Y.M., Jeong, U., Moon, H.C.: Balancing the concentrations of redox species to improve electrochemiluminescence by tailoring the symmetry of the AC voltage. ChemElectroChem **5**, 2836–2841 (2018)

47. Lee, H., et al.: Communication—screen-printed silver electrodes for enhanced performance in light-emitting devices based on electrochemiluminescence. ECS J. Solid State Sci. Technol. **8**, R146 (2019)
48. Oh, H., Seo, D.G., Moon, H.C.: Performance improvement of yellow emitting electrochemiluminescence devices: effects of frequency control and coreactant pathway. Org. Electron. **65**, 394–400 (2019)
49. Hirata, S., et al.: Improvement of electroluminescence performance of organic light-emitting diodes with a liquid-emitting layer by introduction of electrolyte and a hole-blocking layer. Adv. Mater. **23**, 889–893 (2011)
50. Kubota, K., et al.: Liquid carbazole substituted with a poly (ethylene oxide) group and its application for liquid organic light-emitting diodes. Chem. Lett. **41**, 934–936 (2012)
51. Hirata, S., et al.: Improved device lifetime of organic light emitting diodes with an electrochemically stable π-conjugated liquid host in the liquid emitting layer. Jpn. J. Appl. Phys. **51**, 041604 (2012)
52. Kasahara, T., et al.: Fabrication and performance evaluation of microfluidic organic light emitting diode. Sens. Actuators Phys. **195**, 219–223 (2013)
53. Kobayashi, N., et al.: A wide-energy-gap naphthalene-based liquid organic semiconductor host for liquid deep-blue organic light-emitting diodes. J. Lumin. **200**, 19–23 (2018)
54. Tsuwaki, M., et al.: Fabrication and characterization of large-area flexible microfluidic organic light-emitting diode with liquid organic semiconductor. Sens. Actuators Phys. **216**, 231–236 (2014)
55. Ishimatsu, R., et al.: Kinetics of excimer electrogenerated chemiluminescence of pyrene and 1-pyrenebutyricacid 2-ethylhexylester in acetonitrile and an ionic liquid, triethylpentylphosphonium bis(trifluoromethanesulfonyl)imide. J. Phys. Chem. B **123**, 10825–10836 (2019)
56. Ishimatsu, R., Tashiro, S., Nakano, K.: Monomer and excimer emission in electrogenerated chemiluminescence of pyrene and 2,7-di-*tert*-butylpyrene associated with electron transfer distance. J. Phys. Chem. B **127**, 9346–9355 (2023)
57. Kasahara, T., et al.: Multi-color microfluidic organic light-emitting diodes based on on-demand emitting layers of pyrene-based liquid organic semiconductors with fluorescent guest dopants. Sens. Actuators B Chem. **207**, 481–489 (2015)
58. Kobayashi, N., et al.: Microfluidic white organic light-emitting diode based on integrated patterns of greenish-blue and yellow solvent-free liquid emitters. Sci. Rep. **5**, 14822 (2015)
59. Itoh, N.: Electrochemical light-emitting gel. Materials **3**, 3729–3739 (2010)
60. Moon, H.C., Lodge, T.P., Frisbie, C.D.: Solution-processable electrochemiluminescent ion gels for flexible, low-voltage, emissive displays on plastic. J. Am. Chem. Soc. **136**, 3705–3712 (2014)
61. Hong, K., et al.: Self-supporting ion gels for electrochemiluminescent sticker-type optoelectronic devices. Sci. Rep. **6**, 29805 (2016)
62. Moon, H.C., Lodge, T.P., Frisbie, C.D.: Electrochemiluminescent displays based on ion gels: correlation between device performance and choice of electrolyte. J. Mater. Chem. C **4**, 8448–8453 (2016)
63. Soulsby, L.C., et al.: Colour tuning and enhancement of gel-based electrochemiluminescence devices utilising Ru(ii) and Ir(iii) complexes. Chem. Commun. **55**, 11474–11477 (2019)
64. Kwon, D.-K., Myoung, J.-M.: Ion gel-based flexible electrochemiluminescence full-color display with improved sky-blue emission using a mixed-metal chelate system. Chem. Eng. J. **379**, 122347 (2020)
65. Kwon, D.-K., Myoung, J.-M.: Wearable and semitransparent pressure-sensitive light-emitting sensor based on electrochemiluminescence. ACS Nano **14**, 8716–8723 (2020)
66. Lee, S., et al.: Water washable and flexible light-emitting fibers based on electrochemiluminescent Gels. ACS Appl. Mater. Interfaces **14**, 17709–17718 (2022)

Microscopic Imaging of Redox and Electrochemical Reactions with Light

Fabien Miomandre and Vitor Brasiliense

Abstract This chapter is devoted to the description of optical microscopic techniques used to investigate electrochemical reactions. The first section is focused on fluorescence microscopy and its main configurations in order to image or control an electrochemical reaction involving emissive or fluorogenic species. In a second section, other optical techniques based on reflection, transmission, or scattering modes are described with practical examples of studies. In both parts, a special attention is brought to spatial and time resolution limits as well as sensitivity issues.

Keyword Fluorescence · Microscopy · Scattering · SECM · Electrofluorochromism

In complement to ECL, where light is produced by a combination of redox and electrochemical reactions, photoluminescence and more generally optical techniques based on scattering, reflection, absorption, or interference can also be used to provide information on the mechanisms and location of redox and electrochemical processes. The two past decades have seen the emergence of imaging techniques with an ever-higher resolution and/or sensitivity, which is of course one of the main challenges in the imaging field. The aim of this chapter is to provide guidelines to understand the basics of these approaches and describe some general trends observed in the recent literature.

F. Miomandre (✉) · V. Brasiliense
PPSM—CNRS—ENS Paris-Saclay, Gif Sur Yvette, France
e-mail: mioman@ens-paris-saclay.fr

F. Paolucci et al. (eds.), *Electrogenerated Chemiluminescence Microscopy and Imaging*,
https://doi.org/10.1007/978-3-032-14602-1_13

1 Imaging Electrochemical Reactions with Fluorescence

1.1 Introduction

The idea of using fluorescence as an optical readout to image electrochemical reactions has been thought since the 1980s [1], that is to say at the same time the now popular scanning electrochemical microscopy (SECM) was invented by Bard [2]. While SECM is a powerful scanning probe technique to analyze the electrochemical reactivity of surfaces, it does not provide real images as an optical microscope does. Fluorescence microscopy, an imaging technique that focuses on the analysis of emitted light from excited species, can fill this gap, of course with a spatial resolution which remains diffraction limited. Nevertheless, it took almost 20 years to see the combination of SECM with fluorescence microscopy appear in the literature [3, 4]. In parallel to the development of new instruments, an intensive effort started to find high yield fluorescent dyes able to report redox events with the best contrast. Historically, the first examples in the literature were based on indirect reports through pH changes, which were translated into a dramatic change of emission intensity using pH-sensitive dyes. Later on, electrofluorochromic compounds, presenting direct changes in their emission properties according to their redox state [5], have emerged and become the topic of growing interest. This way, electrochemical reactions occurring onto or in the near vicinity of the electrode surface could be directly imaged through fluorescence intensity variations.

This short milestone collection reveals how both instrumentation and chemical advances are essential to meaningfully describe redox reactivity through imaging process. In this section, we shall describe some of the most popular instrumental approaches used in emission-based opto-electrochemical setups, besides providing elements to guide the choice of pertinent molecular reporters.

1.2 Various Configurations of Instrumental Setup

Fluorescence microscopy is a widespread technique to image and report a variety of processes occurring at the micron scale. The essential elements of a fluorescence imaging setup are a light source (lamp, LED, or laser), an imaging device (such as a microscope equipped with large numerical aperture objectives), and a method for rejecting the excitation wavelength with respect to the emission. Most often, the emitted light is collected through the same objective, separated from the excitation by filters and a dichroic mirror, and collected on a CCD camera, a photomultiplier, or a spectrometer. The coupling with an electrochemical cell is relatively straightforward but variations exist depending mainly on the choice of the working electrode (Fig. 1): it can be either the substrate itself (typically an optically transparent semiconductor like ITO or FTO or a semi-transparent ultra-thin metallic layer, in each case coated on a glass slide or coverslip) or a metallic disk or tip dipped in the electrolyte solution.

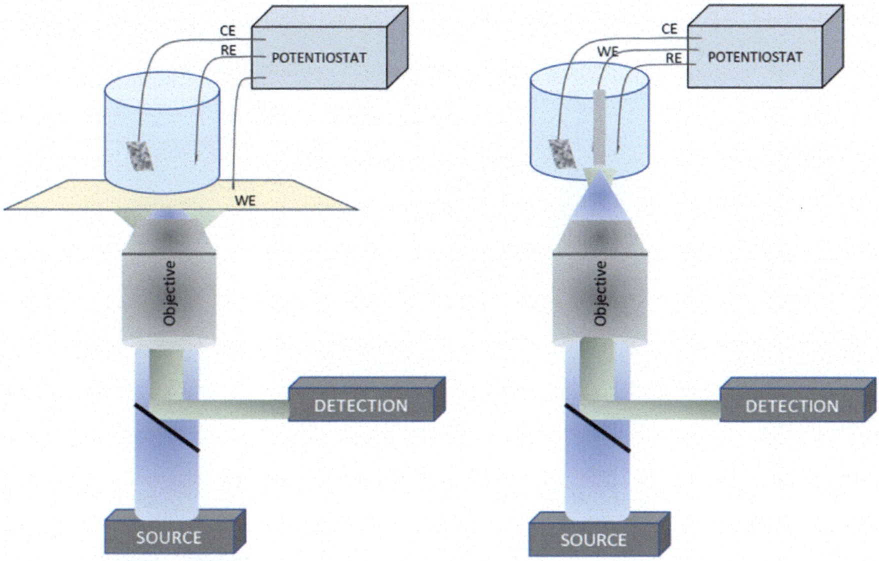

Fig. 1 Fluorescence microscope coupled to an electrochemical cell using either the substrate (left) or an independent (micro)disk (right) as the working electrode

In the latter case, objectives with long working distances are required to focus the beam on the working electrode surface.

The use of the substrate as the working electrode offers the advantage of requiring smaller volumes of electrolyte solution but restricts the choice of the electrode material to (semi)-transparent materials. Moreover, since conductive layers are rarely as transmissive as glass, some fraction of the signal is lost by absorption or reflection in the interface. The second configuration (Fig. 1 right) requires to design an electrochemical cell with a very thin optical window at the bottom. In such case, quartz is often preferred over other glasses due to its wider spectral window. This configuration might be more complex to implement, but offers the advantage of using all the panel of electrode materials and geometries.

Fluorescence microscopy can be performed in several configurations (Fig. 2): either in wide field by illuminating the substrate perpendicularly or in Total Internal Reflection Fluorescence (TIRF) by illuminating the substrate with an angle above the critical angle of total reflection. In that case, the excitation operates through the evanescent wave propagating across the interface, thus decreasing the probed area to typically a few hundreds of nanometers, resulting in a higher spatial resolution. Another popular strategy to limit the excitation area, minimizing parasitic signal, is through the use of focused laser in confocal mode. In this approach, a pinhole is used to limit the excitation zone, while the sample is scanned in three dimensions allowing better-resolved images to be collected at the expense of time resolution. In addition, this configuration allows one to selectively probe a given plane at a certain

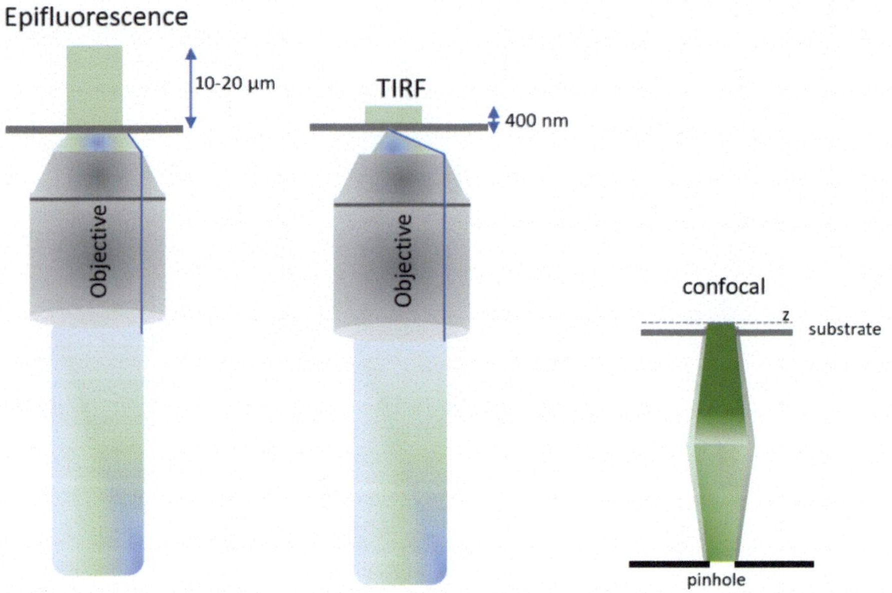

Fig. 2 Various configurations of fluorescence microscopy. Excitation beam in light blue and emission beam in green

distance z of the substrate allowing a 3D reconstruction of the diffusion layer when z is gradually changed (see Sect. 1.3).

Confocal Light Scanning Microscopy (CLSM) was originally used with an electrochemical cell for the first time by Unwin in 2002 [6], providing direct images of the reduction of benzoquinone at the surface of a platinum microdisk. Then, other types of setup have been designed to image electrochemical reactivity using fluorescence: One can cite the coupling of SECM with fluorescence microscopy developed in our group [7] with a unique molecular probe for the electrochemical and fluorescence detections, following earlier attempts in the early 2000s [3, 4] and the application of SECM-Fluorescence microscopy coupling to investigate biochemical issues [8, 9]. The idea of using fluorescence to report the reaction at the counter electrode rather than at the working electrode emerged in the 2020s [10], based on a concept well known in bipolar electrochemistry [11, 12] consisting in spatially offsetting the place where the phenomenon is triggered from the one where it is reported. The general principle is illustrated in Fig. 3: The electrochemical reaction occurring on one pole is reported by the fluorescence signal on the other pole. This technique was named Fluorescence-Enabled Electrochemical Microscopy (FEEM) and developed among others by Zhang with arrays of close bipolar electrodes [12]. Application to the detection of various analytes like glucose [13] or hydrogen peroxide [11] can be found in the recent literature.

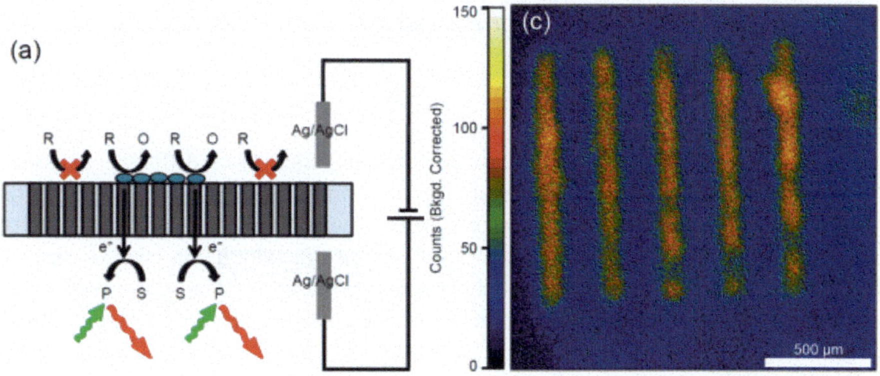

Fig. 3 Electrochemical reactions in bipolar electrochemistry reported by a fluorescence signal (reproduced with permission from ref. [12])

1.3 Direct Visualization of the Diffusion Layer

The original goal of coupling fluorescence microscopy with electrochemistry is to use the optical technique to image the electrochemical reaction. This can be done by probing either directly the surface when the electroactive species are immobilized on it or by probing the diffusion layer of the electrode as it forms in real time ("operando" as we say now).

An example of surface confined reactions probed by fluorescence microscopy can be found in the adsorption/desorption process of self-assembled monolayers (SAMs) on gold reported 20 years ago by Bizzotto [14]. The desorption process which is triggered by an electrochemical reduction can be followed by the instantaneous increase of fluorescence intensity of the probe attached to the monolayer, as long as it moves away from the quenching metallic surface. A similar concept was implemented to investigate other types of SAMs by the same group [15, 16].

Later on, Bouffier et al. used confocal microscopy coupled to electrochemistry to image and reconstruct in 3D the diffusion layer of an electrode [17]. Confocal microscopy is indeed a technique of choice to select the emission coming from a precise vertical distance from the substrate which correlates with the concentration of one redox state monitored by the potential of the electrode. This concept was applied to decipher more complex mechanisms involving chemical reactions coupled to electron transfer [18] or enzymatic reactions [19]. At approximately the same period, Compton et al. also used confocal microscopy to image the reaction layer [20] or heterogeneities on electrode surfaces [21], exploiting the high spatial resolution of this technique and an ultimate chemical confinement of a pH-sensitive fluorophore.

1.4 The Concept of Electrochemical Fluorescence Switch

The concept of the electrochemically controlled fluorescence switch is generally called electrofluorochromism [22] due to its similarity with electrochromism. It encompasses all cases where at least two redox states have different emission properties (generally emission intensity but in some cases emission wavelengths as well [23]). A subgroup of these electrofluorochromic compounds, generally called fluorogenic, corresponds to non-emitting species which can be activated by a redox process with special interest in bioanalysis [24]. Basically, the term electrofluorogenic is often used as a synonym of electrofluorochromic in many reports [25].

From a molecular point of view, there is a quite large choice of organic or organometallic compounds possessing this property [26, 27]. Polycyclic aromatics having heteroatoms in their backbone like rhodamine, boron dipyrromethene (Bodipy), or perylenediimide (PDI) belong to this category. Indeed, the polycyclic skeleton ensures a high rigidity which is acknowledged to be favorable for efficient fluorescence properties, while the presence of heteroatoms and electron donating or withdrawing groups brings electroactivity in an accessible potential range. This is also the case of tetrazine derivatives which are strongly electron-deficient aromatic compounds with for some of them fluorescence properties triggered by excitation in the visible [28]. Another molecular system widely used for its electrofluorogenic properties is the resazurin/resorufin one [17]. Resazurin is a weakly fluorescent phenoxazine derivative that can be converted into the strongly fluorescent resorufin by an irreversible two-electron reduction. The fluorescence of resorufin can be in turn reversibly quenched by its transformation into dihydroresorufin by a two-electron reduction in basic aqueous conditions (Scheme 1).

Being able to control the emission state of a molecule or a material through a potential signal opens the way toward various application fields like making high-contrast displays [29] or anticounterfeiting systems [30]. Indeed, electrofluorochromic compounds are very often electrochromic as well, allowing one to associate the "output" state (emissive or not, plus the color) to a combination of several "input" values like the electrode potential and different excitation wavelengths (typically in UV or visible).

1.5 Pushing the Resolution to the Limit

From the point of view of imaging chemical or biochemical processes, several developments have been highlighted recently in the literature allowing to reach very low detection thresholds. Some are based on the use of bipolar electrodes in relation with a fluorescence switch as already mentioned. The use of fluorescence microscopy in bipolar electrochemistry can be implemented with simple redox reactions involving protons coupled with fluorophores whose emission is strongly dependent on the local

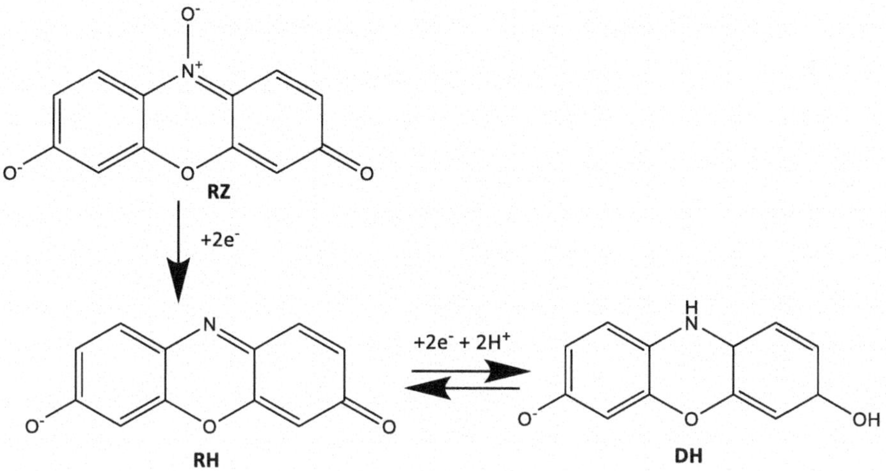

Scheme 1 Electrochemical reactions involving resazurin (RZ), resorufin (RH), and dihydroresorufin (DH)

pH (like fluorescein for example) [31]. Hence, according to this principle, cancer cells glycoproteins can be detected on a surface with a high sensitivity (10 fM) [32].

Another interesting development of electrochemical fluorescence switch is the concept of EC-STORM recently proposed by Gooding [33]. STochastic Optical Reconstruction Microscopy (STORM) is a well-known technique for imaging objects with improved, subdiffractive, resolutions in which UV light is used to quench the emission of luminophores, enabling their localization as individual emitters when they stochastically return to the emissive states. EC-STORM applies the same concept but replacing the UV excitation by an electrochemical signal, and its use to detect single molecules was recently demonstrated [34]. The precise control of the luminescence states and the rapid switch between them are important factors to make the detection quantitative and very well spatially resolved. Applying the concept to the famous Green Fluorescent Protein (GFP) demonstrated its interest in biological cell imaging with a high signal-to-noise ratio and much less cell damage than with all optical super-resolution techniques [35].

1.6 Time-Resolved Detection and Transient Reactions

One of the main advantages of optical techniques is the improved time resolution they typically offer. However, when dealing with time resolution, one must first discriminate between time-resolved fluorescence techniques on one hand and investigation of electrochemical reactions in the transient regime on the other hand.

Time-resolved fluorescence may help to decipher the mechanisms involved in the electrochemical fluorescence switch, as demonstrated by our group [36].

Indeed, simple conversion processes can be unambiguously distinguished from those involving the electrogenerated species as a possible quencher of the excited state of the fluorophore. Fluorescence imaging of transient events in (electro)chemical reactions remains somehow challenging if one wants to investigate short time regimes (sub-millisecond). The difficulties to overcome lie in the necessary large number of photons to be collected in a very short time range and the possible signal loss related to the diffusion of emitters. The first condition is strongly related to the brilliance of the dye, which reports the ability to absorb a large number of photons and a high yield for emitting photons from the excitation process. When the event is directly reported by an electrofluorochromic probe whose emission is related to its redox state, mitigation of diffusion can be obtained by increasing the medium viscosity or even suppressed by attaching the probe to the surface. Furthermore, the fluorescence switching rate can be enhanced by chemical confinement as demonstrated by Zhang using resorufin [13]. The principle is illustrated in Fig. 4: The presence of a chemical reductant like glucose creates a feedback loop avoiding the accumulation of the dye in its fluorescent state near the electrode, thus enhancing the switching rate. The use of indirect pH-dependent fluorescent probes can also lead to the real-time follow-up of transient electrochemical reactions as demonstrated more recently by Chovin et al. [10], where 100 ms potential pulses applied to generate ferrocene oxidation at the working electrode could be translated in real time into fluorescent peaks at the counter electrode. Beyond Faradaic process, other transient responses related to electric field driven phenomena like electroosmosis can also be tracked using charged pH-sensitive fluorescent dyes in nanofluidic channels [37].

These examples show that an adequate choice of the electrofluorochromic probe and experimental conditions allow imaging transient phenomena down to the millisecond regime.

1.7 Comparing Fluorescence Microscopy with ECLM

ECL and fluorescence are often put in competition to report chemical events in microenvironments. A well-known advantage of ECL over photoluminescence in general is the absence of background signal due to the absence of light source in ECL. Furthermore, ECL is by essence spatially resolved since light emission only occurs in the vicinity of the electrode surface where the first reactant of the light-generating reaction is produced. However, as underlined by Zhang in a recent review [38], ECL suffers from an electron-to-photon efficiency which is generally quite low. This might become a critical issue when very small amounts of molecules are involved compared to photoluminescence for which the emitted to absorbed photon quantum yield may be quantitative. ECL relies on the production of radicals which are usually very reactive to avoid any diffusion before their reaction with the electrogenerated species. While this is an advantage in terms of spatial resolution, this also means that competitive reaction pathways (e.g., with the solvent) might limit the overall yield. Another point to be considered is the time response, as mentioned above.

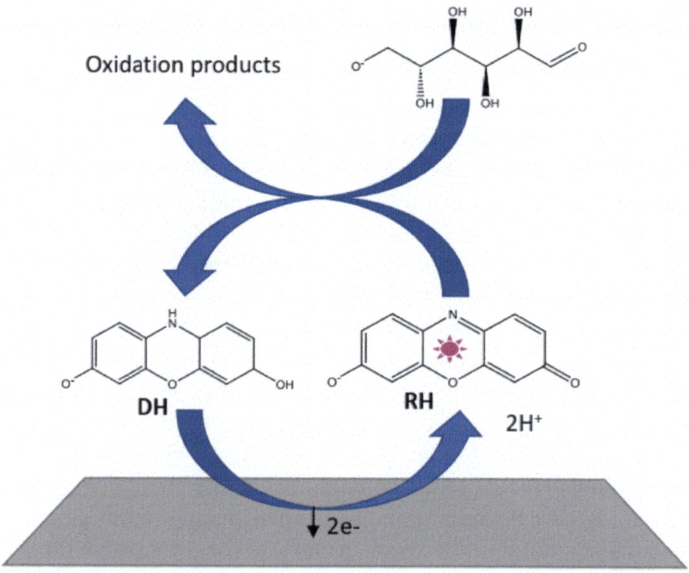

Fig. 4 Chemical confinement allowing to enhance the switching rate of fluorescent resorufin (RH) to detect glucose (adapted from Oja et al. [13])

If one wants to shorten it to image transient events, the direct consequence is the collection of fewer photons and thus requires high yield for their production, which is generally the case with fluorophores. When this high yield is associated with an optical technique with high spatial resolution like TIRF or confocal microscopy, short time responses become accessible. Short-time detection also overcomes the limitation of photoluminescence techniques due to photobleaching, a phenomenon that renders electrochemical fluorescence switches for a small collection of emitters generally difficult to observe [39].

One can also mention that efficient ECL is often obtained with a relatively narrow panel of emitters/coreactant pairs, which can restrict its use in terms of color or electrolytic medium. Electrofluorochromic probes are more numerous, although some of them attract more attention due to their high quantum yield and stability in aqueous medium, e.g., resazurin [17], resorufin [11, 40], or rhodamine [41]. Furthermore, an alternative to direct electrofluorochromic dyes can be found in using pH-sensitive fluorophores, as electrochemical reactions in aqueous solutions very often involve protons. This is the case of fluorescein, which is much more fluorescent in basic conditions and can thus serve as a reporter of pH gradients in electrochemical reactions [42, 43]. More recently, it has been demonstrated the same trend with derivatives of coumarins [44].

2 Imaging Electrochemical and Redox Reactions with Other Optical Techniques

While emission-based methods remain extremely powerful tools for imaging redox processes, they pose a fundamental requirement on the probed system: the analyzed materials and molecules must be emissive. Other optical techniques are therefore becoming increasingly popular for imaging redox reactions with high spatial and temporal resolutions and have been used to analyze redox reactivity in a variety of contexts [45], described briefly in the second part of this chapter.

Optical microscopy and spectroscopy encompass a wide family of configurations and techniques, many of which have been coupled to electrochemistry. Depending on the sought spatial and time scales, different setups or configurations can be designed for providing the required redox information from optical observables. As a result, a growing plethora of methods are under development, providing complementary information about the redox activity of surfaces, individual nanoobjects, or molecular layers. Rather than attempting to exhaustively present all existing opto-electrochemical methodologies, we instead chose to focus on iconic literature examples that illustrate how different optical setups and techniques can be used to obtain redox information, and how they have been coupled to electrochemical experiments. Interested readers are referred to excellent review papers on the topics [46–49].

We broadly present three categories of optical setups, based on successively reflection, scattering, and transmission, focusing on examples that illustrate different strategies to couple electrochemistry and optics in different contexts, and the kind of redox information that can be extracted. Whenever possible, we have chosen to illustrate each configuration with: (i) straightforward methods that require relatively little adaptation to commercial microscopes; (ii) advanced methodologies, often based on home-made setups, enabling precise quantification of the underlying redox processes; (iii) spectroscopically based methods, which emphasize the extraction of chemical characteristics of the analyzed systems and their evolution with redox reactions.

2.1 Reflection-Based Methods

Electrodes used in classical electrochemical setups are often metallic surfaces. Therefore, it should be of little surprise that optical techniques exploiting light reflection were among the first opto-electrochemical strategies to be developed [50, 51]. Reflectivity is perhaps the most straightforward manner to control the surface-state of an electrode with high resolution. It requires analyzing the evolution in the light reflected by the electrode surface; therefore, it can be readily implemented at most laboratories equipped with an optical microscope. Most commonly, reflectance experiments are set as indicated in Fig. 5a and thus pose little requirements in terms of the geometry and dimensionality of the probed system. Instrumentally, most microscope objectives have limited working distances ($\sim$ millimeters), so either thin closed cells are used, or an immersion objective is required. Such implementation simplicity favors its interfacing with different complementary systems (scanning probes [52], spectroscopy, etc.), with a flexibility allowing the extraction of redox information from different samples at different scales. While normal incidence is by far the most common methodology, examples using oblique illumination also exist [53].

Reflectivity-based setups have been used in different electrochemical contexts. For instance, as indicated in Fig. 5b, the reflectance of electrocatalytic surfaces was applied to differentiate various crystallographic orientations, and to identify both increase in catalytic activity and deactivation phenomena, such as the onset of corrosion [54, 56–58]. Quantification of catalytic phenomena at the nanoscale is also possible, notably at single nanoobjects. Such approach can be carried out at different levels of sophistication, e.g., using either the direct reflectivity signal [59], comparing the response at different color (RGB) channels, or through interferometric approaches [60–62].

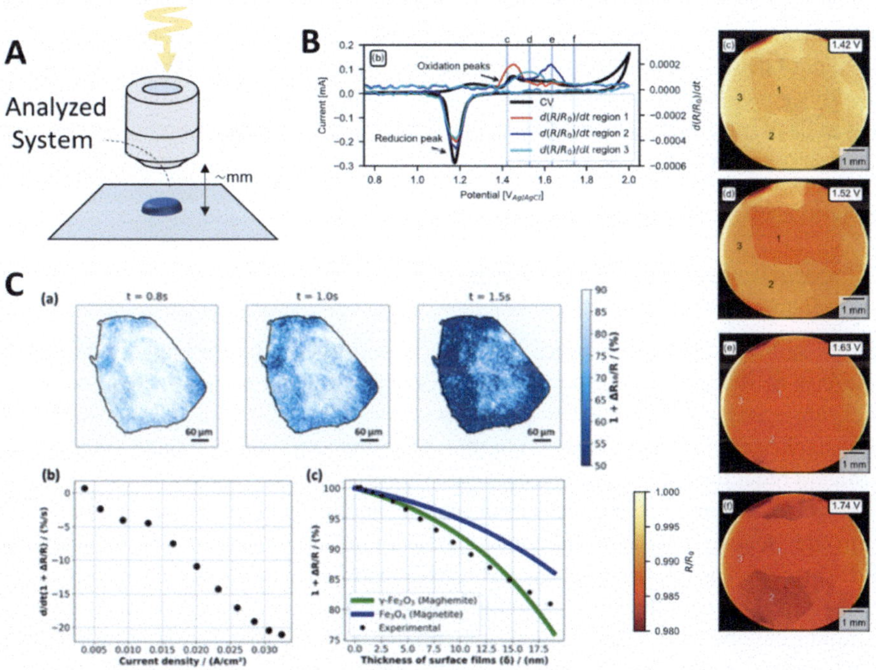

Fig. 5 **a** Schematic view over reflection-based microscopy techniques. **b** Analysis of reflectivity evolution of different Au crystallographic domains at different anodic potentials. As indicated in the inset, the spatially resolved reflectivity results can be directly compared to the electrochemical current, revealing the electrochemical behavior of different domains. Reprinted with permission from Pfaff et al. [54]. **c** Opto-electrochemical analysis of corrosion of iron surfaces, attempting to quantify the extent of surface modification through reflectance. Reprinted with permission from Makogon et al. [55]

Quantitative approaches have also been developed, to analyze the electrodeposition and growth of subdiffractive objects, such as catalytic particles, or to monitor the reactivity of micrometric ones, such as graphene flakes [63] or lithium dendrites. Deposition of thin films in operando conditions can be readily analyzed, providing in situ information about the layer thickness and homogeneity [64]. Interestingly, the ionic transport in dilute solution triggered by electrochemical reactions has been recently shown to lead to measurable reflectance variations, and therefore can be used to reveal concentration gradients [65, 66].

Reflectance microscopy also finds diverse applications in corrosion science. Primarily, it provides a visual guideline for identifying the onset of different corrosion modes and verifying the integrity of coatings. More recent uses have emphasized its potential as a quantitative tool, particularly in the case where the chemical identity (and hence optical properties) of the formed phases are known. The corrosion of aluminum and iron surfaces [55, 67], for example, has been carefully analyzed quantitatively using reflectance based microscopy (as indicated in Fig. 5c), in some cases aided by advanced image analysis and segmentation schemes. As with other optical techniques, automated image analysis tools are becoming fundamental, enabling the behavior of complex materials to be studied. For instance, in battery electrochemistry, reflectivity measurements have been used to reveal complex reactivity phenomena, such as Li plating [68], transport and intercalation, active material precipitation, and electrodeposition [69].

2.2 *Scattering-Based Techniques*

Scattering-based optical strategies are among the most popular for imaging the redox reactivity of isolated scattering objects, such as individual nanoparticles. With the emerging interest in single-entity electrochemistry [70–72], optical scattering techniques are becoming increasingly popular, enabling extraction of small signals due to drastic reduction of the background.

The most common configuration for measuring the scattering from nanoparticles is through the use of a dark field condenser, as indicated in Fig. 6a [73]. Condensers for most commercial microscopes are readily available, and work by illuminating the sample at high angle, larger than the collecting angle of the objective, defined by its numerical aperture. Only if local scatterers (e.g., nanoparticles) are present, will light be significantly detected, minimizing the background and favoring the detection of small signals. Such a technique is popularly used to analyze the electrodeposition of different metals on transparent electrodes (e.g., ITO [74], FTO, or metallic thin films) and for monitoring electrochemically induced nanoparticle growth [75].

While easily implemented, the use of commercial condensers imposes limits upon the system numerical aperture, thus limiting the access to the system, precluding interfacing with local probes, for example. In such cases, off-axis illumination schemes enabled by optical fibers are useful alternatives that have been, for example, used for

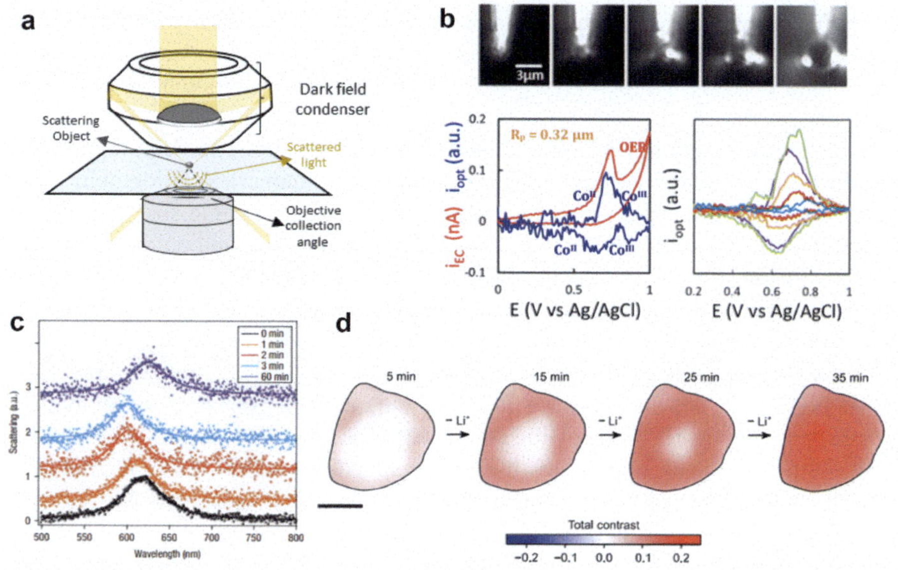

Fig. 6 **a** Schematic view over classical implementation of single object scattering imaging using a dark field condenser. **b** Operando analysis of single NP electrodeposition over a carbon nanoelectrode, using off-axis scattering strategy. Adapted with permission from Brasiliense et al. [76]. **c** Variation in the plasmonic resonance of individual Au nanoparticles during reduction reaction, due to the accumulation of charge on the particle. Reprinted with permission from Novo et al. [85]. **d** Evolution of the lithiation distribution of an NMC microparticle using interferometric scattering (iScat). Reprinted with permission from Xu et al. [100]

analyzing particles on the tip of nanoelectrodes and nanopipettes [76–78], as indicated in Fig. 6b. In a slightly less general setting, when electrodes which can support propagating plasmonic modes are used, scattering of propagating plasmonic modes becomes powerful tools for detecting small signals, coming from nanoparticles, molecular clusters, or single molecules [79–81].

Owing to their enhanced scattering cross sections, plasmonic nanoobjects are particularly fit for scattering experiments, remaining the most popularly studied systems using these techniques. This is however not a fundamental limitation, as the analysis of the electrochemical response of materials with no (or little) plasmonic character such as transition metal oxides [76, 78], Prussian blue analogues [82], hybrid materials [83], or Pd [84] have also been demonstrated. Dark field imaging setups can be readily interfaced with more advanced spectroscopic analysis. For instance, while analyzing electrocatalytic processes, coupled dark field and fluorescence microscopies are able to provide information about the reaction efficiency, while monitoring the NP catalysts scattering signal. Moreover, upon modification of the detection path, single particle spectroscopic analysis is possible, enabling more chemical insight into physicochemical processes. For instance, monitoring the plasmonic resonance of individual NPs has enabled charge injection from a chemical reductant to be precisely controlled, as illustrated in Fig. 6c [85]. Upon adsorption on the electrodes different chemical redox processes can similarly be monitored, such as Ag NP conversion [86–88], redox reactions [89], non-Faradaic electrochemical tuning [90, 91], NP dealloying [92, 93], or delicate underpotential deposition processes.

Interferometric approaches based on the scattering signal are also appealing, enabling for example 3D nanoparticle tracking, as in numerical holography [94]. Such techniques have been used for estimating the hydrodynamic size of colloidal nanoparticles, before analyzing their coupled electrochemical and optical responses upon collision with an electrode. Besides enabling tracking, interference-based methods are interesting due to increased sensitivity, as the detection efficiency is proportional to the product of the scattering and reference fields. For a scattering particle, the scattered amplitude is proportional to the particle volume. Therefore, while pure scattering intensity (as in dark field) is proportional to the amplitude square, generating a V^2 signal for small particles, interferometric signals which scale with V become more sensitive for small particles [95]. Several emerging optical techniques exploit such linearity [96], including interferometric scattering microscopy (Iscat), an emerging and readily implementable technique that allows detection of single subdiffractive objects such as proteins and tracking even small (10 nm) nanoparticles [97] over long distances. Within electrochemistry contexts, this technique has been used, for example, to image Li transport in battery materials, and to monitor lithium deposition and stripping on Cu surfaces [98–100].

2.3 *Transmission-Based Techniques*

Transmission-based techniques are another class of optical methods that can be readily implemented and can be used for analyzing redox phenomena. Its usage in electrochemical frameworks is a bit less general, requiring some degree of light transmission through the sample. This limits the objects that can be analyzed with transmission techniques to either thin layers, semi-transparent objects, or small subdiffractive objects, as discussed below.

Absorption differences between two redox states are the main properties through which transmission mode microscopy can be used to extract electrochemical information about samples [101]. Several methodologies have been developed based on this principle, normally applied in thin cells or being used to measure changes in thin layers. At thin-layer levels, such techniques are usually coupled with spectroscopic methods, such as UV–Vis absorption, enabling electrochemical process to be quantified while informing about the chemical identity of the generated/consumed species.

Thin objects, such as 2D materials, are also good candidates for optical analysis in transmission mode; even highly absorbing materials enable light transmission, making 2D materials, such as MoS_2 [102] or graphene [103, 104], well adapted to transmission-based optical techniques. It is important to highlight that both redox and non-Faradaic mechanisms can generate absorption differences in such thin materials.

For instance, MoS_2 transmission can be modulated through Faradaic processes, such as in lithium intercalation (Fig. 7b), or through electrochemical charging (Fig. 7c).

Thick materials are relatively difficult to analyze in transmission modes, as little light is actually transmitted, but such difficulty can be occasionally circumvented by varying the observation wavelength. Such principle was elegantly illustrated by employing NIR light to analyze battery systems in transmission mode [105].

Sub-diffractive individual particles can also be observed in transmission modes, but require some amplification mechanism to extract the signal of single objects from the background: for example, photothermal microscopy methods [106, 107], in which a pump beam is used to induce local refractive index changes due to absorption. Such a technique has been used to analyze individual absorbers such as single nanoparticles or single molecules, but its use in redox contexts is still limited.

While too opaque objects are difficult to analyze quantitatively using transmission microscopy implementation, objects absorbing too little are also challenging. One solution in such a case is to increase the transmission signal by promoting several light passages through the object of interest. For instance, the local oxidation kinetics of micrometric particles in flow batteries was imaged through this approach, using a SECM electrode as a microscale mirror to increase the transmission path [108]. Alternatively, phase-based methods can be used, such as digital holography or wave-front sensing, that measure differences in the optical path. Becoming widely popular in nanosciences [109, 110] and biology [111], relatively few examples of quantitative phase imaging methodologies exist so far for analyzing redox reactions. Our group has analyzed photoinduced generation and grafting of aryl-radicals from diazonium salts, as indicated in Fig. 7c [112, 113]. We have shown that local surface modification can be quantitatively measured, using a common-path interferometric approach that can be implemented with a thin diffuser [114]. The same operando technique was later adapted to control the irradiation rate, enabling precise (5 attoliter) control over the volume of the formed layers [115].

3 Conclusion

Optical techniques are becoming increasingly popular for tracking redox activity and reactivity. In this chapter, we have collected emerging strategies that materialize such potential, along with indications of popular implementations of the different techniques.

For emissive molecules, as described in the first part of the chapter, the development of fluorescence microscopy coupled with electrochemical techniques has become a very efficient tool to image electrochemical reactions with high spatial and temporal resolutions. One takes advantage of the high sensitivity of fluorescence and the ability of electrochemistry to rapidly switch the emission on and off reversibly, opening new routes toward super-resolution allowing single-molecule detection. Efforts are still necessary if one wants to enhance the long-term stability and shorten the time scale of the investigated events.

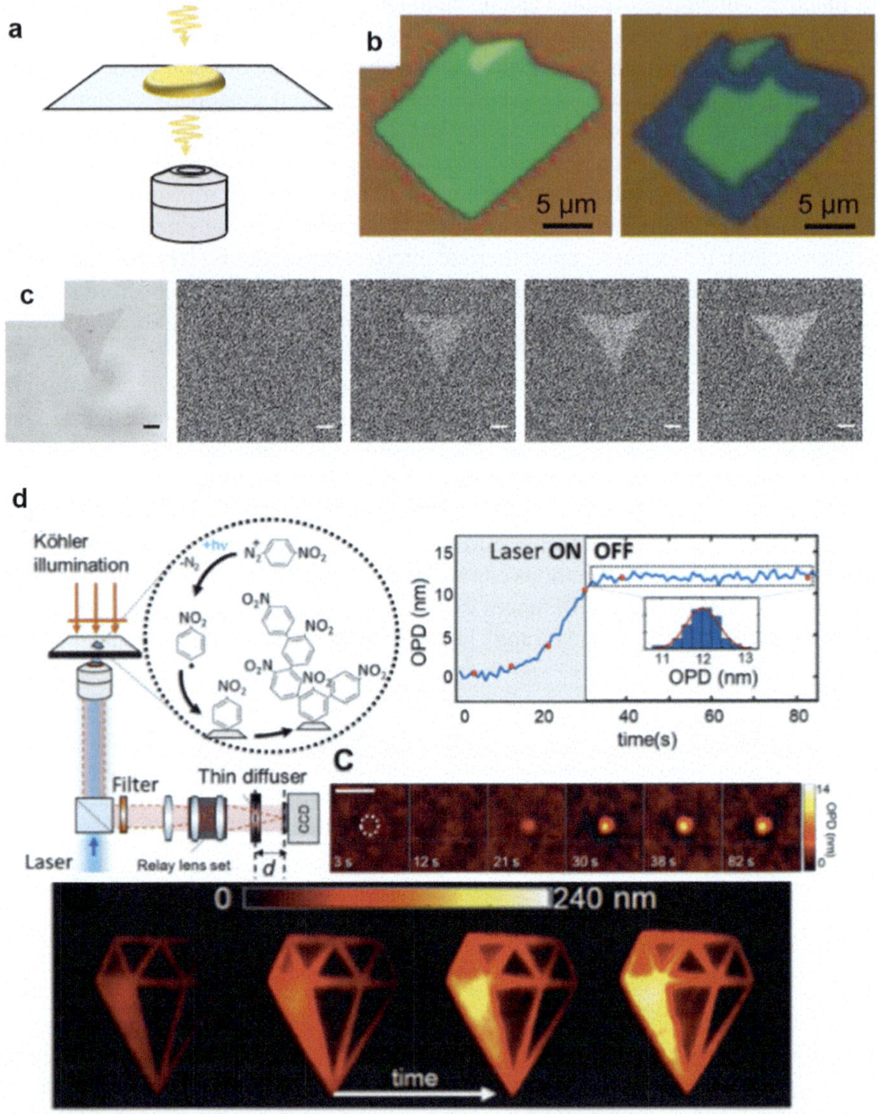

Fig. 7 Different approaches for analyzing redox approaches in transmission-based optical techniques **a** typical scheme in such techniques, which analyzes the light transmitted through the sample. **b** Optical transmission microscopy of MoS_2 nanoflakes before (left) and during (right) electrochemical Li intercalation process. Reprinted with permission from Xiong et al. [116]. **c** Transmission contrast of an atomic flat MoS_2 layer in transmission mode, and differential transmission images at different potentials. Reprinted with permission from Zhu et al. [102]. **d** Quantitative phase imaging for monitoring the formation of thin layers due to photoactivation of diazonium salts over a glass substrate. The setup is shown in the inset, next to a sequence of images showing the monitoring of the layer thickness on a single diffraction-limited spot (as measured by the optical path difference). The bottom sequence shows the evolution of a more complex pattern, generated with structured light. Reprinted with permission from Brasiliense et al. [112]

For non-emissive systems, several other optical imaging modalities are available, as briefly summarized in the second part of the chapter. Depending on the system dimensionality, and the time and spatial scales of interest, we have presented different examples of how reflectance, scattering, or transmission optical setups have been designed to detect, quantify, and analyze redox processes.

We have thus showed that imaging approaches can be relatively easily implemented and coupled to electrochemical modulation, allowing several redox information to be accurately and locally measured. Several quantitative approaches can be developed, enabling high-throughput characterization of different redox processes, while revealing spatial correlations which are difficult to obtain through traditional methods. We therefore anticipate that the current trend of coupling optical and electrochemical methods will still continue to grow, becoming a standard methodology for electrochemists of different horizons.

References

1. Patterson, M.L., Allen, C.S.: Fluorescence detection of polishing alumina on glassy carbon electrode surfaces. Anal. Chem. **57**(13), 2751–2752 (1985)
2. Liu, H.Y., Fan, F.R.F., Lin, C.W., Bard, A.J.: Scanning electrochemical and tunneling ultramicroelectrodes microscope for high-resolution examination of electrode surface in solution. J. Am. Chem. Soc. **108**(13), 3838–3839 (1986)
3. Boldt, F.M., Heinze, J., Diez, M., Petersen, J., Börsch, M.: Real-time pH microscopy down to the molecular level by combined scanning electrochemical microscopy/single-molecule fluorescence spectroscopy. Anal. Chem. **76**(13), 3473–3481 (2004)
4. Mukhopadhyay, R.: SECM meets fluorescence microscopy. Anal. Chem. **76**(13), 224A (2004)
5. Illos, R.A., Harlev, E., Bittner, S.: A novel all-organic chemical and electrochemical fluorescent switch. Tetrahedron Lett. **46**(48), 8427–8430 (2005)
6. Cannan, S., Douglas Macklam, I., Unwin, P.R.: Three-dimensional imaging of proton gradients at microelectrode surfaces using confocal laser scanning microscopy. Electrochem. Commun. **4**(11), 886–892 (2002)
7. Guerret-Legras, L., Audibert, J.F., Dubacheva, G.V., Miomandre, F.: Combined scanning electrochemical and fluorescence microscopies using a tetrazine as a single redox and luminescent (electrofluorochromic) probe. Chem. Sci. **9**(27), 5897–5905 (2018)
8. Salamifar, S.E., Lai, R.Y.: Use of combined scanning electrochemical and fluorescence microscopy for detection of reactive oxygen species in prostate cancer cells. Anal. Chem. **85**(20), 9417–9421 (2013)
9. Kim, J., Izadyar, A., Shen, M., Ishimatsu, R., Amemiya, S.: Ion permeability of the nuclear pore complex and ion-induced macromolecular permeation as studied by scanning electrochemical and fluorescence microscopy. Anal. Chem. **86**(4), 2090–2098 (2014)
10. Djoumer, R., Chovin, A., Demaille, C., Dejous, C., Hallil, H.: Real-time conversion of electrochemical currents into fluorescence signals using 8-hydroxypyrene-1,3,6-trisulfonic acid (HPTS) and amplex red as fluorogenic reporters. ChemElectroChem **8**(12), 2298–2307 (2021)
11. Stefano, J.S., Conzuelo, F., Masa, J., Munoz, R.A.A., Schuhmann, W.: Coupling electrochemistry with a fluorescence reporting reaction enabled by bipolar electrochemistry. J. Electroanal. Chem. **872**, 113921 (2020)
12. Guerrette, J.P., Percival, S.J., Zhang, B.: Fluorescence coupling for direct imaging of electrocatalytic heterogeneity. J. Am. Chem. Soc. **135**(2), 855–861 (2013)

13. Oja, S.M., Guerrette, J.P., David, M.R., Zhang, B.: Fluorescence-enabled electrochemical microscopy with dihydroresorufin as a fluorogenic indicator. Anal. Chem. **86**(12), 6040–6048 (2014)
14. Shepherd, J.L., Kell, A., Chung, E., Sinclar, C.W., Workentin, M.S., Bizzotto, D.: Selective reductive desorption of a SAM-coated gold electrode revealed using fluorescence microscopy. J. Am. Chem. Soc. **126**(26), 8329–8335 (2004)
15. Murphy, J.N., Cheng, A.K.H., Yu, H.Z., Bizzotto, D.: On the nature of DNA SAM on Au: measuring surface heterogeneity with electrochemical in situ fluorescence microscopy. J. Am. Chem. Soc. **131**, 4042–4050 (2009)
16. Meunier, A., Triffaux, E., Bizzotto, D., Buess-Herman, C., Doneux, T.: In Situ fluorescence microscopy study of the interfacial inhomogeneity of DNA mixed self-assembled monolayers at gold electrodes. ChemElectroChem **2**(3), 434–442 (2015)
17. Doneux, T., Bouffier, L., Goudeau, B., Arbault, S.: Coupling electrochemistry with fluorescence confocal microscopy to investigate electrochemical reactivity: a case study with the resazurin-resorufin fluorogenic couple. Anal. Chem. **88**(12), 6292–6300 (2016)
18. de Poulpiquet, A., Goudeau, B., Garrigue, P., Sojic, N., Arbault, S., Doneux, T., Bouffier, L.: A snapshot of the electrochemical reaction layer by using 3 dimensionally resolved fluorescence mapping. Chem. Sci. **9**(32), 6622–6628 (2018)
19. Tassy, B., Dauphin, A.L., Man, H.M., Le Guenno, H., Lojou, E., Bouffier, L., de Poulpiquet, A.: In Situ fluorescence tomography enables a 3D mapping of enzymatic O_2 reduction at the electrochemical interface. Anal. Chem. **92**(10), 7249–7256 (2020)
20. Yang, M.J., Batchelor-McAuley, C., Kätelhön, E., Compton, R.G.: Reaction layer imaging using fluorescence electrochemical microscopy. Anal. Chem. **89**(12), 6870–6877 (2017)
21. Pruchyathamkorn, J., Yang, M.J., Amin, H.M.A., Batchelor-McAuley, C., Compton, R.G.: Imaging electrode heterogeneity using chemically confined fluorescence electrochemical microscopy. J. Phys. Chem. Lett. **8**(24), 6124–6127 (2017)
22. Audebert, P., Miomandre, F.: Electrofluorochromism: from molecular systems to set-up and display. Chem. Sci. **4**(2), 575–584 (2013)
23. Qing, Z., Audebert, P., Clavier, G., Meallet-Renault, R., Miomandre, F., Tang, J.: Bright fluorescence through activation of a low absorption fluorophore: the case of a unique naphthalimide-tetrazine dyad. New J. Chem. **35**(8), 1678–1682 (2011)
24. Guille-Collignon, M., Delacotte, J., Lemaître, F., Labbé, E., Buriez, O.: Electrochemical fluorescence switch of organic fluorescent or fluorogenic molecules. Chem. Record **21**(9), 2193–2202 (2021)
25. Sojic, N., Bouffier, L.: Interplay between electrochemistry and optical imaging: the whole is greater than the sum of the parts. Curr. Op. Electrochem. **34**, 101007 (2022)
26. Miomandre, F.: How molecular electrochemistry may shine light by designing electrofluorochromic compounds. Curr. Op. Electrochem. **24**, 56–62 (2020)
27. Guille-Collignon, M., Delacotte, J.M., Lemaître, F., Labbé, E., Buriez, O.: Electrochemical fluorescence switch of organic fluorescent or fluorogenic molecules. Chem. Record **21**(9), 2193–2202 (2021)
28. Gong, Y.-H., Miomandre, F., Meallet-Renault, R., Badre, S., Galmiche, L., Tang, J., Audebert, P., Clavier, G.: Synthesis and physical chemistry of s-Tetrazines: which ones are fluorescent and why? Eur. J. Org. Chem. **35**, 6121–6128 (2009)
29. Nakamura, K., Kanazawa, K., Kobayashi, N.: Electrochemically-switchable emission and absorption by using luminescent Lanthanide(III) complex and electrochromic molecule toward novel display device with dual emissive and reflective mode. Displays **34**(5), 389–395 (2013)
30. Lin, L., Kaojin, W., Haozhong, W., Anjun, Q., Ben Zhong, T.: Simultaneously achieving high capacity storage and multilevel anti-counterfeiting using electrochromic and electrofluorochromic dual-functional AIE polymers. Chem. Sci. **12**, 7058–7065 (2021)
31. Bouffier, L., Doneux, T., Goudeau, B., Kuhn, A.: Imaging redox activity at bipolar electrodes by indirect fluorescence modulation. Anal. Chem. **86**(8), 3708–3711 (2014)

32. Tian, Z., Mi, L., Wu, Y., Shao, F., Zou, M., Zhou, Z., Liu, S.: Visual electrofluorochromic detection of cancer cell surface glycoprotein on a closed bipolar electrode chip. Anal. Chem. **91**(12), 7902–7910 (2019)
33. Yang, Y., Ma, Y.Q., Berengut, J.F., Lee, L.K., Tilley, R.D., Gaus, K., Gooding, J.J.: Electrochemically controlled blinking of fluorophores for quantitative STORM imaging. Nature Photon. **18**(7), 713–720 (2024)
34. Yang, Y., Ma, Y.Q., Gooding, J.J.: The electrochemical modulation of single molecule fluorescence. Farad. Discuss. **257**, 333–343 (2025)
35. Yang, Y., Fan, S.J., Webb, J.A., Ma, Y.Q., Goyette, J., Chen, X.Q., Gaus, K., Tilley, R.D., Gooding, J.: Electrochemical fluorescence switching of enhanced green fluorescent protein. Biosens. Bioelectron. **237**, 115467 (2023)
36. Guerret-Legras, L., Audibert, J.-F., Gonzalez-Ojeda, I.M., Dubacheva, G.V., Clavier, G., Miomandre, F.: Time-Resolved fluorescence microscopy combined with scanning electrochemical microscopy: a new way to visualize photo-induced electron transfer quenching with an electrofluorochromic probe. J. Phys. Chem. C **124**(43), 23938–23948 (2020)
37. Scida, K., Eden, A., Arroyo-Currás, N., MacKenzie, S., Satik, Y., Meinhart, C.D., Eijkel, J.C.T., Pennathur, S.: Fluorescence-Based observation of transient electrochemical and electrokinetic effects at nanoconfined bipolar electrodes. ACS Appl. Mater. Int. **11**(14), 13777–13786 (2019)
38. Lewis, T.H., Zhang, B.: Fluorescence for spatially and temporally resolved electrochemical imaging. Curr. Op. Electrochem. **40**, 101334 (2023)
39. Guerret-Legras, L., Maillot, B., Audibert, J.F., Dubacheva, G.V., Galmiche, L., Lang, P., Miomandre, F.: Electrofluorochromism of surface-confined tetrazines investigated on the monolayer scale. J. Phys. Chem. C **123**(48), 29255–29261 (2019)
40. Oja, S.M., Zhang, B.: Imaging transient formation of diffusion layers with fluorescence-enabled electrochemical microscopy. Anal. Chem. **86**(24), 12299–12307 (2014)
41. Cízková, M., Cattiaux, L., Mallet, J.M., Labbé, E., Buriez, O.: Electrochemical switching fluorescence emission in rhodamine derivatives. Electrochim. Acta **260**, 589–597 (2018)
42. Rudd, N.C., Cannan, S., Bitziou, E., Ciani, I., Whitworth, A.L., Unwin, P.R.: Fluorescence confocal laser scanning microscopy as a probe of pH gradients in electrode reactions and surface activity. Anal. Chem. **77**(19), 6205–6217 (2005)
43. Xu, W., Ma, C., Bohn, P.W.: Coupling of independent electrochemical reactions and fluorescence at closed bipolar interdigitated electrode arrays. ChemElectroChem **3**(3), 422–428 (2016)
44. Liu, X.Q., Savy, A., Maurin, S., Grimaud, L., Darchen, F., Quinton, D., Labbe, E., Buriez, O., Delacotte, J., Lemaitre, F., Guille-Collignon, M.: A dual functional electroactive and fluorescent probe for coupled measurements of vesicular exocytosis with high spatial and temporal resolution. Angew. Chem. Int. Ed. **56**(9), 2366–2370 (2017)
45. Chen, B., Zhang, H., Xuan, J., Offer, G.J., Wang, H.: Seeing is believing: in situ/operando optical microscopy for probing electrochemical energy systems. Adv. Mater. Techn. **5**(10), 2000555 (2020)
46. Lemineur, J.-F., Wang, H., Wang, W., Kanoufi, F.: Emerging optical microscopy techniques for electrochemistry. Ann. Rev. Anal. Chem. **15**, 57–82 (2022)
47. Wang, H.-W., Lu, S.-M., Chen, M., Long, Y.-T.: Optical-facilitated single-entity electrochemistry. Curr. Op. Electrochem. **34**, 100999 (2022)
48. Sojic, N., Bouffier, L.: Interplay between electrochemistry and optical imaging: the whole is greater than the sum of the parts. Curr. Op. Electrochem. **34**, 101007 (2022)
49. Wang, W.: Imaging the chemical activity of single nanoparticles with optical microscopy. Chem. Soc. Rev. **47**(7), 2485–2508 (2018)
50. Feinleib, J.: Electroreflectance in metals. Phys. Rev. Lett. **16**(26), 1200–1202 (1966)
51. McIntyre, J.D.E.: Electrochemical modulation spectroscopy. Surf. Sci. **37**, 658–682 (1973)
52. Godeffroy, L., Ciocci, P., Nsabimana, A., Miranda Vieira, M., Noël, J.-M., Combellas, C., Lemineur, J.-F., Kanoufi, F.: Deciphering competitive routes for nickel-based nanoparticle electrodeposition by an operando optical monitoring. Angew. Chem. Int. Ed. **60**(31), 16980–16983 (2021)

53. Larsson, A., Vorobyova, M., Pfaff, S., Abbondanza, G., Pan, J., Zetterberg, J., Lundgren, E.: Operando surface optical reflectance microscopy study of corrosion film growth on a Ni–Cr–Mo Alloy during anodic polarization. J. Phys. Chem. C **127**(44), 21871–21877 (2023)
54. Pfaff, S., Larsson, A., Orlov, D., Harlow, G.S., Abbondanza, G., Linpé, W., Rämisch, L., Gericke, S.M., Zetterberg, J., Lundgren, E.: Operando reflectance microscopy on polycrystalline surfaces in thermal catalysis, electrocatalysis, and corrosion. ACS Appl. Mater. Int. **13**(16), 19530–19540 (2021)
55. Makogon, A., Noël, J.-M., Kanoufi, F., Shkirskiy, V.: Deciphering the interplay between local and global dynamics of anodic metal oxidation. Anal. Chem. **96**(3), 1129–1137 (2024)
56. Onderwaater, W.G., Taranovskyy, A., van Baarle, G.C., Frenken, J.W.M., Groot, I.M.N.: In situ optical reflectance difference observations of CO oxidation over Pd(100). J. Phys. Chem. C **121**(21), 11407–11415 (2017)
57. Weber, T., Ortmann, T., Escalera-López, D., Abb, M.J.S., Mogwitz, B., Cherevko, S., Rohnke, M., Over, H.: Visualizing potential-induced pitting corrosion of ultrathin single-crystalline IrO_2(110) films on RuO_2(110)/Ru(0001) under electrochemical water splitting conditions. ChemCatChem **12**(3), 855–866 (2020)
58. Pfaff, S., Larsson, A., Orlov, D., Rämisch, L., Gericke, S.M., Lundgren, E., Zetterberg, J.: A polycrystalline Pd surface studied by two-dimensional surface optical reflectance during CO oxidation: bridging the materials gap. ACS Appl. Mater. Int. **16**(1), 444–453 (2024)
59. Lemineur, J.-F., Stockmann, T.J., Médard, J., Smadja, C., Combellas, C., Kanoufi, F.: Optical nanoimpacts of dielectric and metallic nanoparticles on gold surface by reflectance microscopy: adsorption or bouncing? J. Anal. Test. **3**(2), 175–188 (2019)
60. Kadoguchi, N., Uesugi, Y., Nagasako, M., Kobayashi, T., Kozawa, Y., Sato, S.: Nanoprocessing of self-suspended monolayer graphene and defect formation by femtosecond-laser irradiation. Nano Lett. **23**(11), 4893–4900 (2023)
61. Lemineur, J.-F., Noël, J.-M., Courty, A., Ausserré, D., Combellas, C., Kanoufi, F.: In Situ optical monitoring of the electrochemical conversion of dielectric nanoparticles: from multistep charge injection to nanoparticle motion. J. Am. Chem. Soc. **142**(17), 7937–7946 (2020)
62. Godeffroy, L., Lemineur, J.-F., Shkirskiy, V., Miranda Vieira, M., Noël, J.-M., Kanoufi, F.: Bridging the gap between single nanoparticle imaging and global electrochemical response by correlative microscopy assisted by machine vision. Small Methods **6**(9), 2200659 (2022)
63. Li, W., Wojcik, M., Xu, K.: Optical microscopy unveils rapid, reversible electrochemical oxidation and reduction of graphene. Nano Lett. **19**(2), 983–989 (2019)
64. Kitta, M., Murai, K., Yoshii, K., Sano, H.: Electrochemical surface plasmon resonance spectroscopy for investigation of the initial process of lithium metal deposition. J. Am. Chem. Soc. **143**(29), 11160–11170 (2021)
65. Utterback, J.K., King, A.J., Belman-Wells, L., Larson, D.M., Hamerlynck, L.M., Weber, A.Z., Ginsberg, N.S.: Operando label-free optical imaging of solution-phase ion transport and electrochemistry. ACS En. Lett. **8**(4), 1785–1792 (2023)
66. Sanchez, T., Kurchavova, E., Shkirskiy, V., Światowska, J., Vivier, V., Volovitch, P.: Detection and quantification of defect evolution at buried metal-oxide-polymer interface on rough substrate by local electrochemical impedance mapping. Electrochim. Acta **388**, 138467 (2021)
67. Homborg, A.M., Olgiati, M., Denissen, P.J., Garcia, S.J.: An integral non-intrusive electrochemical and in-situ optical technique for the study of the effectiveness of corrosion inhibition. Electrochim. Acta **403**, 139619 (2022)
68. Chen, Y., Chen, K.-H., Sanchez, A.J., Kazyak, E., Goel, V., Gorlin, Y., Christensen, J., Thornton, K., Dasgupta, N.P.: Operando video microscopy of Li plating and re-intercalation on graphite anodes during fast charging. J. Mater. Chem. A **9**(41), 23522–23536 (2021)
69. Steiger, J., Kramer, D., Mönig, R.: Mechanisms of dendritic growth investigated by in situ light microscopy during electrodeposition and dissolution of lithium. J. Power. Sources **261**, 112–119 (2014)
70. Baker, L.A.: Perspective and prospectus on single-entity electrochemistry. J. Am. Chem. Soc. **140**(46), 15549–15559 (2018)

71. Albrecht, T., MacPherson, J., Magnussen, O., Fermin, D., Crooks, R., Gooding, J., Hersbach, T., Kanoufi, F., Schuhmann, W., Bentley, C., Tao, N., Mitra, S., Krischer, K., Tschulik, K., Faez, S., Nogala, W., Unwin, P., Long, Y., Koper, M., Tian, Z., Alpuche-Aviles, M.A., White, H., Brasiliense, V., Kranz, C., Schmickler, W., Stevenson, K., Jing, C., Edwards, M.: Electrochemistry of single nanoparticles: general discussion. Farad. Discuss. **193**, 387–413 (2016)
72. Zhang, L., Wahab, O.J., Jallow, A.A., O'Dell, Z.J., Pungsrisai, T., Sridhar, S., Vernon, K.L., Willets, K.A., Baker, L.A.: Recent developments in single-entity electrochemistry. Anal. Chem. **96**(20), 8036–8055 (2024)
73. Jing, C., Reichert, J.: Nanoscale electrochemistry in the "dark-field." Curr. Op. Electrochem. **6**(1), 10–16 (2017)
74. Hill, C.M., Pan, S.: A dark-field scattering spectroelectrochemical technique for tracking the electrodeposition of single silver nanoparticles. J. Am. Chem. Soc. **135**(46), 17250–17253 (2013)
75. Brasiliense, V., Noël, J.-M., Wonner, K., Tschulik, K., Combellas, C., Kanoufi, F.: Single nanoparticle growth from nanoparticle tracking analysis: from monte carlo simulations to nanoparticle electrogeneration. ChemElectroChem **5**(20), 3036–3043 (2018)
76. Brasiliense, V., Clausmeyer, J., Berto, P., Tessier, G., Combellas, C., Schuhmann, W., Kanoufi, F.: Monitoring cobalt-oxide single particle electrochemistry with subdiffraction accuracy. Anal. Chem. **90**(12), 7341–7348 (2018)
77. Suvira, M., Ahuja, A., Lovre, P., Singh, M., Draher, G.W., Zhang, B.: Imaging single H2 nanobubbles using off-axis dark-field microscopy. Anal. Chem. **95**(43), 15893–15899 (2023)
78. Brasiliense, V., Clausmeyer, J., Dauphin, A.L., Noël, J.-M., Berto, P., Tessier, G., Schuhmann, W., Kanoufi, F.: Opto-Electrochemical in situ monitoring of the cathodic formation of single cobalt nanoparticles. Angew. Chem. Int. Ed. **56**(35), 10598–10601 (2017)
79. Shan, X., Patel, U., Wang, S., Iglesias, R., Tao, N.: Imaging local electrochemical current via surface plasmon resonance. Science **327**(5971), 1363–1366 (2010)
80. Wang, Y., Shan, X., Wang, H., Wang, S., Tao, N.: Plasmonic imaging of surface electrochemical reactions of single gold nanowires. J. Am. Chem. Soc. **139**(4), 1376–1379 (2017)
81. Shan, X., Díez-Pérez, I., Wang, L., Wiktor, P., Gu, Y., Zhang, L., Wang, W., Lu, J., Wang, S., Gong, Q., Li, J., Tao, N.: Imaging the electrocatalytic activity of single nanoparticles. Nature Nanotech. **7**(10), 668–672 (2012)
82. Jiang, D., Sun, L., Liu, T., Wang, W.: Thin-Film electrochemistry of single prussian blue nanoparticles revealed by surface plasmon resonance microscopy. Anal. Chem. **89**(21), 11641–11647 (2017)
83. Jing, C., Gu, Z., Xie, T., Long, Y.-T.: Color-coded imaging of electrochromic process at single nanoparticle level. Chem. Sci. **7**(8), 5347–5351 (2016)
84. Nguyen, T.N., Cappillino, P.J., Chang, W.-S.: Revealing enhanced size uniformity of the electrochemical deposition of palladium nanoparticles via single-particle dark-field scattering imaging. J. Phys. Chem. C **129**(13), 6477–6485 (2025)
85. Novo, C., Funston, A.M., Mulvaney, P.: Direct observation of chemical reactions on single gold nanocrystals using surface plasmon spectroscopy. Nature Nanotech. **3**(10), 598–602 (2008)
86. Brasiliense, V., Berto, P., Combellas, C., Kuszelewicz, R., Tessier, G., Kanoufi, F.: Electrochemical transformation of individual nanoparticles revealed by coupling microscopy and spectroscopy. Farad. Discuss. **193**, 339–352 (2016)
87. Smith, J.G., Yang, Q., Jain, P.K.: Identification of a critical intermediate in galvanic exchange reactions by single-nanoparticle-resolved kinetics. Angew. Chem. Int. Ed. **53**(11), 2867–2872 (2014)
88. Nguyen, M.-C., Berto, P., Valentino, F., Lemineur, J.-F., Noel, J.-M., Kanoufi, F., Tessier, G.: 3D Spectroscopic tracking of individual brownian nanoparticles during galvanic exchange. ACS Nano **16**(9), 14422–14431 (2022)

89. Jing, C., Rawson, F.J., Zhou, H., Shi, X., Li, W.-H., Li, D.-W., Long, Y.-T.: New insights into electrocatalysis based on plasmon resonance for the real-time monitoring of catalytic events on single gold nanorods. Anal. Chem. **86**(11), 5513–5518 (2014)
90. Byers, C.P., Hoener, B.S., Chang, W.-S., Link, S., Landes, C.F.: Single-Particle plasmon voltammetry (spPV) for detecting anion adsorption. Nano Lett. **16**(4), 2314–2321 (2016)
91. Liu, T., Liu, S., Jiang, W., Wang, W.: Tracking sub-nanometer shift in the scattering centroid of single gold nanorods during electrochemical charging. ACS Nano **13**(6), 6279–6286 (2019)
92. Hu, S., Yi, J., Zhang, Y.-J., Lin, K.-Q., Liu, B.-J., Chen, L., Zhan, C., Lei, Z.-C., Sun, J.-J., Zong, C., Li, J.-F., Ren, B.: Observing atomic layer electrodeposition on single nanocrystals surface by dark field spectroscopy. Nature Commun. **11**(1), 2518 (2020)
93. Al-Zubeidi, A., Stein, F., Flatebo, C., Rehbock, C., Hosseini Jebeli, S.A., Landes, C.F., Barcikowski, S., Link, S.: Single-Particle hyperspectral imaging reveals kinetics of silver ion leaching from alloy nanoparticles. ACS Nano **15**(5), 8363–8375 (2021)
94. Brasiliense, V., Patel, A.N., Martinez-Marrades, A., Shi, J., Chen, Y., Combellas, C., Tessier, G., Kanoufi, F.: Correlated electrochemical and optical detection reveals the chemical reactivity of individual silver nanoparticles. J. Am. Chem. Soc. **138**(10), 3478–3483 (2016)
95. Young, G., Kukura, P.: Interferometric scattering microscopy. Ann. Rev. Phys. Chem. **70**, 301–322 (2019)
96. Cheng, C.-Y., Liao, Y.-H., Hsieh, C.-L.: Dynamic signal of live biological cells under interferometric scattering (iSCAT) microscopy and its impacts on single-particle tracking. J. Physics D: Appl. Phys. **54**(36), 364001 (2021)
97. Taylor, R.W., Sandoghdar, V.: Interferometric scattering microscopy: seeing single nanoparticles and molecules via rayleigh scattering. Nano Lett. **19**(8), 4827–4835 (2019)
98. Feng, G., Shi, Y., Jia, H., Risal, S., Yang, X., Ruchhoeft, P., Shih, W.-C., Fan, Z., Xu, W., Shan, X.: Progressive and instantaneous nature of lithium nucleation discovered by dynamic and operando imaging. Sci. Adv. **9**(21), 6813 (2023)
99. Merryweather, A.J., Schnedermann, C., Jacquet, Q., Grey, C.P., Rao, A.: Operando optical tracking of single-particle ion dynamics in batteries. Nature **594**(7864), 522–528 (2021)
100. Xu, C., Merryweather, A.J., Pandurangi, S.S., Lun, Z., Hall, D.S., Deshpande, V.S., Fleck, N.A., Schnedermann, C., Rao, A., Grey, C.P.: Operando visualization of kinetically induced lithium heterogeneities in single-particle layered Ni-rich cathodes. Joule **6**(11), 2535–2546 (2022)
101. Garoz-Ruiz, J., Perales-Rondon, J.V., Heras, A., Colina, A.: Spectroelectrochemical sensing: current trends and challenges. Electroanalysis **31**(7), 1254–1278 (2019)
102. Zhu, H., Zhang, F., Wang, H., Lu, Z., Chen, H.-Y., Li, J., Tao, N.: Optical imaging of charges with atomically thin molybdenum disulfide. ACS Nano **13**(2), 2298–2306 (2019)
103. Viinikanoja, A., Wang, Z., Kauppila, J., Kvarnström, C.: Electrochemical reduction of graphene oxide and its in situ spectroelectrochemical characterization. PCCP **14**(40), 14003–14009 (2012)
104. Walker, E.K., Vanden Bout, D.A., Stevenson, K.J.: Spectroelectrochemical investigation of an electrogenerated graphitic oxide solid-electrolyte interphase. Anal. Chem. **84**(19), 8190–8197 (2012)
105. Wang, X., Wang, S.-C., Ma, J., Xie, R.-C., Wang, W.: Near-infrared visualisation of single microparticle electrochemistry for batteries. Chem. Sci. **15**(22), 8536–8544 (2024)
106. Hou, L., Adhikari, S., Tian, Y., Scheblykin, I.G., Orrit, M.: Absorption and quantum yield of single conjugated polymer Poly[2-methoxy-5-(2-ethylhexyloxy)-1,4-phenylenevinylene] (MEH-PPV) molecules. Nano Lett. **17**(3), 1575–1581 (2017)
107. Adhikari, S., Spaeth, P., Kar, A., Baaske, M.D., Khatua, S., Orrit, M.: Photothermal microscopy: imaging the optical absorption of single nanoparticles and single molecules. ACS Nano **14**(12), 16414–16445 (2020)
108. Moghaddam, M., Godeffroy, L., Jasielec, J.J., Kostopoulos, N., Noël, J.-M., Piquemal, J.-Y., Lemineur, J.-F., Peljo, P., Kanoufi, F.: Scanning electrochemical microscopy meets optical microscopy: probing the local paths of charge transfer operando in booster-microparticles for flow batteries. Small **20**(36), 2309607 (2024)

109. Baffou, G., Bon, P., Savatier, J., Polleux, J., Zhu, M., Merlin, M., Rigneault, H., Monneret, S.: Thermal imaging of nanostructures by quantitative optical phase analysis. ACS Nano **6**(3), 2452–2458 (2012)
110. Baffou, G., Quidant, R., García de Abajo, F.J.: Nanoscale control of optical heating in complex plasmonic systems. ACS Nano **4**(2), 709–716 (2010)
111. Park, Y., Depeursinge, C., Popescu, G.: Quantitative phase imaging in biomedicine. Nature Photon. **12**(10), 578–589 (2018)
112. Brasiliense, V., Audibert, J.-F., Wu, T., Tessier, G., Berto, P., Miomandre, F.: Local surface chemistry dynamically monitored by quantitative phase microscopy. Small Methods **6**(1), 2100737 (2022)
113. Maillot, B., Johnson, M., Audibert, J.F., Miomandre, F., Brasiliense, V.: Operando surface optical nanometrology reveals diazonium salts' visible photografting mechanism. Nanoscale **15**(19), 8754–8761 (2023)
114. Berto, P., Rigneault, H., Guillon, M.: Wavefront sensing with a thin diffuser. Opt. Lett. **42**(24), 5117–5120 (2017)
115. Maillot, B., Audibert, J.F., Miomandre, F., Brasiliense, V.: Nanometrology based control: taming radical grafting reactions with attoliter precision. Nanoscale **16**(15), 7594–7602 (2024)
116. Xiong, F., Wang, H., Liu, X., Sun, J., Brongersma, M., Pop, E., Cui, Y.: Li intercalation in MoS_2: in situ observation of its dynamics and tuning optical and electrical properties. Nano Lett. **15**(10), 6777–6784 (2015)

Correction to: Electrochemiluminescence Nano-Emitters and Electrochemiluminescence Microscopy-Based Single Nanoparticle Analysis

Zejing Xing, Zhichen Zhang, Jingjuan Xu, and Jun-Jie Zhu

Correction to:
Chapter 9 in: F. Paolucci et al. (eds.), *Electrogenerated Chemiluminescence Microscopy and Imaging*,
https://doi.org/10.1007/978-3-032-14602-1_9

In the original version of Chapter 9, the author name Zhichen Zhang was incorrectly spelled as "Zicheng Zhang" which has now been corrected. The chapter has been updated with the changes.

The updated version of this chapter can be found at
https://doi.org/10.1007/978-3-032-14602-1_9

F. Paolucci et al. (eds.), *Electrogenerated Chemiluminescence Microscopy and Imaging*,
https://doi.org/10.1007/978-3-032-14602-1_14

Correction to: Super-Resolution Electrochemiluminescence Microscopy

Wenxin Zhu, Pinlong Zhao, and Jiandong Feng

Correction to:
Chapter 6 in: F. Paolucci et al. (eds.), *Electrogenerated Chemiluminescence Microscopy and Imaging*, https://doi.org/10.1007/978-3-032-14602-1_6

In Chapter 6, the figures 5 and 6 were swapped mistakenly. It is now corrected and placed the appropriate figures.

The updated version of this chapter can be found at
https://doi.org/10.1007/978-3-032-14602-1_6

F. Paolucci et al. (eds.), *Electrogenerated Chemiluminescence Microscopy and Imaging*,
https://doi.org/10.1007/978-3-032-14602-1_15

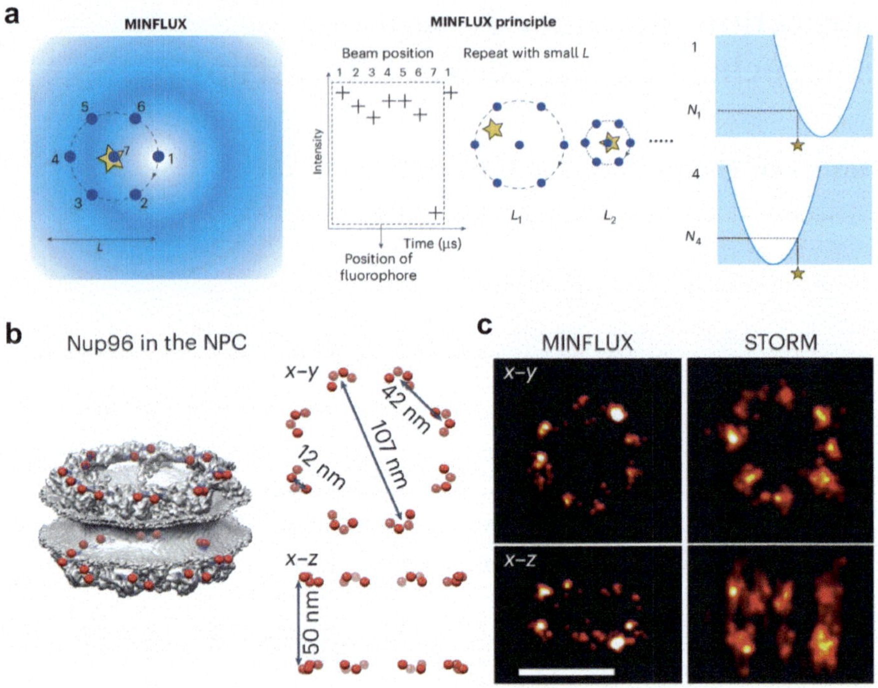

Fig. 5 Principles and performance of MINFLUX [32]. **a** Illustration of super-resolution principles of MINFLUX. **b** Detailed structures of nuclear pores. **c** Comparison of the reconstructed super-resolved images of nuclear pores obtained by MINFLUX and STORM. (**a–c** are all adapted with permission from Ref. [32]). Copyright © 2025, Springer Nature

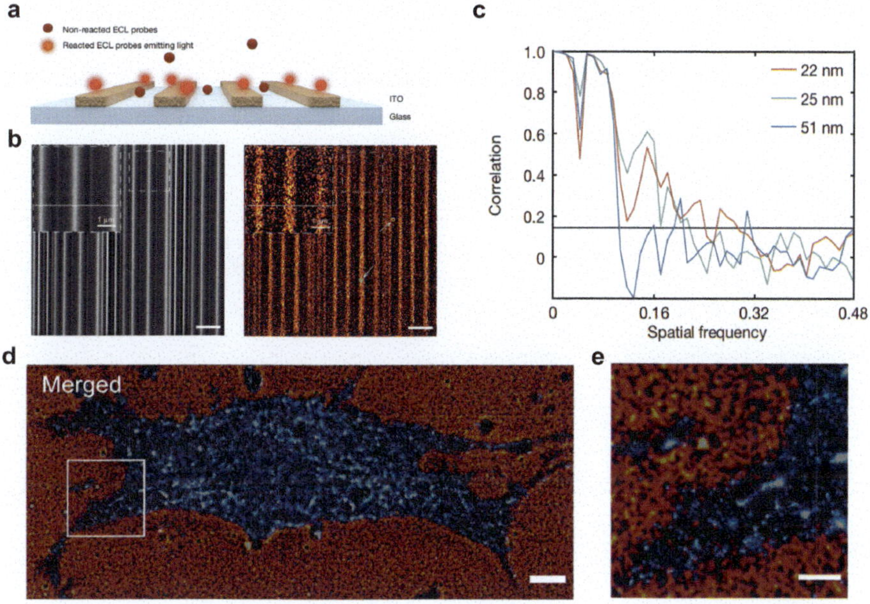

Fig. 6 Localization-based super resolution ECL imaging [35]. **a** Representative single-molecule collision on the electrode surface. **b** Correlation of super-resolution ECL reconstruction and scanning electron microscopy. **c** Fourier ring correlation analysis of selected region of the interest in "b". **d** Merged super-resolution ECL reconstruction and STORM image of cell adhesion. **e** Enlarged white box region in "d". (**a**–**e** are all adapted with permission from Ref. [35]). Copyright © 2021, Springer Nature

The manufacturer's authorised representative in the EU is Springer Nature Customer Service Centre GmbH, Europaplatz 3, 69115 Heidelberg, Germany. If you have any concerns regarding our products, please contact ProductSafety@springernature.com

Printed and bound by CPI Group (UK) Ltd, Croydon, CR0 4YY
07/10/2026
02229350-0009